TERRESTRIAL PLANT ECOLOGY

SECOND EDITION

TERRESTRIAL PLANT ECOLOGY

SECOND EDITION

Michael G. Barbour
University of California, Davis

Jack H. Burk
California State University, Fullerton

Wanna D. Pitts
San Jose State University

The Benjamin/Cummings Publishing Company, Inc.

Menlo Park, California • Reading, Massachusetts
Don Mills, Ontario • Wokingham, U.K. • Amsterdam • Sydney
Singapore • Tokyo • Madrid • Bogota • Santiago • San Juan

Sponsoring Editor: Andrew Crowley

Production Editor: Julie Kranhold/*Ex Libris*

Production Coordinator: Mimi Hills

Cover and Consulting Designer: Sara B. Hunsaker

Cover Art: F. French from *Happy Hunting Grounds* by W. H. Gibson. New York: Harber & Bros. 1887.

Part Opening Art: Parts I, III, and IV: Brine, M. D. 1886. *From Gold to Grey: Being Poems and Pictures of Life and Nature.* New York: Cassell & Company, Limited. Part II: Frye, A. E. 1899. *Introductory Geography.* Boston, MA: Ginn & Company.

The basic text of this book was designed using the Modular Design System, as developed by Wendy Earl and Design Office Bruce Kortebein

Library of Congress Cataloging-in-Publication Data

Barbour, Michael G.
    Terrestrial plant ecology.

    Bibliography: p.
    Includes index.
    1. Botany—Ecology. 2. Botany—North America—
Ecology. I. Burk, Jack H., 1942-   . II. Pitts,
Wanna D., 1932-   . III. Title.
QK901.B345 1987   581.5′264   86-26812
ISBN 0-8053-0541-6

13 14 15 16-MA-97 96

The Benjamin/Cummings Publishing Company, Inc.
2727 Sand Hill Road
Menlo Park, California 94025

PREFACE
To The First Edition

Our purpose in writing this book has been to condense the literature pertaining to the ecology of wild plants into a digestible survey for the undergraduate student. Most readers will be in a one quarter or one semester course at the upper division level and will have had previous courses in basic biology, botany, and possibly in general ecology or environmental science.

This is not a detailed, thorough, unbiased reference work, nor does it take us to a new level of understanding. Rather, it is a basic synthesis that we hope will be a useful starting point for instructors and students. The present science of plant ecology is so broad as to encourage specialization and to narrow one's perspective of the entire field. We believe it is essential to maintain a broad outlook, and to examine the important aspects of all the specialized branches of plant ecology, if we are to grasp the essentials of the interactions between plants and their environment.

This book is divided into four parts. The first part introduces the broad aspects of plant ecology and gives an historical overview of the development of the science. Part II discusses autecology, the responses of populations to their environment; and Part III focuses on synecology, that is, community attributes. Part IV describes the role of five environmental factors (light, temperature, fire, soil, and water) at both the autecological and the synecological levels. The concluding chapter describes the major vegetation types of North America, attempting to show how the many factors and responses covered individually in earlier sections interact to produce the vegetation of this continent. Except for this last chapter, which has its own list of references, the references cited are listed in one section at the end of the book. We hope these references offer a good starting point for the reader who wishes to go further with some topic of interest. Most chapters conclude with a relatively detailed summary.

We believe that the study of plant ecology may provide lessons that take us beyond the conventional, narrow limits of "science." It is possible that our social and psychological evolution has been influenced by vegetation. Natural settings have generated negative feelings in western (Judeo-Christian) societies, as described by Roderick Nash:*

> . . . appreciation of the wilderness must be seen as recent, revolutionary, and incomplete. . . . Ambivalence, a blend of attraction and repulsion, is still the most accurate way to characterize the present feeling toward wilderness. . . . Wilderness has risen far on the scale of man's priorities. But the depth and intensity of previous antipathy suggests that it still has a long way to go.

The poet-anthropologist Loren Eisely reminds us how important our development in natural, wild vegetation will continue to be to us despite these negative feelings. "Man was born and took shape among earth's leafy shadows," and now he is to embark upon a new and difficult journey:†

> At the climactic moment of his journey into space he has met himself at the doorway of the stars. And the looming shadow before him has pointed backward into the tangled gloom of a forest from which it has been his purpose to escape. Man has crossed, in his history, two worlds. He must now enter another and forgotten one, but with the knowledge gained on the pathway to the moon. He must learn that, whatever his powers as a magician, he lies under the spell of a greater and a green enchantment which, try as he will, he can never avoid, however far he travels. The spell has been laid on him since the beginning of time—the spell of the natural world from which he sprang.

Many persons, some anonymous, have read all or part of the manuscript, and provided valuable insights and corrections, thereby helping to maintain the intended focus of the book. We thank these reviewers, who include: Jerry M. Baskin, Gary L. Cunningham, Andrew M. Greller, H. Thomas Harvey, Gordon L. Huntington, Subodh K. Jain, Robert P. McIntosh, Jack Major, Bruce M. Pavlik, Robert W. Pearcy, Otto T. Solbrig, Robin Smith, James P. Syvertsen, Irwin A. Ungar, John L. Vankat, and Lynn D. Whittig. Final decisions concerning approach, presentation, and content were our own and we take full responsibility for any oversights or inaccuracies contained herein.

Several other individuals have contributed long hours in manuscript preparation and have tolerated and understood the sometimes intolerant authors. Special thanks go to Heather Stout, who worked many long hours organizing the Literature Cited, securing permissions, and preparing the index. We also thank Patricia Burk, Maggie Hummer, Jerry Pitts, and the many ecology students who "tried out" sections of this book as a text in courses during the past two years. We thank you for your help and patience.

M.G.B., J.H.B., W.D.P.

---

*Nash, R. 1973. *Wilderness and the American Mind.* Yale University Press.
†Eisely, L. 1970. *The Invisible Pyramid.* Scribners.

PREFACE
To The Second Edition

P lant ecology has changed substantially since the first edition of this text
was completed in 1979. At that time, we had no expectation of address-
ing our readers for a subsequent edition. We are pleased that our broad approach
to plant ecology has found an accepting audience—an audience large and persistent enough to warrant a second edition.

While several important organizational changes have been made in this edition, we remain committed to an integrated approach to plant ecology. Our overall format is the same because we feel that separation of physiological, population, community, and ecosystem perspectives into mutually exclusive sections of
the book would be misleading to the student. As a result, material on a specific
topic may appear in more than one section, and in different contexts. To assist
the reader in finding all information pertinent to one topic, we have provided a
very detailed index.

We have written the chapters as self-contained units, which may be assigned
for reading in any order. Each chapter concludes with a summary.

Plant population ecology has undergone dramatic increases in knowledge
and activity. Under the leadership of Dr. John L. Harper, plant population ecology
has moved from an orphan science to a rapidly expanding body of knowledge,
where enough basic information is available to support the development of models
and generalizations typical of mature branches of science. Our text has expanded
accordingly.

Other important advances in plant ecology have been stimulated by rapid
technological development. The equipment used to measure plant response,
measure the microenvironment, and analyze data has improved markedly, leading to new insights and discoveries in most subdisciplines of plant ecology. These
recent developments have dictated many modifications throughout the text.
Material on classification and ordination, mineral cycles, light, temperature, and
water have been significantly reorganized and rewritten to make the flow of
topics more logical and their coverage more complete and current.

We have added other new sections on such diverse topics as: antiherbivore chemical defense; ecotype research; mycorrhizal classification; theoretical aspects of community stability, diversity, and degree of integration; remote sensing; acid deposition; forest herb phenology; the historical development of plant ecology; and new ordination techniques. At the same time, we have tried to prune the first edition, so that overall length is not appreciably increased.

Our goal throughout the text has been to cite works that summarize or provide a particularly broad or clear view of a topic, rather than cite tediously long lists of more narrow contributions. Nevertheless, we have listed over 1300 references in the "Literature Cited" section at the back of the book.

We are particularly appreciative of the excellent help provided by reviewers. We have accepted most of their suggestions. Reviewers include: Rex Cates, Bruce Mahall, Robert Pearcy, Marcel Rejmanek, William Schlesinger, Maureen Stanton, Stephen Stephenson, John Vankat, and Susan Ustin. We are also grateful to Gayle Gutierrez, who handled the process of requesting permissions and several other time-consuming jobs. We appreciate the continuing support of Patricia Burk, Jerry Pitts, and Valerie Whitworth who, along with our colleagues and students, made sacrifices in the interest of *Terrestrial Plant Ecology*.

Michael G. Barbour
Jack H. Burk
Wanna D. Pitts

CONTENTS

TERRESTRIAL
PLANT
ECOLOGY

SECOND EDITION

PART I

BACKGROUND AND
BASIC CONCEPTS

The origins of informal plant ecology go back hundreds of years to plant geographers, taxonomists, and naturalists who were people of energy and insight, such as Humboldt, De Candolle, and Darwin. In contrast, the science of ecology has had a very brief, intensive history, beginning less than a century ago. In that brief period, certain energetic men and women have had such a profound effect on the developing science that many current research projects and viewpoints are simply extensions of their work.

In this part, we will examine some of that history and some of those personalities, and we will present a common language to use in our study of plant ecology. There are many approaches to plant ecology, but each one attempts to answer the same basic question: How do plants cope with their environment? Some of the approaches we will discuss are paleoecology, phytosociology, community dynamics, systems ecology, and autecology.

1

CHAPTER 1

INTRODUCTION

Plant ecologists try to discover an underlying order to vegetation. They do this at finer and finer levels for the same reasons that biologists, chemists, and physicists pursue their worlds to the level of DNA, hydrogen bonds, and subatomic particles: There seems to be a human need to know the complete story, to explain the past, and to predict the future. What threads link plants to each other and to their environment? How flexible are these threads, and how intermeshed? How do plants "solve" the problems of dispersal, germination in a suitable site, competition, and the acquisition of energy and nutrients? How can they withstand unfavorable periods of fire, flood, or storm?

What can plants tell us by their presence, vigor, or abundance about the past, present, and future course of their habitat? Can plants be used as a scientific tool to analyze the intricacies of the environment or to test hypotheses about evolution?

What can plants tell us about our management hopes for the land? Once a forest is cut, what will replace it, how long will the process take, and how can we most efficiently manipulate this process? Once domestic livestock graze at a given density for a given time, what will the vegetation look like and how many animals will it then be able to support? Once topsoil is removed by strip mining in semiarid regions, what plants should be introduced to stabilize the remodeled landscape? Once brushfields have been sprayed, burned, and replanted as grasslands, what will happen to watershed quality, soil nutrient levels, and the rate of siltation behind nearby dams? What is the residence time of herbicides in soils and are there side effects on nontargeted organisms? How many backpackers can use an alpine trail without altering adjacent vegetation? If fire or flood is a natural catastrophe that must recur with a certain frequency to maintain particular vegetation types, how can we incorporate such regular disasters into state and federal park management plans?

All of these questions, and more, are being investigated by plant ecologists. Some researchers are more interested in generating basic information that has to do with the description of vegetation or the biology of component species. Others are more interested in applying that basic information to management problems. Applied plant ecologists may be called range managers, foresters, or agronomists, but they are all plant ecologists and they all share the same pleasure in discovering the subtle ways in which plants are adapted to their environment. Their objective is very close to a formal definition of **ecology**: the study of organisms in relation to their natural environment.

Ecology, Environment, and Vegetation

The word ecology was coined little more than 100 years ago by the German zoologist Ernst Haeckel. He spelled the word "oekologie," but ecologists soon dropped the first o, to the annoyance and confusion of some purists. The word comes from the Greek roots *oikos*, home, and *logos*, the study of. Thus, oekologie translates loosely as: the study of organisms in their home, their environment.

Environment is the summation of all **biotic** (living) and **abiotic** (nonliving) factors that surround and potentially influence an organism; it is the organism's habitat. Examples of biotic factors include competition, mutualism, allelopathy, and other interactions between organisms, which are described in Chapters 6 and 7. Abiotic factors include all chemical and physical aspects of the environment that influence a plant's growth and distribution.

The environment can be divided into two parts: the macroenvironment and the microenvironment. The **macroenvironment** is the prevailing regional environment and the **microenvironment** is the environment close enough to an object to be influenced by it (Figure 1-1). The microenvironment may be quite different from the macroenvironment. For example, the microenvironment beneath a forest canopy is different from the macroenvironment above it in such traits as humidity, wind speed, and light intensity; the microenvironment beneath a rock in desert soil may be cooler and moister than other parts of the macroenvironment; the microenvironment just 1 mm above a leaf surface may differ in wind speed, humidity, and temperature from the macroenvironment 10 mm away. Each organ or part of a plant is exposed to a different microenvironment, as shown in Figure 1-2. Obviously, the microenvironment is what a plant responds to, and so the microenvironment will be emphasized in this book.

Plant ecology is concerned not only with individual plants and plant species, but also with vegetation. **Vegetation** consists of all the plant species in a region (the flora) and the ways those species are spatially or temporally distributed. If the region is large, its vegetation will consist of several prominent plant communities. Each vegetation type is characterized by the **growth form** of its dominant plants (the largest, most abundant, characteristic plants). Examples of growth forms include annual herbs, broadleaf evergreen trees, drought-deciduous shrubs, plants with bulbs or rhizomes, needleleaf evergreen trees, perennial bunchgrass, and dwarf shrubs (Figure 1-3). Growth forms may include any or all of the following, depending on the context: (a) the size, life-span, and woodiness of a taxon, for example, herb, annual, perennial, herbaceous perennial, woody per-

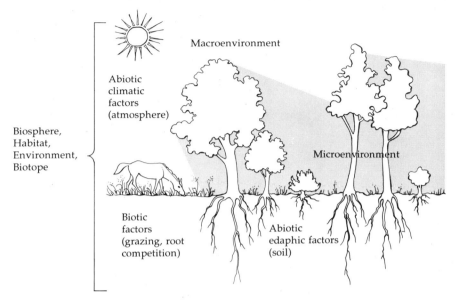

Figure 1–1. Components of the environment.

Figure 1-2 Temperature of the microenvironment (°C) near different parts of an alpine plant in sun and shade on a July day when the temperature of the macroenvironment was 11.7°C. (Redrawn from Tikhomirov 1963; also cited in Larcher 1975.)

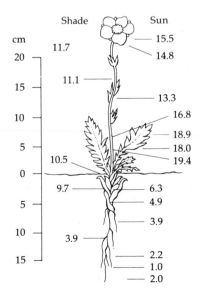

(a)

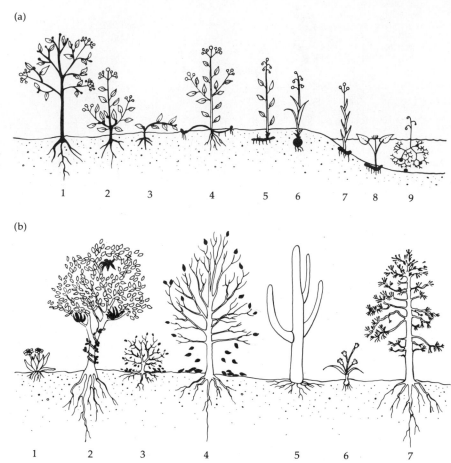

(b)

Figure 1-3   Examples of life forms (growth forms). (a) A scheme of classification by Raunkiaer (1934) based on the location of overwintering, perennating parts, such as buds, bulbs, or seeds, shown in dark, solid areas: (1) phanerophyte, (2–3) chamaephytes, (4) hemicryptophyte, (5–9) cryptophytes (also called geophytes). (b) Life forms based on other criteria, such as length of life, succulence, and leaf traits: (1) annual herb, (2) broadleaf evergreen tree with a liana (a woodly vine) and an epiphyte (a plant which grows on another plant but uses it only for mechanical support, not as a source of nutrients), (3) drought-deciduous shrub, (4) broadleaf deciduous tree, (5) stem succulent, (6) bulbous herbaceous perennial, (7) needleleaf evergreen. ((a) Redrawn from C. Raunkiaer, 1934. *The Life Forms of Plants and Statistical Plant Geography.* Courtesy of Clarendon Press, Oxford.)

ennial, shrub, tree, or vine; (b) the degree of independence of a taxon, for example, green and rooted to the ground, parasitic, saprophytic, or epiphytic; (c) the morphology of a taxon, for example, stem succulent, leaf succulent, rosette form, spinescent, or pubescent; (d) the leaf traits of a taxon, for example, large, small, sclerophyllous, evergreen, winter-deciduous, drought-deciduous, needleleaf, or broadleaf; (e) the location of perennating buds, as defined by Raunkiaer (1934); and (f) **phenology,** the timing of life-cycle events in relation to environmental cues.

Vegetation is also characterized by the architecture of its canopy layers. Different forest types have one to four tree canopy layers. Architecture and life form both contribute to the **physiognomy** (outer appearance) of vegetation, and each vegetation type has its own characteristic physiognomy. A vegetation type that extends over a large region is called a **formation**. For example, a tropical rain forest is a formation dominated by broadleaf evergreen trees and is characteristic of thousands of square kilometers in humid tropical regions on several continents. Part of the formation classification scheme adopted by UNESCO (1973) appears in Table 1-1 at the end of this chapter.

Formations may be subdivided into **associations**. An association is the collection of all plant **populations*** coexisting in a given habitat. As formally defined in an international botanical congress early this century, an association has the following attributes: (a) it has a relatively fixed floristic composition, (b) it exhibits a relatively uniform physiognomy, and (c) it occurs in a relatively consistent type of habitat. The same species tend to recur together wherever a particular habitat repeats itself. Associations are usually named by their dominant or most characteristic taxa; thus the red fir forest in California's Sierra Nevada, which occurs on well-drained, undisturbed sites between 1800 and 2400 m, is a red fir (*Abies magnifica*) association in a conifer forest formation.

Typically, several to many associations may belong to the same formation, all sharing a similar physiognomy but each differing qualitatively or quantitatively in species composition. A fascinating ecological phenomenon is the similarity of vegetation types in similar macroenvironments scattered around the world. It is as though a particular physiognomy has been selected for in similar but isolated habitats. Evidently there has been convergent evolution among vegetation types (see Chapter 8).

Specializations Within Plant Ecology

Synecology (Community Ecology)

One large segment of plant ecology has followed directly from plant geography (Figure 1-4). This is **synecology**, and some of its many synonyms include community ecology, phytosociology, geobotany, vegetation science, and vegetation ecology.

---

*A population is a group of individuals of the same species occupying a habitat small enough to permit interbreeding among all members of the group. Some populations do not interbreed, being self-pollinated or reproducing only asexually, but they exist in a habitat small enough to permit the *potential* exchange of genes. Very restricted species, limited to a small area, may consist of only a single population, but most species include many populations. Consequently, the terms species and population are not usually synonymous.

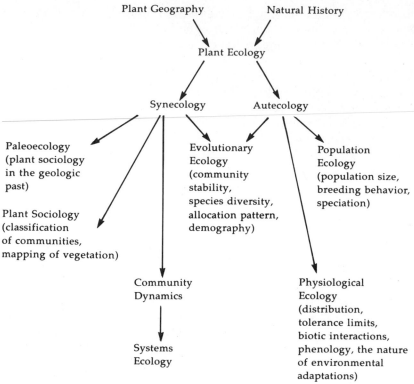

Figure 1-4   The relationship between, and origins of, specializations within the field of plant ecology.

One phase of synecology is **plant sociology,** the description and mapping of vegetation types and communities. (**Community** is a general term that can be applied to any vegetation unit, from the regional to the very local.) In the past 50 years there has been a proliferation of standard methods for sampling vegetation and treating and analyzing sampling data. With these standard methods, valid conclusions can be drawn and vegetation from all over the world can be compared on an equal basis. (The description of *past* vegetation types and associations, as they have existed through geologic time, is part of the field called **paleoecology**.)

Another phase of synecology is the examination of **community dynamics,** which includes processes such as the transfer of nutrients and energy between members, the antagonistic or symbiotic relationships between members, and the process and causes of **succession** (community change over time). The study of community dynamics can be abstracted to a mathematical level, where complex formulae and computer programs summarize, simulate, or model the particular dynamic system being examined. This type of research is called **systems ecology**.

A third phase of community ecology tries to deduce evolutionary themes that determine the fundamental nature of communities. What determines the number of species that may coexist in a habitat? How may communities be described in terms of function rather than as taxa? What determines the stability or fragility of a community? How is physiognomy selected for? How may species be described in terms of adaption patterns? How have plants and animals coevolved in the complex, gradual formation of the communities that exist today? This phase is called **evolutionary ecology,** and it overlaps with autecology and population ecology.

Autecology

Another large segment of plant ecology deals with the adaptations and behavior of individual species or populations in relation to their environment; this is **autecology**. Subdivisions of autecology include demecology (speciation), population ecology and demography (the regulation of population size), physiological ecology or ecophysiology, and genecology (genetics). Autecologists try to explain the *why* of a given species' distribution: What phenological, physiological, morphological, behavioral, or genetic traits seem most important to the species' continued success in a given habitat? They try to illustrate the pervasive influence of the environment at the population, organismic, and suborganismic levels. There is some overlap with evolutionary ecology when autecologists attempt to summarize all this as a species' adaptive pattern for survival. Autecology flows easily into other specialties outside the field of ecology proper, such as physiology, genetics, evolution, biophysics, and biosystematics (a part of taxonomy).

A Final Word on Specialization

Plant ecology itself may be thought of as a specialization within ecology. Some scientists and educators criticize the division of ecology into plant and animal ecology, arguing that the division is artificial and damaging to an understanding of the interdependence that permeates ecosystems. (An **ecosystem** is the sum of the plant community, animal community, and environment in a particular region or habitat.) Plant ecology as a discrete science is partly an artificial creation, but no more so than any other science. There are many ways in which chemistry is linked to physiology, or physiology to behavior, or soils to geology, or mathematics to economics, yet these fields are accepted as reasonable specializations. We all specialize, and in this manner progress is made. One person cannot master all of plant ecology, let alone plant and animal ecology both. In addition, we think a case could be made that the inherent differences in structure, behavior, and function between plants and animals are so profound that many principles of plant ecology cannot be translated into principles of animal ecology, and vice versa.

The answer to the charge that we artificially fragment ecology is not that we should specialize less, but that we should communicate more. The excellent reviews of the history of general ecology in North America by Egerton (1976) and McIntosh (1976) are a good beginning in this direction.

Table 1-1   Part of the formation classification scheme adopted by UNESCO (1973).

Formation	Description
Closed forest formation class	Dominants 5^+ m tall, crowns interlocking
Mainly evergreen forest	Individual trees may shed leaves, but canopy as a whole remains green
Tropical ombrophilous forest (= tropical rain forest)	Dominants mainly broadleaved, evergreen, with drip tips; neither cold- nor drought-resistant
Tropical and subtropical evergreen seasonal forest	Number of drought-deciduous trees intermediate between above and below
Tropical and subtropical semideciduous forest	Most upper canopy trees drought-deciduous; many understory trees evergreen and sclerophyllous; leaves without drip tips
Subtropical ombrophilous forest	Local variant, grading into tropical rain forest
Mangrove forest	Intertidal location in tropics and subtropics; dominated by evergreen sclerophyllous broadleaved trees with stilt roots or pneumatophores; vascular epiphytes rare
Temperate and subpolar evergreen ombrophilous forest	Occurs in extremely oceanic, frostfree climates of Southern Hemisphere, as in *Nothofagus* or *Podocarpus* forests of New Zealand
Temperate evergreen seasonal broadleaved forest	Dominated by hemisclerophyllous evergreen trees; rich in herbaceous undergrowth, but few epiphytes and lianas; grades into above and below
Winter-rain evergreen broadleaved sclerophyllous forest	Dominated by sclerophyllous, evergreen trees with little understory but some lianas
Tropical and subtropical evergreen needleleaved forest	Dominated by needleleaved or scaleleaved evergreen trees; vascular epiphytes and lianas absent
Temperate and subpolar evergreen needleleaved forest	As above, but to the north
Mainly deciduous forest	Majority of trees shed foliage simultaneously in connection with an unfavorable growing season

Table 1-1 *continued*

Formation	Description
Tropical and subtropical drought-deciduous forest	Foliage shed during dry season (usually winter)
Cold-deciduous forests with evergreen trees	Foliage shed during frost season; deciduous trees dominant but evergreens present as in hemlock-hardwood forest
Cold-deciduous forests without evergreen trees	Deciduous trees absolutely dominant, vascular epiphytes absent
Extremely xeromorphic forest	Dense stands of xeromorphic trees, with succulents and xeromorphic shrubs, often grading into woodland (below)
Woodland formation class	Dominants 5^+ m tall, crowns not usually touching, but canopy cover 40^+%; herbaceous layer may be present
Mainly evergreen woodland	Dominants evergreen
Mainly deciduous woodland	Dominants variously deciduous
Extremely xeromorphic woodland	Similar to xeromorphic forest, but trees less dense
Scrub formation class	Dominants shrubs or dwarf trees 0.5–5 m tall
Mainly evergreen scrub	Includes chaparral
Mainly deciduous scrub	Includes riparian thickets
Extremely xeromorphic (subdesert) shrubland	Very open stands of shrubs with xerophytic adaptations; some plants with thorns
Dwarf-scrub and related communities	Dominants less than 0.5 m tall; includes arctic-alpine tundra, bogs, heaths
Herbaceous vegetation	Dominated by graminoids or forbs; more or less continuous cover; woody synusia less than 40% cover
Tall graminoid vegetation	Dominant graminoids 2^+ m tall when in flower; forb cover less than 50%
Tall grassland with tree synusia 10–40% cover	An open woodland with graminoid cover greater than 50%
Tall grassland with tree synusia less than 10% cover	Savannas, sometimes with shrubs
Medium tall grassland	Dominant graminoids 0.5–2 m tall when in flower; forb cover less than 50%
Short grassland	Dominant graminoids less than 0.5 m tall when in flower; forb cover less than 50%; includes meadows, some types of tundra
Forb vegetation	Forb cover greater than 50%; graminoid cover less than 50%

CHAPTER 2

A BRIEF HISTORY
OF PLANT ECOLOGY

The route a science takes in its development is determined by the personal traits, interests, culture, and social surroundings of certain people along the way, perhaps as much as it is determined by the hypotheses, facts, or approaches that each person has contributed. Good science may be impersonal, dispassionate, and unbiased, but most scientists are not. Consequently, this brief history of plant ecology is flavored not only by people's ideas, but also by the people themselves.

Foundations in Plant Geography

Prior to the Nineteenth Century

Plant ecology has been studied informally and pragmatically since the beginning of the human race. Gatherers and hunters mastered a knowledge of the distribution of the wild food and forage plants that sustained them and the prey they hunted. Shamans (medicine men) learned the narrow habitat requirements of rare species that had healing, narcotic, or hallucinogenic qualities; no doubt being able to find such plants was as important as knowing how to use them. Aristotle and Theophrastus, ca. 300 B.C., may have been the first to write about plant geography and plant ecology, but illiterate peoples throughout the world must certainly have had an understanding of these topics well before that time.

Some of the earliest formal ecology papers, dating back to the seventeenth century, were concerned with the succession of communities surrounding gradually filling lakes and bogs, and the term succession was used in its modern context by the beginning of the nineteenth century (Clements 1916). However, the real development of plant ecology came through books on plant geography written by people trained as taxonomists or general botanists. World exploration, especially in the nineteenth century, molded an increasingly rigorous ecological viewpoint. Carl Ludwig Willdenow (1765–1812) was the pioneer of this line of thought. He was an early plant geographer who noted that similar climates produce similar vegetation types, even in regions thousands of kilometers apart, such as southern Africa and Australia.

The Nineteenth Century

Willdenow's teaching at the University of Göttingen, in Germany, greatly influenced a wealthy young Prussian student, Friedrich Heinrich Alexander von Humboldt (1769–1859) (Figure 2-1), who studied botany to round out his education in higher mathematics, natural sciences, and chemistry. Not long after graduating, Humboldt met Johann Forster, who had accompanied James Cook on his world voyage of discovery. Forster's stories made Humboldt determined to visit the new world tropics. A decade later Humboldt met a young French botanist named Aimé Bonpland, who had a similar desire. Plans crystallized, and they received the permission and protection of King Carlos IV of Spain to travel in what is now Latin America. They took with them the best equipment of the day for measuring latitude, elevation, temperature, humidity, and other physical factors. For five years they traveled from steamy lowland rain forest to cold alpine paramo and from arid desert to thorn scrub. They explored Cuba, Venezuela, Ecuador, Peru, Mexico, and the Orinoco and Amazon Rivers; they climbed nearly to the top of Mt. Chimborazo (to 5900 m); they collected 60,000 plant specimens. On his way back to Europe in 1804 Humboldt was a house guest of President Jefferson in Washington. Jefferson himself was keenly interested in plant responses to climate and studied the phenology of garden plants along latitudinal gradients. Jefferson had only recently sent Lewis and Clark west, and he questioned Humboldt on the nature of his discoveries (Billings 1985). Summaries of Humboldt's travels can be found in Botting (1973) and McIntyre (1985).

Humboldt returned to France and began to write his monumental 30-volume work, *Voyage aux regions equinoxiales*. The first 14 volumes were devoted to botany, and in those he coined the term **association**, described vegetation in terms of physiognomy, correlated the distribution of vegetation types with environmental factors, and described the synergistic effects of some physical factors (for example, elevation, latitude, and temperature). His statement, "In the great chain of causes and effects no thing and no activity should be regarded in isolation," is a striking preview of our modern view of interdependence within communities and ecosystems. Near the end of his life he wrote a five-volume encyclopedia, *Kosmos*, which attempted to describe and explain the entire universe. Humboldt was one of the last Renaissance people, attempting to master all the knowledge of his time.

Humboldt's study of plant geography was furthered by Schouw, De Can-

Figure 2-1 Friedrich Heinrich Alexander von Humboldt, 1769–1859. (Courtesy of the Hunt Institute for Botanical Documentation, Carnegie–Mellon University, Pittsburgh, PA.)

Figure 2-2 Johannes Eugenius Bülow Warming, 1841–1924. (Courtesy of the Hunt Institute for Botanical Documentation, Carnegie–Mellon University, Pittsburgh, PA.)

dolle, Kerner, and Grisebach, among others. J. F. Schouw (1789–1852), a professor at the University of Copenhagen, methodically described the effects of major environmental factors on plant distribution in an 1823 book, emphasizing the role of temperature. This search for single, most important factors is still with us today, but more and more we understand the interdependence of all factors. Schouw popularized the procedure of naming associations by combining the dominant genus with the suffix *-etum*. Thus, *Quercetum* is an oak woodland association (*Quercus* + *etum*) and *Pinetum* is a pine forest association (*Pinus* + *etum*). Some modern schemes of association nomenclature still use this concept.

Anton Kerner von Marilaun (1831–1898) studied medicine at the University of Vienna but gave up practice after he experienced a cholera epidemic as an intern. He turned to botany as a less traumatic career, became Professor of Botany at the University of Innsbruck, and then was commissioned by the Hungarian government to describe the vegetation in parts of eastern Hungary and Transylvania. The book that resulted, *Plant Life of the Danube Basin* (1863), has fortunately been translated into English, so a larger audience can now appreciate the beauty of his vegetation descriptions and his clear understanding of succession. Later, Kerner became one of the first experimental ecologists. He established several transplant gardens at various elevations in the Tyrolean Alps, from Vienna at 180 m to his villa at 1200 m, and up to the alpine zone at 2200 m. In each garden he grew alpine and lowland forms of more than 300 species together. He found that some variations within a species were fixed and heritable, while other variations were nonheritable modifications induced by the environment (Kerner 1895).

August Grisebach (1814–1879) traveled widely and described more than 50 major vegetation types in very modern physiognomic terms, relating their distribution to various climatic factors. He succeeded Willdenow as Professor of Botany at Göttingen.

Alphonse Luis Pierre Pyramus De Candolle (1806–1893) was an herbarium taxonomist, an armchair plant geographer, but his access to vast plant collections led him to try to "discern the laws of plant distribution." Like Schouw, he chose to study temperature. He summed temperatures according to a formula so useful that his 1874 data later became the basis for Köppen's famous classification of climate, published half a century later (see Gates 1972 and Ackerman 1941).

Other biologists who contributed to the development of plant ecology in the mid- to late-nineteenth century include Oscar Drude (1890 and 1896), Adolf Engler (1903), George Marsh (1864), Asa Gray (1889), and Charles Darwin. The theory of evolution and the evidence which Darwin marshalled to support it in his book, *Origin of Species* (1859), are inherently ecological (see Harper 1967). It is fitting that Darwin's explorations and accounts were inspired by Humboldt's plant ecology work.

The Establishment of Plant Ecology Apart from Plant Geography

The plant geography period culminated with the publication of several extremely important, innovative books between 1895 and 1916. At the end of those 20 years, plant ecology was firmly established as a science in its own right, and most of our modern research directions had been initiated. The major contributors were Warming, Schimper, Ramensky, and Paczoski in Europe, and Merriam, Cowles, and Clements in America.

Johannes Eugenius Bülow Warming (1841–1924)

As a young man in Holland, Johannes Warming (Figure 2-2) was offered the chance to help a paleontologist conduct a field study near Lagoa Santa, Brazil. It was an area of tropical woodland-savanna, 42 days' travel northwest of Rio de Janeiro. He spent three years there, and 30 years later, when the last of his 2600 plant specimens had finally been identified or described, he wrote a book (1892) on the vegetation. Its classical ecological organization would serve as a model for any vegetation study done today. Introductory chapters on geology, soils, and climate are followed by sections on each of the major vegetation types and communities. He discussed dominants and subdominants, the adaptive value of various life forms, the effect of fire on community composition and succession, and the phenology of communities and taxa.

In the meantime, he had returned to teach at the University of Copenhagen as a successor to Schouw. He organized what may have been the world's first ecology course and was recognized as an outstanding teacher. He published his lecture notes in 1895, in Danish, as the world's first plant ecology text. The book had an immediate impact on many botanists throughout the Western world. It was later translated into German, Polish, Russian, and finally English (1909). Warming synthesized plant morphology, physiology, taxonomy, and biogeography into a coherent science for the first time. He concluded that soil has more of an effect on vegetation than climate, and he emphasized both moisture and temperature as prime climatic factors. He coined such useful, and still commonly used, terms as **halo-**, **hydro-**, **meso-**, and **xerophyte**, meaning plants of saline, wet, moist, and dry habitats, respectively.

Andreas Franz Wilhelm Schimper (1856–1901)

Andreas Schimper was born into a celebrated family of German botanists. He studied geology and botany at the University of Strassburg, and later taught plant histology, physiology, ecology, and geography at the University of Bonn. He traveled extensively in the tropics and concluded that the basic work of plant geography and taxonomy would soon be completed; he was not quite correct. Based on his conclusion, however, the goal he set for himself was to explain the *causes* of regional differences in floras and vegetation. Near the end of his short life, he published *Plant Geography on a Physiological Basis* (1898 and later in English in 1903). The book's title is somewhat misleading; he stressed morphological features of presumed adaptive value and presented little of plant physiology in the modern sense. However, many of his semiphysiological conclusions are still accepted today, and Billings (1985) calls his work the real beginning of plant physiological ecology.

Like Warming, Schimper gave weight to both climatic and edaphic (soil) factors, and among climatic factors he emphasized temperature and moisture. He appeared to borrow heavily from Warming's text and figures (Warming's book had been translated into German two years before), but nowhere did he give Warming so much as a footnote of credit. It is possible that Schimper came to Warming's conclusions independently, but some of Warming's supporters still suspect Schimper of plagiarism (Goodland 1975).

Jozef Paczoski (1864–1941) and Leonid Ramensky (1884–1953)

The spread of Jozef Paczoski's (pronounced pach · ós · ky) ideas suffered because they were published in a Slavic rather than a western European language, and they were not quickly translated (see Maycock 1967). He has only belatedly been credited in the U.S.S.R. and North America as the father of **phytosociology** (which he defined in his 1891 and 1896 papers as all the sociological relationships of plants). He showed how plants modify the habitat, creating their own microenvironment. He discussed the role of competition, the causes of succession, the role of fire, the interdependence of species in a community, the continuum nature of community boundaries, and such physiological adaptations as shade tolerance. He later published a text on phytosociology (1921) and founded the world's first phytosociology department at the University of Poznan, Poland. Many current phytosociology practices, terms, and concepts in Europe were conceived by him long before their current popularity.

Ramensky (also written Ramenskii) has also only recently been appreciated by Western plant ecologists (Rabotnov 1953 and 1978, Soboleve and Utekhin 1978, McIntosh 1983a). His concepts of the individuality of species and the continuum of vegetation predate those of Gleason and Whittaker by half a century. He developed methods of gradient analysis and, like Paczoski, showed how communities graded into one another (he coined the term **phytocoenosis**). He also expressed community composition in a table form much like that later popularized by Braun-Blanquet. Ramensky also prefigured the autecological *C-S-R* and *r-K* categories of Grime, MacArthur, and Pianka by dividing plants into three groups: "violent" (competitors, *K*-strategists), "patient" (stress-tolerators), and "exploring" (ruderals, *r*-strategists).

Clinton Hart Merriam (1855–1942)

Clinton Merriam received an M.D. degree from Columbia University, but later devoted himself to biology (especially zoology) and served as Chief of the U.S. Biological Survey from 1885 to 1910. During that time he visited many parts of the western United States, writing with a naturalist's eye about new species of mammals and the distribution of vegetation types, and with an anthropologist's eye about Indian groups in California. He was a founder of the National Geographic Society.

As his large expeditions with long supply trains moved slowly through the West, he was impressed with the similarities of elevational zones of vegetation from one mountain range to another. He believed that temperature, especially warmth in the growing season, was the determining environmental factor, and he developed formulae to sum degrees of warmth much as Grisebach had done. He then correlated each vegetation type to values of warmth, enlarged the vegetation types into life zones (which he named Boreal, Transitional, Canadian, Hudsonian, Tropical, etc.) and extrapolated the distribution of his life zones across the entire North American continent (1890, 1894, and 1898; Figure 2-3 is a small example).

Merriam had a strong measure of self-confidence, for he later wrote, "in its broader aspects the study of the geographic distribution of life in North America is completed. The primary regions and their subdivisions have been defined and mapped, the problems involved in the control of distribution have been solved, and the laws themselves have been formulated." Although some naturalists and biology books still use his life-zone terminology and concept, many of his assumptions, calculations, and conclusions have been rightfully challenged (see,

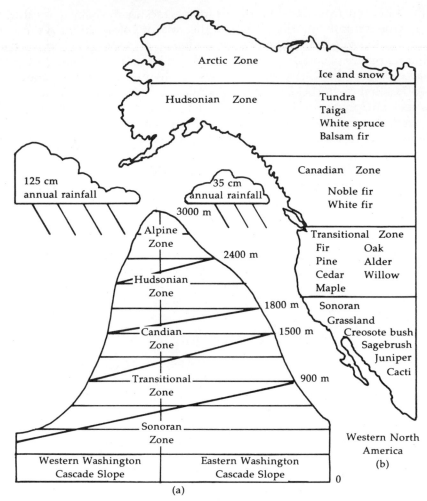

Figure 2-3   Example of Merriam's life zones (a) in the Cascade Mountains of Washington and (b) diagrammatically extended across the western states. (Adapted from *Living Systems* by J. Ford and J. Monroe. Copyright © 1971 by Ford and Monroe. Reprinted by permission of Harper and Row, Publishers, Inc.)

for example, Daubenmire 1938), and his latitudinal extrapolations are not widely accepted. Nevertheless, Merriam had a strong impact on American plant ecology.

Henry Chandler Cowles (1869–1939)

Henry Cowles was a geologist-turned-botanist who investigated plant succession on sand dunes around Lake Michigan from the perspective of a geologist. A leading botanist of that time, John Coulter, was his major professor. Cowles had a gentle, amiable personality that attracted many students to him.

Gleason wrote in 1940, "he was a man of infectious gaiety and high spirits, of infinite humanity and humor." The University of Chicago, where he taught, became a center of plant ecology during the first 20 years of this century. He was an organizer of the Ecological Society of America in 1915 and served as its president three years later. Most professional ecologists in the United States today belong to this society. It publishes research journals and sponsors scientific meetings at which ecologists can exchange information and ideas.

His Ph.D. dissertation and later papers on dune succession, from 1898 to 1911, emphasized the dynamic nature of vegetation, as Warming and Schimper somehow had not. This dynamic aspect of ecology attracted botanists who until then had thought of plant ecology as merely the mapping and description of static pieces of vegetation. Cowles had an impact on American plant ecology mainly through his students, who adopted his emphasis on succession and applied it throughout the midwestern and eastern United States. The pioneer animal ecologists Charles C. Adams and Victor Shelford were colleagues of Cowles at Chicago and they were strongly influenced by his work. However, the University of Chicago "school" of plant ecology was soon supplanted as a center of plant ecology by the Nebraska "school" of Clements.

Frederick Edward Clements (1874–1945)

Frederick Clements (Figure 2-4) provided a geographic balance to Cowles; he was born, raised, and educated in Nebraska and traveled widely in the western United States. Clements' life and contributions have been well described by Tobey (1981). During his student days at the University of Nebraska, he was prodded by Professor of Botany C. E. Bessey to expand his classical undergrad-

Figure 2-4   Frederick Edward Clements, 1874–1945. (Courtesy of the Hunt Institute for Botanical Documentation, Carnegie–Mellon University, Pittsburgh, PA.)

uate education by adding a large dose of field botany. Together with some other students, Clements attempted to collect, identify, and describe the distribution of every plant species in the state, from algae to oaks. He collected hundreds of plants and identified many new species of fungi, ultimately co-authoring a book in 1898, *The Phytogeography of Nebraska*, which received wide recognition.

He married a university student, Edith Schwartz; the next year she switched her Ph.D. studies from German to botany. Long after (1960), she wrote a charming book about her life with Clements, and it is clear that she was a great help to him in his career, functioning alternately as driver, secretary, photographer, translator, and sometimes co-author.

Clements described much of the vegetation of North America, naming regional formations and associations, local variants, and seral stages with great authority, if not always with great precision. He wrote about the causes of succession, the use of certain species as environmental indicators, and the methods for documenting succession or in identifying associations. His classical background meshed with his philosophical, precise, rigid personality to produce large, comprehensive books (e.g., 1916 and 1920) filled with many new terms and with conclusions that have a dogmatic tinge. He defined a plant association in terms of an organism in order to illustrate the interdependent nature of an association's component species (see Chapters 8 and 11). The wealth of information, the new terms, the philosophical sweep of completeness, and the assertive conclusions confused some readers and offended others, and thus many of his ideas have been criticized unjustly. Much of his plant community work retains validity today.

In 1917 he was hired by the Carnegie Institution of Washington, D.C., to direct research at a coastal laboratory at Santa Barbara, California, during the winter and an alpine laboratory at Pikes Peak, Colorado, during the summer. He devoted the rest of his life to full-time research on the causes of plant distribution, experimenting with transplant gardens much as Kerner had done before him, yet arriving at completely different conclusions, which are not considered valid today (see Clements and Hall 1921). The Carnegie Institution was an important source of support for the young science of plant ecology (McIntosh 1983b, Billings 1985). Frederick Coville, who did pioneering botanical work in Death Valley, convinced the Institution to establish a desert laboratory for the study of southwestern desert vegetation. In 1903 the laboratory was erected just west of Tucson, Arizona. Very productive researchers, such as William Cannon, Daniel MacDougal, Forrest Shreve, and Heinrich Walter, worked out of that laboratory during the same years that Clements worked at the other two Carnegie laboratories.

Clements performed valuable public service work with the Soil Conservation Service during the Dust Bowl of the mid-1930s, and he wrote one of the first American plant ecology textbooks with John E. Weaver in 1929. He was actively engaged in research until a few weeks before his death at the age of 70.

His personality was quite different from that of Cowles. It was "powerful" according to Tansley (1947), "decidedly puritan, even ascetic . . . and his manner was apt to be tinged with a certain arrogance . . . [Nevertheless] he had that best of all senses of humour which enables a man to laugh at himself. . . . [He was] by far the greatest individual creator of the modern science of vegetation."

Plant Ecology Since 1925

By 1925 there were many plant ecologists in Europe and America, contemporaries of the above pioneers, each asking his or her own particular questions and expanding the field of plant ecology in a particular direction.

William Cooper, Edgar Transeau, and Emma Lucy Braun, all students of Cowles, described forests and bogs and paths of succession in the Midwest. Cooper also traveled long distances, for those days, to examine succession behind retreating glaciers in Alaska, on sand dunes along the Pacific coast, and in California chaparral. Transeau introduced ecology into general botany texts and foreshadowed our current interest in ecosystem productivity and energy transfer by working out an energy budget of a cornfield in 1926. E. Lucy Braun became famous for her descriptions of the virgin deciduous forests of eastern North America. She and her sister traveled extensively through the Appalachian Mountains during Prohibition and the Depression, gaining acceptance by the local people when male strangers might have been rebutted.

In 1926, Henry Gleason wrote a widely read paper that attacked Clements' basic assumptions about the nature of associations (see Chapter 8). As a result, he wrote in 1953, "to ecologists I was anathema. Not one believed my ideas; not one would even argue the matter. . . . For ten years, or thereabout, I was an ecological outlaw, sometimes referred to as 'a good man gone wrong'." Gleason's ideas have since become standard parts of current ecology books, now often taken with as much faith as Clements' ideas had been taken before.

Forrest Shreve and William Cannon labored in the deserts, Cannon relating morphology and anatomy to habitat in the tradition of Schimper and Warming, and Shreve describing the major communities of the warm deserts of North America as a research associate of the Carnegie Institution from 1908 to 1945. Shreve drew one of the first published vegetation maps of North America in 1917, and he was a member of the small group Cowles brought together to found the Ecological Society of America. His wife, Edith Shreve, wrote some excellent articles on plant-water relations.

Ecophysiological research expanded in the 1940s and 1950s (Billings 1985). Frits Went supervised the construction at Cal Tech of an elaborate growth chamber called a phytotron, which permitted research on the roles of temperature and photo-period on plant behavior. Gas exchange chambers, capable of measuring whole plant or leaf photosynthesis, were developed in the 1950s, and the first portable one was taken into the mountains of Wyoming in the summer of 1958 by Harold Mooney, Ed Clebsch, and Dwight Billings (Billings et al. 1966). Within a few years, several groups in the United States and Europe had mobile laboratories for the measurement of transpiration, plant-water stress, and photosynthesis.

Major synecological contributions were made by Robert H. Whittaker from the 1940s through the 1970s (see Figure 2-5). Raised during the Depression, Whittaker developed a strong work ethic. He turned his keen intellect and energy to a wide diversity of topics: the classification of communities; development of techniques such as ordination and gradient analysis, which permitted complex vegetation patterns to be related to equally complex environmental factors; the measurement of species diversity and an assessment of its significance; studies of the process and driving forces of succession; comparative studies of biomass and plant productivity; and an analysis of the roles of inhibitory metabolic compounds (allelochemics) in the ecosystem. He also proposed a five-kingdom classification for organisms, an approach since widely adopted by biology texts and instructors.

Whittaker concentrated on zoology and entomology in his undergraduate work but combined botany and zoology as a graduate student at the University of Illinois, working under both Vestal (a botanist) and Kendeigh (a zoologist). Interestingly, the Botany Department would not admit him as a botany graduate student because it was felt he lacked appropriate background courses. (Such mistakes can be made anywhere.) He completed his Ph.D. in two and one-half years, surely a modern record and far below today's normal period of five to six years. For a first-person account of his difficult times as a graduate student, read his essay in Jensen and Salisbury's 1972 textbook. Whittaker was aggressive in his challenge of many ecological ideas accepted as dogma at the time (especially Clementsian views on vegetation). Perhaps because of the unease this raised in colleagues who felt challenged, Whittaker was released from his first academic position without tenure. He went on, however, to serve at several universities, completing his career at Cornell, where he died prematurely of cancer in 1980. In view of his enormous importance to plant ecology on this continent, many view him as a second Clements. Unlike Clements, however, his ideas were accepted worldwide, and Whittaker did much to bring together plant ecologists of Europe and North America in a common approach to the study of vegetation.

In Europe, Christen Raunkiaer succeeded Warming as Professor of Botany at Copenhagen. He developed a life form classification that is still widely used (see Figure 1-3a) and a quantitative method of sampling vegetation whereby data could be treated statistically without bias. In Sweden, Göte Turesson experimentally brought plants together in a common garden to examine the genetic basis of variation. His method was not much different from those of Bonnier (1895) and Kerner (1895) before him, but his conclusions were much more general, powerful, and well documented. He described the ecological variation within species as ecotypes (see Chapter 3), and this new concept of the "ecological species" triggered a long series of experiments by others on the subtle ways in which genotypes are selected by local environments. In the United States, one of Clements' associates, Hall, began to establish a transect through California for the purpose of studying plant variation. After his death, Clausen, Keck, and Heisey, from the Carnegie Institution, collaborated to extend that work into a classic ecotype study that further generalized Turesson's ideas.

Sir Arthur Tansley investigated the vegetation of Britain, founded the British Ecological Society, coined the term *ecosystem*, called for more physiological investigations in field studies, and later led a conservation movement in England decades before a movement equally strong developed in the United States. In his first presidential address to the British Ecological Society (1914), he was not

Figure 2-5 Robert H. Whittaker, 1920–1980. (W. E. Westman and R. K. Peet. Reproduced by permission from Dr. W. Junk, Publisher. *Vegetatio* 48:97–122. 1982.)

Figure 2-6 Josias Braun-Blanquet, 1884–1980. (Courtesy of the Hunt Institute for Botanical Documentation, Carnegie–Mellon University, Pittsburgh, PA.)

hesitant to put the importance of the new science of ecology above the older, traditional fields of study. Tansley contributed a great deal to the philosophy and process of ecological research, and he called for a more experimental approach even though he himself was not adept at such an approach. He encouraged an international exchange of views and fostered that with field trips on three continents between 1911 and 1930 (Cooper 1957).

In Wales, beginning in the 1950s, John Harper developed the area of plant demography into an exciting science that attracted many autecologists throughout the world. His examination of the dynamics of plant populations focused on weedy species, but the concepts he formulated in his 1977 book have been applied by others to wildland plants in natural habitats. As a result of his work, weed science has turned in a more biological, ecological direction. "Integrated pest management," rather than reliance on herbicides or mechanical methods, has gained acceptance (Radosevich and Holt 1984).

In Switzerland and later in France, Josias Braun-Blanquet (Figure 2-6), following the path laid down by Kerner, developed his methods of community sampling, data reduction, and association nomenclature that dominate much of plant community ecology today. Eduard Rübel made many other early contributions in Switzerland, and cooperation between Rübel and Braun-Blanquet led to the development of an approach to plant synecology called the Zurich-Montpellier School of Phytosociology.

Braun-Blanquet began his career as a bank clerk, but his strong interest in alpine plants drove him to become a self-made plant ecologist. Winning the acceptance and esteem of more traditional botanists by the quality of his work (and to some extent by the pleasant force of his personality), he eventually received a doctorate and in 1930 founded a research station at Montpellier, France, called Station Internationale de Géobotanique Méditerrènne et Alpine (SIGMA). SIGMA remains a center for vegetation research to this day. Although Braun-Blanquet's methods of vegetation analysis and classification are widely accepted throughout the world, other schools or approaches to synecology have developed in Czechoslovakia, Germany, Scandinavia, and the Soviet Union (see Whittaker 1962).

Internationalism characterized two decades of novel research alliances promoted by the International Biological Program (IBP), created in 1960. Dozens of scientists—animal ecologists, plant ecologists, mathematicians, biochemists, climatologists, systematists—studied different aspects of a single ecosystem in the hope that the pieces could be put together and large questions answered. IBP projects were located in arctic tundra, boreal forests, deserts, grasslands, and marine seabeds, with study sites on several continents. The IBP formally ended in 1974, but publications resulting from its work continued to appear into the 1980s.

The IBP emphasized the **biome**—the sum of plant and animal communities coexisting in the same region. This perspective stimulated research on the role of plants in energy transfer, nutrient cycling, ecosystem stability, ecosystem modeling, and evolutionary ecology. One of the most exciting aspects of evolutionary ecology to receive recent attention is the nature of plant–herbivore relationships. Hard data on co-evolved intricacies are being generated slowly, but a vast body of speculative theory is growing fast (see Chapters 6 and 7).

Where to Now?

Because of the amount of information and data that have accumulated on plant ecology, the diversity of questions that have interest and importance, and the sophistication of modern research methods, professional ecologists today typically concentrate their efforts in a single area of plant ecology. The Ecological Society of America ended its first year with 300 members; today there are more than 5000. Just as we lost our Renaissance people in the nineteenth century, we lost our general plant ecologists in the twentieth.

Another area that will continue to be accented is ecology's contribution to land use planning. In 1969, passage of the National Environmental Policy Act (NEPA) by the U.S. Congress required that future federal activities that would have a significant effect on the environment be postponed until an adequate Environmental Impact Statement could be written and commented on by the public and other agencies. Since then, many states have adopted similar legislation for nonfederal projects. Some impact statements are written by federal, state, or local agencies, but many others are written by private consulting companies who hire ecologists and other specialists. New magazines, technical journals, and many books have been published for this applied, professional audience.

The response and tolerance of vegetation to pollution and to other disturbances will continue to be investigated. In 1971, the Ecological Society of America
helped to create The Institute of Ecology (TIE) to deal with large environmental
policy questions that are beyond the scope of individuals or single institutions.
Its objective is to educate the public and government on environmental issues
and to generate research that will improve the ecological soundness of regional
decisions about land, air, and water use. The National Academy of Sciences has
participated in similar activities.

Major forms of pollutants that will surely receive continued attention include:
increasing carbon dioxide concentration in the atmosphere, acid deposition from
SO_x and NO_x, ozone and hydrocarbons from internal combustion engines, and
toxic waste. In addition, the continued loss of tropical forests and other vegetation types have major ecological impact that will require assessment. The biological consequences include the loss of gene sets (species) as valuable genetic
resources. Apparently, we still need to generate and adopt a life ethic— a credo
that all living organisms are important because they share the unique anomaly
we call life, quite apart from any homocentric uses we may attribute to them.

Basic ecological research will also continue and should be encouraged. Such
research into areas, species, or concepts has no immediate known practical objective; it simply aims to increase our understanding of the biological world. Applied
ecology depends on continuing work in basic ecology, however, and basic research
does often lead to practical applications (McIntosh 1974). Basic research in all the
aspects of plant ecology described in this chapter should continue. None is unimportant; none is completely understood. As Pielou (1981) has cautioned us, conceptual models of the ecological world are still of limited use; they can be improved
only by incorporating additional data based on observations of, and experiments
with, nature.

PART II

THE SPECIES AS
AN ECOLOGICAL UNIT

An autecological question is asked in this part: How do the members of just one species in a community cope with their environment? An incidental, secondary question is also asked: What is an ecological species? A taxonomic species is made up of individuals and populations that may be genetically heterogeneous. An ecological species, however, is a genetically more homogeneous collection of plants adapted to one particular set of microenvironmental conditions.

Part of an organism's environment is composed of adjacent plants and animals, which may or may not be members of the same species. Interactions between pairs of organisms lie along every possible part of a continuum between obligate and incidental, and between mutually beneficial and mutually detrimental. These interactions, whether mediated by chemical or physical factors, can affect the spatial distribution of individuals; in fact, a striking distribution pattern may be an ecologist's first clue to the existence of an interaction.

CHAPTER 3

THE SPECIES IN
THE ENVIRONMENTAL
COMPLEX

The surface of the earth is a network of environmental factors that vary in both space and time. These environmental factors determine the direction of evolution of plant species and may be correlated with the patterns of plant life on the planet. We will consider the ways that environmental extremes and the gradients between them are related to the physiological tolerances of species, as well as how plant evolution reflects variations in and predictability of the environmental complex. We will see the plant species as a dynamic set of reactions able to respond to an ever-changing environment. We will consider examples of how these sedentary organisms are able to adjust to variations in the environment, not only as individuals but also on the level of organs, such as leaves, which react physiologically and developmentally.

Environmental Factors and Plant Distribution

Later chapters will review the importance of certain environmental factors one by one, but there are some general principles that we will discuss in this chapter: the law of the minimum, the theory of tolerance, and the holocoenotic concept of the environment.

The "Law" of the Minimum

In 1840, the agriculturist and physiologist Justus von Liebig wrote that the yield of any crop depended on the soil nutrient most limited in amount. This conclusion has come to be called the **law of the minimum**. The factors covered have been expanded, so that a loose definition of the law would now be: The

29

growth and/or distribution of a species is dependent on the one environmental factor most critically in demand.

The validity of the law has been shown in many parts of the world. The poor growth of some clover pastures in parts of Australia, for example, was shown to be a result of deficiencies in the micronutrients copper, zinc, or molybdenum. Addition of only 6–8 kg ha^{-1} of copper or zinc sulfate every 4 to 10 years increased plant growth 300% and increased the wool harvest from sheep grazing the vegetation even more. As little as 140 g ha^{-1} of sodium molybdate, applied every 5 to 10 years, increased pasture yield six to sevenfold (Moore 1970). In England, the range of certain calcicoles (plants on calcium-rich soil with basic pH) abruptly ends when soil pH drops below pH 5 (Grubb et al. 1969; Rorison 1969), and in California the abundance of typical chaparral shrubs declines (sometimes to zero) when the substrate changes to serpentinite, which has an exceptionally low level of calcium (Walker 1954).

Two limitations to the law of the minimum have become evident, however. First, organisms have an upper tolerance limit to every factor as well as a lower limit. Second, most factors act in concert rather than in isolation; a low level of one factor can sometimes be partially compensated for by appropriate levels of other factors, or the influence of one factor may be magnified as other factors reach their maximum or minimum limits.

The Theory of Tolerance

The pioneer American animal ecologist Victor Shelford (1913) noted weaknesses in Liebig's concept and proposed a modification, which has come to be called the **theory of tolerance**. Ronald Good, a plant geographer, later elaborated on it (1931, 1953), as follows. Each and every plant species is able to exist and reproduce successfully only within a definite range of environmental conditions (Figure 3-1). In general importance, Good rated climatic factors above edaphic factors, and both of these above biotic factors (competition, etc.). Tolerance ranges may be broad for some factors and narrow for others, and they may vary in their relative width according to the phenological stage of the species. Tolerance ranges cannot be determined from morphology; they are related to physiological features that must be measured experimentally. Tolerance ranges may change in the course of evolution, but this process is so slow that environmental change is typically accompanied by plant migration rather than a change in tolerance.

Some ecologists disagree strongly with Good's second conclusion, believing that edaphic or biotic factors are more important than climatic ones, depending on the species under discussion. There have been a few experiments that dramatically show how the tolerance range or optimum for a physical factor is modified by competition (Harper 1964; Ellenberg 1958). For example, when the common weedy annuals wild radish (*Raphanus raphanistrum*) and spurrey (*Spergula arvensis*) are grown in separate pots in controlled conditions, their growth curves exhibit similar pH tolerance ranges and optima. Radish exhibits optimum growth at pH 5, and spurrey shows an optimum at pH 6 (Figure 3-2). When grown

Tolerance ranges for:

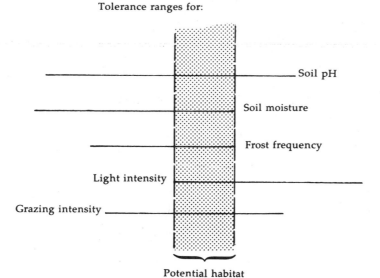

Figure 3-1 According to the theory of tolerance, the range of habitats for a given species is the sum of tolerance limits for each environmental factor. The extent of each horizontal line represents the tolerance range for that factor. The stippled area is the region of overlap for all tolerance ranges, which represents the potential habitat.

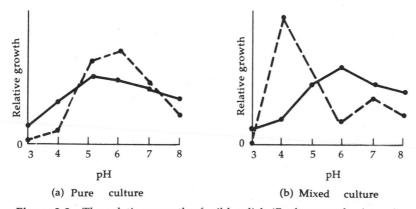

Figure 3-2 The relative growth of wild radish (*Raphanus raphanistrum*) (solid line) and spurry (*Spergula arvensis*) (dashed line) as a function of substrate pH, (a) when grown alone and (b) when grown in competition with each other. (From H. Ellenberg, "Bodenreaktion (einschieplich kalfrage)." In W. Ruhland (editor), *Handbuch der pflanzenphysiologie*, 1958. vol. 4. By permission of Springer-Verlag.)

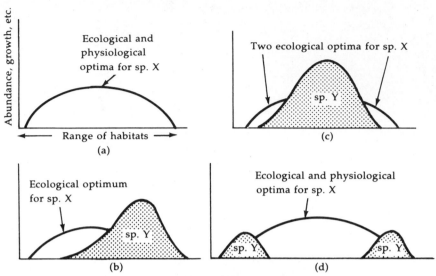

Figure 3-3   Examples of how the ecological optimum may be different from the physiological optimum due to competition. In (a), species X grows alone and laboratory experiments determine that its physiological optimum lies in the center of the curve, as shown. In nature, it faces competition from other species which displace it from habitats it could grow in alone; consequently its ecological optimum is shifted far to the left of its physiological optimum, as shown in (b). In (c), species X competes poorly and is displaced from the middle of its range; it appears to have two ecological optima. In (d), species X competes poorly at the range extremes, so that the ecological optimum still coincides with the physiological optimum, but the range is narrowed. (From H. Walter, *Vegetation of the Earth in Relation to Climate and the Eco-physiological Conditions*, 1973. By permission of Springer-Verlag.)

together, however, the optimum of spurrey shifts to pH 4 and its range for good growth becomes narrow, while the optimum for radish shifts slightly towards pH 6 and its tolerance range remains much as when grown alone.

The conditions under which a species can exist and grow best in isolation are its **potential (physiological) range** and **potential (physiological) optimum**, respectively. These may differ from its observed **ecological range** and observed **ecological optimum** in nature, where the species grows in competition with other species (Figure 3-3). The role of competition in plant distribution is very important. Undoubtedly, other biotic interactions, such as herbivory and pollination, also affect the distribution of species.

The Holocoenotic Concept of the Environment

Roughly 100 years after Humboldt wrote that everything was somehow interconnected and interdependent, Karl Friederich gave that ecosystem attribute a name: **holocoenotic** (Billings 1952) (holocoen is a synonym for ecosystem). The holocoenotic concept is a natural climax to the modifications others have applied to Liebig's law. The holocoenotic concept states that it is impossible to isolate the importance of single environmental factors to the distribution or abundance of a species, because the factors are interdependent and synergistic. Therefore, single-factor ecology (the search for one all-important environmental factor that best determines plant distribution) is short-sighted and naive according to this concept.

This concept does not mean that all factors are necessarily equal—only that they are interactive. Certain factors in any ecosystem are of overriding importance, such as moisture in desert scrub, fire in chaparral scrub, or moving sand in coastal dune scrub. Billings (1970) calls these important factors **trigger factors**.

The Taxonomic Species

Although the species is at the heart of our classification scheme, there is no unanimity on a working definition for species. We will use the following definition, synthesized from several sources. A **species** consists of groups of morphologically and ecologically similar natural populations that may or may not be interbreeding but that are reproductively isolated from other such groups.

Three aspects of classification are combined in this definition: (1) external appearance (morphology), (2) breeding behavior, and (3) habitat distinctiveness. The last aspect is clearly third in importance to most taxonomists, and morphology or reproductive isolation receives the most weight in deciding what a species is. Traditional taxonomists weigh morphology heavily, but biosystematists give more weight to reproductive isolation.

Many biologists believe that living organisms do not vary continuously over the whole range but fall into more or less well-defined groups, which are commonly called species. Traditional taxonomists adhere to this philosophy of discrete species. A traditional taxonomist examines primarily plant morphology, searching for a few conservative, genetically controlled traits that consistently allow the separation of plants into well-defined groups. Traditional taxonomists do not limit themselves to the examination of only a few traits at the start of a study. They examine many morphological, anatomical, and chemical characters from many specimens but eventually select only a few morphological characters to serve in defining the various species. The characters selected are those that show discontinuities and thus are most helpful in separating species. However, the selection process is somewhat subjective.

Biosystematists—in contrast to traditional taxonomists—are interested in determining natural biotic units: populations of plants that maintain their distinctiveness because of biological barriers that genetically isolate them from other populations. These isolating barriers may be due to breeding behavior (time of flowering, type of pollinator), habitat or geographic isolation, or inability to form fertile hybrids with closely related groups (perhaps because of chromosomal differences or pollen incompatibility with the stigma and style).

Conflicts arise between biosystematists and traditional taxonomists because the natural biotic units do not always correspond to well-defined groups. Two populations of the same traditional species may prove, upon crossing in the greenhouse, to yield no offspring or infertile offspring; thus they belong to two different biosystematic species. The traditionalist argues that such nonvisible traits as crossability are theoretically important, but they cannot be interpreted in the field with a hand lens and thus are of no practical importance. Also, greenhouse crosses may not be a valid imitation of crossing frequency in nature. One further difficulty with the biosystematic approach is that crossability is seldom all or nothing, so subjective decisions still have to be made. For example, if populations A and B are 78% interfertile, are A and B in the same species?

Other biosystematists choose to recognize natural biotic units by a mathematical averaging of an entire collection of traits, including breeding behavior, DNA chemistry, external morphology, and fossil records. Traits are defined in such a way that either a taxon possesses it (1) or lacks it (0). Computers are then programmed to manipulate the data, calculating degrees of similarity between all possible pairs of taxa and creating charts or graphs that show existing relationships between taxa (phenograms) or possible evolutionary/phylogenetic relationships (cladograms). Although this method and the resulting graphs appear to confer mathematical objectivity on taxonomy, we should realize that humans have chosen the traits to include in the mathematical pot and devised the rules by which the computer program runs. Ultimately, numerical taxonomy is as biased as traditional taxonomy. Its use of many more traits on which to base conclusions may, or may not, result in a closer approximation of the real world and its evolutionary past.

Conclusions

First, the process of identifying and defining species differs from taxonomist to taxonomist, but there is one facet that all the diverse approaches share to some degree: The result is arbitrary. Species are partly natural and partly human artifacts.

Second, habitat characteristics are seldom important taxonomic criteria. Consequently, some species have such enormous ranges that it is doubtful they are genetically homogeneous entities. Quaking aspen (*Populus tremuloides*), for example, has the widest distribution of any North American tree, covering 110° of longitude and 47° of latitude. Red maple (*Acer rubrum*) not only occurs throughout the eastern deciduous forest, but within that area it can be as abundant on sandy, dry sites as it is on wet, bottomland sites. Subalpine fir (*Abies lasiocarpa*) extends south from sea level in the Yukon Territory to 3600 m at timberline near the Mexican border, and east from the maritime climate of coastal British Columbia to the harsh continental climate along the front range of the Rocky Mountains. Clearly, such taxonomic species are not ecological tools; that is, they are not very precise indicators of a relatively narrow set of environmental ranges. Some wide-ranging species have been subdivided into regional variants (subspecies or varieties), but even then the ranges often cover heterogeneous areas.

Other taxonomic species, with restricted ranges, might serve as excellent indicators of a certain environment, but these are seldom common or dominant species. Coast redwood (*Sequoia sempervirens*) is restricted to a relatively narrow coastal strip in California, perhaps 90 km^2 in extent. Within that area it is a dominant species and serves as an indicator of summer fog and a certain moderate range of annual and diurnal temperature fluctuation, among other factors. At an extreme, the cypress *Cupressus stephensonii* is restricted to a single grove in southern California about 0.15 ha in size. This species may completely characterize its unique site, but the relationship has no practical, predictive value because of the plant's absence from the rest of the world.

Can the average, somewhat arbitrary, taxonomic species be redefined or subdivided to make it a better ecological tool? The answer is yes in theory, and often no in practice, as discussed in the next section.

The Ecological Species

The ultimate objective of any science is to be able to make accurate predictions or inferences about a given system, be it chemical, physical, or biological. Plant ecologists would like to use species as deductive tools with which to understand ecosystems. If the ecological requirements of species A are known, and the resource allocation pattern is understood, then the presence and degree of vigor of species A anywhere permits us to make many inferences about the environment, such as soil depth, soil nutrient levels, frequency of frost, light intensity, length of growing season, frequency of disturbance, and the presence or absence of other plants and animals that may interact with that species. This type of analysis is our objective. In this section we ask: Is it theoretically possible to use plant species as deductive tools?

Ecotypes

Linnaeus and taxonomists after him recognized that some taxonomic species were not homogeneous: Their member plants varied in height, leaf size, flowering time, or other attributes, with change in light intensity, latitude, elevation, or other site characteristics. It was thought that such differences within a species were plastic, not heritable responses. The early transplant gardens of Kerner (1895) in the Tirol supported this view.

This conclusion was challenged by the experiments of the botanist Göte Turesson in the early twentieth century. He hypothesized that many variations within a species were heritable and were of adaptive value to particular habitats within the species' range. During the 1920s he attempted to arrive at, as he put it, "an ecological understanding of the Linnaean species." At first he examined plants from Sweden only, but later he studied species that ranged all over Europe. For each species (usually a perennial), he brought back vegetative material or seeds from different habitats or regions and grew the plants in a test garden near

Table 3-1   Some morphological and phenological traits of hawkweed (*Hieraceum umbellatum*) ecotypes, as revealed in Turesson's uniform garden. (From Briggs and Walters 1969, *Plant Variation and Evolution*, Cambridge University Press, Cambridge.)

	Ecotypes		
Traits	**Woodland**	**Field**	**Dune**
Habit	Erect	Prostrate	Intermediate
Leaves	Broad	Intermediate	Narrow
Pubescence	Absent	Present	Absent
Autumn dormancy	Present	Present	Absent

his home at Åkarp, Sweden. He reasoned that if the morphological or pheno-logical differences noted in the field were retained in the garden, then the traits were heritable and genetically based.

Table 3-1 summarizes a typical result, using the herbaceous perennial hawk-weed (*Hieraceum umbellatum*), which in southern Sweden grows on coastal sand dunes, on rocky headlands, and in inland fields and woodlands. The field dif-ferences were retained in the test garden. Were these genetically distinct types of plants different species? Turesson made all the crosses and found all the types to be interfertile. Therefore, the types were technically part of only a single species.

Turesson called these entities ecotypes. An **ecotype** is the product of genetic response of a population to a habitat. It is a population or group of populations distinguished by morphological and/or physiological characters, interfertile with other ecotypes of the same species, but usually prevented from naturally inter-breeding by ecological barriers. (Although Turesson wrote in English, it is a very dense form of English and is best read in the "translation" written by Turrill in 1946.) Turesson also coined the terms ecospecies (approximately the equivalent of the biosystematist's natural biotic unit) and coenospecies (the equivalent of a genus with few species, or a section of a larger genus), but these terms and concepts have not been widely used since.

It is important to outline all the elements that were part of ecotypes as Turesson saw them: (a) they were genetically based, (b) their distinctiveness could be morphological, physiological, phenological, or all three, (c) they occurred in distinctive habitat types, (d) the genetic differences were adaptations to the dif-ferent habitats, (e) they were potentially interfertile with other ecotypes of the same species, and (f) they were discrete entities, with clear differences separating one ecotype from another. Our modern concept of the ecotype does not agree with all these elements, as will be shown later in the chapter.

At the same time that Turesson was describing ecotypes in over 50 common European species, three biologists were reporting similar results with western North American perennial plants. In 1922, Jens Clausen, a geneticist and cytol-

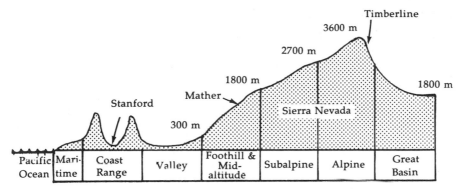

Figure 3-4 Profile, from west to east, of California at approximately 37°30'N latitude, showing the location of the three principal transplant gardens used by Clausen, Keck, and Hiesey. (From Clausen et al. 1940. Courtesy of the Carnegie Institution of Washington.)

ogist, David Keck, a taxonomist, and William Hiesey, a physiological ecologist, established a 323 km long study transect in California, supported by the Carnegie Institution. The transect extended from near sea level at Stanford University (just south of San Francisco), across the Coast Ranges, through California's Central Valley, up the gradual west face of the Sierra Nevada to timberline at 3000 m elevation, and partly down the steep, relatively arid east face of the range (Figure 3-4). Despite great environmental diversity along this transect, Clausen, Keck, and Hiesey were able to find about 180 species whose ranges extended over much or all of the transect.

Each species was collected at a variety of locations along the transect, brought back to greenhouses at Stanford, cloned (literally torn into parts and each part induced to root separately so that many genetically identical plants were produced), grown for 6 months, then transplanted to test gardens along the transect. Initially there were 11 gardens, but the number was soon reduced to three: Stanford, near sea level; Mather, in the mid-elevation Sierra Nevada; and Timberline. Table 3-2 summarizes the garden environments. About 60 species were rugged

Table 3-2 The environments of the three principal transplant gardens used by Clausen, Keck, and Hiesey. (From Clausen et al. 1940. Courtesy of the Carnegie Institution of Washington.)

Garden name	Elevation (m)	Surrounding vegetation	Growing season (mo)	Annual ppt. (cm)	Ppt. as snow (cm)	Mean max. temp. (°C)	Mean min. temp. (°C)
Stanford	30	Oak woodland, chaparral	12.0	30	None	36	− 3
Mather	1400	Mixed conifer forest	5.5	95	90	36	−11
Timberline	3050	Subalpine forest	2.0	73	485	23	−22

Table 3-3 Ecotypes (subspecies) of cinquefoil (*Potentilla glandulosa*) as revealed by the research of Clausen, Keck, and Hiesey. (From *Plants and Environment*. Daubenmire. Copyright © 1974 by John Wiley and Sons, Inc. Reprinted by permission.)

Coenospecies	Ecospecies	Ecotypes	Environment
		nevadensis	Alpine and subalpine, 1600–3500 m
		hanseni	Mid-elevation wet meadows, 250–2200 m
Potentilla, section Drymocallis	*glandulosa*		
		reflexa	Mid-elevation dry slopes, 250–2200 m
		typica	Coast Range

enough to survive this initial handling, and their growth, phenology, and mortality were followed for as long as 16 years.

The herbaceous perennial cinquefoil (*Potentilla glandulosa*) can serve as an example of their garden results (Table 3-3). Based on morphology, phenology, physiology, and habitat, there appeared to be four ecotypes in the species (Clausen, Keck, and Hiesey chose the more conservative, taxonomic term subspecies, but ecotype is synonymous in this case). Ecotype *typica* was a lowland form. It grew best at Stanford, survived at Mather, but did not last a year at Timberline. Its herbage was relatively frost tolerant, and at Stanford it was able to grow all year round, but at Timberline it could not withstand the hard winter frosts (Figure 3-5). Ecotypes *reflexa* and *hanseni* were mid-elevation forms, distinctively taller than *typica*. One occurred on dry sites, the other on wet meadow sites, and there were subtle morphological differences between them. Both flowered late in summer, so late that when grown at Timberline they often were hit by frost before flowering or seed set could be completed. Their herbage was not frost tolerant, and cold temperatures induced them to enter dormancy. They survived at all locations, but grew best at Mather. Ecotype *nevadensis*, a timberline form, was the shortest of the four. It flowered early in the season and its herbage was very frost tolerant. Its winter dormancy may have been induced by short day length rather than temperature, for it was winter dormant even in the mild climate at Stanford. A period of winter cold may still have been physiologically necessary, however, for growth at Stanford declined in subsequent years. The herbage seemed more susceptible to disease at lower elevations than at Timberline. Crossing experiments showed that all the ecotypes were interfertile.

Clausen, Keck, and Hiesey concluded that most species are composed of an assemblage of ecotypes, each ranging in size from a single population to a regional

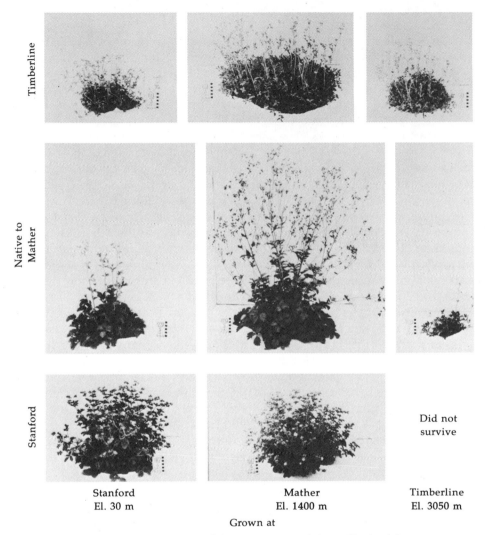

Timberline

Native to Mather

Stanford

Stanford	Mather	Timberline
El. 30 m	El. 1400 m	El. 3050 m

Did not survive

Grown at

Figure 3-5 The appearance of three ecotypes of *Potentilla glandulosa* as grown at three different locations. The top row shows ecotype *nevadensis*, the middle row is ecotype *reflexa*, and the bottom row shows ecotype *typica*. (Courtesy of Dr. William M. Hiesey.)

group of many populations; the wider the species' range is, the more ecotypes there are within the species. Ecotype research by many other investigators, on many different species throughout the world, has corroborated this conclusion. The terms *race*, *genecotype*, and *ecological race* are sometimes used as synonyms for ecotype. Random genetic variants (individuals or groups of individuals) within ecotypes are called *biotypes*. Populations whose uniqueness in nature is due to nongenetic plasticity are called *ecophenes* or *phenecotypes*, to distinguish them from ecotypes.

Ecoclines

The ecotype concept of Turesson may seem to give us the ecological species tool we seek, but more recent research has shown its limited practicality.

In Scotland, J. W. Gregor (1946) closely examined what at first appeared to be two ecotypes of the coastal plantain, *Plantago maritima*. One ecotype inhabited salt marshes regularly flooded by high tide with a soil salinity of approximately 2.5%. These plants had short leaves, small seeds, and thick, short, somewhat decumbent* flowering stalks. The other ecotype inhabited nonsaline meadows farther inland, and these plants had longer leaves, larger seeds, and thinner, taller, more upright flowering stalks. Gregor collected seeds of each and sowed them in a test garden. In addition, he collected and sowed seeds from plants growing in an **ecotone**, an intermediate habitat. The resulting plants showed that field differences were genetically fixed, but more importantly, they showed a continuous gradation from one extreme, one ecotype, to the other. There were no discrete boundaries between ecotypes or even between ecotypes and plants from the ecotone (Table 3-4). For example, 3.9% of the salt marsh ecotype plants had a habit grade score of 3, which is well within the range of meadow ecotype plants, and 2% of the meadow ecotype plants had a habit grade score of 2, which is well within the range of salt marsh ecotype plants. Ecotone plants completely overlap the habit grades for both salt marsh and meadow ecotype plants.

Olaf Langlet (1959), a compatriot of Turesson, brought seeds of the pine *Pinus sylvestris* from 580 sites throughout Sweden to a test garden. When he examined saplings for growth rate and morphological features, he found the extremes to be quite different, but a **cline**, a continuum of variation, connected the extremes. There were no sudden breaks in the range of variation where one could say ecotype A ended and ecotype B began.

Cavers and Harper (1967*a* and *b*) also found significant variation within what had been called a discrete ecotype of the weedy annual curly dock (*Rumex crispus*). They used germination response as an indicator of genetic heterogeneity. They first collected seeds from separate populations (seeds pooled from many plants in each population), then from separate plants in the same population, then from separate inflorescences (clusters of flowers and later of seeds) of the same plant, and finally from upper and lower portions of the same inflorescence. In every case, the range of germination response was very large from population to population, from plant to plant, etc. The implication was that other genetic, adaptive traits could fluctuate equally widely. Again, here was an ecotype which was as heterogeneous in certain traits as a wide-ranging species.

The ecotype concept remains useful because it emphasizes the genetic heterogeneity of taxonomic species and the pervasive influence of the local environment at morphological, physiological, and successively more subtle levels of plant behavior. The ecotype concept has practical importance to reforestation or revegetation projects in general. Seed or vegetative material from a wide-ranging species must be taken from that portion of the species' range most similar to the environment of the revegetation site, or the success of the project may be reduced.

---

*Lying or growing along the ground but erect at or near the apex.

Table 3-4 Results of Gregor's cultivation experiments with plantain (*Plantago maritima*) ecotypes. Habit grade 1 includes the most prostrate, succulent plants, with shortest leaves and smallest seeds, habit grade 5 is the other extreme, and habit grades 2–4 are intermediate. (From Briggs and Walters 1969, *Plant Variation and Evolution*, Cambridge University Press, Cambridge.)

Habitat	Mean scape length (cm)	% of plants in habit grades 1–5				
		1	2	3	4	5
Salt marsh, soil salinity 2.5%	23.0	74.5	21.6	3.9		
Upper marsh edge, ecotone	38.6	10.8	20.6	66.7	2.0	
Nonsaline meadow above marsh, soil salinity 0.25%	48.9		2.0	61.6	35.4	1.0

However, the ecotype concept typically does not permit one to recognize homogeneous populations or groups of populations in nature for use as ecological tools. Ecotypes are not more discrete and distinct from each other than are species. Turesson's stairstep concept of ecotypes (Figure 3-6) must be replaced with an ecocline concept. An **ecocline** is a gradation in the attributes of a species (or community or ecosystem) associated with an environmental gradient. Sometimes ecocline is used to refer to the environmental gradient itself (Hanson 1962). Turesson, and to some extent Clausen, Keck, and Hiesey, thought of ecotypes as discrete because their sampling method prejudiced their results: Plant material was selected from widely separate places, and ecotones were ignored.

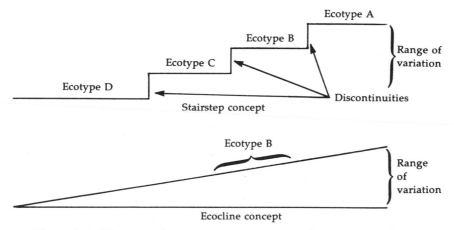

Figure 3-6 Diagrams that contrast two concepts of ecotypes. In the stairstep view, ecotypes are distinct and separated from each other by discontinuities in their morphology or other attributes (the vertical lines indicate discontinuities). In the ecocline view, there are no discontinuities; there is just a continuum of variation.

Sometimes habitats are discrete, with sudden changes from one habitat to the other (usually because of steep terrain or edaphic factors), but typically habitats intergrade, and plant populations vary with the same subtlety. An ecotype, then, is only an arbitrary segment of an ecocline, which may be convenient to recognize for reference purposes.

Genoecoclinodemes

Since the work of Turesson and Clausen, Keck, and Hiesey, ecotypes and ecoclines have been described in many species, but the rigor of ecotype definition has been lost. Many investigations of ecotypes have not tested breeding behavior, some have not grown ecotypes together to test for genetic differentiation, and a few have implied habitat uniqueness without quantifying it. All of these missing items are important parts of Turesson's six criteria for identifying ecotypes.

Gilmour devised an entire system of nomenclature that eliminates the need to use the term ecotype and that gives precision to the results of autecological and biosystematic research (Gilmour and Heslop-Harrison 1954). All the terms are prefixes to a neutral suffix, -deme, which never stands alone but can be defined as a group of closely related individuals. As more information becomes known about any group, a series of single or multiple prefixes can be added, and these prefixes convey exactly what is known about the group, no more, no less.

For example, a topodeme is a group of individuals co-existing in a given locale; a gamodeme is a group of naturally interbreeding individuals (equivalent to the term population); an ecodeme is a group in a specific, unique habitat; a genodeme is a group which is genetically distinct from others; a phenodeme is a group whose differences are not yet known to be genetic; a plastodeme is a group whose differences are known to be nongenetic. A genoecodeme is a group whose differences are genetically fixed and which exists in a unique habitat; a genoecoclinodeme is a group whose differences are genetically fixed, which exists in a unique habitat, and which is part of a continuous cline of gradation. In short, a genoecoclinodeme is the equivalent of an ecotype.

Not surprisingly, this system of multiple prefixes has been slow to gain acceptance, even though its objectivity and precision have much to recommend it. We will continue to use the term ecotype in this book, but its synonym, genoecoclinodeme, reminds us of the real meaning of ecotype.

Ecotype Research at the Physiological Level

One aspect of recent ecotype research has revealed the subtle physiological, metabolic basis of plant adaptation to the local habitat. That aspect will be illustrated by a series of investigations that takes us successively closer to the ultimate control of adaptations, the genes themselves.

Harold Mooney and Dwight Billings (1961) published a classic study on the perennial herb alpine sorrel (*Oxyria digyna*) (Figure 3-7), for which they received the 1962 George Mercer award from the Ecological Society of America for the most outstanding ecological paper published by young ecologists in the previous

Figure 3-7 Alpine sorrel (*Oxyria digyna*) above the timberline in Wyoming. Note the heart-shaped leaves and the profusion of flowers, both traits of the alpine ecotype. (Courtesy of W. D. Billings.)

2 years. *Oxyria* has a circum-boreal distribution in the treeless arctic tundra and extends south in alpine tundra along several mountain chains. In the conterminous United States it is found at high elevations in the Sierra Nevada and Rocky Mountains.

Arctic tundra vegetation and alpine tundra vegetation share many similarities, but there are important environmental differences. Alpine areas in the temperate zone experience higher light intensity and greater extremes of temperature in summer than do arctic areas.

Just as there are two environmental extremes, there are two ecotypic extremes of *Oxyria*. Mooney and Billings showed that the arctic and alpine ecotypes differed morphologically and phenologically even when grown together from seed in controlled chambers that simulate a natural, uniform environment (the modern test garden) (Table 3-5). Mooney and Billings also showed that the metabolism of the two ecotypes differed.

Table 3-5 Some morphological, biochemical, and phenological traits of arctic and alpine ecotype extremes of alpine sorrel (*Oxyria digyna*). (From "Comparative physiological ecology of arctic and alpine populations of *Oxyria digyna*" by H. A. Mooney and W. D. Billings, *Ecological Monographs* 31:1-29. Copyright © 1961 by the Ecological Society of America. Reprinted by permission.)

Trait	Arctic	Alpine
Rhizomes	Present	Absent
Intensity of flowering	Low	High
Leaf anthyocyanin level (red color)	Low	High
Critical photo-period for flowering	24 hr at 70°N	15–17 hr at 38–48°N
Leaf shape	Ovate	Heart-shaped

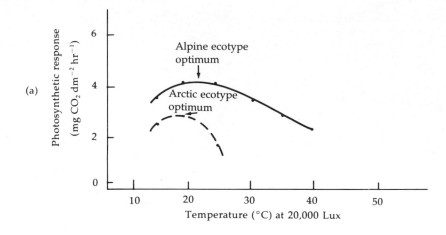

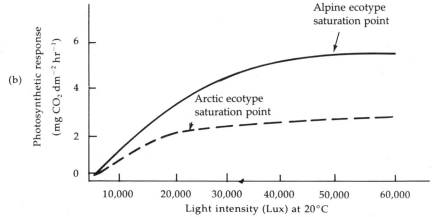

Figure 3-8 (a) Photosynthetic response of arctic (dashed line) and alpine (solid line) ecotypes of *Oxyria digyna* to temperature at 20,000 lux. The arctic ecotype temperature optimum for photosynthesis is lower than that of the alpine ecotype. (b) Photosynthetic response to light at 20°C. (From "Comparative physiological ecology of arctic and alpine ecotypes of *Oxyria digyna*" by H. A. Mooney and W. D. Billings, *Ecological Monographs* 31:1-29. Copyright © 1961 by the Ecological Society of America. Reprinted by permission.)

By placing potted plants in a small plexiglass chamber whose microenvironment could be controlled, then measuring the CO_2 content of air passing into and out of the chamber, the rate of photosynthesis per unit leaf area or weight could be determined. As shown in Figure 3-8a, the alpine ecotype had a significantly higher temperature optimum for photosynthesis. The amount of light required to saturate the photosynthesis system, beyond which increasing light fails to increase the rate of photosynthesis very much, was greater for the alpine ecotype (Figure 3-8b), again correlating well with the environmental differences in nature. These physiological differences most likely correlated with enzymatic and other biochemical factors, but Mooney and Billings did not pursue their investigation to that level.

One species of cattail, *Typha latifolia*, is widely distributed in the northern hemisphere. McNaughton (1966) collected dormant rhizomes from such disparate habitats as the cool, maritime, foggy Pacific coast at Point Reyes, California and the relatively hot, arid Sacramento Valley near Red Bluff, California, more than 100 km inland. He potted the rhizomes and placed them in a greenhouse regulated at 30/25°C day/night, where they broke dormancy, produced shoots and grew for three months. He then took samples of leaf tissue, extracted their enzymes, and subjected the extract to a heat stress of 50°C for periods of time up to 30 minutes. This simulated the leaf temperature that Red Bluff plants might experience in nature. Point Reyes plants probably do not experience a leaf temperature in nature higher than 30°C.

After each period of heat stress, he tested the activity of three important respiratory enzymes: malate dehydrogenase, glutamate oxaloacetate transaminase, and aldolase. He reasoned that the heat tolerance of the Red Bluff ecotype must inherently lie in the heat stability of some or all of its enzymes. As shown in Figure 3-9, malate dehydrogenase was significantly more heat stable in the Red Bluff ecotype than in the Point Reyes ecotype; the other two enzymes showed no difference. Enzymes are complex macromolecules with tertiary or quarternary structure and the reaction site is a relatively small part of their total architecture and chemistry. Thus, malate dehydrogenase could differ in the two ecotypes in many ways that might increase stability or activity (McNaughton 1972). Why only one of the enzymes, rather than all three, showed ecotypic differentiation is not clear.

In another paper, McNaughton (1967) showed that cattail ecotypes from Point Reyes at sea level and Wyoming at 1980 m elevation differed in their photosynthetic efficiency. The high elevation, short growing season ecotype exhib-

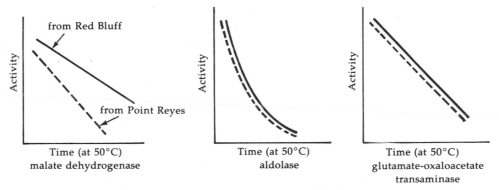

Figure 3-9 Enzyme response to incubation at 50°C. Enzymes were extracted from coastal (Point Reyes, California) (dashed line) and inland (Red Bluff, California) (solid line) ecotypes of *Typha latifolia*. (From McNaughton 1966. Thermal inactivation properties of enzymes from *Typha latifolia* L. ecotypes. *Plant Physiology* 41:1736–1738. By permission of the American Society of Plant Physiologists.)

ited about twice the photosynthetic rate of the Point Reyes ecotype. He was able to trace the difference to higher reducing activity (an important process in the light reactions of photosynthesis) in the chloroplasts of the Wyoming ecotype.

Respiration is also attuned to elevation at the enzymatic level, as Klikoff (1966) showed for populations of the grass squirrel tail (*Sitanion hystrix*) from different elevations of the Sierra Nevada. Isolated mitochondria showed higher oxidative rates at lower temperatures with increasing elevation of the parent plant.

Some species are differentiated into sun ecotypes (those that germinate and develop in the open) and shade ecotypes (those that develop beneath the canopy of other plants). A European species of goldenrod (*Solidago virgaurea*), a perennial herb, has such ecotypes. Olle Björkman, in a series of papers culminating in 1968, examined the differences between the two goldenrod ecotypes at successively more subtle levels. He collected vegetative material from plants growing in exposed heath vegetation in Norway and from plants growing beneath oak forests in Sweden, then cloned the material and grew plants in controlled environments.

Among other differences, he showed that the light saturation curve of the sun ecotype was different from the light saturation curve of the shade ecotype. The sun ecotype had a higher light saturation point and showed a higher rate of photosynthesis at that saturation point. To find the reason for the difference, he first searched at the morphological level, asking if sun ecotype leaves perhaps absorbed more light; the answer was no. He then searched at the cellular level, asking if the chlorophyll concentration was higher in sun leaves; the answer was no. Finally he examined the enzyme level. The enzyme responsible for fixation of CO_2 into the dark reaction pathway of photosynthesis is ribulose bisphosphate carboxylase (also called carboxydismutase). When he assayed for the concentration (activity) of this important enzyme, he found that it was two to five times greater in the sun ecotype, which was sufficient to account for the five times higher rate of photosynthesis of the sun ecotype at high light intensity.

Ultimately, this line of research should be able to show us which enzymes are most important to basic types of climatic or edaphic ecotypes, how these enzymes differ (in amount, as Björkman showed for one enzyme, or in structure, as McNaughton implied for another), and what the genetic basis for the difference is (how many loci and how many genes at each locus). When we get to that point, the information will have great practical importance for the breeding of ecotypes for use in reclaiming terrain such as nutritionally poor road cuts, toxic mine spoil, saline or coarse dredge spoil, strip mine areas, dunes that are mobile and low in nitrogen, eroded clear-cuts, etc. At present, we have yet to reach this level.

Acclimation

Acclimation (also called acclimatization) is a plastic, temporary change in an organism caused by an environment to which it has been exposed *in the past*. Matthaei (1905) may have been the first to document such a phenomenon in

Table 3-6 The effect of acclimation temperature on the optimum temperature for photosynthesis of alpine and arctic ecotype extremes of alpine sorrel (*Oxyria digyna*). (Adapted from Billings et al. Reproduced with permission of the Regents of the University of Colorado from *Arctic and Alpine Research* 3:277–289, 1971.)

| Population site | Ecotype | Optimum temperature for photosynthesis at each acclimation regime (°C) | | | |
		Warm (32/21°C)	Medium (21/10°C)	Cold (12/4°C)	Shift (warm–cold)
Sonora Pass, California	Alpine	28	21.5	17	11
Pitmegea River, Alaska	Arctic	21	20.5	20	1

plants, and the effect of past temperatures on photosynthesis and respiration rates has been reviewed by Semikhatova (1960).

Billings et al. (1971) conducted an experiment that provides a good example of acclimation, and once again alpine sorrel proved to be a good study plant. Alpine sorrel (*Oxyria digyna*) seeds were collected from a range of habitats, germinated and grown in a uniform greenhouse for 4 months, then subdivided into three growth chamber environments: warm (32/21°C day/night), medium (21/10°C), and cold (12/4°C). After 5 to 6 months in the chambers, replicates of each collection were measured for net photosynthesis at a range of temperatures, from 10 to 43°C, and the optimum temperature for photosynthesis was noted. Table 3-6 shows that representatives of arctic and alpine ecotypes possessed different acclimation capacities. The optimum temperatures for alpine plants shifted as much as 11°C, depending on the temperatures they had been growing at before the photosynthesis measurement, but the optimum temperatures for arctic plants shifted only 1°C.

Similar effects of preconditioning, or acclimation, have been shown for plants as diverse as pine trees (Rook 1969) and desert shrubs (Mooney and West 1964; Mooney and Harrison 1970; Strain and Chase 1966).

The relationship between plant and environment can thus be written:

$$\text{phenotype} = \text{genotype} + \frac{\text{prevailing}}{\text{environment}} + \frac{\text{past}}{\text{environment}}$$

How distant a past environment can influence phenotype? The influential past environment may have been uncomfortably and unaccountably long ago, perhaps reaching back to parent generations, according to Rowe (1964). Seeds of groundsel (*Senecio vulgaris*) were germinated at different temperatures. The seedlings were immediately transferred to a common environment and allowed to grow for 80 days, then the shoots were weighed. The sevenfold difference in shoot weights shown in Table 3-7 is hard to explain except as a result of temperature differences at the time of germination, 80 days earlier. Plants of the weedy annual prickly lettuce (*Lactuca scariola*), subjected to different day lengths or appli-

Table 3-7 The effect of germination temperature on subsequent growth of groundsel (*Senecio vulgaris*). (From "Environmental preconditions with special reference to forestry" by J. S. Rowe, *Ecology* 45:399–403. Copyright © 1964 by the Ecological Society of America. Reprinted by permission.)

Germination temperature (°C)	Subsequent growing conditions, next 80 days	Final plant weight (mg)
10		147
14	All grown together at 17°C,	775
23	16 hr photo-period	1078
30		390

cations of growth regulators, produced progeny that differed in germination, seedling growth, and time of flowering (Gutterman et al. 1975). Other examples of parental environment carrying over to germination have been reported by Baskin and Baskin (1973), Quinn and Colosi (1977), and Hume and Cavers (1981).

Highkin (1958) grew pure-line peas under two sets of conditions: 24/14°C day/night and 26°C constant. Pollen from plants in either regime was transferred to the stigmas of other pure-line peas that had been kept separate in uniform conditions. The seeds were harvested, kept separate according to the temperature regime of the pollen donor, germinated, and grown in uniform conditions. The two types of progeny differed significantly in height and number of nodes, and there was a diminishing but still measureable carryover effect for several generations.

A non-Lamarckian, genetic explanation for acclimation is possible, but our objective has been only to show that acclimation can occur; the genetic basis for it need not concern us here. The significance of past environments on plant behavior is insufficiently recognized by plant autecologists, even though it appears that acclimation is important to an understanding of plant distribution.

Integrated Approaches to Ecotype Research: A Case Study

An excellent example of recent ecotype research that utilizes a wide assortment of techniques, including population biology and physiological ecology, is a study of *Dryas octopetala* ecotypes from the Alaskan tundra (McGraw and Antonovics 1983, McGraw 1985a and b). The authors combined such techniques as growth chamber studies in the Duke University phytotron, field transplants, competition trials, pollination ecology, demographic observations, environmental manipulation *in situ*, photosynthesis measurements, and determination of photosynthate allocation patterns to various plant organs. This rich mixture made their conclusions ecologically powerful and important.

Dryas octopetala is a perennial dicot herb of the rose family (Figure 3-10). It grows in two distinct habitats, which may be very near one another, and several traits distinguish the plants of each habitat (Table 3-8). Fellfield sites are on ridges

Figure 3-10 *Dryas octopetala* ssp. *octopetala*. (Courtesy of James B. McGraw.)

or slopes that experience dry summer soils and minimal winter snow cover. Snowbed sites are in protected swales with late-melting snow; as a result, summer aridity is much less pronounced. Plant cover is dense on snowbed sites but rather open on fellfields. McGraw and Antonovics selected a study site where fellfield *Dryas* grew only 150 m away from, and 30 m above, snowbed *Dryas*.

First they demonstrated that the two forms (which taxonomists had distinguished as subspecies) were fully interfertile. Both are outcrossers, and there was no evidence that either could self-pollinate or produce seeds by apomixis. Hybrids in nature were often encountered, though they accounted for less than 1% of all plants. When grown together in a common garden (the phytotron) or when reciprocally transplanted in the field, some of the morphological differences in Table 3-8 became less pronounced, but they still remained distinct. Generally, the snowbed ecotype showed more phenotypic plasticity than did the fellfield ecotype: Its petiole length, leaf area, leaf margin, and specific leaf weight approached those of the fellfield ecotype more than vice versa.

If the two are fully interfertile and are obligate outcrossers, how did this genetic differentiation occur? Observations of insect pollinator flight paths re-

Table 3-8 Some genetically fixed differences between *Dryas octopetala* ecotypes. The fellfield ecotype, taxonomically, is ssp. *octopetala*; the snowbed ecotype is ssp. *alaskensis*.

Fellfield	Snowbed
Plants long-lived (>100 yr)	Plants may be shorter-lived
Rhizomatous spread modest	Rhizomatous spread extensive
Flowers early in growing season	Flowers up to 2 mo later
Leaves deciduous	Leaves evergreen
Leaves with orange scales and branched hairs	Leaves with glands
Leaves small, short	Leaves large, long
Fewer marginal teeth on leaves	More marginal teeth on leaves
Low specific leaf weight (g cm^{-2})	High specific leaf weight
Petioles short	Petioles long

Table 3-9 Relative survival of *Dryas* ecotypes at various life cycle stages, when planted from seed in two habitats and followed for several years. A dash indicates maximum survival of the one ecotype relative to the other (or, if both ecotypes have a dash, no significant difference in survival between them). Numbers show significantly decreased survival of the one ecotype relative to the other (1.00 = total mortality). P&S = pollination and seed set. (From McGraw and Antonovics, 1983.)

Ecotype	Habitat	Germina-tion	After first winter	After first summer	End of first yr	Adult	P&S
Fellfield		—	0.89	—	0.85	—	—
	Fellfield						
Snowbed		0.85	—	—	—	—	—
Fellfield		—	0.89	0.74	0.97	0.50	—
	Snowbed						
Snowbed		—	—	—	—	—	—

vealed, first of all, that more than 99% of all pollen was transferred within ecotypes rather than between them. Second, the onset of flowering for snowbed plants was later (up to 2 months later, depending on the year) than that for fellfield plants (late snow covered the snowbed plants, and wet soil warms more slowly than the drier fellfield soil).

Field manipulations of the environment revealed that the most significant microenvironmental factors for differential survival of the two ecotypes were light and nutrients. Fellfield plants beneath shade screens declined in growth more than snowbed plants. Addition of nitrogen and phosphorous stimulated snowbed plant growth, whereas it eventually had a negative effect on fellfield plant growth (mainly due to the increased growth of fellfield grasses that over-topped *Dryas*). Habitat differences in degree of wind exposure and soil moisture were not important.

Competition trials in the phytotron between the two ecotypes demonstrated that root competition was more critical than shoot competition and that the fellfield ecotype was consistently the "loser"—that is, it experienced a greater reduction of growth in the presence of interecotype competition than did the snowbed ecotype. The authors concluded that the snowbed ecotype "won" because of more efficient nutrient uptake and a plasticity in leaf form that permitted less self-shading in crowded conditions.

Finally, by following reciprocal transplants started from seed for several years, they were able to determine at which life cycle stages the ecotypes were sorted by the different microenvironments. They devised a simple scale of comparative survival success which ranged from 0 (equivalent survival of the two ecotypes) to 1 (zero survival for one ecotype). Departures from 0 were statistically analyzed; Table 3-9 shows only the significant departures.

Note that selection has not yet perfectly matched each ecotype to its usual habitat. The fellfield ecotype germinated equally well in both habitats but survived less well during its first year in both habitats, then survived best as an adult in its usual fellfield habitat. The snowbed ecotype germinated poorly in the fellfield but could survive to adulthood equally well in both habitats. Pollination success (seed set) was equivalent for both ecotypes in both habitats. Thus, only in the process of completing an entire life cycle will each ecotype ultimately be favored by the microenvironment of its usual habitat.

Summary

Plant ecologists would like to use species as deductive tools, as rather precise indicators of certain levels of environmental factors. This may not be a realistic objective, for two reasons. First, plants respond to a complex of climatic, edaphic, and biotic factors, and the impact of single factors is difficult to isolate. The tolerance range of a species to factor X may be modified by factors Y or Z. Good's version of the theory of tolerance recognizes the special confounding effect of competition on tolerance ranges. Second, taxonomic species, whether recognized on morphological, biological, or statistical grounds, are partially artifacts of the human desire to classify. Wide-ranging species, which occur in many different habitats, are not genetically homogeneous and thus cannot serve as ecological indicators.

Turesson searched for "an ecological understanding of the Linnaean species." He discovered that taxonomic species were composed of ecologically important subunits, which he called ecotypes. He defined ecotype as the genetic response of a population (or group of populations) to a habitat, distinguished by morphological and/or physiological characters, yet interfertile with other ecotypes of the same species. Most wide-ranging species are now known to be made up of many ecotypes, but the practical utility of the concept was diluted when it was discovered that ecotypes are just as heterogeneous, with boundaries just as vague, as species. Turesson's stairstep concept of ecotypes must be replaced with an ecocline concept, and the equivalent of the term ecotype in Gilmour's -deme terminology, genoecoclinodeme, properly emphasizes that fact.

The ecotype concept is still important from the standpoint of basic science, because it has led to research that shows the pervasive influence of the environment at all levels of plant behavior, from morphology and phenology to subtle levels of physiology, metabolism, and genetics. Recent ecological research has emphasized these subtle levels, but it has not yet reached the point where we know which enzymes are most important to basic types of climatic or edaphic ecotypes, nor how these enzymes differ from ecotype to ecotype, nor how many genes are involved, nor how we may breed for certain ecotypic features as we currently breed for seed yield or disease resistance. A confounding factor in ecotype research is acclimation: An environment to which an organism has been exposed in the past (sometimes a generation or more in the past) may cause a physiological change in the organism. Indeed, the capacity for acclimation may itself be an ecotypic trait.

CHAPTER 4

POPULATION STRUCTURE AND PLANT DEMOGRAPHY

P lants are not evenly distributed in nature. Differences in environmental conditions, resources, neighbors, and disturbance are but a few of the factors that influence the population dynamics and pattern of plants. Different sets of environmental conditions not only modify the distribution and abundance of individuals but are likely to change the growth rate, seed production, branching pattern, leaf area, root area, and size of the individuals. Distribution, survival, and patterns of growth and reproduction reflect the plant's adaptations to a particular environmental regime and thus are a critical part of plant ecology.

The arrangement of plants in space and time presents special problems for the population ecologist. Unlike most animals, many plants produce new individuals asexually and can drop or add new sets of organs (flowers, leaves, roots, and branches) in response to changes in the external environment. The material of the plant population ecologist is, therefore, not limited to the distribution and dynamics of individuals in a population but includes the dynamic growth of the ever-changing plant body.

Our goals in this chapter are twofold. First, we will examine population density and distribution patterns of individuals in a species and discuss their possible causes. Second, we will consider plant population dynamics by discussing plant demography as it applies to individuals and modules of plant growth.

Density and Pattern

Density: Definitions and Methods

Density is the number of individuals per unit area, such as 300 sugar maples (*Acer saccharum*) per hectare in a Michigan deciduous forest or 3000 creosote

Figure 4-1 Placement of twelve 2 m² circular quadrats by (a) a random method and (b) a stratified random method (three quadrats in each section). In each case, 1% of the total area has been included in the quadrats. One of the quadrats in (a) has been located by the random numbers 5 and 15, indicating that the investigator should locate the quadrat center 5 paces up from the base line and 15 paces down the base line. The other quadrat centers were located by using different pairs of random numbers.

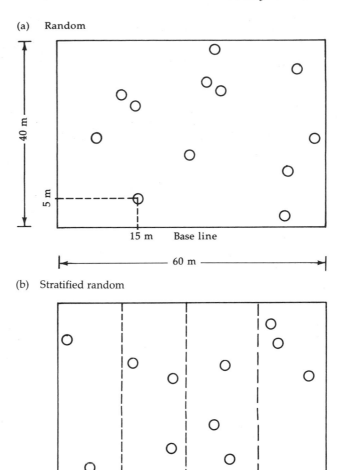

(a)   Random

40 m

5 m

15 m   Base line

60 m

(b)   Stratified random

bushes (*Larrea tridentata*) per hectare in a New Mexico desert scrub. It is not necessary to count every individual in a large area to arrive at a density value. Random sampling with quadrats whose combined area is perhaps only 1% of the overall area can give a close estimate.

A **quadrat** is an area of any shape that can be delimited in vegetation so that cover can be estimated, plants counted, or species listed. It is usually small enough that one person, standing at one point along its edge, can easily survey its extent, but quadrats for tree samples may be 10–50 m on a side, so taking a census can require more than one person.

Quadrats can be located randomly by constructing two imaginary axes along the edges of the large area being sampled, dividing the axes into units, and picking pairs of units from a table of random numbers. For the example shown in Figure 4-1, the random pair (5,15) meant that the investigator placed the center of the circular quadrat at a point 5 m up and 15 m over from the origin of the axes. Altogether, 12 quadrats of 2 m² each were placed in the 2400 m² area, so that 1% of the total area was included in the quadrats.

By chance, of course, all the random quadrats might be clustered in one section of the area. To avoid this possibility, the area can be subdivided into roughly equal areas and each section randomly sampled with fewer quadrats. This is called stratified random sampling, or restricted random sampling (Figure 4-1b). This is especially important when the ecologist has reason to believe that some important environmental parameter varies systematically across the sampled area. An example would be the upper and lower sections of a slope.

Density can also be estimated by distance methods, which do not use quadrats. Random spots are picked, from which the investigator begins to walk through the area; in some versions the path is a straight line, in others it is not. At intervals, the distance to the nearest individual is measured. Density can be calculated based on certain geometrical and statistical assumptions, knowing only the average distance from point to plant. The assumptions and statistical machinations are detailed in books such as Cox (1985), Greig-Smith (1964), Kershaw (1973), Mueller-Dombois and Ellenberg (1974), and Phillips (1959); Pielou (1977) requires a good mathematical background.

Pattern: Definitions and Methods

Density alone is a static measure. It does not reveal the dynamic interactions that may exist among members of the same species. The **pattern,** or spatial distribution, of the 300 sugar maples or the 3000 creosote bushes gives additional information about the species.

The same number of plants in an area can be arranged in three basic patterns: random, clumped, or regular (Figure 4-2). In a **random** pattern, the location of any one plant has no bearing on the location of another of the same species. In a **clumped** pattern (also called aggregated or underdispersed), the presence of one plant means there is a high probability of finding another of the same species nearby. A **regular,** or overdispersed, pattern is similar to the pattern of trees in an orchard. In regularly distributed populations, where one individual is found, there is a lower probability of finding another than would be expected by the assumption of randomness.

Members of most species seem to be clumped, and there are at least two reasons for this. The first has to do with reproduction: Seeds or fruits tend to fall close to a parent, or runners or rhizomes produce vegetative offspring near a parent. The second reason concerns the microenvironment: The habitat is homogeneous at a macroenvironmental level, but at a finer level it consists of many different microsites that permit the establishment of a species with varying degrees of success. Those microsites most suitable for a species will tend to become more densely populated with that species.

There are many ways to measure pattern; details can be found in the books cited earlier. One method utilizes random quadrats. The number of rooted individuals of species A is tallied in each quadrat and summarized in table form (Table 4-1). These are the observed data. The expected data—if members of species A were distributed at random—are generated by a rather simple formula, the Poisson distribution, which requires only that we know the average number

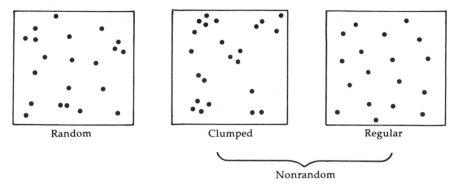

Figure 4-2 Random and nonrandom distribution patterns of individuals. Nonrandom distribution patterns may be clumped or regular. Dots represent individual plants as viewed from above.

of plants per quadrat. Discrepancy between the observed data and the expected data is evaluated by a chi-square calculation. In the example shown in Table 4-1, the chi-square value is greater than would be expected by chance, so the conclusion is that members of species A are not distributed at random. Are they, then, clumped or regular? Inspection of the table shows that fewer quadrats than expected

Table 4-1 Poisson analysis of quadrat data for a species with a nonrandom plant distribution. In the formulas, m is the average number of plants per quadrat, 1.56 in this case, and e^{-m} is 0.21. To employ this test, each category must have an expected value of $> 5\%$ of the total quadrats; to achieve this, category 5 had to be lumped together with category 4. The χ^2 sum is thus based on 5 numbers, and the degrees of freedom is $5 - 2 = 3$. At the 99% level of significance for 3 degrees of freedom, any $\Sigma\chi^2$ value > 11.34 suggests that the null hypothesis be rejected. The null hypothesis here is that the plants are distributed randomly. Therefore, the plants are distributed nonrandomly.

No. of plants per quadrat (x)	Observed no. of quadrats with x plants	Expected no. of quadrats with x plants $=$ $(e^{-m})\left(\dfrac{m^2}{x!}\right)(100)$	$\chi^2 =$ $\dfrac{(\text{observed} - \text{expected})^2}{\text{expected}}$
0	13	21.0	3.0
1	51	32.8	10.1
2	23	25.6	0.3
3	3	13.3	8.0
4	$\left.\begin{matrix}0\\10\end{matrix}\right\}10$	$\left.\begin{matrix}5.2\\1.6\end{matrix}\right\}6.8$	1.5
5			
Totals	100	99.5	$\Sigma\chi^2 = 22.9$

had either zero or more than one plant, and more than expected had one plant. By deduction, then, we can say that the members of species A were distributed regularly. There are other, shorter, treatments of the data that permit the type of nonrandom distribution to be calculated, rather than deduced. For a discussion of problems and alternatives to the random quadrat method see Kershaw (1973).

Distance methods (nonquadrat methods) can also be used to detect distribution patterns. In this case, the distance between adjacent members of the same species is tallied (see Chapter 8). Frequency is another measure used to assess pattern. Frequency is the fraction of all quadrats that contain a given species. If 50 quadrats are placed and species A is noted as present in 10 of them, the frequency of A is 0.20, or 20%. If density is high but frequency is low, one could conclude that species A is clumped; if the reverse is true, then one could conclude that species A is regular. Density and frequency are usually independent measures; knowing one does not help predict the other unless the plants are distributed at random. In that case, $100 - \%$ frequency $= e^{-m}$, where e^{-m} is the expected number of quadrats with no plants, the first entry in the Poisson distribution (Table 4-1) (Blackman 1935).

Frequency values are highly dependent on quadrat size. If the quadrats are too large, most species will have 100% frequency; if they are too small, many species will have close to 0% frequency. Daubenmire (1968a) suggested that the appropriate quadrat size for sampling a given life form be small enough that only one or two species show 100% frequency, but Blackman (1935) suggested that the maximum frequency should be 80%. It is difficult to compare frequency values in different studies unless the sampling methods used were identical.

Plant Demography

Demography is the study of changes in population size with time. By determining birth and death rates of individuals of each age in a population, the demographer projects how long an individual is likely to live, when it will produce offspring and how many, and the overall change in population numbers with time. Collecting data for demographic study seems a simple problem of counting individuals, as one would do with animals. However, the great plasticity and morphological complexity of plants and their ability to reproduce asexually mean that the individual plant can take on different forms, depending on circumstances. Because of these complications, plant population ecology did not develop in consort with animal ecology; it has become a field of active research only during the past two decades. The approaches we use to deal with plant demography were developed or stimulated primarily by John L. Harper (1977).

One approach to plant demography is to describe the various stages of the life history of a plant and quantify the numbers present at each stage. For example, Figure 4-3 is a diagrammatic representation of the stages in the life history of a plant population. The seeds present in the soil are referred to as the **seed pool** (or seed bank). Some of these seeds germinate to become seedlings. The environment acts as a sieve, as some seedlings become established and other

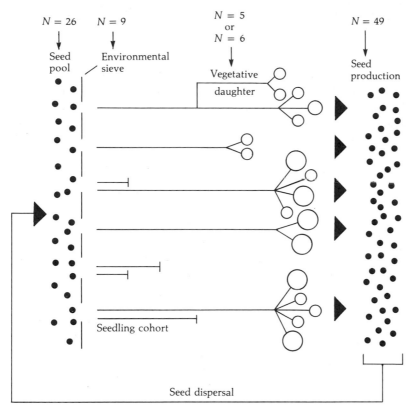

Figure 4-3 An idealized plant life history. Near the middle of the growing period, N is 5 or 6, depending on whether the vegetative daughter is considered an individual or part of the parent plant. (Copyright © 1977 by J. L. Harper, *Population Biology of Plants.*)

seeds remain in the seed pool. Some of the plants die before reaching reproductive maturity, and still others form new vegetative daughters by vegetative reproduction. Near the end of the growing season, new seeds are produced, and another seed pool is available for the next generation. The vegetative recruits are initially attached to the parent plant; the population ecologist must decide whether a vegetative rosette, tiller, stem, or the whole plant is to be considered a new population unit. In plant demography, the population unit is not always the individual formed by the germination of a seed. Vegetatively produced population units are referred to as **ramets.** This comes from the Latin root meaning a branch, as in ramification. A **genet** refers to a population unit arising from a seed. Thus a group of ramets—even if they have been severed from the parent plant, or the parent plant is no longer living—are considered a single genet because they were produced asexually by the same plant and are genetically identical. A group of ramets produced vegetatively from a single parent may also be referred to as a **clone.** Ramets themselves may reproduce vegetatively, increasing the size of a clone.

Two plants of the same age that have great differences in size and shape because of environmental circumstances have different impact as part of the population. For example, a large plant may produce many more seeds than a small plant. Therefore, it is often necessary to identify **modules** of growth and conceptualize a plant as a metapopulation or a population of modules that make up the morphological structure of a genetic individual (White 1979). We can get an idea of the dynamic complexities of a plant population by considering a hypothetical population of trees. The growth and reproduction of the trees depends on environmental conditions. The population of trees responds to these conditions and to time in one of two ways. First, it may increase by producing seeds and thus new genets. The population density then changes, and the age structure of genets changes because a new genetic individual has been added to the population. Genets may then adjust either physiologically or morphologically in response to changes in the environmental circumstances caused by additional individuals. These adjustments can take the form of acclimation, production of allelochemics, increased or decreased sexual reproductive output, and changes in the shape and size of the individual by adding or shedding modules of the plant body, such as leaves, branches, or roots. Thus, the second way a population of genets may respond to changes in environmental conditions is by changing the overall shape and orientation of the plant body. Proliferation of modules may utilize the available resources and impede the recruitment of new individuals from seeds.

One could hardly say, though, that the population is static because no new seedlings are being established. This plasticity of module numbers in response to changing environmental conditions presents a level of population dynamics superimposed on the changing of population size by establishment or death of trees. Studying the dynamics of the tree population includes following both the birth and death of individual trees and of the modular components of the plant body. Repeating modules can serve as units of demographic study. These modules may be a branch with associated buds, leaves, and flowers, individual leaves, or any other module of plant growth that repeats as a plant becomes larger.

Unlike the animal kingdom, where the distribution of body size is typically characterized by a few large and a few small individuals, with most clustered around the mean (a normal distribution), mature plant populations typically have an L-shaped frequency distribution (Figure 4-4). In plant populations, not all individuals add modules at the same rate because of limitations on resources and competition between individuals. Many individuals remain small, while a few grow large and occupy an inordinate amount of the habitat space. Size in these plants is not necessarily proportional to age.

Population Growth Models

A thorough mathematical treatment of population growth is beyond the scope of our discussion. Any good general ecology or population ecology text-

Figure 4-4 A typical frequency distribution of individual plant weight for a plant population experiencing intraspecific competition.

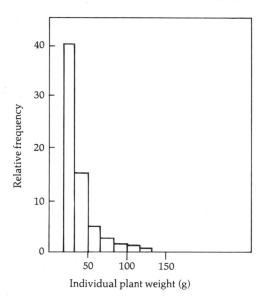

book can be consulted for background (e.g., Harper 1977; Pianka 1983; Silvertown 1982; Krebs 1985).

Using a **continuous-time model,** we can determine the number of plants expected at some time in the future (N_{t+1}) by adding the number of plants present at a certain time (N_t), the number established from seed produced by the plants present (B), and those dispersed onto the site (I), then subtracting the number that have died (D) and the number of seeds dispersed out of the area (E) during the time period t to $t + 1$. In equation form:

$$N_{t+1} = N_t + B + I - D - E \hspace{4cm} \text{(Equation 4-1)}$$

Since we are seldom able to make complete counts of the births and deaths for an entire population, data are usually expressed as individual birth (b) and death (d) rates. Ignoring for the moment immigration and emigration, we can calculate the instantaneous rate of increase (r) per individual (also called the intrinsic rate of natural increase) in the population as:

$$r = b - d \hspace{5cm} \text{(Equation 4-2)}$$

We can now calculate the instantaneous rate of change in population numbers using the differential equation

$$dN/dt = rN \hspace{5cm} \text{(Equation 4-3)}$$

where N is the number of individuals in the population at time t (Lotka 1925). This equation shows geometric growth (or decrease if $b < d$), plotted as curve (a) in Figure 4-5. We know that no population grows for long without some restric-

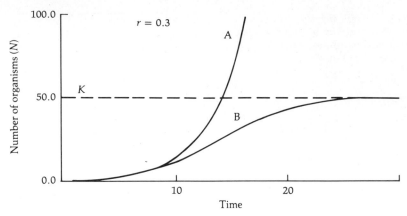

Figure 4-5   Geometric (line A) versus logistic (line B) population growth over time. *K* is the carrying capacity of the habitat for a population showing density-dependent logistic growth (line B).

tion due to lack of resources, space, or other limitation. We can assume, therefore, that the environment has a **carrying capacity** *(K)*; that is, the number of individuals of a species that can be supported by the environment. As the population numbers approach *K*, the environment restricts population growth, resulting in a logistic population growth curve (curve (b) in Figure 4-5). The classical Verhulst-Pearl equation for logistic population growth is:

$$dN/dt = rN(K - N/K) \qquad\qquad \text{(Equation 4-4)}$$

The Verhulst-Pearl equation is sometimes inadequate for plant populations, because the carrying capacity of vascular plants depends not only on the number of individuals but also on the biomass of the individuals. Plasticity of growth makes it possible for a few large or many small individuals to have the same influence on population recruitment. This problem is dealt with in more detail in the next section.

The continuous-time population growth models discussed above are appropriate for populations with continuous growth and in cases where birth rates, death rates, and size are correlated with age, such as in many annual plants and populations of leaves. However, plant populations usually reproduce for only a short period during the year, and not all plants reach reproductive maturity on a predictable time scale. Also, indeterminate growth in plants makes the number of individuals a poor indicator of population resource demands. **Matrix models**, which allow the determination of population growth in plants by dealing with discrete time periods and identifiable phases of the plant's life history (based on a similar model developed by Leslie (1945)), are available from a variety of sources (e.g., Lefkovitch 1965; Goodman 1969; Lewis 1972; Sarukhan and Gadgil 1974; Werner and Caswell 1977; Harper and Bell 1979; Silvertown 1982).

Before we begin our discussion of a matrix model, it is necessary to define some basic terms that apply to matrices and their manipulation. A matrix consisting of a single column is referred to as a **column matrix**. We could construct a column matrix that represents the number of individuals in each of three stages of development. For example, the number of seeds (N_s) in a seed pool, the number of plants in rosette form (N_r), and the number of plants in the flowering phase (N_f) would appear in matrix form as:

Column matrix of three growth stages   (Matrix 4-1)

$$\begin{bmatrix} N_s \\ N_r \\ N_f \end{bmatrix}$$

A **transition matrix** for three growth stages is square and consists of a group of probability values that represent the chance that a plant in a given stage of development will arrive at a different (or stay at the same) developmental stage during the time between population census dates. The transition matrix for three growth stages would appear as follows:

This census   (Matrix 4-2)

		Seed	Rosette	Flower	
Next census	Seed	a_{ss}	a_{rs}	a_{fs}	Transition matrix of three growth stages
	Rosette	a_{sr}	a_{rr}	a_{fr}	
	Flower	a_{sf}	a_{rf}	a_{ff}	

In the transition matrix, a_{sr}, for example, represents the probability that a seed from this census will develop to the rosette stage by the next census, and a_{rf} is the probability that a plant in the rosette stage at this census will be in flower at the next census. The columns represent the fate of, from left to right, seeds, rosettes, and plants in flower. The data are obtained by observing the timing of transition from one stage to another. For example, by following the development of rosettes during the time period under consideration, values for the relative number of rosettes that remain rosettes or flower become the probability values. Assume for the sake of illustration that 60% of the rosettes present at this census die before the next census, leaving 40%, of which 75% flower and 25% remain rosettes. The probability of a single rosette flowering (a_{rf}) would then be $0.4 \times 0.75 = 0.3$, and the probability of a single rosette remaining a rosette (a_{rr}) would be $0.4 \times 0.25 = 0.1$. The values for the remainder of the matrix could be calculated from similar data on the other growth stages.

In the theoretical model of Matrix 4-2, any growth stage has the capacity to become any growth stage at the next census. However, when we consider the biological constraints, certain transitions are not possible. For example, most perennial plants remain vegetative for one or more years after germinating, so seeds produce only rosettes in the next generation of growth stages. Annual plants, on the other hand, have a different set of biological constraints and would have a different set of potential values. For example, an annual plant in the rosette phase at the end of the season would produce no individuals in the next generation because it produces no seeds. Thus, the rosette column in a matrix for an annual would contain zero in all cells.

End of growing season,
year 1   (Matrix 4-3)

		Seed	Rosette	Flower	
End of growing season, year 2	Seed	a_{ss}	0	a_{fs}	Transition matrix for a hypothetical annual plant (census taken yearly at the end of the growing season)
	Rosette	a_{sr}	0	a_{fr}	
	Flower	a_{sf}	0	a_{ff}	

The frequency with which the census of the populations is taken also changes the matrix. If the census of an annual plant population were to be taken in mid-growing season and again at the end of the growing season, we would obtain a second matrix and therefore more information. Assuming only a single germination period, the end-of-season matrix would appear as follows:

Middle of growing season,
year 1   (Matrix 4-4)

End of growing season, year 1

$$\begin{bmatrix} a_{ss} & 0 & a_{fs} \\ 0 & a_{rr} & 0 \\ 0 & a_{rf} & a_{ff} \end{bmatrix}$$

End-of-season transition matrix for a hypothetical annual plant (census taken mid-season)

A transition matrix for a perennial in a yearly census would have the following characteristics:

Year 1   (Matrix 4-5)

Year 2

$$\begin{bmatrix} a_{ss} & 0 & a_{fs} \\ a_{sr} & a_{rr} & a_{fr} \\ 0 & a_{rf} & a_{ff} \end{bmatrix}$$

Yearly transition matrix for a perennial plant

The transitional probabilities a_{ss}, a_{rr}, and a_{ff} are the probability that the growth stage of a particular individual does not change during the year. In the case of a_{rr}, an individual may not change or may generate new rosettes without sexual reproduction.

The column matrix, established by counting the number of individuals in each growth stage, and the transition matrix are then multiplied to estimate the number of individuals of each growth stage in the next generation. For example, the number of seeds expected in the next generation is found by summing the products of each cell of the column matrix and the corresponding cell in the first row of the transition matrix. The result of matrix multiplication would be as follows:

$$
\begin{array}{cccccccc}
A & \times & B_1 & = & & B_2 & & \text{(Matrix 4-6)}
\end{array}
$$

$$
\begin{bmatrix} a_{ss} & 0 & a_{fs} \\ a_{sr} & a_{rr} & a_{fr} \\ 0 & a_{rf} & a_{ff} \end{bmatrix} \times \begin{bmatrix} N_s \\ N_r \\ N_f \end{bmatrix} = \begin{bmatrix} (N_s a_{ss}) & + & 0 & + & (N_f a_{fs}) \\ (N_s a_{sr}) & + & (N_r a_{rr}) & + & (N_f a_{fr}) \\ 0 & + & (N_r a_{rf}) & + & (N_f a_{ff}) \end{bmatrix}
$$

By summing the rows, B_2 becomes a new column matrix of N_s, N_r, and N_f for the second generation. Multiplying B_2 by A will provide estimates of the population for generation three. Continuing to multiply the column matrix for each generation by the transition matrix (A) gives the population growth estimates for the population through time. After a few generations (if the values used to construct the transition matrix remain constant and $r>0$), the relative number of individuals in each growth stage will remain constant; the age structure is then stable.

The net reproductive rate (R_0) of the population is a measure of whether the population is increasing, decreasing, or stable. To determine R_0 for a population, calculate the ratio of the number of individuals in a particular growth stage in two succeeding generations in a population that has reached a stable age structure. For example, the number of rosettes in generation ($t + 1$) divided by the number of rosettes in generation (t) equals the net reproductive rate for the population if sufficient generations have passed to reach a stable age distribution. If $R_0 = 1.0$, the population numbers are constant; if $R_0 < 1.0$, the population is decreasing; and if $R_0 > 1.0$, the population numbers are increasing.

The matrix model is of great advantage when the population units move from one identifiable stage of growth to another and when the ecologist is interested in the effect of different transition probabilities, such as would be present in contrasting habitats or in changing environments. Werner and Caswell (1977) conducted a study of teasel (*Dipsacus sylvestris*), a short-lived perennial that dies after flowering. After germination, teasel forms a rosette that may exist for up to 5 years before flowering. Vernalization is required for flowering, so no reproductive activity is possible in the first year. Teasel was studied in an open field and in an old field where much of the ground was shaded by shrubs. The transition matrices are shown in Table 4-2. The instantaneous rate of increase per individual (r) was 0.957 in the open field and -0.465 in the shrub-covered field. Since r measures environmental impact at the population level and integrates the effects of birth and death rates, which are both related to population dynamics and population fitness, we can use r values calculated from the transition matrix to draw conclusions about the population. Teasel populations in the open field were expanding geometrically ($r > 0$), and the populations in the shrub-covered field were doomed to extinction ($r < 0$), assuming transition probabilities did not

Table 4-2 Transition matrices for *Dipsacus sylvestris* (teasel) in open old field and in shrub-covered old field. Data for stable size distribution are in percent. The number in the "Flowering" column is average number of seeds produced per flowering plant. (Modified from P. Werner and H. Caswell, "Population growth rates and age versus stage distribution models for teasel (*Dipasacus sylvestris* Huds)," *Ecology* 58:1103–1111. Copyright © 1977 by the Ecological Society of America. Used with permission.)

Open field

	Seeds	Seeds 1 yr	Seeds 2 yr	Rosette small	Rosette medium	Rosette large	Flowering
Seeds	—	—	—	—	—	—	635
Seeds 1 yr	.634	—	—	—	—	—	—
Seeds 2 yr	—	.974	—	—	—	—	—
Rosette small	.013	.017	.011	.000	—	—	—
Rosette medium	.109	.004	.002	.077	.212	—	—
Rosette large	.006	.003	.000	.038	.281	.000	—
Flowering	—	—	—	.000	.063	1.000	—

Shrub-covered field

	Seeds	Seeds 1 yr	Seeds 2 yr	Rosette small	Rosette medium	Rosette large	Flowering
Seeds	—	—	—	—	—	—	476
Seeds 1 yr	.423	—	—	—	—	—	—
Seeds 2 yr	—	.987	—	—	—	—	—
Rosette small	.024	.009	.006	.007	—	—	—
Rosette medium	.044	.000	.000	.050	.158	—	—
Rosette large	.001	.000	.000	.002	.008	.000	—
Flowering	—	—	—	.000	.000	.250	—

Stable size distribution for open field teasel

Seeds	Seeds 1 yr	Seeds 2 yr	Rosette small	Rosette medium	Rosette large	Flowering
71.45	17.39	6.50	0.50	3.30	0.56	0.29

change. The population in the open field cannot continue to increase indefinitely. The finite amount of resources in the environment might cause the teasel population to reach carrying capacity, or the environment may change, reducing or even causing negative *r* values. The evidence gathered by Werner and Caswell shows that if shrubs invade the old field, we might expect a serious reduction in the teasel population. Since shrub invasion is the expected trend in old field succession, we have evidence that teasel is a fugitive species that occupies the habitat only temporarily during succession. The stable size distribution of teasel plants in the old field is shown at the bottom of Table 4-2. Remember that these values depend on the unlikely occurrence of a constant transition matrix.

The continuous-time model is useful when continuous data are available on age-specific birth and death of individuals or population modules. Recently, Kirkpatrick (1984) has constructed continuous-time demographic models based on size, not age.

Density Dependent Versus Crowding Dependent Population Regulation

We noted earlier that carrying capacity may be determined not only by the number of individuals in the population but also by the size and growth rate of individuals in the population. The influence of density on total biomass accumulation (yield) is exemplified by the data on clover in Figure 4-6a. Seedling density was varied from 6 to 32,500 plants m^{-2} and biomass of mature plants per unit area measured at each density. There was no mortality. At maturity, yield is independent of seedling density over a very wide range of densities. Thus, differences in density are largely compensated for by differences in individual plant size. Below 1500 plants m^{-2}, interplant distance is great enough that growth is not influenced by neighbors. Therefore, at low density, yield is dependent on the number of plants present. Where plant density is high enough for intraspecific interference to be important, yield is predictable regardless of plant density. This result is consistent for many plant species and has been referred to as the **"law" of constant yield** (Kira et al. 1953).

The magnitude of plant yield at a site depends on the availability of resources. For example, the yield of *Bromus unioloides* was constant over a wide range of densities but varied with nitrogen availability (Figure 4-6b). The environment sets a limit on the amount of plant biomass that can be supported at a site. Thus the concept of carrying capacity, which relates to the number of individuals that can be supported in a given environment, must be expanded to include a yield or biomass component.

All individuals in a plant population would need to be identical for each individual in the population to be equally inhibited by its neighbors. Either because of genetic or microhabitat differences, some individuals gain more than their share of resources and grow faster than plants of equal size. The result is a gradual thinning of very dense populations as individual plants die and other individuals dominate the stand. Plant mortality due to competition in crowded even-aged stands follows a predictable pattern, described by the self-thinning rule.

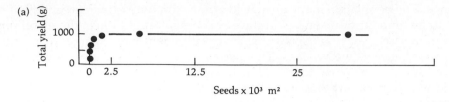

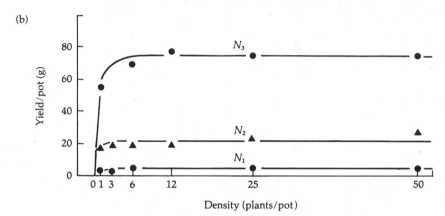

Figure 4-6   The relationship between yield (dry weight per unit area) and population density for mature populations of *Trifolium subterraneum* (a) and mature populations of *Bromus unioloides* (b) grown at three levels of nitrogen fertilization. (From C. M. Donald, 1951. *Australian Journal of Agricultural Research* 2:355–376.)

Population size in increasing populations, as predicted by most population growth models, depends on density dependent changes in survival or reproductive rates as population numbers get greater. As N gets closer to K, r decreases until it averages zero. The description of the population depends only on variations in N and is therefore density dependent. We know from the "law" of constant yield that plants respond to crowding caused not only by density but also by size of individuals. It is clear that the state of a plant population cannot be described by biomass alone, nor by density alone. It is more accurate to say that plant populations are crowding dependent rather than density dependent.

The self-thinning rule was proposed by Yoda et al. (1963) as a means of describing the interacting influences of density and biomass on plant population dynamics. Figure 4-7 shows the relationship between log biomass and log N that has been tested and found to fit more than 100 plant species (White 1980; Gorham 1979). The generalization may apply to most if not all plants. The *B-N* diagram shows the self-thinning line that describes the mortality of plants as a function of the rate at which biomass accumulates. The line is described by the equation:

$$B = CN^{-1/2} \tag{Equation 4-5}$$

B is biomass per unit area, N is number of survivors per unit area, and C is a

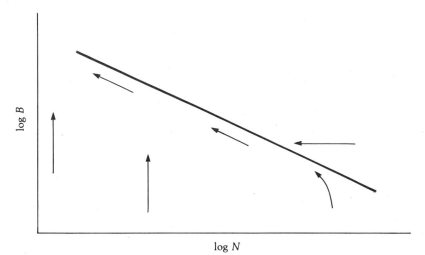

Figure 4-7 The *B-N* diagram. The line of slope −1/2 is the self-thinning line. Arrows indicate the trajectories that stands follow at different biomass-density combinations. (From *American Naturalist* 118:581–587, M. Westoby. Copyright © 1981 by University of Chicago Press. Reprinted by permission.)

constant.* The slope of the thinning line is −1/2. Where stands have biomass and density values that place them below the thinning line, biomass will increase until the line is approached. Individuals will then die in relation to biomass at a rate that depends on the rate of biomass accumulation. The slope of the line is the same for herbs, shrubs, and trees. The slope is probably due to the amount of nonphotosynthetic tissues necessary to support larger plants. A stand contains more biomass when the individuals are larger and fewer in number. There are some interesting speculations concerning the applicability of the rule to vascular plants in general (White 1981).

The self-thinning rule provides a mathematical means of determining the state of a population where the variables include both biomass and numbers. We can refer to the concept as carrying capacity where population regulation depends only on density or as crowding capacity where regulation depends on both density and biomass (as with most vascular plants) (Westoby 1981; 1984).

Stage Versus Age

Classical demographic theory uses age as the basis for estimating fecundity (*b* in Equation 4-2) and survivorship. However, age may not be an indicator of reproductive status in plants. There are two primary reasons for this. First, size

---

*The original formulation of the self-thinning rule was $w = KD^{-3/2}$, where w is average weight per plant, K is a constant, and D is density. We have followed Westoby (1981) in using biomass per unit area and N as the variables because these are more easily relatable to conventional population growth models.

is not necessarily correlated with age, and second, many plants flower when they reach a particular size, regardless of their age. For example, some plants listed as biennials in the floras (i.e., forming a vegetative rosette one year then flowering and dying the next) will flower in the second year only if the rosette has reached a critical size. If the rosette is too small, the plant may survive in the rosette stage until photosynthetic area and carbohydrate storage are sufficient to induce flowering (Werner 1975; Werner and Caswell 1977). Conversely, in an optimal environment, the necessary size and stored carbohydrates may be accumulated quickly, and flowering may occur in the first year of growth. A tree seedling may remain small for years when growing in the dense shade of a forest. It is the stage of development that determines the demographic status of the individual, not its age. In certain circumstances, a plant may even revert back to an earlier developmental stage (e.g., a plant may flower one year and revert to a vegetative state for one or more succeeding years) (Rabotnov 1978). Age also becomes a meaningless term in plant demography when dormancy interrupts the life cycle for a period of time. For example, a seed may remain dormant for years, during which time its population status does not change. The same is true for a bulbous herbaceous desert perennial, which may remain dormant during several years of drought. Many plants have a great deal of morphological plasticity; complete demographic analysis thus requires data on both stage of development and age.

Life Tables

Life tables were originally constructed by insurance companies as a means of determining the relationship between age and the potential for a client surviving to pay sufficient insurance premiums to keep the company solvent. These insurance life tables provide some basic information on survival for demographic studies but ignore the process of birth. By expanding life tables to include information on fecundity (birth rate) and age, the ecologist has an effective means of organizing demographic data.

There are two kinds of life tables, depending on the life-span of the individuals in the population. One kind of life table is referred to as a cohort or dynamic life table. A **cohort life table** is used when the observer can follow all of the seedlings germinating at a particular time (a cohort) until all individuals die. Such a life table is commonly used for plants that live for short periods of time as compared to the life-span or tolerance of the plant ecologist. Thus, annuals and short-lived perennials are usually studied from cohort life tables. Trees and shrubs often live longer than plant ecologists and are therefore studied using a static or time specific life table. In a **static life table**, the age structure of a population consisting of multiple cohorts is used to estimate the survival patterns of the various age groups. Figure 4-8 illustrates the difference between cohort and static life tables.

Table 4-3 is an example of a cohort life table and fecundity schedule for an annual plant, *Phlox drummondii.* Age is used here as a demographic parameter because these annual plants show nearly simultaneous germination and the

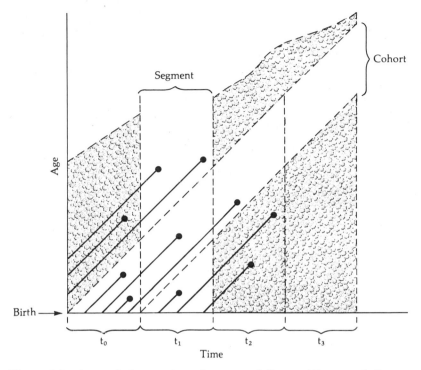

Figure 4-8 A population portrayed as a set of diagonal lines, each line representing the life "track" of an individual or a module. As time progresses, each individual ages and eventually dies. Three individuals are born prior to t_0, four during t_0, and three during t_1. To construct a cohort life table, a "searchlight" is directed onto the cohort of individuals born during t_0 and the subsequent development of the cohort is monitored. Two of the four individuals have survived to the beginning of t_1; only one of these is alive at the beginning of t_2; and none survives to the start of t_3. To construct a static life table, the searchlight is directed onto the whole population during a single segment of time (t_1). The ages (or stages) of the seven individuals alive at some time during t_1 may be taken as an indication of the age (stage)-specific survival rates if we assume that the rates of fecundity and survival are constant. (From Skellam. 1972 In J. N. R. Jeffers (ed.), *Mathematical Models in Ecology,* British Ecological Society Symposium No. 12, pp. 13–29. Blackwell Scientific Publications, Oxford.)

developmental stages are related directly to age. Age (x) is presented as an interval ($x - x'$) in days and the **age specific mortality rate** (q_x) as the percentage of the population dying during a particular age interval (in this case, 1 day). By knowing the number dying during the age interval (d_x) and the number of seeds in the seed pool (996 in Table 4-3), we can calculate the number surviving to the beginning of the next age interval (N_x). The proportion of the population surviving at the beginning of the age interval is referred to as **survivorship** (l_x). Survivorship is set at 1.0 for the original cohort and is equal to the proportion

Table 4-3 A cohort life table for *Phlox drummondii*. (From W. S. Laverich and D. A. Levin. *American Naturalist* 113:881–903. Copyright © 1979 by the University of Chicago Press.)

Age Interval (days) $x - x'$	No. surviving to day x N_x	Survivorship l_x	No. dying during interval d_x	Average mortality rate per day q_x
0– 63	996	1.0000	328	.0052
63–124	668	.6707	373	.0092
124–184	295	.2962	105	.0059
184–215	190	.1908	14	.0024
215–231	176	.1767	2	.0007
231–247	174	.1747	1	.0004
247–264	173	.1737	1	.0003
264–271	172	.1727	2	.0017
271–278	170	.1707	3	.0025
278–285	167	.1677	2	.0017
285–292	165	.1657	6	.0052
292–299	159	.1596	1	.0009
299–306	158	.1586	4	.0036
306–313	154	.1546	3	.0028
313–320	151	.1516	4	.0038
320–327	147	.1476	11	.0107
327–334	136	.1365	31	.0325
334–341	105	.1054	31	.0422
341–348	74	.0743	52	.1004
348–355	22	.0221	22	.1428
355–362	0	.0000		

of the original cohort surviving to the beginning of each age interval. Survivorship thus goes from one to zero during the life-span of the longest-lived individual in the cohort.

Plotting the log of the number of survivors at each age interval against time results in a **survivorship curve** (Figure 4-9). Deevey distinguished three types of survivorship curves that represent the extremes of population response. A type I survivorship curve is characteristic of organisms with low mortality in the younger stages and rapid mortality in the older age classes; type II is a straight line, where the probability of dying is essentially the same at any age; and type III is typical of organisms having high rates of juvenile mortality followed by a period of low mortality. Forest trees frequently have very high rates of seed mortality because of frugivory and granivory, then a period of more or less constant mortality as self-thinning occurs, and finally an extended period of low

Figure 4-9 Hypothetical
survivorship curves. (E. S. Deevey, Jr.
1947. Life tables for natural
populations of animals. *Quarterly
Review of Biology* 22:283–314. © The
Stony Brook Foundation, Inc. Used
with permission.)

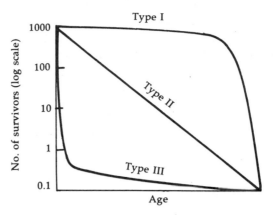

mortality as the surviving trees occupy positions in the canopy. The overall form
of the survivorship curve is a Deevey type III (Figure 4-10a). Annual plants
without seed dormancy, growing in open sites, may have a Deevey type I curve,
since most seedlings are able to survive to reproduction (Figure 4-10b). On less
open sites, intraspecific competition may result in mortality prior to reproduction, thus showing a Deevey type II curve. The survivorship curves of annual

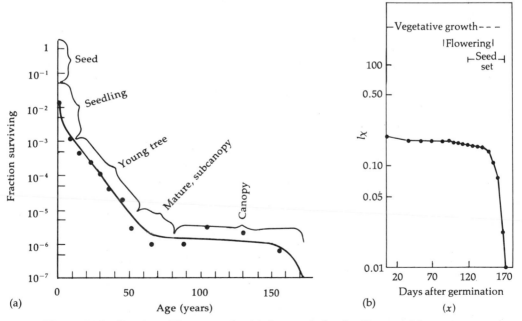

Figure 4-10 Survivorship curves for (a) the tropical palm *Euterpe globosa*.
(Reprinted by permission from L. Van Valen, 1975. *Biotropica* 7:260–269.)
(b) *Phlox drummondii*. (Modified from Leverich and Levin. *American
Naturalist* 113:881–903. Copyright © 1979 by University of Chicago Press.
By permission.)

plants may vary from year to year and from site to site but are usually type I or II. Herbaceous perennials and shrubs have various forms of survivorship curves that, depending on circumstances, span the entire range from type I to type III.

The length of time that seeds survive in the seed pool is related to growth form and environment. Annual plants in very harsh environments and weeds tend to have extremely long periods as viable seeds and very short postgermination life-spans. Seeds of long-lived perennials in moderate environments usually survive for only short periods in the seed pool. In general, we can say that seed longevity is proportional to environmental severity and inversely proportional to adult life-span.

It is interesting to contrast the survivorship curves for plants and animals. In animals, the larger, longer-lived species have type I survivorship curves, and the smaller, more rapidly reproducing species have type III curves. The opposite is true of plants. The relationship between survivorship and life history characteristics is discussed more fully in Chapter 5.

Fecundity (b_x, also called the age specific individual birth rate or natality) is measured by counting the total number of seeds produced by the cohort during each age interval and dividing by the number of surviving individuals in the cohort. Fecundity is thus the average number of seeds produced by an individual in the population at time or age interval x. If the plants are dioecious (male and female flowers on separate plants), only the female plants are considered in the life table. Multiplying survivorship (l_x) by fecundity (b_x) and summing over the life-span of the cohort gives an estimate of the **net reproductive rate** of the cohort (R_0). Symbolically:

$$R_0 = \sum_{x=0}^{\infty} l_x b_x \qquad \text{(Equation 4-6)}$$

On the average, every plant that dies is replaced by a new seed in the seed pool when $R_0 = 1$—i.e., the population is not changing in size. When R_0 is < 1.0, fewer seeds are being placed into the seed pool than are necessary to replace the population. The population of *Phlox drummondii* of Table 4-3 is increasing, since $R > 1.0$.

The success of a colonizing population or the survival of an established population depends on the ability of existing individuals to contribute offspring to future generations. **Reproductive value** (V_x) is a measure of the relative contribution an average individual of age x will make to the seed pool before it dies. Reproductive value is the sum of the average number of seeds produced by individuals of age x (b_x) and the total number of seeds produced by individuals older than x (b_{x+1}) times the probability that an individual of age x will survive to each older age category (l_{x+1}/l_x).

$$V_x = b_x + \sum_{x}^{\infty} (l_{x+1}/l_x) b_{x+1} \qquad \text{(Equation 4-7)}$$

Reproductive value is generally low in the early stages of growth because of the relatively high probability of death before reproduction. However, when the

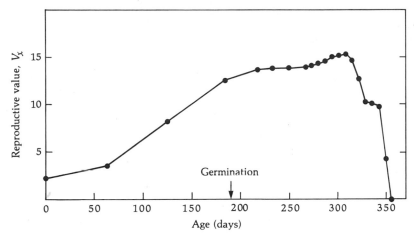

Figure 4-11   Reproductive values V_x for *Phlox drummondii*. Age is in number of days after seed dispersal. (Modified from Leverich and Levin. *American Naturalist* 113:881–903. Copyright © 1979 by University of Chicago Press. By permission.)

plant survives to reproduction, it maintains a high reproductive value until it reaches senility. The reproductive values for the *Phlox drummondii* cohort are plotted in Figure 4-11.

Age and Stage Structure

Regardless of the life-span, one can recognize eight important **stages** in an individual plant or a population (Rabotnov 1969): (1) viable seed, (2) seedling, (3) juvenile, (4) immature, vegetative, (5) mature, vegetative, (6) initial reproductive, (7) maximum vigor (reproductive and vegetative), and (8) senescent. If a perennial population shows only the first four or five stages, it is obviously invading and is part of a seral community. If a population shows all eight stages and if no further changes occur in the age structure, it may be stable and replacing itself at the site. If it shows only the last four stages, the population may be in decline or replaced by temporally infrequent cohorts. It takes only one surviving seedling per adult to maintain the population; successful reproduction once in a lifetime is sufficient.

The age distribution of a population can be used as a predictive tool in community ecology. Figure 4-12 shows an age distribution of longleaf pine (*Pinus palustris*) in a northern Florida forest. (Actually, trunk diameter at breast height (dbh) is shown, not age, but in this case age correlates closely with dbh.) The pine population has many mature trees but no seedlings or saplings; it appears to be a senescent population that is not reproducing itself and is part of a seral community. What will replace it? The same graph shows two hardwood populations combined, water oak (*Quercus nigra*) and wax myrtle (*Myrica cerifera*), which have many seedlings and saplings but no large (mature) individuals. The

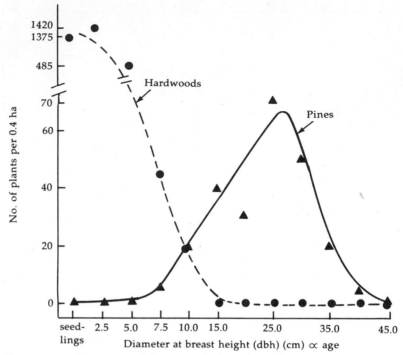

Figure 4-12 Summary of the ages of all trees in a forest near Gainesville, Florida. Succession is progressing from a pine community (solid line) to a hardwood community (mainly oak and wax myrtle) (dashed line). (Based on data from "The relation of fire-to-stand composition of longleaf pine forests" by F. Heyward, *Ecology* 20:287–304. © 1939 by the Ecological Society of America and from Plant Communities by R. Daubenmire © 1968 by Harper and Row, Publishers, Inc. New York.)

hardwoods are the invading populations, part of a community that will replace the pine forest.

In contrast, Figure 4-13 shows the age distribution of red spruce (*Picea rubens*) in virgin forests of the White Mountains of New Hampshire. Here we have a reverse J-shaped curve, with many seedlings and saplings and fewer larger, older trees. Many ages, and all age stages, are represented, and the high density of young trees means there is a high probability of the population maintaining itself as part of a climax community.

Not all stable, climax populations will show this same kind of age distribution curve, however. Woody perennials are long-lived and so can successfully maintain themselves even if their seedlings become established only sporadically in time. For instance, creosote bush (*Larrea tridentata*) is an evergreen shrub that dominates the warm deserts of North America. It seems to have all the attributes of a climax species that has maintained itself in the region for thousands of years

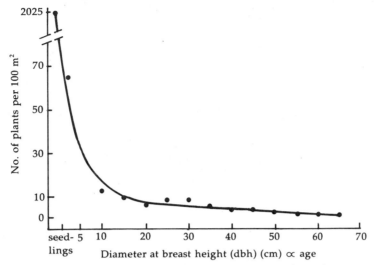

Figure 4-13 A reverse J-shaped age distribution curve for red spruce
(*Picea rubens*) in four stands in the White Mountains of New Hampshire.
(Based on data from "A comparison of virgin spruce-fir forest in the
northern and southern Appalachian system" by H. J. Oosting and W. D.
Billings, *Ecology* 32:84–103. Copyright © 1951 by the Ecological Society of
America. Reprinted by permission.)

(Vasek et al. 1975b; Laudermilk and Munz 1938). However, the species has
remarkably mesic requirements for good germination and seedling development
(abundant moisture, neutral pH, low salinity, and moderate temperature) (Barbour, Cunningham, Oechel, and Bamberg 1977), and such a combination of
conditions may occur only occasionally in nature. When a suitable year for establishment does arrive, we can expect an age distribution curve to show a peak.
Robert and Alice Chew (1965) aged a *Larrea* stand in southeastern Arizona and
found most shrubs to be 15–20 years old at the time (Figure 4-14a). No reproduction had taken place in the past 4 years, and the oldest shrub was 65; this is
a relatively young stand. Barbour, MacMahon, Bamberg, and Ludwig (1977) sampled older stands, relating age to height; stands on more arid sites often showed
several age peaks, with no recent reproduction (Figure 4-14b). Evidently, mass
germination and establishment in this species is a periodic phenomenon. There
is no reverse J-shaped age curve, and only a few age classes are represented, yet
the populations still maintain themselves.

In the mid-elevations of the Sierra Nevada Mountains, some groves of giant
sequoia (*Sequoiadendron giganteum*) have many young trees and a reverse J-shaped
age curve, while others do not (Figure 4-15). By correlating trunk diameter at
breast height to age, Rundel (1971) showed that reproduction had not occurred
for the past 500 years at sites such as Ponderosa Grove, and he predicted that
these groves would become extinct in the next 1000 years as the old trees died

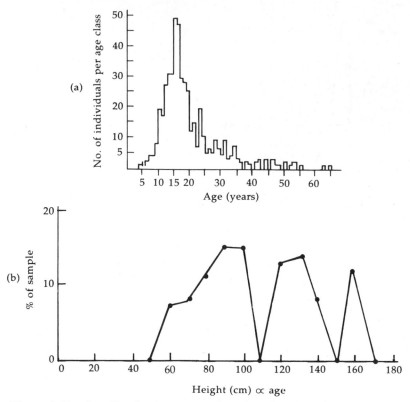

Figure 4-14   Age distribution curves for a long-lived woody perennial, creosote bush (*Larrea tridentata*). (a) A relatively young stand that shows a peak of establishment 15–20 years ago and an absence of reproduction during the past 4 years. (b) An older stand in a more xeric (arid) environment, showing several past peaks of reproduction and an absence of reproduction for several years. (From "The primary productivity of a desert shrub (*Larrea tridentata*)" by R. M. and A. E. Chew, *Ecological Monographs* 35:353–375. Copyright © 1965 by the Ecological Society of America. Reprinted by permission.)

and were not replaced with their own kind. Yet the life-span of the giant sequoia is 2000–3000 years (Hartesveldt et al. 1975), so if a good year for establishment comes only once in a millennium, that should be sufficient to maintain the population.

Knobcone pine (*Pinus attenuata*) is a closed cone conifer that typically occurs in single-aged (even-aged) stands that date back to periodic destructive fires (Vogl 1973). The reproduction of this species is dependent on fire melting a resin that otherwise seals cones shut and prevents seed dispersal. Apparently, conifers in the Canadian boreal forest also show single-aged populations, and these date back to severe storms, fires, or other natural disasters (Jones 1945). The frequency of these disasters is sufficient to maintain populations even though their age structure is not typical for climax species (Loucks 1970).

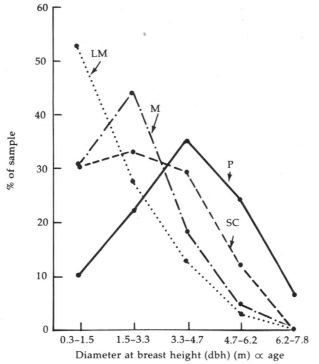

Figure 4-15 Age distribution curves for several stands of giant sequoia (*Sequoiadendron giganteum*) in the Sierra Nevada Mountains. Grove names are: LM, Long Meadow; M, Muir; SC, South Calaveras; and P, Ponderosa. Only LM shows a reverse J-shaped curve. (From Rundel et al. 1977. In *Terrestrial Vegetation of California.* Edited by Barbour and Major. Copyright © 1977 John Wiley and Sons, Inc. Reprinted by permission.)

Summary

The measurement and description of plant population structure and dynamics are the central focus of this chapter. The factors that influence population dynamics and distribution of plants include environmental conditions, resource availability, competitors, and disturbance. These conditions are expressed in the plant population by changes in reproductive output, growth, branching pattern, and biomass. Plant population ecologists strive to understand and predict the responses of plant populations to internal and external conditions.

The density of plants, expressed as the number of individuals per unit area, is an important quantity used to describe populations. When density is combined with measures of spatial distribution, we can deduce more about habitat preference, competitive dynamics, and microhabitat distribution than with density alone. Clumping is the most common distribution pattern in plants. This is because seeds and vegetative offspring tend to concentrate near the parent and because plants tend to cluster in areas with suitable microenvironment.

Plant demography is the study of changes in plant populations through time. Plant populations increase or decrease not only by birth and death of individuals but also by indeterminate growth over a large range of potential sizes. Complete knowledge of plant population dynamics thus requires information about the number of genetically distinct individuals (genets), the number of vegetatively reproduced individuals (ramets), and the number of growth modules present on an individual. Growth modules may be individual leaves, terminal buds with associated lateral meristems, or individual branches, depending on what growth regions form repeating modules throughout the individual. Understanding the dynamics of seeds in the soil (the seed pool) provides important insights into plant demography.

Plant demography is studied using either continuous-time models or matrix models to reveal the consequences of variations in birth rate and death rate for the population. Continuous-time models are used for populations with continuous growth where birth, death, and size are correlated with age. Studies of leaf demography and dynamics of annual plant populations are frequently based on continuous-time models. Most plants have a less predictable growth rate, age at death, and age at reproduction than is necessary for use of the continuous-time models. Matrix models deal with discrete time periods and use identifiable phases of the plant's life history, rather than age, as the basis for prediction.

The number of individuals of a species that can be supported by a unit of habitat (carrying capacity) and the age of individuals at particular stages of development are complicated as population parameters by the growth and development plasticity of plants. The total amount of biomass present at a site depends on the availability of resources and, once the resources are fully utilized, is constant over a wide range of densities (the "law" of constant yield). As high-density populations mature, population density declines in a predictable fashion, described by the self-thinning rule. Reproduction in many plants depends on size or accumulation of stored reserves rather than on age. Therefore, stage is a better population parameter in most plants than is age.

The data for demographic studies are organized into cohort or time specific life tables, depending on the life-span of the plants. Life tables provide information on age specific mortality rate, survivorship, fecundity, net reproductive rate, and reproductive value for a population.

The static age and stage structure of plant populations is used as a predictive tool to deduce trends in population replacement with time, to reconstruct the periodicity of reproductive success of a population, and to reconstruct population response to periodic fire or other disturbances.

CHAPTER 5

LIFE HISTORY PATTERNS AND RESOURCE ALLOCATION

P lants have patterns of growth, reproduction, and longevity that are related to specific demands for survival in a particular place at a particular time. We refer to these patterns of growth, reproduction, and longevity as **life history patterns.** In this chapter, we attempt to deduce the survival value of variations in life history patterns. The variations can be traced to differences in the timing and magnitude of energy and nutrients allocated to growth, reproduction, and survival. The life history pattern that is successful in a given situation is one that enhances reproduction, survival, and/or growth. A successful pattern may involve remaining small and allocating most available carbohydrates and nutrients to reproduction; another may be to grow rapidly and tie up resources (light, water, nutrients, space, etc.), to the detriment of competitors; a third may be to extend life by growing slowly and allocating resources to maximize resistance to herbivores, parasites, disease and abiotic conditions. Winning in this existential game is a product of both short-term success and the ability to continue playing for as long as possible (Slobodkin and Rapoport 1974).

The life history pattern determines at what place and time a plant can successfully carry out the processes of germination, growth, and reproduction. For example, a plant that allocates most resources to reproduction and remains small is more likely to occupy a disturbed habitat than a more stable habitat. In a stable habitat, a tree, which allocates resources to growth rather than reproduction (at least in the early stages of development), would be more successful. Life history pattern is a reflection of the **niche** a species occupies in a community.

The term *strategy* has been applied to the concept of life history variation. This term must be used cautiously, however, since it implies that the plant has the ability to plan a scheme of resource allocation that maximizes fitness. Patterns of resource allocation may change through the process of natural selection, but resource distribution is determined by response to environmental factors within the confines of a plant's genotype.

Plant Economics

Plant function is somewhat analogous to the operation of a business, or perhaps to personal finance. In fact, it is instructive to apply economic theory and principles to ecological systems, and vice versa (see Bloom et al. 1985). Intuitively, the species or individual that has the greatest profit (produces the most offspring) is at an advantage. There are, however, costs associated with

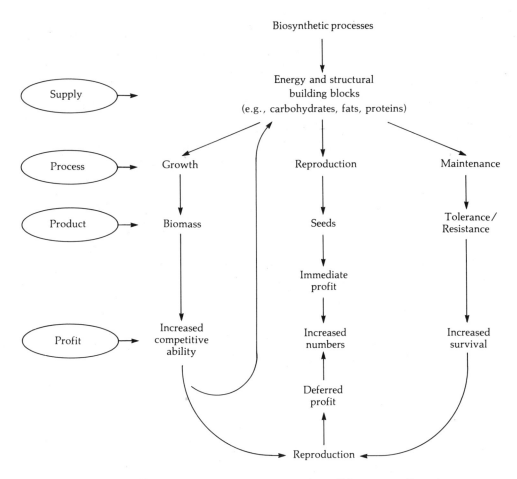

Figure 5-1   A diagrammatic representation of possible energy allocation pathways in plants. Items circled at the left are analogous terms from economics.

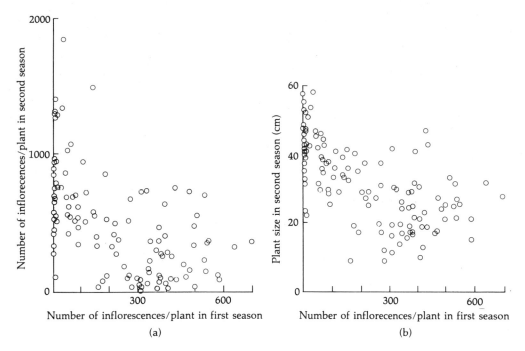

Figure 5-2 Scatter diagrams of number of inflorescences per plant in the first season and (a) number of inflorescences per plant in the second, and (b) plant size in the second season. (From Law, *American Naturalist* 113:3–16. Copyright © 1979 by University of Chicago Press. By permission.)

massive seed production. These costs are realized in decreased vegetative growth and survival (Figure 5-1). In other words, growth, reproduction, and survival represent competing demands for the finite supply of carbohydrates, fats/waxes, and proteins produced by synthetic processes in the plant. Gadgil and Bossert (1970) developed a theoretical model, based on cost/benefit analysis, which shows that early or large allocations of energy to reproduction lead to decreased probability of survival and/or reduced potential for future reproduction. The Gadgil and Bossert model has been confirmed by several studies. *Poa annua* completes the life cycle within one or more years, depending on conditions (Law 1979). High levels of reproduction in the first season were correlated with lower levels of reproduction in the second season (Figure 5-2a). Plants with high first-season reproductive output were smaller in the second season than their associates that had little or no early reproductive output (Figure 5-2b). There is also evidence that early, heavy reproduction increases the subsequent risk of mortality in plants. Pinero et al. (1982) reported that the probability of a tropical palm (*Astrocaryum mexicanum*) surviving for 15 years was essentially zero when a plant produced 40 to 50 fruits per year (Figure 5-3).

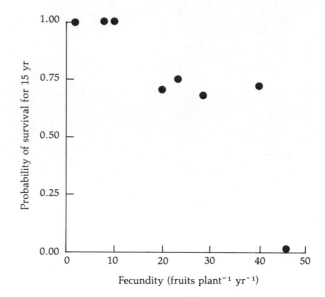

Figure 5-3 The probability of surviving for 15 years as a function of fecundity in ten-year-old plants of *Astrocaryum mexicanum* in Veracruz, Mexico. (From Pinero et al. *Journal of Ecology.* 70:473–481.)

Life-Span and Reproduction

The periodicity and intensity of reproduction are intimately related to plant life history and demography. Some species have a single period of reproductive effort followed by death, whereas others have repeated periods of low to moderate reproductive activity. Plants that reproduce only once and die are termed **monocarpic** (Greek "one" "fruit") or **semelparous** (Latin "at the same time" "birth"). In a lighter vein, monocarpy is sometimes referred to as "big bang reproduction." **Polycarpic** (Greek "many" "fruit") or **iteroparous** (Latin "again" "birth") plants have repeated periods of reproduction.

Assuming the plant that produces the greatest number of seeds during its lifetime has the highest fitness, we can consider the circumstances where either iteroparity or semelparity confer the greatest advantage (Figure 5-4). The important parameters are seed output (b_x) and the probability of reproducing in the future (residual reproductive value = RRV) when allocating a particular fraction of the available resources to reproduction. If b_x responds rapidly to increases in RA (the proportion of resources allocated to reproduction), the resulting curve is concave (Figure 5-4a); if response is slow, the curve is convex (Figure 5-4b). RRV is the product of the reproductive value of an average individual one year older (V_{x+1}; see Equation 4-7) and the probability that a plant of age x will survive to reproduce at age $x+1$ (l_{x+1}/l_x):

$$RRV = V_{x+1} (l_{x+1}/l_x) \qquad\qquad \text{(Equation 5-1)}$$

When reproduction causes a rapid decrease in residual reproductive value, the RRV versus RA curve is concave (Figure 5-4c). Where reproduction causes only

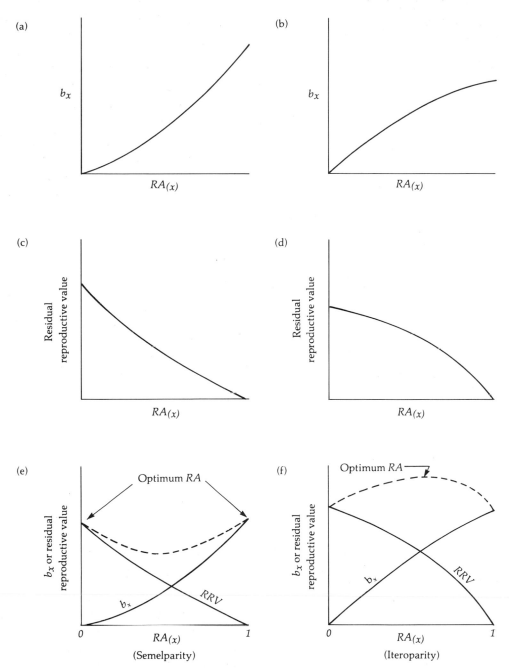

Figure 5-4 Seed production (a, b) at year x (b_x) and residual reproductive value (c, d) $V_{(x+1)}(l_{x+1}/l_x)$ as a function of the proportion of resources allocated to reproduction (RA). The total of b_x and residual reproductive value (RRV) (dashed line) is highest at $RA = 0$ or 1 (semelparity) in (e) and at intermediate RA values (iteroparity) in (f). See text for details.

a small decrease in *RRV*, the curve is convex (Figure 5-4d). Adding the residual reproductive value and b_x for all possible *RA* values from 0 (no reproduction) to 1 (allocating all resources to reproduction) gives a series of numbers, the maximum of which is associated with the *RA* of maximum fecundity. When both the residual reproductive value and b_x are concave (Figure 5-4e), maximum fecundity is attained either by not reproducing in year *x* or by committing all resources to reproduction (semelparity). Where both b_x and the residual reproductive value are convex (Figure 5-4f), maximum fecundity will result from allocating an intermediate amount of resources to reproduction (iteroparity).

Life-Span and Environment

A basic classification considers plants annual, biennial, or perennial, depending on length of life and age at first reproduction. **Annuals** are semelparous and go from seed to seed in less than 12 months. **Biennials** are semelparous, with a season of vegetative growth followed in the second season by seed production and death. **Perennials** live for more than two years and can be either semelparous or iteroparous. In general, postponing reproduction increases reproductive output, leading to speculation that the perennial habit would have the greatest fitness (Schaffer and Gadgil 1975). Why then do we see so many species of annuals in some habitats? One answer involves the chances that an adult plant can survive long enough to gain the advantage of postponed reproduction. The annual growth form should have the greatest fitness in situations where the probability that a seed will establish an adult plant exceeds the probability that the adult plant will survive to another growing season. The annual growth form also has maximum fitness in habitats where high values of *r* are an important asset. This is because being annual is associated with the possibility of reproduction at an early age and thus high potential rates of population growth.

Situations that reduce adult survival and therefore favor the annual life cycle may be related to disturbance and the existence of temporary habitats. Disturbance removes competing vegetation, makes resources readily available, creates a temporary habitat, and thus favors rapid growth and reproduction. Species adapted to temporary habitats are commonly described as weeds because they are well adapted to life in habitats disturbed by human activity. Correctly used, **weed** refers to any plant (annual to perennial) that is growing where it is either undesired or unappreciated.

Disturbance is not the only phenomenon that favors the annual habit. For example, over 90% of the desert flora of Death Valley, California, is annual. The extreme levels of physiological stress associated with drought and high temperature reduce the probability of survival of perennating organs (organs capable of perennial activity, such as buds, tubers, corms, and bulbs) of perennials. Seeds are potentially much more resistant to extreme conditions than perennating structures. Resource availability is highly unpredictable in the desert. For example, Went and Westergaard (1949) reported that annual precipitation in Death

Valley varied from 19 mm to 94 mm. Since about 60 mm of precipitation is required within a few hours for germination of many species, one or more years may pass between germination episodes; even then, the length of time that resources are available may be very short, and reproduction must occur quickly. For example, *Boerrhavia repens* of the Sahara Desert goes from seed to seed in as few as 10 days (Cloudsley-Thompson and Chadwick 1964). The annual plants that occupy habitats with temporally and spatially unpredictable resources are properly referred to as **ephemerals.**

Not all plants showing the annual growth form are obligate annuals—death may be due to external rather than internal factors. Obligate annuals die after reproduction because all resources are committed to reproduction and/or the growth meristems are the source of reproductive tissue, leaving no meristematic regions for further vegetative growth. Other annuals maintain the potential for continued growth and reproduction until some environmental factor surpasses the plant's range of tolerance. For example, *Perityle emoryi* grows as a semelparous ephemeral in the southwestern deserts and as an iteroparous perennial in island and coastal habitats of California.

The classification of plants as annual, biennial, and perennial is a useful convention when considering plants in near-optimal (or at least predictable) conditions, such as might be found in horticultural or agricultural settings. In natural circumstances, however, there may not be any plants with an obligate biennial life cycle (Harper 1977). Many plants classified as biennials are really short-lived semelparous perennials that are induced to flower when they attain a certain size or carbohydrate storage level (Werner 1975). Flower induction may typically take two years; under better conditions flowering may occur in the first season; and under worse conditions it may be delayed to 3 or more years. In some biennials, flowering the first year is impossible, as a vernalization period (period of cold temperature) is required to induce flowering.

Short-lived semelparous perennials (biennials) typically occupy sites disturbed periodically but not every year, where they could exist as annuals and produce seeds yearly; why then is reproduction postponed? Postponing reproduction may be an advantage in areas with a short growing season. For example, *Melilotus alba* (white sweet clover) is an annual at lower elevations but postpones reproduction to the second year at high elevations, where the growing season is contracted (Smith 1927). Overwintering is costly in terms of prereproductive mortality, but the possibility of a very large seed crop in the second year as a result of the large stored reserves may compensate for the increased mortality and the doubling of generation time. Harper (1967) suggests that a biennial pattern of growth and reproduction will be favored if the second-year seed set is the square of that possible at the end of the first year. *Verbascum thapsus* (mullein) is an early successional, short-lived semelparous perennial that has the characteristics of a biennial life form (Gross 1980; Semenza et al. 1978).

Longer-lived perennials can be either semelparous or iteroparous and thus offer material for study of the relative merits of suicidal versus repeated reproduction. Semelparity has the advantage of potentially larger reproductive struc-

Figure 5-5   Examples of large inflorescences of the perennial semelparous *Frasera speciosa* (a) and *Yucca whipplei* (b) and, in the background, *Agave deserti*.

tures and thus may gain a greater share of the pollinators in a competitive system. For example, some species of *Agave, Yucca,* and *Frasera* (Figure 5-5) produce a large inflorescence with copious amounts of nectar—an energetic cost for reproduction that would be impossible in an iteroparous species. *Frasera speciosa* populations show both temporal and spatial flowering synchrony, produce large amounts of nectar for an extended period, and have inflorescences up to 2 m tall. As a result, a small number of individuals (relative to the total number of individuals of all species competing for pollinators) receive up to 80% of the floral visits made in a sampling area of the Colorado Rocky Mountains (Beattie et al. 1973).

At first glance it seems that semelparity would be a disadvantage in habitats where reproductive success is highly variable. Where age determines reproductive maturity, plants from seeds produced all at once by a semelparous individual would flower synchronously, and in a bad year extinction of that genome could occur. *Frasera speciosa* is a semelparous plant that lives in an environment where reproductive success is variable. However, because flowering depends on factors other than age, not all offspring from one individual will flower together. Thus the *Frasera* population has an advantage that one would expect to find in iteroparous species, because the offspring of any one individual may reproduce at different seasons, over an extended period of time (Taylor and Inouye 1985).

Occasional massive reproduction may produce more seeds than can be consumed by granivores, thus increasing the probability of some seeds surviving. *Phyllostachys bambusoides* (Chinese bamboo) lives about 120 years before flowering. Synchronous flowering increases the chance of outcrossing in this wind-pollinated species. Flowering is physiologically controlled and is not stimulated by external environmental cues. The seeds are vulnerable because they are not protected chemically, contain more nutrients than rice or wheat, and disperse close to the parent plant, often reaching a depth of 15 cm below the parent. However, granivore populations are apparently unable to predict reproduction in bamboo, and so their populations are not sufficient to consume all of the seeds. Individuals that flower out of synchrony, or in areas isolated from the main group of flowering individuals, lose essentially all of their seeds to granivores (Janzen 1976).

Some iteroparous woody perennials reproduce heavily on some years (**mast years**), followed by one or more years of little or no reproduction. Mast years are common in temperate trees and serve to increase the chances of seeds escaping granivores, much as semelparity does in Chinese bamboo. Smith (1970) studied seed predation in conifer species of the Pacific Northwest and found that escape from predation was related to mast years when the squirrel populations were satiated. Mast years may be unpredictable when the reproductive cycle responds to external rather than physiological changes. Alternatively, irregular reproduction may be caused by severe resource depletion during the mast year. The period of time necessary to replace the energy reserves is often unpredictable because it varies with environmental conditions. Predictable mast years are found in conifers where the reproductive process takes a predictable time to complete.

Seed Dormancy and Dispersal

Well-timed periods of seed dormancy and suitable patterns of seed dispersal are necessary components of the successful life history pattern. **Dormancy** is a mechanism to avoid seasonally harsh conditions and, in the case of seeds, may be a means of spreading the risk of failure over several periods of potential growth. Arrival in new habitats and making the best use of already occupied habitats depend on appropriate patterns of **dispersal.**

In predictable environments, seeds are frequently of low density in the immediate vicinity of an isolated parent, of high density in the area adjacent to the parent, and of declining density further away (*Verbascum thapsus* in Figure 5-6). Such a seed distribution pattern would serve to reduce intraspecific competition by increasing the distance between parent and offspring, but because most seeds are in the same general area as the parent, the chances that the offspring will establish in a habitat already tried by the parent is high. Where a dense population of parents is present, dispersal (and hence colonization) takes place along an advancing front, where the highest seed density is often under the canopy and decreases rapidly as the distance from the parents increases (*Tussilago farfara* in Figure 5-6).

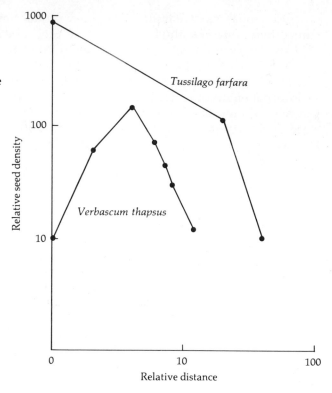

Figure 5-6 The relationship between seed density and the distance of seed dispersal from an isolated plant of *Verbascum thapsus* and from the edge of a dense population of *Tussilago farfara*. (© 1977 by J. L. Harper, *Population Biology of Plants*.)

Long-distance dispersal is typical of plants that occupy patchy environments. Agricultural weeds typically have effective means of dispersal that depend on wind or animal (including man) vectors. Riparian trees grow along the margins of streams and rivers and are thus adapted to widely separated or patchy environments. Seeds of these trees are often dispersed by wind or water. Dispersal distance may also be related to the density of patches in the environment that contain an appropriate combination of conditions for seedling establishment. Microenvironments specifically meeting the germination requirements of a species are called **safe sites**. Maple (*Acer* spp.), Ash (*Fraxinus* spp.), and Tuliptree (*Liriodendron tulipifera*) are wind-dispersed trees that inhabit the eastern deciduous forests of North America and have different requirements for seedling establishment and, therefore, different safe sites. Maple samaras have the shortest dispersal range and the greatest density of safe sites, whereas ash and tuliptree disperse further and have fewer safe sites (Green 1983).

Venable and Lawlor (1980) modeled the interplay between dispersability and seed dormancy in desert annuals. Their model predicts that plants with long-range dispersal will have few dormancy mechanisms and thus quick germination. In contrast, those without adaptations for long-distance dispersal should have delayed germination (see also MacArthur 1972; Cohen 1966). Table 5-1 is a list of species that produce dimorphic seeds, some capable of long-distance dispersal, others not. In most cases the seeds without long-range dispersal had delayed germination, confirming the predictions of the Venable and Lawlor model.

Table 5-1 All species listed here have two types of seeds with differences in both dispersal and germination. N and F signify whether the seed type is dispersed nearer or farther from the parent than the other type. D and Q signify which seed type has delayed or quick germination relative to the other seed type. Dispersal is inferred from presence versus absence of dispersal structures such as barbed or plumed pappus for animal or wind dispersal respectively. If dispersal differences are inferred from substantial differences in size and weight of propagules, symbols are marked with an asterisk (*). (Reprinted from Venable and Lawlor, *Oecologia* 46:272–282, Springer-Verlag, New York.)

Species of Asteraceae	Dispersal		Germination	
	Outer	Inner	Outer	Inner
Dimorphotheca plurialis	N	F	slightD	Q
Xanthocephalum gymnospermoides (= Gutierrezia g.)	N	F	D	Q
Heterotheca latifolia (= H. lamarckii)	N	F	D	Q
Charieis heterophylla	N	F	Q	Q
Bidens bipinnata	similar		D	Q
Sanvitalia procumbens	N	F	D	Q
Verbesina enceliodes	N	F	slightD	Q
Synedrella nodiflora	N	F	D	Q
Heterospermum xanthii	N	F	D	Q
Galinsoga parviflora	N	F	Q	D
Layia platyglossa subsp. campestris (= L. elegans)	N	F	slightD	Q
L. platyglossa	N	F	D	Q
L. heterotricha	N	F	D	Q
Achyrachaena mollis	N	F	D	Q
Chrysanthemum segetum	N*	F*	D	Q
C. coronarium	N*	F*	D	Q
C. viscosum	N*	F*	D	Q
C. frutescens	N*	F*	D	Q
Coleostephus myconis (= Chrysanthemum m.)	N*	F*	D	Q
Chardinia xeranthemoides	N	F	D	Q
Leontodon taraxacoides	N	F	D	Q

Species of Brassicaceae	Dispersal		Germination	
	Upper	Lower	Upper	Lower
Cakile maritima	F	N	Q	D
Rapistrum rugosum				
(with capsule wall)	F	N	D	Q
(without capsule wall)	F	N	Q	D
Sinapis arcense	N	F	D	Q
S. alba	N	F	D	Q
Hirshfieldia incana	N	F	D	Q
Brassica tournefortii	N	F	D	Q

Seed size is often variable within populations and even within individuals (Stanton 1984). Variations in seed size may influence dispersal distance, germination requirements, and the probability of survival upon germination (Harper 1977; Cook 1980). Thus, polymorphism in seeds may lead to a generalist life history pattern with respect to reproduction.

A **seed pool** is an accumulation of living seeds in the soil. **Transient seed pools** are represented by seeds that will either die or germinate within a year, whereas **persistent seed pools** will accumulate seeds over a longer period of time (Thompson and Grime 1979). Transient seed pools have little impact on the population, aside from their numbers and position in safe sites, and are usually formed by annuals growing in predictable habitats where successful reproduction is essentially a sure thing. Persistent seed pools are characteristic of ephemeral plants in unpredictable habitats and of perennial herbs (especially short-lived semelparous perennials) or shrubs that germinate in response to some unpredictable event that opens habitat space in a later successional community. For example, chaparral shrubs add seeds to the seed pool yearly, but they germinate only when fire removes the overstory (J. Keeley 1977). Permanent seed pools contain a multitude of genotypes, produced in past environments and potentially capable of germinating at any time. This accumulation of genotypes from the past increases the chance that a well-adapted genotype will be present following a germination event (Templeton and Levin 1979). This is particularly important for annuals in unpredictable environments.

Freas and Kemp (1983) compared germination of *Pectis angustifolia*, a Chihuahuan desert annual that germinates in response to predictable summer precipitation, with *Lappula redowskii* and *Lepidium lasiocarpum*, which germinate following unpredictable winter and spring rainfall. *Pectis* had no innate dormancy (i.e., 100% germination), whereas *Lappula* and *Lepidium* both showed innate dormancy in a portion of the seeds. Thus, the seeds with innate dormancy are added to the seed pool and presumably will germinate at some future time. An interesting sidelight to their study was that both the winter and summer annuals had germination rates proportional to the amount of rainfall. The seeds of *Pectis* would not all germinate unless there was a sufficient amount of rainfall. This fits the prediction of Venable and Lawlor (1980) that a sensitive, environmentally induced germination mechanism can substitute for innate dormancy in desert ephemerals.

Short-lived semelparous perennials occupy habitats of 2–3 years duration and depend on seed being present in the soil so that growth can begin immediately when habitat becomes available. Typically, seeds of these "biennials" are long-lived enough to survive extended periods when habitat is unavailable and they have few special requirements for breaking dormancy.

Classification of Life History Patterns

The immense number of plant species and the complexity of adaptations they represent have led to an effort to classify plants according to similarities of life history patterns. Several assumptions underlie the classification schemes. First, evolution leads to adaptations that maximize fitness. Second, species growing in similar spatial and temporal microhabitats have similar life history patterns. Since there are spatial and temporal microhabitat differences within any habitat, we expect that a community will contain plants with a variety of different life history patterns and that successional changes can be characterized by more or less predictable changes in life history patterns of the component plants. Third, the expression of characteristics that determine life history patterns may vary from species to species even when, from outward appearances, they are of the same pattern. This is because increased expression of one characteristic may compensate, adaptively, for decreased expression of another (Grime 1982). For example, either increased seed production or earlier reproduction will increase r, so two populations may have equal intrinsic rates of natural increase. One may be due to higher reproductive allocation, the other to earlier reproduction. Next, physiological characteristics reflect life form, habitat, and reproductive patterns. Finally, a life history classification scheme will aid in our effort to understand and communicate information about plant adaptations. Many of these assumptions are under attack by one or more groups of scientists (e.g., Stearns 1977; Hickman 1975; Wilbur 1976).

The life history pattern of an organism can be viewed as a suite of co-adapted characters whose functions are complementary and enhance the organism's reproductive success in a specific environmental setting. Bartholomew (1972) uses the term **adaptive suite** in the same context as life history pattern.

r- and *K*-Selected Life History Patterns

The most widely used classification of life history patterns places organisms on a spectrum between the extremes of resource allocation, r and K (MacArthur and Wilson 1967). r refers to the intrinsic rate of natural increase, as defined in Equation 4-2. Selection acts to maximize r in temporary habitats, where rapid, abundant production of seed and short life-span are adaptive. Desert ephemerals are classified near the r extreme of the r to K spectrum. The K refers to carrying capacity (see Equation 4-4) and represents the number of individuals of a population that can co-exist in a given habitat. Thus, an organism with a K-selected life history pattern allocates a greater proportion of available resources to functions that increase competitive ability and survival and a lesser proportion to reproductive output. Redwood trees are classified near the K extreme of the $r–K$ spectrum because fitness is determined by the ability of individuals to function in populations that remain near carrying capacity. Our discussion of r and K will focus on organisms near the extremes of adaptation, but it is important to remember that most plants fall somewhere between the extremes and that the position of an organism in the spectrum is relative to other organisms, not to some artificial standard. Table 5-2 summarizes the important distinctions between r- and K-selected species.

The r-selected population has a life history pattern that maximizes fitness in unpredictable habitats and/or habitats that exist for only a short time; they expe-

Table 5-2   Some traits that correlate with *r*- and *K*-selection. (Modified from E. Pianka, *American Naturalist* 104:592–597. Copyright © 1970 by University of Chicago Press. By permission.)

Trait	*r*-selection	*K*-selection
Climate	Variable and/or unpredictable; uncertain	Fairly constant and/or predictable; more certain
Mortality	Often catastrophic; density independent	Density dependent
Survivorship	Usually types I and II (see Figure 4-9)	Often type III (see Figure 4-9)
Population size	Variable in time; not in equilibrium; usually well below carrying capacity of the habitat; recolonization each year	Fairly constant in time; in equilibrium; at or near carrying capacity of the habitat; no recolonization necessary
Intraspecific and interspecific competition	Variable; often lax	Usually keen
Life-span	Short, usually less than 1 year	Longer, usually more than 1 year
Selection favors	Rapid development; early reproduction; small body size; single reproduction period in life-span	Slower development; greater competitive ability; delayed reproduction; larger body size; repeated reproduction periods in life-span
Overall result	Productivity	Efficiency

rience significant environmental and population fluctuations (Figure 5-7). Mortality of both seedlings and adults is usually caused by external factors such as drought, frost, or other physical damage and is density independent. Seedling mortality is density independent—when more are produced, more survive—so maximum fitness is attained by producing many small seeds. Small seeds are distributed over more potential safe sites, thus increasing the probability that the new generation will survive to reproduce. This ability to disperse over large areas is important where habitats are temporary. Adult mortality may come about quickly or unpredictably by means of some catastrophic, density independent factor. The life history pattern must then consist of early, explosive reproduction. In terms of plant economics, this means that little or no energy is allocated to growth and survival; *r*-selected plants are small and have short life-spans and generation times. Because habitats are often unpredictable, *r*-selected popula-

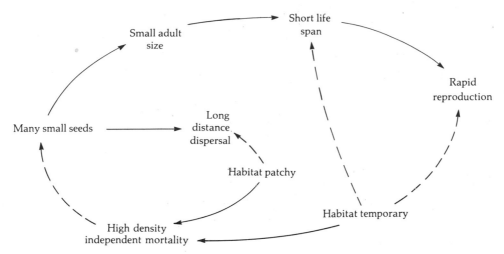

Figure 5-7 A model showing selective forces (dashed arrows) and associated plant–plant and plant–habitat characteristics (solid arrows) for an *r*-selected population.

tions typically have very long life expectancy as seeds. Short life-span and early reproduction, together with little mortality due to intraspecific competition, lead to low postgermination mortality. The survivorship curve is usually Deevey type I or II (see Figure 4-9).

In general, desert ephemerals fit the concept of *r*-selected species, but not all of them occupy the same position on the *r–K* spectrum. For example, Clark and Burk (1980) studied patterns of allocation of structural (biomass) and non-structural (stored carbohydrates) carbon in the co-occurring desert ephemerals *Plantago insularis* and *Camissonia boothii* (Figure 5-8). These species germinate simultaneously in response to cool-season precipitation but have markedly different patterns of growth and reproduction. *Plantago* allocates significantly more structural and nonstructural carbon to reproduction early in the season, completing its life cycle in less than 60 days, whereas *Camissonia* continues both vegetative and reproductive growth for over 100 days. The extended growing season of *Camissonia* is made possible by allocating more energy to the production and maintenance of vegetative tissues and to storage. Presumably, in years when no further rainfall occurs after germination, *Plantago* will have higher reproductive output because of early, heavy commitment of resources to reproduction. However, in years when it rains again later in the growth period, *Camissonia* (by allocating resources to longer life) will realize higher overall reproduction than would *Plantago* in the same year. The two seasonal patterns of precipitation must occur with sufficient frequency to maintain both life history patterns as desert ephemerals, with *Plantago* to the *r*-side of *Camissonia* on the *r–K* spectrum.

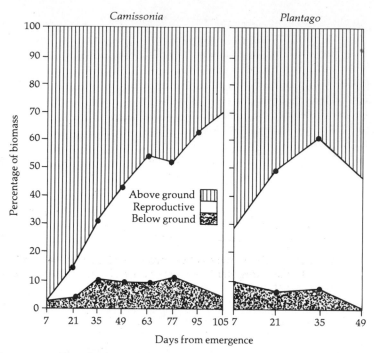

Figure 5-8   Distribution of aboveground, belowground, and reproductive biomass (as % of total biomass) throughout the growing season of the desert ephemerals, *Camissonia boothii* and *Plantago insularis.* (From D. D. Clark and J. H. Burk, *Oecologia* 46:272–282. Copyright © 1980 by Springer-Verlag, New York. Reprinted by permission.)

K-selected populations, on the other hand, have adaptive suites that maximize fitness in constant (or predictable) habitats where an individual exists for an extended period of time (Figure 5-9). *K*-selected populations therefore experience fewer fluctuations in population numbers and are more resistant to environmental fluctuations than are *r*-selected species. Mortality is usually highest when population numbers are high and is therefore density or crowding dependent. Density in adult populations is typically low, but most resources are utilized by the adults, leaving little chance for seedlings to establish. Low reproductive success is associated with low reproductive effort and an extended time to reach reproductive maturity. The successful energy allocation pattern favors growth and maintenance over reproduction. A seedling is more likely to survive if the seed has sufficient energy reserves to allow establishment and growth in highly competitive or resource-poor habitats. Therefore, seeds are generally large and contain high-energy carbohydrates and fats (somewhat analogous to parental care in animals). Consequently, they have poor dispersal capacity. The chance of producing a seedling that will reach reproductive maturity is low for any one growing season, so *K*-selected species tend to be iteroparous. Once a plant is

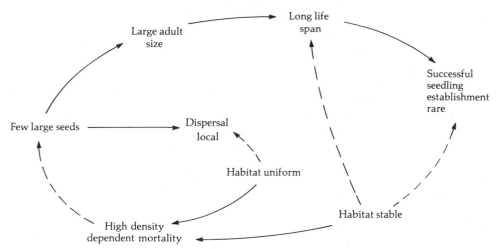

Figure 5-9   A model showing selective forces (dashed arrows) and associated plant–plant and plant–habitat characteristics (solid arrows) for a *K*-selected population.

established, maximum fitness is attained by living a long time (increasing the chance of producing successful offspring) and becoming large. The survivorship curve is typically a Deevey type III (see Figure 4-9), because very few of the seeds are successful but once an individual is established it is likely to survive for an extended period. Existing for a long period with tissues carried over from year to year necessitates morphological and physiological adaptations that do not respond to minor fluctuations in environmental factors—the same factors that cause density independent mortality in *r*-selected populations.

   r- and *K*-selection should result in a continuous distribution of species between hypothetical extreme specialists at each end of the spectrum. We should be able, therefore, to place all plants at their correct position along the continuum by studying the intensity of expression of the various characteristics deemed important in determining the life history patterns. The extreme plasticity of plants makes the absolute characterization of life history patterns very difficult, as one species may have different life history patterns in different habitats. Also, two species that are similar ecologically often have life history patterns that are difficult to distinguish. One adaptation may compensate for another, and we are faced with the impossible task of determining which factors are most important in determining the life history. We are left with a classification of life history phenomena that is strictly qualitative. The value of a qualitative life history classification scheme is in the clarification of general relationships between species and their environment and as an aid in understanding the structure of communities based on the general distribution of life history patterns.

R-, C-, and S-Selected Species

Recently, J. P. Grime (1977; 1979) expanded the *r* and *K* concept of life history classification to include patterns selected in resource abundant temporary habitats (*R*), in resource abundant predictable habitats (*C*), and in resource stressed habitats, (*S*). The three adaptive patterns coincide with the three allocation possibilities diagrammed in Figure 5-1; allocation primarily to reproduction (*R*), primarily to growth (*C*), and primarily to maintenance (*S*). *R* comes from the word **ruderal**, which describes a temporary, usually disturbed habitat. It refers to an adaptive suite identical to that described in the previous section for *r*-selected species. Plants that allocate most available resources to growth are the best competitors (*C*) and are found in habitats with resources readily available only to the most **competitive species**. The final category of life history patterns includes those plants that live in habitats where, because of limited resources or where physiological stress restricts resource utilization, survival depends on allocating most resources to maintenance; these are **stress-tolerant** species (*S*). Table 5-3 contains an overview of the characteristics of competitive, stress-tolerant, and ruderal plants.

We can get a better idea of the relationship between Grime's life history patterns and habitat by considering Figure 5-10, where the relative abundance of species with the indicated life history pattern is plotted against a continuum of resource availability and disturbance. At the left in the graph, disturbance is high and resources are readily available; the adaptive suites of the plants are of the *R* type. As we follow the abscissa from left to right, at (i) disturbance becomes infrequent enough to favor longer-lived species with a competitive life history pattern. Moving further to the right, we find that at (ii) resources become critically scarce for plants with the competitive life history pattern, and survival depends on allocating more and more resources to maintenance. It is the separation of organisms along the *K* side of the spectrum into stress adapted and competitive species that distinguishes the Grime model from MacArthur and Wilson's *r* and *K* model.

Figure 5-10 Diagram describing the frequency (*f*) of ruderal (*R*), competitive (*C*), and stress tolerant (*S*) life history patterns. At point A, disturbance is infrequent enough to favor longer-lived competitive species. At point B, resources become limiting to plants with the competitive life history pattern and favorable to the noncompetitive stress tolerant populations. (Modified from J. P. Grime, *American Naturalist* 111:1169-1194. Copyright © 1977 by University of Chicago Press. By permission.)

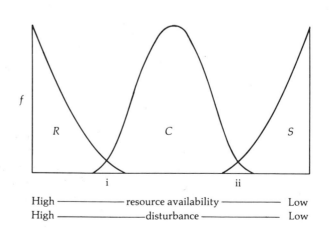

Table 5-3 Some characteristics of competitive, stress-tolerant and ruderal plants. (*Plant Strategies and Vegetation Processes* by J. P. Grime. Copyright © 1979 by John Wiley and Sons, Ltd. Reprinted by permission.)

	Competitive	Stress tolerant	Ruderal
Morphology			
Life forms	Herbs, shrubs, and trees	Lichens, herbs, shrubs, and trees	Herbs
Morphology of shoot	High, dense canopy of leaves Extensive lateral spread above and below ground	Extremely wide range of growth forms	Small stature, limited lateral spread
Leaf form	Robust, often mesomorphic	Often small or leathery, or needlelike	Various, often mesomorphic
Life history			
Longevity of established phase	Long or relatively short	Long–very long	Very short
Longevity of leaves and roots	Relatively short	Long	Short
Leaf phenology	Well-defined peaks of leaf production coinciding with period(s) of maximum potential productivity	Evergreens, with various patterns of leaf production	Short phase of leaf production in period of high potential productivity
Phenology of flowering	Flowers produced after (or, more rarely, before) periods of maximum potential productivity	No general relationship between time of flowering and season	Flowers produced early in the life history
Frequency of flowering	Established plants usually flower each year	Intermittent flowering over a long life history	High frequency of flowering
Proportion of annual production devoted to seeds	Small	Small	Large
Perennation	Dormant buds and seeds	Stress tolerant leaves and roots	Dormant seeds
Regenerative strategies[1]	V, S, W, B_s	V, B_τ	S, W, B_s

(Table continues on next page)

Table 5-3   *continued*

	Competitive	Stress tolerant	Ruderal
Physiology			
Maximum potential relative growth rate	Rapid	Slow	Rapid
Response to stress	Rapid morphogenetic responses (root–shoot ratio, leaf area, root surface area) maximizing vegetative growth	Morphogenetic responses slow and small in magnitude	Rapid curtailment of vegetative growth, diversion of resources into flowering
Photosynthesis and uptake of mineral nutrients	Strongly seasonal, coinciding with long continuous period of vegetative growth	Opportunistic, often uncoupled from vegetative growth	Opportunistic, coinciding with vegetative growth
Acclimation of photosynthesis, mineral nutrition and tissue hardiness to seasonal change in temperature, light, and moisture supply	Weakly developed	Strongly developed	Weakly developed
Storage of photosynthate mineral nutrients	Most photosynthate and mineral nutrients are rapidly incorporated into vegetative structure but a proportion is stored and forms the capital for expansion of growth in the following growing season	Storage systems in leaves, stems, and/or roots	Confined to seeds

Table 5-3 *continued*

	Competitive	Stress tolerant	Ruderal
Miscellaneous			
Litter	Copious, often persistent	Sparse, sometimes persistent	Sparse, not usually persistent
Palatability to unspecialized herbivores	Various	Low	Various, often high

[1]Key to regenerative strategies: V—vegetative expansion, S—seasonal regeneration in vegetation gaps, W—numerous small wind-dispersed seeds or spores, B_s—persistent seed bank, Bт—persistent seedling bank.

Intuitively, it seems that survival in a resource stressed habitat would be enhanced by competitive ability. However, the adaptations necessary to be successful in a highly competitive situation are, for the most part, counterproductive in a habitat with severely restricted resources. An effective competitor must have the ability to extract resources rapidly from the environment when they become available. This ability is enhanced by the potential for rapid growth, uncompromised by more than minimal allocations of energy to maintenance and survival. On the other hand, rapid growth following a short period of resource abundance in a usually resource depressed habitat would leave the plant with vulnerable tissues when stress returns. Therefore, the successful stress tolerant species has a very slow growth rate and produces stress tolerant tissues. Since the norm in resource limiting habitats is stress, these plants do not have the ability to grow rapidly even when ample resources are available. Stress tolerant species have very long life-spans, tissues that are resistant to stress and therefore expensive to produce, a well-developed acclimation potential, reproduction that is often vegetative, and seedlings that may remain small for extended periods of time.

It is important to remember that plants can be adapted to any combination of disturbance, competition, and stress. This means that a classification system must be based on the relative importance of each in determining the life history pattern. To emphasize this point, Grime (1979) presents his ideas in the form of a triangular model (Figure 5-11), in which the competitive, ruderal, and stress tolerant populations described in Table 5-3 occupy the corners and plants with intermediate life history patterns the central area of the triangle.

Competitive ruderals (*CR*) occur in resource abundant habitats where disturbance is severe enough to prevent high-intensity competition but infrequent or moderate enough to prevent noncompetitive ruderals from dominating. *CR* plants can be annual or perennial, characterized by large size because of a relatively long period of vegetative growth before flowering. The vegetation sup-

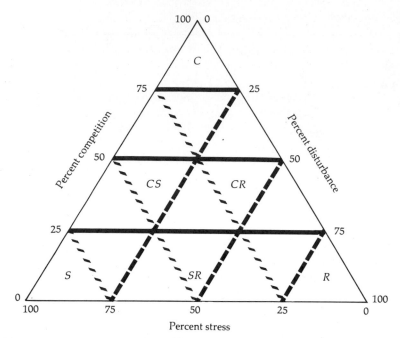

Figure 5-11   Grime's model of life history variation based on % occurrence of competition (solid lines), disturbance (long dashed lines), and stress (short dashed lines). Life history patterns are: *C* = competitive, *S* = stress tolerant, *R* = ruderal, *CS* = stress tolerant competitor, *CR* = competitive ruderal, and SR = stress tolerant ruderal. (From *Plant Strategies and Vegetation Processes* by J. P. Grime. Copyright © 1979 by John Wiley and Sons, Ltd. Reprinted by permission.)

porting *CR* species is a mix of annual and perennial plants, with the competitive ruderals taking advantage of seasons of low productivity of the competitive species or occupying sites occasionally flooded, intensively grazed, or otherwise disturbed.

Stress tolerant ruderals (*SR*) occur in unproductive habitats with intermediate levels of disturbance. These plants are generally small herbaceous annuals or short-lived perennials with restricted time for growth due to resource stress. *SR* plants experience sufficient resource stress during growth to reduce overall size significantly. Desert ephemerals, desert geophytes, arctic/alpine annuals, and many bryophytes are examples of stress tolerant ruderals.

Where the habitat is not subjected to disturbance and sufficient resources are available for moderate growth, the plants are characterized as stress tolerant competitors. Herbaceous stress tolerant competitors include those plants that show moderate growth rates, spread aggressively by vegetative reproduction, and possess evergreen leaves that show dramatic shifts in biomass with the seasons, such as *Carex lacustris* (Bernard and MacDonald 1974; Bernard and Solsky 1977). Many woody plants that grow in unproductive habitats, or during the later stages of succession when many of the resources are tied up in biomass, may also fall into the category of stress tolerant competitors.

A general relationship is found between habitat, life form, time, and life history pattern. The relationship between life form and life history pattern is diagrammed in Figure 5-12. Note that, with the exception of perennial herbs and ferns, the life form groups occupy a restricted area of the Grime triangular model. Life form thus gives preliminary information concerning the habitat and adaptive suite of a plant, and confirms that certain life forms (and life history patterns) are restricted to growing in particular habitats. In addition, changes in species with successional time follow a predictable pattern with respect to the *C*, *S*, and *R* life history patterns. Of course, the specific pattern of replacement depends on environment (see Chapter 11 for details), but we can generalize for a specific region, such as the deciduous forests of eastern North America. In secondary succession, the initial plant community is dominated by plants with a ruderal life history pattern when there is no competition from long-lived species and when resources are maximally abundant in the habitat. Intermediate stages of succession are characterized by competitive species able to maintain themselves in the community only by rapid growth and large size. During the intermediate stages of succession, resources become progressively bound up in biomass. The last stage of succession is therefore dominated by stress tolerant species, which have probably been present for long periods as a seedling bank on the forest floor. These elderly seedlings require little in the way of light and soil nutrients,

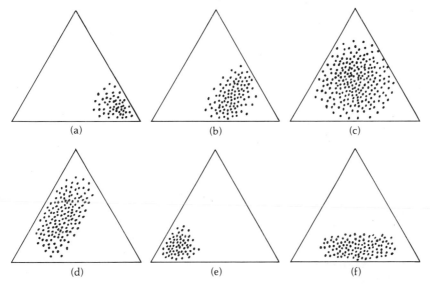

Figure 5-12 Diagram describing the range of life history patterns encompassed by (a) annual herbs, (b) biennial herbs, (c) perennial herbs and ferns, (d) trees and shrubs, (e) lichens, and (f) bryophytes. For the distribution of life history patterns within the triangle, see Figure 5-11. (Modified from J. P. Grime. *American Naturalist* 111:1169–1194. Copyright © 1977 by University of Chicago Press. By permission.)

so they increasingly gain an advantage as resources get tied up in biomass. Their establishment requirements and the very long life-spans of stress tolerant plants allow these species to replace themselves in the habitat and remain for extended periods as "climax" species.

Figure 5-13 shows the trend in life history patterns during secondary succession using Grime's triangle model. The origin of the arrow coincides with the time immediately following disturbance in the ruderal corner of the triangle, and the arrow terminates at climax in the stress corner of the triangle. Relative biomass at various stages is indicated by the size of the circles in the arrow. Note that in the upper line, where potential productivity is high, the biomass of the climax species is greater than for less productive systems. Also note that competition is most important in highly productive communities. Where stress is high, life history patterns remain in the lower part of the triangle, moving only from ruderal to stress tolerant with consistently low biomass.

Any attempt to locate a species in the Grime triangular model will result, as with the *r–K* continuum, in some conflicting characteristics. For example, you may find that a woody plant growing in a resource restricted habitat has all of the necessary qualities of a stress tolerant competitor but produces massive numbers of small seeds yearly. Such conflicts are not unexpected, since such species' adaptive possibilities are limited by factors other than those used to describe life history patterns in Tables 5-2 and 5-3. The idea of an identifiable life history pattern is appropriate only at the general level (Grime 1982). Detailed studies at the physiological and morphological level will reveal an uninterpretable mass of conflicting data. The concepts of life history pattern are instructive only where we can generalize about groups of species in a community that have some ecological similarity. Categorizing species as they co-exist or change temporally or spatially in a community allows the ecologist to simplify the system and make observations, which may lead to a greater understanding of community organization (Whittaker and Goodman 1979). Exceptions to expected life history properties may lead the physiological ecologist to interesting adaptive modes, which can be examined at a more detailed level.

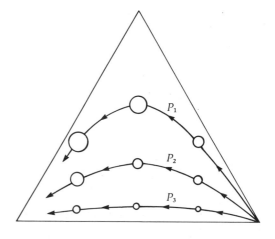

Figure 5-13   Grime's triangular model showing the path of secondary succession under conditions of high (P_1), moderate (P_2) and low (P_3) potential productivity. The size of the plant biomass at each stage of succession is indicated by the circles. The axes are the same as in Figure 5-11. (Modified from J. P. Grime. *American Naturalist* 111:1169–1194. Copyright © 1977 by University of Chicago Press. By permission.)

Summary

Life history patterns describe the patterns of growth, reproduction, and longevity of plants. They are different in different habitats and change through time as succession takes place. The life history pattern of a plant is a matter of economics. There is a finite amount of material that can be produced by a plant, and that material must be allocated preferentially to growth, reproduction, or increasing the chance of survival because there is never enough for optimal alloction to all three. Allocation patterns are somewhat flexible, with different patterns in different environmental conditions, but fixed within the overall adaptive suite of the plant. In other words, an annual can adjust to different situations by changing allocation patterns, but being a successful annual precludes being a good competitor or living a long time.

Life-span and pattern of reproduction during the plant's life are important determinants of life history. Semelparous plants reproduce only once during their lifetime since the cost, in terms of reduction in future reproductive output, is minimized by either not reproducing in a given year or allocating all available resources to reproduction. Iteroparous plants have repeated reproductive periods. Here longevity and thus future expectation of offspring are maximized by less intensive reproduction, scattered over several years.

Life-span is also related to environment. Annuals are semelparous and have a short life as a means of survival in temporary, usually disturbed habitats. Biennials are semelparous and usually live into the second season before reproducing. Postponing reproduction presumably has benefits in terms of future reproductive success that outweigh the mortality cost associated with longer life. Perennials are either semelparous or iteroparous and are best suited to more permanent habitats. There are some advantages of semelparity, including a large flower display to attract pollinators and massive seed production to satiate predators and increase chances of successful reproduction.

The successful life history pattern has well-timed periods of dormancy and seed dispersal. Dormancy is a means of avoiding harsh environmental conditions and, for plants with long-lived seeds, spreads the risk of failure over several growing seasons. Dispersal is the mechanism for efficiently utilizing a habitat and moving to other habitats. Plants in patchy environments tend to have long-distance dispersal capabilities and few dormancy mechanisms, whereas plants from uniform environments typically have lower dispersal distance. Annuals, in unpredictable environments, tend to have effective dormancy mechanisms and extremely long seed life, forming persistent seed pools in the soil. In predictable environments where successful reproduction is essentially assured, annuals have few dormancy mechanisms and form transient seed pools.

There are as many life history patterns as there are species, but, because species with similar patterns tend to occupy similar habitats in space and time, they can be classified to reduce the complexity. One classification system places life history patterns along a continuum from the extreme transient *r*-selected species to the extreme perpetual competitor or *K*-selected species. *r*-selected species have short life-spans, allocate most resources to reproduction, have high dispersal distances, small size, little competitive ability, and live in temporary habitats. *K*-selected species live longer, are larger, allocate more resources to increase longevity and competitive ability, and occupy more permanent habitats. The life history classification of Grime defines three nodes of life history patterns, rather than only two as in the *r–K* classification. *R* (ruderal) species are identical to *r*-selected species, *C* (competitive) species are those that allocate most of their energy to rapid growth and competitive ability, and *S* (stress tolerant) species are those that allocate most of the available resources to increase longevity. Stress tolerant species live in resource poor habitats, competitive species in resource rich permanent habitats, and ruderal species in resource rich temporary habitats. Many species can be placed clearly in one of these categories; others may be intermediate. Those with intermediate life history patterns occupy intermediate habitats. In general, there is a shifting of life history pattern during secondary succession. Early arrivals after disturbance are *R*-selected, intermediate stages of succession are dominated by *C*-selected species, and the climax species tend to be stress tolerant.

CHAPTER 6

SPECIES INTERACTIONS
Competition and Amensalism

F or the most part, Chapters 3, 4, and 5 dealt with plant species and populations as though they existed in isolation. In nature, most communities consist of more than one plant population. In addition, they show the influence of nonplant populations, such as decomposers (bacteria and fungi) in the soil, parasitic pathogens, and herbivorous animals. Interactions between these diverse populations modify the genetic potential of each species (its physiological optimum and range) to yield communities based on ecological optima and ranges. Harper (1964) has written a powerful review that demonstrates how the biology of organisms grown in isolation does not compare to their biology when grown in mixtures.

Many ecologists believe that the associated organisms in a community are somehow interdependent, that they are not associated by chance, and that disturbance of one organism will have consequences to all organisms. Clements (1916), and especially his followers, took this view to an extreme. They equated climax communities with superorganisms and considered their component populations as interdependent as the cells, tissues, or organs of a single organism. The objectives of this chapter and Chapter 7 are to survey the kinds of interactions that may occur between members of a community. This will lead, in Chapter 8, to an assessment of Clementsian views, and more moderate views, of community interdependence and integrity.

Table 6-1 catalogs some possible interactions according to a symbolic scheme developed by Burkholder (1952) (see also Odum 1971 and Malcolm 1966). Each interaction is described by its effect on two populations or organisms, A and B, when they are in contact (the interaction is "on") and when they are apart (the interaction is "off"). As an example, consider herbivory. When a herbivore and its food plant are together, the herbivore is stimulated (its growth, reproduction, or general success is improved), and the plant is depressed (its growth, reserves,

Table 6-1   A partial list of biologically possible types of interactions according to Burkholder (1952). When organisms A and B are close enough to participate in the interaction, the interaction is "on"; otherwise it is "off." Stimulation is symbolized as +, no effect as 0, and depression as −. (From Burkholder 1952. Reprinted by permission of *American Scientist*, journal of Sigma Xi, The Scientific Research Society.)

Name of interaction	On		Off	
	A	B	A	B
Neutralism	0	0	0	0
Competition	−	−	0	0
Mutualism	+	+	−	−
Unnamed	+	+	0	−
Protocooperation	+	+	0	0
Commensalism	+	0	−	0
Unnamed	+	0	0	0
Amensalism	0 or +	−	0	0
Parasitism, predation, herbivory	+	−	−	0

reproduction, or general success declines). When the two are apart, the herbivore is depressed and the plant remains stable. In Table 6-1, herbivory, parasitism, and predation are identical, but the subtle, important differences for other interactions are apparent. Mathematically, there are 81 possible interactions with this symbolism, but Burkholder concluded that only the ten shown in Table 6-1 are logically possible. Among those ten, three are sufficiently rare or at least unexamined that they have no names. Neutralism is included for purposes of comparison and completeness, but it too may be a rarity in nature. This chapter will survey two of the remaining six interactions: competition and amensalism. Protocooperation, commensalism, mutualism, and herbivory will be discussed in Chapter 7.

Some interactions symbolized in Table 6-1 are negative (one or another partner is inhibited, as in competition or amensalism), and others are positive (one or another partner is stimulated, as in commensalism or mutualism). The existence of the interactions can only be conclusively demonstrated by elaborate experimentation. Rather simple field sampling, however, can provide the initial evidence for an interaction.

The sampling is based on the premise that positive interactions will produce positive spatial relationships between the partners; where one partner is found, the probability is high that the other will be found nearby. The two populations attract one another and exist in a nonrandom, clumped pattern (Figure 6-1a).

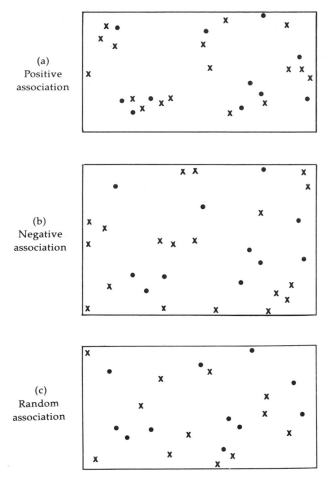

Figure 6-1 An overhead, diagrammatic view of two species (dots and x's) that are (a) positively associated, (b) negatively associated, and (c) randomly associated.

Similarly, negative interactions will produce negative spatial relationships; the two populations appear to repel one another and exist in a nonrandom, regular pattern (Figure 6-1b). If there is no interaction between the populations, then the location of one individual has no influence on the location of others; the two populations are said to be randomly distributed with respect to each other (Figure 6-1c).

One way by which pattern can be revealed is by sampling vegetation with random quadrats of an appropriate size. In each quadrat, the presence or absence of any two (or all) species is noted, then summarized in a contingency table (Table 6-2). This constitutes the observed data. Expected data, assuming a com-

Table 6-2 Contingency table for analysis of association between two species, A and B. In this example, 100 random quadrats were tallied for the presence or absence of each species.

Symbol and description	Observed no. of quadrats	Expected no. of quadrats	$\chi^2 = \dfrac{(observed - expected)^2}{expected}$
a = A and B present	30	$\dfrac{(a + b)}{100} \times (a + c) = 38$	1.7
b = A present, B absent	29	$(a + b) - 38 = 21$	3.0
c = A absent, B present	35	$(a + c) - 38 = 27$	2.4
d = A and B absent	6	$100 - (38 + 21 + 27) = 14$	4.6
Totals	100		$\Sigma\chi^2 = 11.7$

pletely random distribution of the two taxa to be examined, can be compared to observed data by a chi-square formula. The example in Table 6-2 yielded a chi-square value higher than expected by chance, so the two species (A and B) are not distributed randomly with respect to each other. Inspection of the table shows that there were more quadrats with only A or B than expected by chance, and there were fewer quadrats with both species present than expected by chance. The conclusion is that A and B are negatively associated. The reason for the association would then have to be determined by ecological observations and experimentation; the statistical treatment is merely the first step and does not constitute proof of a biological interaction.

Another method of determining pattern is by measuring distances between randomly chosen plants and their nearest neighbors. The dispersion index (R) of Clark and Evans (1954) is then calculated as the ratio of actual mean distance and expected distance, based on a random spatial pattern:

$$R = \frac{\text{mean measured distance}}{0.5\sqrt{\text{density}}} \qquad \text{(Equation 6-1)}$$

where density of plants per unit area can be estimated from quadrats. The departure of R from 1 indicates regularity ($R>1$) or patchiness ($R<1$).

It is important, in all these methods, to realize that pattern may change as plants age (increase in size). A recent review of Sonoran and Mojave Desert plants showed that small shrubs tended to be clumped, medium-sized shrubs tended to be random, and large shrubs could sometimes be regularly dispersed (Phillips and MacMahon 1983). The authors suggested that the data indicate competition increases in intensity as plants grow. Sampling design, consequently, should take size into account.

Competition

Competition results in mutually adverse effects to organisms that utilize a common resource in short supply. An excellent study showing the importance of competition and the mechanism by which it can work, was conducted by Grant Harris (1967) in the northern intermountain region of the United States. Before the middle of the nineteenth century, the dominant plant of this grassland area was bluebunch wheatgrass (*Agropyron spicatum*), a perennial. Then an annual grass called cheatgrass (*Bromus tectorum*) was accidentally introduced to the area from Europe. From that time to the present, ranchers have noticed an enormous increase in the abundance of cheatgrass and an equally impressive decrease in the abundance of bluebunch wheatgrass (Mack 1981). What caused the shift?

Both species have a similar life cycle. They germinate (or break dormancy if perennial) in fall, grow slowly during winter, grow rapidly in spring, form flowers in early summer, and die in June (or begin dormancy in mid-July if perennial). Harris studied the growth and survival of these two species during a year, starting from seed for both. He found that the presence of cheatgrass greatly reduced the growth and survival of bluebunch wheatgrass. In one field trial, bluebunch wheatgrass was sown in October in plots with different densities and cover of cheatgrass (0–100 plants m^{-2}, 0–100% cover). Seedlings were tallied the following June and again after summer drought in October, 7 and 12 months after sowing, respectively. Cheatgrass density had little effect on the number of wheatgrass seedlings still present in June, but it had a great effect on survival and biomass of those seedlings through the summer drought. Harris pointed out that his maximum experimental cheatgrass density of 100 plants m^{-2} was low compared to the entire region, where it could reach 3000–10,000 plants m^{-2}, and there is no establishment of bunchgrass where cheatgrass density is greater than 1000 plants m^{-2}. Clearly, cheatgrass is an important competitor, but how is it competing, and what happens during the summer months to cause the competitive effect?

Harris planted seeds of each species in long glass tubes filled with soil. The tubes were inserted in the field, level with surrounding soil. Every month the tubes were lifted and the depth of rooting was measured. Root growth during winter was much greater for cheatgrass than for wheatgrass (Figure 6-2). At the start of rapid spring growth in April, cheatgrass roots had penetrated 90 cm, in contrast to 20 cm for wheatgrass. This difference in depth allowed cheatgrass to absorb water from a much greater part of the soil profile than wheatgrass could. Consequently, when the upper soil became dry in early summer, with both species drawing water from it, only wheatgrass suffered, for water still remained

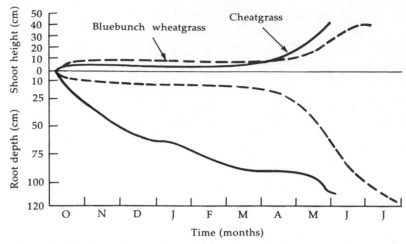

Figure 6-2 Elongation of shoots (above the 0 line) and roots (below the 0 line) of cheatgrass (*Bromus tectorum*) (solid lines) and of bluebunch wheatgrass (*Agropyron spicatum*) (dashed lines) in a field experiment. Plants were grown from the time of germination, in October, until the time of maturity of the annual cheatgrass. (From "Some competitive relationships between *Agropyron spicatum* and *Bromus tectorum*" by G. A. Harris. *Ecological Monographs* 37:89–111. Copyright © 1967 by the Ecological Society of America. Reprinted by permission.)

available at greater depths, where only cheatgrass roots had penetrated. Harris measured soil water availability in the two soil regions at this time and found water potential was below −1.5 MPa in the upper region (near the wilting point of many species), but was only −0.1 MPa in the lower region (essentially still saturated).

The recent success of cheatgrass, then, seems due to its winter root growth, which gives it a spring and early summer advantage over wheatgrass that started from seed at the same time. Each year cheatgrass increases in density, and each year this results in greater competition for moisture in the upper soil and greater stress on wheatgrass.

An equally powerful experimental technique of measuring or demonstrating competition is manipulation of plants in the field. Fonteyn and Mahall (1978 and 1981), for example, manipulated shrub density in a uniform Mojave Desert community of *Larrea* (creosote bush) and *Ambrosia* (burrow bush). In a series of 100 m² plots, one species or the other was removed in several patterns (Figure 6-3). Then the water status of remaining plants was monitored over a period of several months as a measure of release from competition for soil moisture. The results indicated that the three possible types of competition (*Larrea–Larrea*, *Ambrosia–Ambrosia*, *Larrea–Ambrosia*) were not equal in intensity. Release of *Larrea* from surrounding *Larrea* had little effect on plant water status, whereas release of *Ambrosia* from surrounding *Ambrosia* and release of interspecific competition did have significant effects on plant–water status.

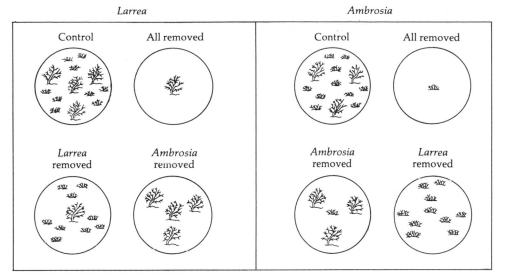

Figure 6-3 The eight treatments of plant removal in *Larrea–Ambrosia* desert scrub. Each circle represents a 100 m² test plot. Large shrubs are *Larrea*, small ones are *Ambrosia*. In the "all removed" plots, the center of the plot was occupied by one remaining plant. Controls were not manipulated. (Redrawn from P. S. Fonteyn and B. E. Mahall 1981. *Journal of Ecology* 69:883–896.)

Using a different manipulation technique, Del Moral (1983) measured the importance of competition in subalpine meadows of Washington. He transplanted species into cleared and uncleared patches of meadow communities that differed in summertime soil moisture and plant cover. He then followed the mortality of the transplants and found that many species from dry or disturbed habitats could survive in cleared plots of mesic habitats but not in uncleared plots. He concluded that mesic sites were characterized by a few highly productive, competitive species, whereas xeric or disturbed sites were characterized by many moderately productive or even slow-growing, weakly competitive species.

In general, it appears that many plants endemic to severe habitats are restricted to them because they are poor competitors on less extreme sites. Severe habitats have very low plant cover, so there is little competition. Plants in severe habitats tolerate the sorts of environmental extremes that most other plants cannot, but they would grow better elsewhere, in the absence of competition. The *r–K* literature, and Grime's (1979) *S-C-R* scheme, discussed in an earlier chapter, incorporate these generalities. Research with serpentine endemics and halophytes illustrates this concept especially well.

Serpentine Endemics

Serpentine is a metamorphic, magnesium silicate rock, often green in color and slippery to the touch, which has a number of traits inimical to plant growth. It is low in such essential nutrients as N, Ca, K, and P; its pH may be far from neutrality (either acidic or basic); and it is high in such toxic elements as Ni and Cr. Soils derived from serpentine rock are sterile, support unusual, endemic floras, and are covered with vegetation whose physiognomy differs from that of surrounding vegetation on nonserpentine soil. Serpentine outcrops are found throughout the world (Europe, Rhodesia, Japan, Southeast Asia, Canada, and the United States; Whittaker 1954), but they are especially common in the Pacific Coast states.

Arthur Kruckeberg, of the University of Washington, experimented with serpentine and nonserpentine ecotypes and species (1954). He reported that herbaceous serpentine endemics became established from seed better and grew faster on nonserpentine soil, providing they were free from interspecific competition. When sown with typical nonserpentine species on nonserpentine soil, they became etiolated and did not survive. On serpentine soil, only the serpentine endemics survived, but there was considerable bare ground and the plants grew slowly.

This is not to say that all serpentine taxa have ranges limited by competition. Tadros (1957) found that toxins released by the higher densities of certain soil microorganisms on nonserpentine soil prevent the herbaceous, serpentine endemic *Emmenanthe rosea* from establishing on normal soil. McMillan (1956) showed that some serpentine taxa grow as well on serpentine soil as on nonserpentine soil, even where interspecific competition is not a factor. Thus there is more than one way in which plants have adapted to serpentine, but tolerance as a means to avoid competition may be the major strategy.

Halophytes

Halophytes (literally, salt plants) are capable of growing in soil with more than 0.2% salt concentration. (Barbour (1970*b*) pointed out that this is a conservative limit, and some researchers put the limit at 0.25% or even 0.5%.) **Glyco-phytes** (literally, sweet plants) are intolerant of salinity above that necessary to supply essential nutrients, approximately 0.1% salt. Certainly, above 0.2% salinity their growth is severely reduced.

All halophytes are not equally tolerant of salt, and a great number of terms have been coined to describe them (Waisel 1972). We will follow Ingram (1957) and use only three terms: **intolerant**, **facultative**, and **obligate**.

Intolerant halophytes show maximum growth at low salinity and declining growth with increasing salinity (Figure 6-4). Facultative halophytes show maximum growth at moderate salinity and diminished growth at both low and high salinity. Obligate halophytes show maximum growth at moderate or high salinity and are unable to survive at low salinity (perhaps below 0.1% salinity, although there is no agreement about this lower limit). As shown by the growth curves in Figure 6-4, all three halophytes may have similar optima and upper tolerance limits; they differ mainly in performance at low salinities.

Most halophytes are intolerant, whether we look at their germination, growth, or reproduction, and whether they are mangroves, coastal salt marsh herbs, beach plants that receive salt spray, or salt desert herbs and shrubs. A few halophytes are facultative. Perhaps there are no obligate halophytes (Barbour 1970*b*),

Figure 6-4 Response to increasing salinity of glycophytes (dashed line) and different types of halophytes: (a) intolerant, (b) faculative, and (c) obligate. (From M. Ingram in *Seventh Symposium of the Society for General Microbiology.* Copyright © 1957 by Cambridge University Press. Reprinted by permission.)

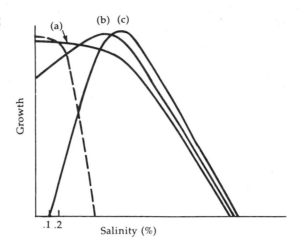

a conclusion that can be reached from both field observations and manipulative experiments in growth chambers.

Competition and Niche

Since competition involves two organisms utilizing the same resource, it is obvious that competing organisms must have, to some extent, overlapping niches. Intraspecific competition is necessarily more intense than interspecific competition, because niche overlap is greater. Current evolutionary theory holds that selection pressure drives species within a community to partition the environment (utilize different parts of it), with the result that competition is minimized.

The concept of "one niche, one species" stemmed from laboratory experiments performed by the Russian microbiologist G. F. Gause (1934). He placed pairs of closely related protozoan or yeast species together in homogeneous microenvironments and noted population growth rates (dN/dt). Initially, growth rates of both species were depressed compared to rates when grown in isolation, indicating that competition was occurring. Ultimately, however, there was a "winner" and a "loser," the winner coming to dominate the mixture and the loser reaching extinction. He concluded that species that coexist in nature must evolve ecological differences, or else they would not be able to coexist. The results of a competition experiment between two species of duckweed (Harper 1961) are shown in Figure 6-5.

The one niche, one species concept is often called **Gause's competitive exclusion principle**, even though zoologists before him had published much the same conclusion (Krebs 1972). Laboratory work and field observations since his time, with animals of more complex life cycles than microbes and with environments more heterogeneous than culture bottles, have shown that minor niche differences exist even between closely related taxa, and these differences are sufficient to permit coexistence and release from some competition (see, for example, Ayala 1969 and MacArthur 1958).

Figure 6-5 Results of a competition experiment performed between two species of duckweed (*Lemna gibba* and *Spirodela polyrhiza*). (Reprinted by permission from Harper 1961.)

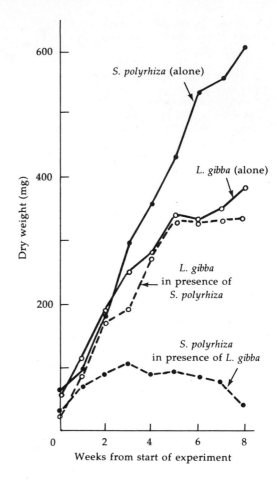

Measurement of Competition

Recall, from Chapter 4, the logistic equation of population growth for a species in isolation:

$$dN/dt = rN\left(\frac{K-N}{K}\right) \qquad \text{(Equation 6-2)}$$

where r is the natural, intrinsic rate of population growth, N is population size, and K is population size at saturation (the environment's carrying capacity). Multiplying out, we can see that the equation already contains an **intraspecific competition** component:

$$dN/dt = rN - \frac{rN^2}{K} \qquad \text{(Equation 6-3)}$$

where rN is the unlimited rate of growth and rN^2/K is the decline caused by crowding (competition).

If we introduce a second species, we must introduce a second negative component to account for **interspecific competition**. The equation for species 1 is:

$$dN_1/dt = r_1 N_1 \frac{(K_1 - N_1 - \alpha N_2)}{K_1} \qquad \text{(Equation 6-4)}$$

where α is the inhibiting (competitive) effect on N_1 for every individual of N_2. Similarly, the equation for species N_2 is:

$$dN_2/dt = r_2 N_2 \frac{(K_2 - N_2 - \beta N_1)}{K_2} \qquad \text{(Equation 6-5)}$$

where β is the inhibiting effect on N_2 for every individual of N_1.

This pair of equations was developed independently by Lotka (1925) and Volterra (1926). They have a number of limitations but still form a starting point for quantifying competition. Appropriate experiments have permitted some zoologists to calculate the components of the Lotka-Volterra equations, for example, α and β (Krebs 1972). If α and β are small, the interspecific components of the equations can be virtually ignored, and the two populations can coexist. Coexistence will also result if $\alpha < K_1/K_2$ and $\beta < K_2/K_1$.

C. T. de Wit (1960, 1961) pioneered in the application of zoological concepts of competition to plants. One of his techniques to measure plant competition is to calculate input/output ratios for each species in a mixture, where:

$$\text{input ratio} = \frac{\text{propagules planted of species A}}{\text{propagules planted of species B}} \qquad \text{(Equation 6-6)}$$

$$\text{output ratio} = \frac{\text{units produced of species A}}{\text{units produced of species B}} \qquad \text{(Equation 6-7)}$$

Propagule can mean seeds, seedlings, potato eyes, bulbs, or lengths of rhizomes; a good synonym is diaspore. Units produced can mean seeds, number of shoots or tillers, length of new stolons, or plant weight. Whatever definition of propagule or unit is chosen, it must be the same for both species. Two species are planted in a replacement series, where the total number of initial propagules is a constant, but the ratio of the two species is changed from pure species A, to an even mix of A and B, and on to pure species B.

Figure 6-6 shows the four possible outcomes of any such experiment. First, species A can "win," in which case the input/output line is parallel to but above the equilibrium line. Second, species B can "win," in which case the ratio line is parallel to but below the equilibrium line (Figure 6-6a). Finally, species A and B can coexist, in which case the ratio line crosses the equilibrium line; that is, the slope is no longer 45° (Figure 6-6b). If the slope is less than 45°, the equilibrium point is stable because movement away from the equilibrium point tends to be dampened by the populations' competitive qualities, and the populations drift back to equilibrium. If the slope is greater than 45°, the equilibrium is unstable because movement away from the equilibrium point tends to lead to further departure from stability, and one species will decline to extinction.

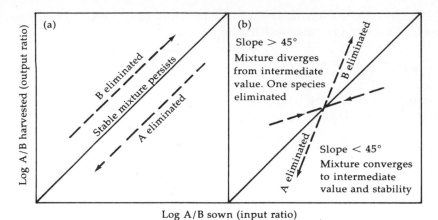

Figure 6-6   Input/output diagrams illustrating the possible outcomes of competition between two species, A and B. (From *Environment and Plant Ecology*. J. R. Etherington. Copyright 1975 by John Wiley and Sons, Inc., LTD, London. Reprinted by permission.

Figure 6-7a–c shows actual data rather than hypothetical data. In Figure 6-7a, the annual grasses barley and oats were sown in a replacement series and output measured as seed production. The results show that barley will outcompete oats at any mixture. Figure 6-7b shows competition between the perennials sweet vernal grass (*Anthoxanthum odoratum*) and timothy (*Phleum pratense*). The input/output (measured as tillers) line lies so close to the equilibrium line that the two species can be expected to coexist at any mixture. Figure 6-7c shows competition between two annual range plants from California grasslands: wild oat (*Avena fatua*) and slender wild oat (*A. barbata*), sown with a combined density of 1380 m^{-2}. The input/output line, with a slope of less than 45°, crosses the equilibrium line at a point corresponding to a mix of about 20% slender wild oat and 80% wild oat. One can conclude that these two species should be able to coexist in nature at this ratio.

Just as Gause's simple competition experiments did not reflect the true complexity of nature, so these replacement experiments may not bring us very close to the real world. Marshall and Jain (1969), for example, who conducted the wild oat study just described, attempted to validate their greenhouse results by examining natural wild oat populations. Did such natural populations at a density of about 1380 m^{-2} show the same sort of 20:80 equilibrum mix that their experiment predicted? No. About two-thirds of such stands in nature were 90–100% pure wild oats or slender wild oats; very few of the mixed stands were of the predicted 20:80 mix. Marshall and Jain believed the discrepancy was due to a far more complex situation in nature than they could imitate with their greenhouse study. For example, the seeds in nature would not be spread uniformly and at a constant depth as they were in the greenhouse study; other competing species from the field were missing in the greenhouse study; and the microenvironment and seed

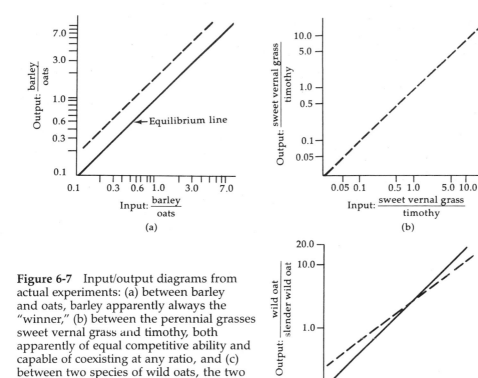

Figure 6-7 Input/output diagrams from actual experiments: (a) between barley and oats, barley apparently always the "winner," (b) between the perennial grasses sweet vernal grass and timothy, both apparently of equal competitive ability and capable of coexisting at any ratio, and (c) between two species of wild oats, the two capable of coexisting at one ratio only. ((a)and (b) from de Wit 1960 by permission. (c) from Marshall and Jain 1969. By permission of the British Ecological Society.)

supply in nature are more heterogeneous than in the greenhouse study. However, de Wit's ideas and methods are an important, essential first step to the goal of quantifying and understanding competition.

A Final Word

It is one thing to be able to document and quantify plant competition; it is another altogether to determine the resource being competed for—the very nature of the competitive interaction. The thoroughness of Harris' intermountain study, described earlier, which showed that competition for soil moisture was the ultimate factor, is unusual in the literature. Most ecologists make an educated guess as to the factors in short supply. If more than one resource is competed for, it may be difficult to demonstrate which is the most important one. As an example of this, consider a study by Hardy Shirley (1945) in north-central Minnesota.

The overstory trees in the Lake states are often the conifers white spruce (*Picea glauca*), white pine (*Pinus strobus*), red pine (*P. resinosa*), and/or jack pine (*P. banksiana*), yet these commercially valuable species do not reproduce in their own shade. Instead, hardwood seedlings and saplings grow up beneath the canopy. Why are conifer seedlings such poor competitors with the hardwoods? Shirley examined the effect of light on conifer seedlings by planting seedlings beneath screens that reduced full sun in steps from 2% to 89%. He found that survival and growth over a 4-year period increased as light intensity increased. Generally, growth was not satisfactory when light intensity was reduced more than 65% (Figure 6-8a). He also examined the combined effect of shade and root competition (Figure 6-8b). Results, after 4 years, were very complex. In dry areas, shade actually improved conifer seedling survival, though not seedling growth. In more mesic areas, removal of the overstory stimulated conifer seedling growth, weeding of ground vegetation further improved it, and trenching around the

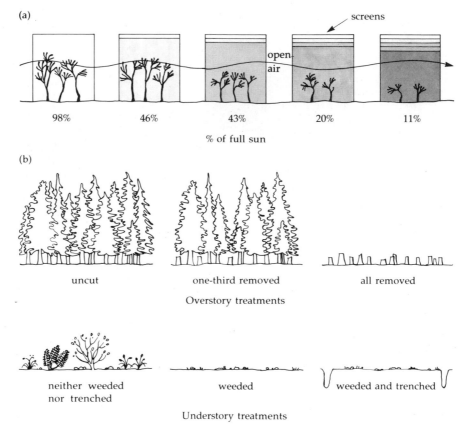

Figure 6-8 Diagrammatic summary of Shirley's experiments to determine the relative importance of competition for light and soil moisture to pine seedlings. (a) Pine seedlings were grown beneath different layers of screens to achieve different levels of sunlight. (b) Seedlings were grown beneath different overstories (closed canopy, one-third removed, all removed) and at the same time in three different understories (control, all other plants removed, all other plants removed and trenches dug around plot to sever the roots of adjacent plants).

plot improved it even further. The conclusion is that competition for shade and soil moisture are intertwined, and a single factor can not be pinpointed as most important.

Furthermore, competition may be affected by other interactions in the real world, so the competitive potential of a species may not be expressed. Mack and Harper (1977) studied the intraspecific and interspecific abilities of four dune annuals in Wales and reached several important conclusions. One species, *Vulpia fasciculata*, showed strong competitive superiority in greenhouse experiments, yet in nature it did not exclude the other species and grow in pure stands. The reason for the anomaly may be herbivory by rabbits, which preferentially graze on *Vulpia*. Another conclusion was that the typical response to competition by these annuals was a plastic reduction in size and seed production, rather than mortality (thinning).

Finally, on theoretical grounds, interspecific competition should be no more important, or likely, than intraspecific competition. Goldberg and Werner (1983) cite three reasons to support this idea of equivalence of competitors:

1. Most plants require the same handful of basic resources in similar amounts, so the possibilities of niche partitioning are very limited.

2. Most communities consist of a rich mixture of species, where the probability of contacts between many pairs of species is high, so selection pressure for minimal niche overlap between any one pair of species is not expected to be high.

3. Plant size typically confers superiority, and size depends on age or growth form rather than on species identity.

Amensalism

Amensalism is an interaction that depresses one organism while the other remains stable. One example of amensalism is **allelochemic interaction**, the inhibition of one organism by another via the release of metabolic by-products into the environment. These by-products are selectively toxic, affecting some species but not others. Allelochemics are viewed by some biologists as simply the mechanism for an aggressive form of competition, but there is an important distinction to be made. Competition is an interaction brought about by the *removal* of a resource from the environment, and allelochemic interactions are brought about by the *addition* of a substance to the environment. A subset of allelochemic interactions involving plant species only is called **allelopathy**. In the 50 years since the word allelochemics was coined, a large body of literature has developed on the subject, seeming to demonstrate the pervasive chemical links between organisms within a community—decomposers, producers, and consumers (Sondheimer and Simeone 1970; Rice 1974; Whittaker and Feeny 1971).

However, the ecological significance of allelochemics is still not clear (Barbour 1973*a*). Before an interaction can be considered ecologically significant, a series of steps should be followed. In the first step, a correlation must be documented. For example, a negative association between two species of plants must be shown. The second step involves experimentation to try to determine cause and effect. If a chemical is involved, what is its identity and how does it affect the species involved? Much of this work can be done in the laboratory, but laboratory conditions should attempt to imitate field conditions. The third step returns to the field situation. Do the factors discovered in the laboratory operate in nature? Can the compounds be detected, and in what concentration? Can they remain viable in the soil system for long periods of time? The pitfalls in carrying a suspected interaction through all three steps are numerous and well illustrated in the literature.

Allelochemic Interactions at the Producer-Decomposer Level

To a large degree, decomposers in the soil and litter beneath a community are affected by the species of plants shedding the litter and penetrating the soil with roots. Soils beneath northern conifer forests are largely acidic, because conifer litter is acidic and its decomposition influences soil pH. As a result, fungi dominate the soil microflora, whereas bacteria dominate more neutral soil beneath deciduous forests (Eyre 1963). There are also differences even within one conifer forest. Pine needles are much more acidic than spruce needles, and the soil beneath most pine species has less decomposer activity and is almost devoid of earthworms, in comparison with soil beneath spruce species.

Understory plants can further modify the soil microenvironment. Tappeiner and Alm (1975) showed that such features as soil pH, litter decay rate, and soil nutrient status were dependent not only on overstory tree species, but also on whether the understory was dominated by hazel shrubs (*Corylus cornuta*), a mixture of other shrubs and herbs, or was largely bare (Table 6-3). Since the understory was often a mosaic of the three types, the authors concluded that there must be considerable spatial variation in the forest soil microenvironment.

Although forest understory herbs are notably patchy, and overstory tree seedlings may be positively or negatively associated with the herbs (Maguire and Forman 1983), it is unlikely that amensalism is usually involved. Instead, herb pattern is often determined by the physical quantity and quality of litter, by uneven distribution of soil nutrients, and by microenvironmental patterns of soil drainage (Sydes and Grime 1981; Rogers 1982 and 1985).

Apart from pH and litter decay products, plants affect soil chemistry by passively contributing a variety of inorganic and organic compounds to the soil. Apparently, plants are very leaky systems. Carlisle et al. (1966) analyzed the nutrient content in rainwater falling directly on the ground and in rainwater falling through the leaf canopy of sessile oak (*Quercus petraea*). Except for nitrogen, the rainwater falling through the oak leaf canopy contained a higher concentration of nutrients (Table 6-4). Tukey (1966) has shown that larger molecules can also be leached from leaves. He grew seedlings and cuttings of 150 species in nutrient culture with certain radioisotopes in it, then leached the plants by atomized mist or immersion in pure water for up to 24 hr. The leachate was

Table 6-3 Effect of overstory and understory plants on soil pH, bulk density (g cm^{-3}), weights of nutrients in the top 3 cm (kg ha^{-1}), and turnover rates (in years) for all dry litter and for selected nutrients. (From "Undergrowth vegetation effects on the nutrient content of litterfall and soils in red pine and birch stands in northern Minnesota" by J. C. Tappeiner and A. A. Alm, *Ecology* 56:1193-1200. Copyright © 1975 by The Ecological Society of America. Reprinted by permission.)

Overstory/ understory	pH	Bulk density (g cm^{-3})	Ca	N	K	Mg	P	Litter	Ca	K
			Weight in soil					Turnover time (years)		
Red pine/none	4.1	0.66	121	470	29	13	7	5.0	4.9	3.9
Red pine/hazel	4.1	0.54	111	535	29	14	7	3.2	2.5	3.1
Red pine/ herb-shrub	4.2	0.62	128	524	30	14	8	4.1	3.8	2.1
Birch/hazel	5.0	0.44	304	617	47	43	8	1.7	1.4	0.2
Birch/ herb-shrub	5.1	0.48	337	602	50	44	9	2.3	1.9	0.3

Table 6-4 Nutrient content in rainwater that falls directly to earth (rainfall) and in rainwater that falls through an oak (*Quercus petraea*) canopy (throughfall). Values are expressed in kg ha^{-1} yr^{-1}. (From Carlisle et al. 1966. By permission of the British Ecological Society.)

Nutrient	In rainfall (kg ha^{-1} yr^{-1})	In throughfall (kg ha^{-1} yr^{-1})
N	9.54	8.82
P	0.43	1.31
K	2.96	28.14
Ca	7.30	17.18
Mg	4.63	9.36
Na[a]	35.34	55.55
Total	60.20	120.36

[a]an essential nutrient for some plants

channeled through an anion-cation exchange resin and analyzed. It contained 14 elements, including such essential nutrients as iron, calcium, phosphorus, potassium, nitrogen, and magnesium, seven sugars, some pectic substances, 23 amino acids, and 15 organic acids, including virtually all the acids in the Krebs cycle of respiration. In nature, these and larger molecules would enter the soil, and they or their decomposition products would accumulate there.

Table 6-5 Root exudate materials (kg ha^{-1} yr^{-1}) released by young roots of the three principal tree species in a northern hardwood forest in New Hampshire. (From "Character and significance of forest tree root exudates" by W. K. Smith, *Ecology* 57:324–331. Copyright © 1976 by the Ecological Society of America. Reprinted by permission.)

Category of material	Birch (*Betula alleghaniensis*)	Beech (*Fagus grandifolia*)	Maple (*Acer saccharum*)
Sugars	1.66	1.27	0.25
Amino acids/amides	0.22	0.29	0.05
Organic acids	2.50	4.18	0.62
Cations	23.56	18.52	4.90
Anions	5.64	2.14	0.65
Total	33.58	26.40	6.47

Chemicals also leak out of roots. Smith (1976) collected root tips from mature trees in a deciduous forest of New Hampshire, leached them in distilled water much as Tukey treated shoots, and analyzed the leachate. Table 6-5 summarizes the impressive results. (Smith was careful to state that they are conservative estimates of root exudation, because older roots were not examined, nor were many types of soluble organic materials searched for, and volatile substances were ignored.) In nature, these substances would diffuse from the roots into the soil, and thus the soil beneath each species would be distinctive.

Plant products may affect the activity of soil bacteria. Abandoned crop land in Oklahoma reverts to prairie through a sequence of successional communities that may require 30 or more years (Figure 6-9). Elroy Rice and his co-workers (Rice 1964, 1965; Blum and Rice 1969; Olmstead and Rice 1970; Wilson and Rice 1968) examined whether or not the course of succession could be directed by chemical interactions—in particular, interactions affecting nitrogen-fixing and nitrifying bacteria. They made aqueous extracts of flowers, leaves, roots, and stems of 14 pioneer weed species and tested them for inhibitory effects on strains of *Azotobacter*, *Nitrobacter*, and *Rhizobium*. All of the extracts proved to be inhibitory to one or more of the bacterial strains. They identified the inhibitors in three of the weed species as chlorogenic acid and gallotannin, both polyphenols. Chlorogenic acid is a strong inhibitor of several enzyme systems (phosphorylase, peroxidase, and oxidase), and gallotannins are excellent protein precipitants. Soils beneath two weed species were found to contain large quantities of gallic and tannic acids (over 600 ppm tannic acid; in laboratory tests, less than 300 ppm tannic acid was effective in reducing symbiotic nitrogen fixation).

Rice claimed that the pioneer weed species demanded less nitrogen than later successional species and that the weed stage prolonged itself by inhibiting certain bacteria, thus reducing nitrogen availability in the soil. However, the story is not ecologically complete. Chlorogenic acid and gallotannins are widespread

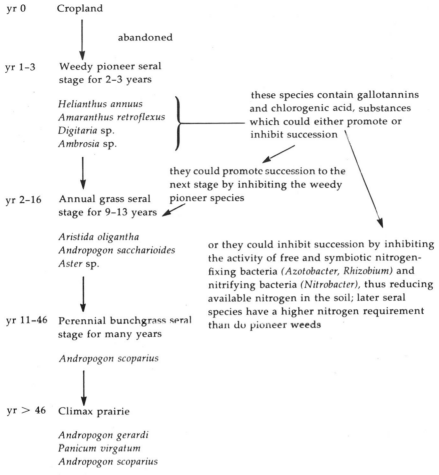

yr 0   Cropland

        abandoned

yr 1–3   Weedy pioneer seral
          stage for 2–3 years

          Helianthus annuus
          Amaranthus retroflexus
          Digitaria sp.
          Ambrosia sp.

these species contain gallotannins
and chlorogenic acid, substances
which could either promote or
inhibit succession

they could promote succession to the
next stage by inhibiting the weedy
pioneer species

yr 2–16   Annual grass seral
           stage for 9–13 years

           Aristida oligantha
           Andropogon saccharioides
           Aster sp.

or they could inhibit succession by inhibiting
the activity of free and symbiotic nitrogen-
fixing bacteria *(Azotobacter, Rhizobium)* and
nitrifying bacteria *(Nitrobacter)*, thus reducing
available nitrogen in the soil; later seral
species have a higher nitrogen requirement
than do pioneer weeds

yr 11–46   Perennial bunchgrass seral
            stage for many years

            Andropogon scoparius

yr > 46   Climax prairie

           Andropogon gerardi
           Panicum virgatum
           Andropogon scoparius

Figure 6-9   The sequence of plant succession in abandoned cropland
of Kansas and Oklahoma, and the effect of allelopathic interactions
detected by Rice. (Based on Booth 1941 and Perino and Risser 1972. By
permission.) A recent study by Collins and Adams (1983) suggests that
the intermediate stages of annual grass and bunchgrass are always not as
predictable as this diagram implies.

in the plant kingdom (Robinson 1967) and might be expected in climax prairie
grasses as well as in pioneer weeds. Indeed, Rice's data show that extracts of the
climax prairie perennial grass *Andropogon scoparius* were just as toxic to the bacteria as many weed extracts. Further, if the inhibitors are stable in soil (and Rice
presented some data to indicate this is so), then why is the weed stage not
prolonged beyond 2–3 years? Possibly, the weed stage is short because chlorogenic acid stimulates the growth of one of the plants in the second community,
wire grass *(Aristida)*, even as it depresses the growth of the weedy pioneers. Not
all plants are equally sensitive to chlorogenic acid. As a final complication, this

grassland phenomenon of declining nitrification with successional time is not universal. Declining nitrification is characteristic of succession in New Hampshire—abandoned cropland to spruce-fir forest (Thorne and Hamburg 1985)—but the reverse is true of dune succession in Indiana and of succession in New Jersey—abandoned cropland to deciduous forest (Robertson and Vitousek 1981). It should be clear, then, that this particular grassland allelochemic story is not yet neatly concluded.

The reverse of plants inhibiting bacteria can also occur: By-products of bacterial metabolism can inhibit higher plants. Brian (1957) listed 38 antibiotics that affect germination and plant growth in low concentration (1–10 ppm); many of these are thought to be formed and released in the normal soil system. Although of relatively high molecular weight, they can be taken up by roots and translocated through a higher plant. Among his list are a number of metabolic inhibitors of great specificity and potency.

Allelopathy

A number of researchers have reported evidence for chemical control of plant distribution, spatial associations between species, and the course of community succession.

One of the most complete studies is that by Muller (1966) on the spatial relationship between coastal sage (*Salvia leucophylla*) and annual grassland in the Santa Ynez valley of southern California. A number of shrub species, including sage, dominate the foothills, and annual grasses and herbs dominate the valley floors. However, patches of sage shrubs may occur in the grassland. Beneath those shrubs, and for 1–2 m beyond the shrub canopy limits, the ground is nearly bare of herbs and grasses (Figure 6-10). Even 6–10 m from the canopy, annuals are stunted. Stunting is not caused by competition for water, since shrub roots do not penetrate that far into the grassland, and stunting is observed even in the wettest times of the year. Soil factors do not seem to be responsible for the negative association either, because major chemical and physical soil factors do not change across the bare zone.

Muller was able to show that *Salvia* shrubs emit a number of volatile oils from their leaves and that some of these (principally cineole and camphor) are toxic to the germination and growth of surrounding annuals. He was able to detect these substances in the field, to demonstrate that they are adsorbed by the soil and can be retained there for months, and to show that they are able to enter seeds and seedlings through their waxy cuticles. However, he was not able to detect the same amounts of oils in natural soils that were necessary to produce inhibition in the laboratory.

Nor was Muller able to eliminate completely other factors as contributors to the maintenance of the bare zones. Bartholomew (1970) examined in detail the influence of mammalian and bird herbivores that reside in the shrub clumps but forage in the grassland. Foraging activity, he reasoned, would increase closer to the shrubs and might be the main cause for maintenance of the bare zone. From seed predation and exclosure experiments, Bartholomew was able to substantiate

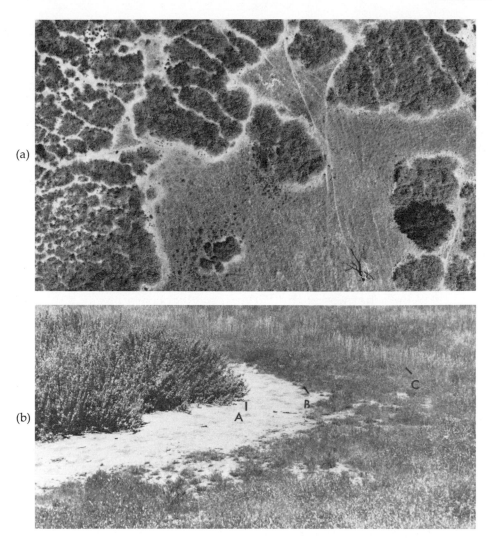

(a)

(b)

Figure 6-10 Nearly bare zones between soft chaparral and annual grassland near Santa Barbara, California. (a) Aerial photograph of *Salvia leucophylla* shrubs and adjoining grassland; the light bands beneath and next to the shrubs are devoid of all but a few species of small herbs. (b) Ground view of the same phenomenon. A indicates soft chaparral, B indicates the edge of the bare zone, and C indicates the grassland. (Courtesy of C. H. Muller.)

that hypothesis. Halligan (1973) discovered a similar influence of herbivores in bare zones around California sagebrush (*Artemisia californica*).

Muller and his students presented a more convincing case for allelopathy in California hard chaparral, a dense, scrubby vegetation dominated by such taxa as chamise (*Adenostoma fasciculatum*). McPherson and Muller (1969) and Christensen and Muller (1975) concluded that substances released from leaves of chamise clearly inhibited the germination and growth of understory herbs because all other environmental factors, such as light, moisture, and soil fertility, were successfully eliminated by experimental manipulation. More recently, however, Kaminsky (1981) has shown that the story is more complex. He demonstrated several anomalies that conflicted with Muller's hypothesis. Soils beneath chamise were inhibitory to herb germination only when wetted, and the inhibition became more intense with time following wetting. The inhibitory effect could be eliminated by fumigating the soil with a bactericidal agent. Also, dry soils typically contained only 100 ppm of known inhibitors (primarily p-coumaric acid), whereas amounts required to induce inhibition in the laboratory were up to four times that concentration. He concluded that chamise merely releases non-toxic material in its litter, which is then converted and concentrated over time by soil microbes into allelochemic substances. The amensal organism, then, is not chamise but must be certain microbial decomposers.

Allelochemics at the Producer–Herbivore Level

For a given herbivore, all species of plants are not equally palatable. Many species are rejected totally, some are eaten preferentially, and others are eaten only when preferential species are absent. Grazing selectivity by large herbivores is easy to observe. It is less obvious, though just as common, in small herbivores such as insects. The attainment of nonpalatability would confer major selective advantage to a plant species. It is quite likely that palatability depends upon the quality and quantity of certain metabolic by-products (Wallace and Mansell 1976).

In lengthy reviews, Brower and Brower (1964) and Ehrlich and Raven (1965) summarized the food preferences of butterfly groups throughout the world. Ehrlich and Raven concluded that many taxonomic groups feed exclusively on one to several families of flowering plants, and that "secondary plant substances [alkaloids, quinones, essential oils, glycosides, flavonoids, and crystals] play the leading role in determining patterns of utilization. This seems true not only for butterflies but for all phytophagous groups and also for those parasitic on plants."

This may be too sweeping a statement. Although the physiological functions of many secondary compounds are indeed unknown, it may not be fair to take them out of the context of metabolic pathways. They may be intermediates in the synthesis of pigments, hormones, or other compounds of known function. There are conflicting reports on their rate of turnover. (A high turnover rate for a substance indicates that it is an intermediate, not an end product; see Seigler and Price 1976.) However, some plant products, whether they are intermediates or stable end products, do affect grazing intensity. At the same time, coevolution on the part of herbivores can include production of enzymes that break down the normally unpalatable or toxic plant products (Krieger et al. 1971).

If the inhibitory chemicals that act as herbivore defenses *are* in fact stable end products rather than intermediates of metabolic pathways then they represent an energetic cost to the plant. Cates (1975), among others, addressed the questions about how much this cost might reasonably be and how herbivores respond. He studied wild ginger (*Asarum caudatum*), an herbaceous perennial that grows in moist conifer forests from British Columbia, Canada, to the Santa Cruz Mountains of California. Within that range, it is fed on by the slug *Ariolimax columbianus*, although slug density is higher in the north than the south. Field observation, coupled with feeding experiments, revealed two ecotypes of wild ginger. One was relatively palatable. In addition, it had a faster growth rate, flowered earlier in the year, produced more seeds per fruit, and grew on moister sites than the other ecotype, which was relatively unpalatable. In laboratory trials, the slug removed more than 60% of the seeds or vegetative matter of the palatable ecotype but less than 25% of the seeds or vegetative matter of the unpalatable ecotype. Cates hypothesized that the slow growth of the unpalatable ecotype was due to the cost of building chemical defenses (although its habitat preference of drier sites would further depress growth). He further predicted that if slug grazing pressure was not an environmental factor, the population density of the unpalatable ecotype would decline because the population density of the palatable ecotype would increase. Using the same slug as a good example of a "generalized" herbivore (one whose food plants cover a broad taxonomic spectrum), Cates and Orians (1975) tested the palatability of 100 plant species from the lowlands of western Washington; some were representative of pioneer or early successional communities (*r*-selected), and others were representative of later seral stages or of climax communities (*K*-selected). Pioneer species proved to be more palatable, supporting their hypothesis that *r*-selected taxa invest relatively little energy in herbivore defense, in contrast to *K*-selected taxa.

Types of Compounds Involved   *r*- and *K*-selected taxa differ in the type of allelochemic substances they contain. Those annuals that do invest energy in chemical herbivore defense appear to do so only with toxins, whereas woody perennials tend to produce tannins (Rhoads and Cates 1976). Toxins are small, relatively inexpensive molecules, of molecular weight 500 g mole^{-1} or less, and they are effective in small concentrations. Toxins usually account for less than 2% of leaf dry weight. They interfere with nerve and muscle activity, with hormone activity, or with liver or kidney functions. They are effective against generalist herbivores; specialist herbivores can develop means to detoxify the substances.

Woody perennials may contain toxins, but they invest considerably more energy into the manufacture of tannins. Tannins are large, complex phenolic molecules of molecular weight 500–3,000, which reduce the digestibility of plant tissue by complexing with plant proteins, carbohydrates, nucleic acids, and herbivore enzymes when the tissue is chewed. Tannin content may exceed 6% of leaf dry weight. Only 17% of all annual plants tested have tannins, but more than 80% of all woody perennials have them. Zucker (1983) has written an extensive review of the chemistry and possible ecological roles of tannins. Two types of tannins are sequestered in different parts of the plant cell. Hydrolizable tannins are in vacuoles, and condensed tannins may be complexed with cellulose, pectin, starch, or alkaloids in or near the cell walls. He hypothesized that the large number of known tannins may make them suitable as a defense against specialist, as well as generalist, herbivores. Hydrolizable tannins may be most effective against chewing herbivores, such as slugs and insects, and condensed tannins may be effective against the cellulase exo-enzymes of fungi and bacteria. Zucker also hypothesized that condensed tannins slow decomposer activity, resulting in a long-term release of nutrients from litter, which would minimize the leaching of nutrients from the root zone.

Terpenes may also function as herbivore deterrents, and these compounds can also be produced in high concentrations by plants in some families, such as the Lamiaceae (mints). Gouyon et al. (1983) have shown how sensitive slug herbivores can be to various terpenes in *Thymus vulgaris* (thyme). Slugs were able to differentiate among four thyme biotypes, each containing a different terpene. Biotypes with two of the four terpenes were most heavily grazed and suffered the highest subsequent mortality. Biotypes with a third terpene exhibited half as much leaf tissue lost to grazing and half as much mortality, and biotypes with a fourth terpene were intermediate in damage. In this particular case, the herbivore was able to judge palatability before tasting a leaf, perhaps by volatile releases of the terpenes from epidermal cells.

The accumulation of tannins, terpenes, or other allelochemic substances represents a significant caloric cost to the plant. Janzen (1974) has hypothesized that the caloric energy invested in defense should be proportional to the metabolic cost of replacing grazed tissue. In other words, one might expect that tissue rich in digestible carbohydrates and inorganic nutrients would also be relatively rich in palatability-reducing defense compounds, and in general this appears to be true. Another slug experiment showed that those lichens that were grazed, from among a cohort of equally available lichens, were those containing the fewest phenolic compounds and the lowest amounts of many essential elements (Lawrey 1983). The lichens that were avoided contained the richest assortment of phenolics and 600% more calcium, 40% more phosphorus, and 25% more potassium. For slow-growing lichens on bare rock, the accumulation of such nutrient concentrations may represent a significant investment, and the protection of that tissue with metabolic by-products fits Janzen's model.

Inducibility of allelochemic substances   It has been known for several decades that inoculation of plants with fungal, bacterial, viral, or nematode extracts can increase plant resistance to subsequent attack by such pathogens or parasites. The substances that accumulate in the plant following inoculation (low molecular weight antibiotics, lignins, tannins, and several other types of compounds) have been called **phytoalexins**.

In the past decade, it has become clear that chemical changes of an amensal nature can be induced by herbivore activity (Rhoades 1979, 1983a, and 1985). At this time, it appears that the plant response is general (not due to a specific herbivore) and may even be generated by mechanical (abiotic) clipping of tissue. Induction can be almost instantaneous or take hours, or it may not become apparent until a full growing season passes (Edwards and Wratten 1985).

Only in the past 5 years has it become clear that the production of allelo-chemic substances can be induced in unattacked plants adjacent to attacked plants. Possible communication routes for a chemical signal to travel from plant to plant include the air (volatile signals) or the soil (mycorrhizal connections or root grafts). Rhoades (1983b) reported that leaves of the willow *Salix sitchensis* declined in palatability following attack by tent caterpillars, and so did leaves of nearby unattacked trees of the same species. Baldwin and Schulz (1983) tore leaves on potted sugar maples (*Acer saccharum*) and poplars (*Populus* x *euroamericana*) in growth chambers and were able to measure a doubling of phenolic content in both the damaged plants and adjacent potted maples or poplars within 52 hours (Figure 6-11).

We should keep in mind that there is variability in antiherbivore chemistry from plant to plant within any population, just as there is variability in any trait. It is likely that future work with antiherbivore chemistry will focus on population variation, as Cates and Redak (1986) recently have done. They placed budworm larvae on a morphologically homogeneous group of Douglas fir trees in a New Mexico forest. When the larvae had pupated 3 weeks later, both pupae and foliage were collected. Foliage was analyzed (qualitatively and quantitatively) for terpene content. Pupae were allowed to incubate until adults emerged, then they were sexed and weighed. Adult weights were found to be inversely related to amounts

Figure 6-11   Change in phenolic content of poplar leaves 52 hr following mechanical damage. *TR* = treated plants, *CC* = communication control plants (untreated but in the same growth chamber as *TR*), and *TC* = true control plants in a separate chamber. Clear bars show total phenolic content per unit of tissue, expressed as percent of pre-treated values. Stippled bars show incorporation of ^{14}C into phenolics per unit of tissue, expressed as percent of pre-treated values. (Based on data from I. T. Baldwin and J. C. Schulz. *Science* 184:753–759. Copyright © 1983 by the American Association for the Advancement of Science. By permission.)

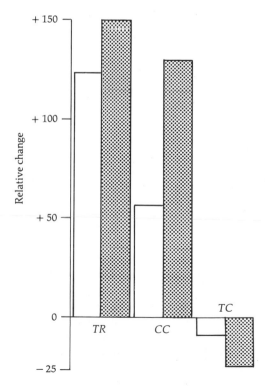

of certain terpenes in tree foliage—suggesting an allelochemic effect by the terpenes—and the amounts of those terpenes varied considerably from tree to tree. Further, the correlations were not consistent from year to year; the most inhibitory terpenes one year were not the most inhibitory the next year. There was much variation in inducible terpene chemistry of this Douglas fir population from tree to tree and in the same tree over time, and selection may in fact favor maximum variation as an optimal population-level defensive attribute.

Hormones as allelochemic agents   The most sophisticated and specific chemical interaction between plants and herbivores may involve hormones. Feeding preferences in some cases may be related to hormonal control rather than simple palatability. Ferns, for example, are generally less extensively eaten by insects than are flowering plants, and it may be no mere correlation that ferns evolved with some insect groups considerably earlier than the flowering plants. Soo Hoo and Fraenkel (1964) commented that larvae of the southern army worm reject ferns in general, and these authors performed some preliminary feeding experiments with water extracts of ground pieces of Boston fern (*Nephrolepis exaltata*). Shortly after, Kaplanis et al. (1967) extracted two major molting hormones from pinnae of bracken fern (*Pteridium aquilinum*): α-ecdysone and β-ecdysone (20-hydroxyecdysone). Ecdysones form a class of compounds similar in structure to cholesterol. They are ordinarily synthesized by the prothoracic gland of larvae, and they promote developmental reactions, such as pupation (Figure 6-12). However, Kaplanis cautioned, "we do not have information on either the significance or function of these steroids in plants. Perhaps . . . these substances . . . interfere with the growth processes of insect predators."

Ecdysone has since been isolated from more than 10 species of conifers, 20 ferns, and 30 flowering plants, out of 1000 species surveyed (Williams 1970). A total of 28 different plant ecdysones are known, the most ubiquitous being β-ecdysone. The ecological significance of β-ecdysone in plants is unclear. It is not toxic when orally ingested (as feeding larvae would obtain it from a food plant), but there is some evidence that it could be a feeding deterrent in concentrations as low as 1 ppb. Perhaps it serves as a steroid base for other compounds once it is in an insect's metabolic system.

Another developmental hormone is the juvenile hormone, which predominates early in larval life (later, ecdysone predominates). It is a methyl ester of the epoxide of a fatty acid derivative, and there is some evidence that its structure differs in different groups of insects. By a series of coincidences, it was discovered that certain paper toweling prevented the European bug, *Pyrrhocoris apterus*, from developing into sexually mature adults. Instead, an extra one or two larval molts ensued, and all eventually died without being able to complete metamorphosis. The juvenility factor was traced to particular conifers used in American paper pulp, mainly *Abies balsamea*, *Tsuga canadensis*, *Taxus brevifolia*, and *Larix laricina*. The active principle was isolated and characterized. It has a structure similar to that of the juvenile hormone and has been named juvabione. It proved to be effective on only one family of insects, the Pyrrhocoridae. Williams (1970), who

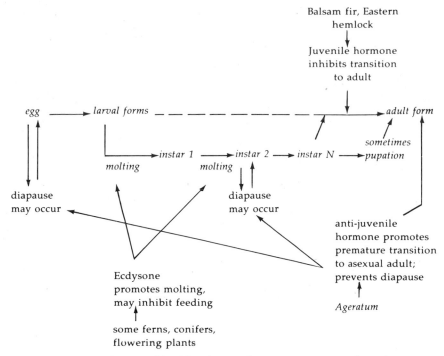

Figure 6-12   The possible role of insect hormones contained in plants, using an idealized insect life cycle.

has summarized the pyrrhocorid story, asked, "Have the plants in question undertaken these exorbitant syntheses just for fun? I think not. Present indications are that certain plants and more particularly the ferns and evergreen trees have gone in for an incredibly sophisticated self-defense against insect predation. . . ."

However, the pyrrhocorid story certainly fails to justify such a hypothesis. *Pyrrhocoris apterus* is a native of Europe, and the only plants that produce the very specific juvabione are natives of North America. The two simply do not occur together. Further, the bug and all members of its family feed by sucking the juices of weak herbs; they are not known to feed on any tree species. What ecological significance is there in the pyrrhocorid story?

Finally, and most recently, an antijuvenile hormone has been extracted from ageratum, a common, annual garden plant (*Ageratum houstonianum*). It has proven capable of inducing precocious molting of immature insects to sterile adults, thus reducing feeding time and reproduction at once (Bowers et al. 1976). Other plants may also possess such a hormone, and its effects on insects could additionally include prevention of diapause (critical resting periods) and prevention of sex pheromone production (Figure 6-12).

Summary

Species interactions can be negative or positive, and the spatial distribution of plants may give a first clue to the presence of an interaction. Burkholder recognized more than nine types of interactions as being biologically and ecologically reasonable. Three of these appear to be poorly known, and one (neutralism) may be rare. Two interactions, competition and amensalism, were discussed in this chapter.

The effect of competition on interacting organisms can be modeled mathematically with the Lotka-Volterra equations, expressed in evolutionary terms with Gause's competitive exclusion principle, or measured with de Wit's experimental designs, which include replacement series plantings and calculation of input/output ratios. However, none of these approaches seems to get us very close to understanding competition in the real world. In addition, many competition studies do not reveal the resource being competed for. Competition may be an important factor in habitat restriction for some plants, such as serpentine endemics and halophytes.

Allelochemic interaction is one form of amensalism, and a review of the literature indicates that it may be an important interaction between plants and soil biota and between plants and herbivores. Allelochemics appear to play a role in plant palatability. Many substances are passively leached from leaves and roots by percolating rainwater; some of these may be allelochemic. Other metabolic products remain stored in plant tissue, including some that mimic insect hormones. The study of allelochemics as a strategy in herbivore defense, with costs and benefits, is currently receiving much research attention, but more investigation is necessary before we can assess the general importance of allelochemics.

CHAPTER 7

SPECIES INTERACTIONS
Commensalism, Protocooperation, Mutualism, and Herbivory

In Chapter 6, two negative interactions resulting in negative spatial associations were examined. In this chapter, we will consider one other negative interaction and several interactions that result in positive spatial associations. These positive interactions have sometimes been lumped together under the term symbiosis. It will be seen that there is a fine line between a positive and a negative interaction, a line that is easily crossed in the course of time (evolution). Herbivory is an example of an interaction that is usually considered to have negative aspects for the plant, but biologists are coming to recognize the complex, subtle, and unexpected ways in which some plants can be stimulated by herbivores.

There is also a fine line between obligate interactions, where both partners fail in the absence of one another, and degrees of facultative interactions, where only one partner is benefited or where both are benefited slightly. This line, too, is easily crossed in the course of evolutionary time; the change is usually from a facultative interaction to an obligate interaction, not vice versa.

Commensalism

Commensalism is an interaction that stimulates one organism but has no effect on the other. One example of commensalism is the growth of epiphytes on host trees (Figure 7-1). Epiphytes may be herbaceous perennials, such as orchids, ferns, bromeliads (such as Spanish "moss"), and cacti, or they may be lower plants, such as true mosses, algae, or lichens. In any event, epiphytes gain no nourishment from the host, using it only as a physical place of anchorage. Some have expanded leaf bases or unusual root surfaces that trap and retain rain

133

(a)

(b)

Figure 7-1 Examples of ephiphytes. (a) The bromeliad *Tillandsia usneoides* (Spanish "moss") in trees along the Gulf Coast. (b) A close-up of *Tillandsia.* (c) The lichen *Ramalina reticulata* (grandfather's beard) on a California oak tree.

(c)

water for later absorption. Others, like Spanish "moss" (*Tillandsia usneoides,* a bromeliad common in the southeastern United States) (Figure 7-1a) or grandfather's beard (*Ramalina reticulata,* a lichen that festoons California oaks) (Figure 7-1c), absorb much of their water from humid air. *Ramalina's* tissue moisture content fluctuates during the day, closely reflecting the trend of the relative humidity

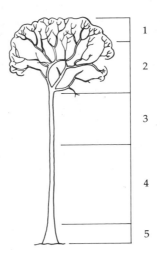

Figure 7-2 Microhabitats of epiphytes within an emergent tree of the tropical rain forest. Small epiphytes are common in zone 1, large epiphytes in zone 2, crustaceous lichen epiphytes in zone 3, and bryophytes in zones 4 and 5. (From *Tropical Rainforest and Its Environment* by K. A. Longman and J. Jenik. Copyright 1974 by Longman, London.)

of the air. Its net photosynthesis is positive only during the early morning hours (Rundel 1974). *Tillandsia*, on the other hand, cannot satisfy its entire water demand from the air even at 80% relative humidity, as Garth (1964) experimentally demonstrated; occasional rain is necessary. Garth found that the distribution of *Tillandsia* correlated very well with the average routes of storms that sweep east and north from Mexico, bringing frequent rain. The range limit of *Tillandsia* corresponds to areas with an average annual relative humidity of 64% or more.

It is likely that competition with other epiphytes and with the canopy of the host tree for space and light has led to very specialized epiphytic species, because censuses of tropical trees have shown that relatively distinct parts of the canopy can be associated with each epiphyte species (Figure 7-2); Longman and Jenik 1974; Janzen 1975). Epiphyte preferences for certain host species seem to reflect such physical attributes as openness of branching and roughness of bark. This is certainly true for Spanish "moss," which festoons cypress and some hardwood trees but avoids pines. Pine avoidance is partly due to the sloughability of pine bark (although bark nutrient content is also important; Schlesinger and Marks 1977).

Epiphytism may easily grade into other types of interactions, such as mutualism, which can result if the epiphyte produces nutrients that are leached by rainwater running down the trunk and enter the soil around host roots. The upper canopy branches of trees in a Colombian rain forest (2700 m elevation, about 250 cm rain a year) are thickly covered with lichen epiphytes. Richard Forman (1975) carefully demonstrated that 86% of these lichens contain the blue-green alga *Nostoc* as the algal symbiont. *Nostoc* is capable of fixing atmospheric nitrogen into organic form, and Forman estimated that it does so in this forest at the rate of $1.5–8.0 \, \text{kg N ha}^{-1} \text{yr}^{-1}$. This is equivalent to the amount of available nitrogen added to the forest from rainwater alone (N_2 is converted to NO_3^{--} during electrical storms, and the nitrate is carried to the earth in rainwater). The nitrogen fixed into lichen epiphytes is probably widely distributed as a result of lichen decomposition and leaching from the living lichen.

Parasitism could conceivably evolve from epiphytism if the epiphyte's roots penetrate beneath the bark into phloem and xylem and develop absorbing organs called haustoria. There seem to be many degrees of parasitism. For example, Hull and Leonard (1964; see also a review by Kuijt 1969) experimented with two genera of green mistletoe that parasitize a variety of conifers in the Sierra Nevada Mountains (Figure 7-3). They exposed host foliage to radioactive $^{14}CO_2$, then took autoradiographs of mistletoe foliage. They concluded that the genus *Phoradendron* derived very little if any carbohydrate from its host, whereas the dwarf mistletoe *Arceuthobium* drew heavily on the photosynthate of its host. (They did not examine water utilization; undoubtedly both genera parasitized host xylem.) Hull and Leonard found that the degree of parasitism closely correlated with chlorophyll content and photosynthetic rate of mistletoe tissue. *Phoradendron*, the hemiparasite, contained 0.93 mg chlorophyll per gram dry weight, about half the value of the host tissue, but *Arceuthobium* contained only 0.37 mg g^{-1}. In turn, the photosynthetic rate of *Arceuthobium* was very low, compared to the rate of the conifer host.

(a)   (b)

Figure 7-3   The mistletoes (a) *Arceuthobium* and (b) *Phoradendron*.

Epiphytes can also become damaging if their size or weight creates strain on the host. For example, the strangler fig (*Ficus aurea* in Florida, other species of *Ficus* elsewhere) germinates in the canopy of a host and begins life as a typical epiphyte. Then, aerial roots grow towards the soil, reach the ground, and thicken, engulfing the host trunk and preventing further growth. At the same time, the canopy of the strangler fig enlarges to overtop the host and deprive it of light. Eventually the host dies (Figure 7-4).

Another example of commensalism is the **nurse plant syndrome.** According to long-term studies by Niering et al. (1963) and Steenbergh and Lowe (1969), virtually all successful saguaro (*Cereus gigantea*) seedlings are found in close prox-imity to a shade-producing object. Occasionally the object is inanimate, but in most cases it is a perennial plant. Observations by these authors, coupled with others by Vandermeer (1980) and Turner et al. (1966), show that the positive effects of the nurse plant are shading (reduces temperature and rate of soil drying), hiding the young cactus from rodent herbivores, and some protection from frost. In the Tucson, Arizona, area, some 15 different species function as nurse plants. The frequency with which they appear as shade plants, however, is an approx-imate indication of their relative abundance and their percentage of cover. In other words, saguaro has no preference for one species over another. Turner et al. showed that soils from beneath different nurse plant species do differ in color, salinity, and nutrient status, but their data indicated that such soil differences had little effect on cactus seedling mortality. Vandermeer presented some evi-dence to indicate that, as the cactus grows, it may compete for soil moisture with at least one of its nurse plant species, the palo verde (*Cercidium floridum*). Dead

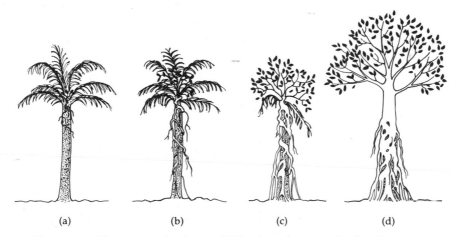

(a)   (b)   (c)   (d)

Figure 7-4 Four stages in the establishment of a strangler fig, *Ficus leprieuri,* on the palm *Elaeis quineensis.* (a) The young fig germinates high up in the palm and sends aerial roots down toward the soil. (b) The roots reach the soil and the fig shoot begins to expand. (c) The fig overtops the palm and the palm begins to senesce. (d) The palm has died but the fig tree remains. (From *Tropical Rainforest and Its Environment* by K. A. Longman and J. Jenik. Copyright 1974 by Longman, London).

Figure 7-5 The nurse plant syndrome: saguaro cactus *(Cereus gigantea)* next to a dead palo verde *(Cercidium floridum).*

palo verdes are more frequently associated with saguaros than not (Figure 7-5). Thus the relationship may shift from a positive one to a negative one over time.

Went (1942), Muller (1953), and Muller and Muller (1956) showed that the nurse plant effect extended to some desert annuals. Species of *Malacothrix* and *Chaenactis,* for example, are positively associated with the canopies of burro bush *(Ambrosia dumosa)* and turpentine-broom *(Thamnosma montana)* shrubs. Apparently the reason for the association is that the growth form of these shrubs is a suitable trap for windblown organic debris. The debris collects beneath the canopies, and this provides a better substrate for the annuals than soil in the open. It may be that seeds of the annuals are also trapped in abundance beneath the canopies. In any case, however, many desert annuals are restricted to association with a shrub. In the Great Basin desert of the western United States, bitterbrush *(Purshia tridentata),* shadscale *(Atriplex confertifolia),* and winterfat *(Eurotia lanata)* seedlings may also require nurse plants (West and Tueller 1971). Perennial bunchgrasses are positively associated with mesquite *(Prosopis juliflora)* canopies in the desert grassland of southern Arizona (Yavitt and Smith 1983).

Protocooperation

Protocooperation is an interaction that stimulates both partners, but it is not obligatory because stable growth continues in the absence of the interaction. One example of protocooperation is root grafts between members of the same or different species. As the roots of some trees grow through a soil and come in contact with each other, a natural graft or union may form. Apparently, this is a more common event than previously supposed. More than 160 tree species are known to form such grafts, and perhaps a fifth of them can form interspecific grafts (Graham and Bormann 1966). Some grafts can even be intergeneric: birch *(Betula alleghaniensis)* with elm *(Ulmus americana),* birch with maple *(Acer saccharinum),* and *Santalum album* with *Eugenia jambolana.* Grafting may occur between trees on relatively mesic sites or on sites as dry as chaparral slopes or woodland with rainfall below 50 cm a year (Saunier and Wagle 1965).

If both partners are equally successful and of the same life form, the relationship is protocooperation, with a mutual, balanced exchange of photosynthate (Figure 7-6a). There is evidence that hormones may also be transferred, resulting in more uniform phenology, such as synchronous spring bud break. If one partner is smaller and suppressed, however, the relationship may be parasitic, with significantly more photosynthate going into the smaller tree from the larger than vice versa (Figure 7-6b). Sometimes, however, the smaller tree merely grows slowly or declines, and so it does not function as a significant parasite.

Woods and Brock (1964) hypothesized that mycorrhizal hyphae (fungal mycelium filaments) may similarly link one tree with another beneath a forest soil surface. They found rapid translocation of labeled (radioactive) ^{45}Ca and ^{32}P from a red maple stump to the leaves of 19 other taxonomically diverse trees and

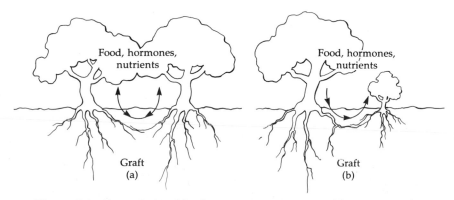

Figure 7-6 Two relationships between trees connected by a root graft, depending on the relative size of the trees. (a) The partners are both overstory trees, and the flow of nutrients is equal in both directions; the interaction is protocooperation. (b) One partner is an understory tree, and it parasitizes the larger tree.

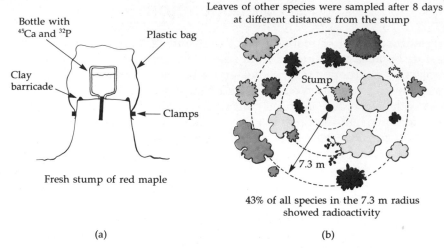

Figure 7-7   The transfer of radioactive ^{45} Ca and ^{32} P from a freshly cut stump of red maple (see detail in (a)) to the leaves of surrounding woody plants. Within 8 days, 43% of all species within a radius of 7.3 m of the stump showed radioactivity, including red maple, Carolina hickory, mockernut hickory, fringe tree, persimmon, ash, holly, juniper, sweet gum, tulip tree, black gum, Virginia creeper, white oak, red oak, black oak, and blackjack oak. (From "Interspecific transfer of ^{45} Ca and ^{32} P by root systems" by F. W. Woods and R. E. Brock. *Ecology* 45:886–889. Copyright © 1964 by the Ecological Society of America. Reprinted by permission.)

shrubs in a North Carolina forest (Figure 7-7). They concluded, "it would seem logical to regard the root mass of a forest . . . as a single functional unit. Inability of a new, invading species to enter into a 'mutual benefit society,' one in which minerals and other mobile materials are exchanged between roots, could have a negative survival value."

Mutualism

Mutualism is an obligate interaction: The absence of the interaction depresses both partners. Closely related organisms do not seem to form such an interaction. Common examples of mutualism are lichens (algae + fungi), mycorrhizae (fungi + higher plants), symbiotic nitrogen-fixation (bacteria or blue-green algae + higher plants), pollination (insects, birds, or mammals + flowering plants), zoochory (animal dispersal of plant propagules), and myrmecophytes (ants + woody plants). We will only describe mycorrhizae, nitrogen fixation, and pollination.

Mycorrhizae

Mycorrhizae (singular: mycorrhiza) are fungal associations with the roots of higher plants. In some cases, the fungus covers the root exterior near the tip of the root with a thick mantle of hyphae. Additional hyphae extend as far as 8 m out in all directions from the root into the soil, and other hyphae penetrate between cortical cells of the host root to form a nutrient-absorbing network (Figure 7-8). These latter hyphae do not form haustoria that penetrate into the cells, but contact with root cells is nevertheless very close, and metabolites are transferred in both directions (Scott 1969). Radioactive $^{14}CO_2$, for example, can be fixed in the leaves of the higher plant by photosynthesis and later be detected in the fungus. Roots passively exude such nutrients as amino acids, and these enter the fungus. Radioactive isotopes of phosphorus, calcium, and potassium can be shown to be taken up in greater amounts by plants with mycorrhizae than by plants without them, and so the growth of the higher plant is stimulated by 25–300%. The relationship, then, is mutualistic. It is not, however, very species-specific. Even along one 10 cm segment of the root of one plant, as many as five different fungal species can form mycorrhizal unions. One tree may host as many as 100 different fungal species (Harley 1969; Hacskaylo 1971; Marks and Kozlowski 1973). Further, hyphae from one mycorrhiza can extend through the soil from one higher plant species to another.

The type of mycorrhizae just described are classed as **ectomycorrhizae.** The biology and ecology of ectomycorrhizae have most recently been reviewed by Harley (1969), Hacskaylo (1971), Marks and Kozlowski (1973), and Smith (1980). Ectomycorrhizae are commonly found on northern temperate zone trees in the oak, pine, willow, walnut, maple, basswood, and birch families. The fungi are typically Basidiomycetes.

Four other kinds of mycorrhizae, sometimes loosely classed under the overall heading **endomycorrhizae,** are ecologically important to an enormous variety of herbs, shrubs, and trees (Table 7-1). These four types do not usually form an external mantle of hyphae but may instead penetrate cortical cells (hence the prefix endo-). The fungi involved may be Phycomycetes, Ascomycetes, or Basidiomycetes. There are very few higher plants that do not form mycorrhizal associations of one of these five types; these exceptions include aquatic vascular plants and members of the families Brassicaceae, Cyperaceae, and Juncaceae.

Ectomycorrhizae in particular appear to permit host plants to grow well in soils that otherwise would not permit good growth. The acidic, leached, nutrient-poor soil beneath the boreal coniferous forest is an example: All the dominant trees are ectomycorrhizal. Revegetation of mine spoil in this country occurs faster if planted conifers are inoculated with the hyphae of particular acid-tolerant mycorrhizae, as revealed by recent work of the U.S. Forestry Sciences Laboratory at Athens, Georgia. Indeed, experiments have shown that the mycorrhizal relationship does not maintain itself unless the environmental conditions are such that they limit the growth of fungi as independent organisms (Scott 1969). Mycorrhizae may also increase drought tolerance (Smith, 1980).

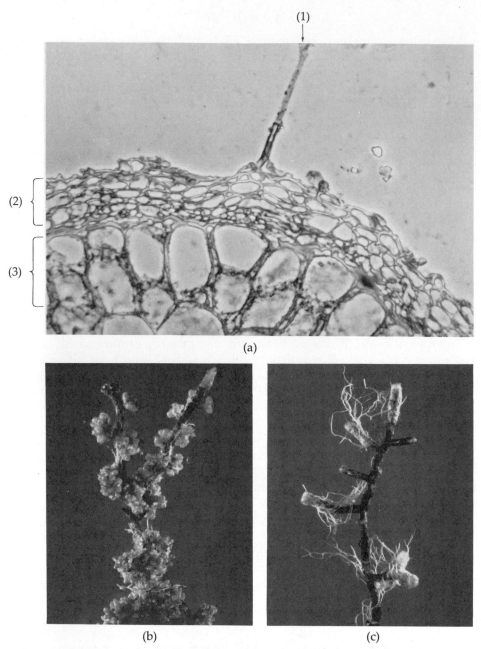

(a)

(b)   (c)

Figure 7-8   Ectomycorrhizae. (a) cross section of a root, showing (1) hyphae branching off into the soil, (2) the fungal mantle coating the outside of the root, and (3) hyphae forming a net between root cortex cells. (b) and (c) show the various shapes that mycorrhizae may take, depending on the species involved. The fungal mantle covers short, club-shaped lateral roots. ((a) Courtesy of F. H. Meyer, (b) and (c) courtesy of B. Zak in Marks and Kozlowski 1973.)

Table 7-1 The five types of mycorrhizae. (Modified from S. S. E. Smith, *Biological Reviews* 55:475–510. Copyright © 1980 by Cambridge University Press.)

Mycorrhizal type	Higher plants involved	Fungi involved
Ectotrophic	Autotrophic trees of temperate zones: conifers, *Eucalyptus*, *Quercus*, *Fagus*	Basidiomycetes and some Ascomycetes
Vesicular-arbuscular	Mainly autotrophic herbs, but also some tropical and temperate deciduous trees	*Glomus, Acaulospora, Endogyne, Gigaspora;* none yet culturable
Ericoid	Autotrophic, sclerophyllous shrubs in the families Ericaceae, Empetraceae, and Epacridaceae	*Clavaria, Pezizella*
Arbutoid	a) Autotrophic, evergreen trees and shrubs of mediterranean climates: *Arbutus, Arctostaphylos*	a) Ectotrophic fungi
	b) Heterotrophic herbs: *Monotropa, Yoania*	b) Ectotrophic fungi are epiparasites in this symbiosis; also *Armillaria*
Orchidaceous	Autotrophic and heterotrophic orchids	*Rhizoctonia* as a saprophyte or epiparasite

In addition to improving host nutrition, mycorrhizae are essential to the normal development of some species. Seedlings of orchids and heaths fail to survive if the fungi are absent, and plantings of pines in areas where they have historically never grown (such as Australia), do poorly unless the mycorrhizal fungi are introduced to the soil. We noted in the last section that mycorrhizae may also serve to link the root systems of all members of a community, in this way evening out soil resources and making it difficult for other species to invade.

The fungal floras of soils are known to change during succession and to differ from vegetation type to vegetation type. There is some evidence that the narrow ecotone between grassland and forest in central North America is maintained, at least in part, by such fungal differences. Tree mycorrhizal fungi are absent from grassland soils, and tree seeds that are blown or carried into the grassland are incapable of producing seedlings that can compete with the grasses without their mycorrhizal association (White 1941; Goss 1960). In the Great Lakes region, stands of red pine *(Pinus resinosa)* can maintain themselves in a macroclimate that favors hardwoods. Tobiessen and Werner (1980) speculated that allelopathic substances from pine litter inhibit the formation of vesicular-arbuscular mycorrhizae between the fungus *Glomus* and the major hardwoods *(Acer rubrum, Fraxinus americana, Tilia americana, Fagus grandifolia)*. These hardwoods did support mycorrhizae and grew at faster rates beneath adjacent plantation stands of the European Scotch pine *(Pinus sylvestris)*. A similar inhibitory phenomenon has been reported beneath *Pinus ponderosa* in Colorado (Kovacic et al. 1984).

Symbiotic Nitrogen Fixation

Nitrogen fixation is the conversion of atmospheric nitrogen gas into organic ammonium. Only certain prokaryotic organisms are capable of this process. Some of these prokaryotes are free-living, and others live in close association with eukaryotes, receiving sugars and other energy-rich molecules from the eukaryotic symbiont. Nitrogen fixation requires energy in the form of ATP and a local anaerobic environment:

$$4N_2 + 6H_2O \xrightleftharpoons[\substack{\text{the enzyme Nitrogenase} \\ \text{no oxygen}}]{\text{ATP} \longrightarrow \text{ADP}} 4NH_3 + 3O_2$$

The association of *Rhizobium* bacteria with the root nodules of legumes is well known, but other symbioses may have an equal or greater ecological importance. Species of the blue-green algae *Nostoc* and *Anabaena* can become associated with bryophyte gametophytes, root nodules of cycads, leaf tissue of the angiosperm *Gunnera*, and leaf tissue of the aquatic fern *Azolla* (Peters 1978). The *Azolla–Anabaena* symbiosis has economic importance in addition to ecological significance. Agricultural trials in California have shown that three-fourths or more of all the nitrogen requirements of rice can be met by cultivating *Azolla* in rice paddies.

Certain soil actinomycetes (filamentous bacteria that resemble fungi) are capable of invading the roots of some higher plants, causing elongate nodules to form. Within the nodules, nitrogen fixation occurs at a rate comparable to that of legume nodules. Some 160 species of woody plants, mainly of the temperate zones, have been shown to possess such nodules (Table 7-2). Many of the plants are pioneers in succession and occur in open habitats on acidic, saline, or sandy soils.

Pollination

Pollination is a very specialized form of mutualism that has developed in flowering plants; it may be the key to much of the variation and specialization in morphology of the angiosperms (Macior 1973). The transfer of pollen from the

Table 7-2 Plants that form root nodules with nitrogen-fixing actinomycetes. The third column shows the fraction of all species in the genus that exhibits nodules. (From J. G. Torrey, *Bioscience* 28:286–292. Copyright © 1978 by the American Institute of Biological Sciences. Reprinted by permission.)

Genus	Family	Nodulated species/Total species
Alnus	Betulaceae	33/35
Casuarina	Casuarinaceae	24/25
Ceanothus	Rhamnaceae	31/55
Cercocarpus	Rosaceae	4/20
Colletia	Rhamnaceae	1/17
Comptonia	Myricaceae	1/1
Coriaria	Coriariaceae	13/15
Discaria	Rhamnaceae	2/10
Dryas	Rosaceae	3/4
Eleagnus	Eleagnaceae	16/45
Hippophae	Eleagnaceae	1/3
Myrica	Myricaceae	26/35
Purshia	Rosaceae	2/2
Rubus	Rosaceae	1/250
Shepherdia	Eleagnaceae	3/3

stamen to the stigma is essential to reproduction in outcrossing species of plants. Even plants that are self-fertile require the transfer of pollen, because the male and female functions are not present in the same tissue. There is a physical space or gap between them that must be spanned for pollination to be effected.

A plant species is usually morphologically adapted to the specific behavioral and morphological characteristics of its pollinator, which is often an insect, bird, or bat. Adapted plants usually exhibit some or all of the following characteristics: (a) petals, sepals, or inflorescence attraction, either visually or olfactorily or both; (b) pollen grains often sculptured or sticky, sometimes massed together; (c) nectar or pollen of nutritive value to the pollinator; and (d) attractants available at blooming time, which is correlated to activity patterns of the pollinator.

In temperate regions, insects are of primary importance as pollinators, especially insects of the order Hymenoptera (Moldenke 1975). Bees, both hive bees and the higher bumblebees (Apidae), have more blossom intelligence than any other insect group. They have a greater capacity for remembering blossom characteristics, discriminating among them, and remaining faithful pollinators (Faegri and van der Pijl 1971).

The significant characteristics of bee-pollinated flowers are: (a) bilateral symmetry; (b) landing platform, mechanically strong flower, often partly closed with sexual organs concealed; (c) bright colors, often blue or yellow, often with nectar guides; (d) moderate quantities of nectar, nectar sometimes partly concealed; and (e) many ovules per ovary and few stamens. The adaptations of the pollinator include: (a) the possession of good color discrimination in the blue, yellow, and ultraviolet ranges (Jones and Buchmann 1974); (b) a relatively high degree of intelligence and a long memory; and (c) a long proboscis capable of probing for nectar.

In the tropics, there is a dearth of insect pollen vectors, and birds assume a more important pollinator function. The sunbirds (Nectarinidae) of Africa and Asia, the honey-creepers (Drepanididae) of Hawaii, and the hummingbirds (Trochilidae) of both Americas are all recognized as important pollinators. Many tropical plants have flowers that are adapted to bird pollination, such as some species in the genera *Eucalyptus* and *Erythrina*. Old World bird-pollinated flowers are borne upon a stout stem or stout inflorescence branch to provide a landing place, because Old World birds that pollinate flowers do not hover. New World hummingbirds do hover and can pollinate pendulous flowers. Table 7-3 presents a summary of those coevolved traits of birds and ornithophilous flowers that are important to their mutualistic interactions.

Table 7-3   A comparison of flowers pollinated by birds and birds that pollinate flowers. (Reprinted with permission from Faegri and van der Pijl, *The Principles of Pollination Ecology.* Copyright 1971 by Pergamon Press.)

Flowers pollinated by birds	Birds that pollinate flowers
Diurnal anthesis	Diurnal activity
Vivid, highly contrasting colors, often scarlet	Visual sensitivity to red
Lip or margin absent or curved back, flower tubate and/or hanging	Too large to alight on the flower itself
Hard flower wall, filaments stiff or united, protected ovary, nectar stowed away but visible	Hard bill
Absence of odors	Scarcely any sense of smell
Nectar very abundant	Large, requiring a large amount of food
Capillary system bringing nectar up or preventing its flowing out	Long bill and tongue
Usually deep tube or spur, wider than in flowers pollinated by butterflies.	Long bill and tongue
Distance between nectar source and sexual organs may be large	Long bill; large body
Nectar guide absent or plain	Intelligent in finding an entrance

The obvious advantage conferred upon animal-pollinated flowers is the possibility of pollen dispersal far from the host anther, allowing for outcrossing and therefore genetic variability. By providing a nutritional source for animals, flowers have influenced the evolution of pollinators and have evolved very specific traits to ensure pollination. The result is a series of mutualistic interactions, each with its own set of special adaptations. An excellent recent review of pollination ecology is a volume edited by Leslie Real (1983).

Herbivory

Herbivory is the consumption of all or part of a plant by a consumer. If the consumer category is looked at very broadly, it includes (a) parasitic microbes or plants, (b) saprophytic microbes that decompose dead tissue, (c) browsing and grazing animals that consume only a part of woody and herbaceous plants, respectively, and (d) animals that consume whole plants (or such a significant part of them that death ensues) or plant propagules (future, potential whole plants). Grazers and browsers are sometimes considered parasites, and whole-plant consumers can be called predators (Slobodkin et al. 1967). Wiegert and McGinnis (1975) additionally suggested that consumers of living tissue be called **biophages** and consumers of dead tissue **saprophages** (Figure 7-9).

Herbivory may also be a part of other interactions already discussed in this chapter. For example, allelochemics may mediate or determine the amount and kind of plant material eaten, and herbivores such as pollinators or ants and rodents that cache seeds, enter into mutualistic relationships with their food sources. Herbivory, incidentally, is a trait that is neither geologically nor taxonomically extensive (Southwood 1985). Fossil evidence of herbivory of the vegetative parts of higher plants does not extend back to the first appearance of vascular plants on the earth's surface but rather comes 70 million years later in the Permian. Extant major taxa of animals that feed on vegetative plant parts are modest in number: Only 27 of 97 orders contain herbivore species.

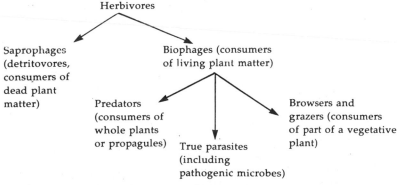

Figure 7-9 Categories of herbivores.

The Impact of Herbivory: Consumption

A great number of studies have investigated the impact of herbivores on plant communities, and their results have been summarized in several publications (Odum 1971; Krebs 1972; Ricklefs 1973; Chew 1974; Larcher 1975; Whittaker 1975; Crawley 1983).

A few definitions are in order. **Gross primary productivity (GPP)** is the total amount of chemical energy fixed by photosynthesis for a given unit of land surface for a given unit of time. Typically, GPP is expressed as calories per square meter per year, but it may be expressed in biomass units also. (There are 4.2 kcal per gram dry weight of plant tissue, on the average.) **Net primary productivity (NPP)** is gross primary productivity minus energy lost through plant respiration and is equivalent to chemical energy stored or accumulated per unit area per unit time. For most terrestrial vegetation, NPP = 30–70% of GPP. Finally, **detritus** is a synonym for litter; it is dead plant material. If we are concerned with an annual plant, then virtually the entire plant body can be considered litter at the end of one year—only the seeds remain living. In the case of a deciduous tree, some bark, twigs, flowers, aborted fruit, and all the leaf matter produced in spring and summer (except the fraction already eaten by herbivores) will be returned to the soil surface as litter.

About 10% of NPP is consumed by biophage herbivores for typical terrestrial vegetation. The percentage varies with vegetation: 2–3% for desert scrub or arctic/alpine tundra, 4–7% for forest, 10–15% for temperate grasslands with minimal grazing, and 30–60% for African grasslands or grasslands managed for domesticated animals. The percentage can also fluctuate from year to year. Episodic outbreaks of herbivores such as tent caterpillars, locusts, budworms, rabbits, voles, lemmings, or pathogenic microbes may raise the consumption to 50–100% of NPP. Limited data show that vegetation may withstand considerable loss to herbivores without lasting damage. For example, tree wood growth rate is not reduced until more than 50% of the canopy's leaf surface is lost. Certainly, however, death can and does result from some herbivore outbreaks, particularly if they persist for more than one growing season.

If NPP is calculated for seed production only, consumption by seed predators is typically well above 10% of NPP, and it often reaches 100%. It may be that herbivores exert their principal effect on vegetation in this way. Janzen (1970) and Connell (1971) concluded that intensive seed predation in the tropics may be the single most important factor regulating tree populations. In the tropical rain forest, a high diversity of tree species coexists, and neighboring trees are usually not the same species. Janzen and Connell proposed that the driving force behind seed dispersal was escape from predators. Predators associate the seed with the parent, and predator populations are high near the parent. Therefore, the closer the seed falls to the parent, the lower is its probability of escaping predation. The optimum for survival is some compromise distance from the parent, where the probability of an animal finding the seed is moderately low, yet the probability of the seed being dispersed that far is moderately high.

The escape hypothesis was proposed with very little empirical support

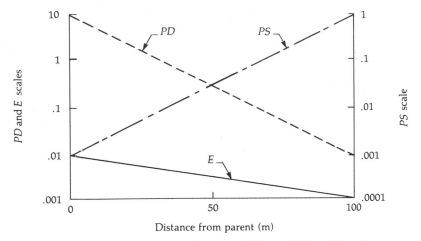

Figure 7-10   The expected distribution of plants *(E)*, based on the probabilities of seed dispersal to distances up to 100 m from the parent *(PD)* and of survival from predators *(PS)*. *E* = *PD* × *PS*. (Redrawn from S. P. Hubbell 1980. Published with permission from Oikos 35:214–229.)

(Hubbell 1980; Howe and Smallwood 1982; Clark and Clark 1984), so it is not surprising that data generated since that time both support the hypothesis and work against it. Hubbell's model of the escape hypothesis, a modification with quantified axes, is shown in Figure 7-10. It permits us to accept data showing that the probability of seedling survival in the tropics *is* measurable (not zero) near a parent and that, as a consequence, adult trees do often occur in clumps. Predation may thus be a contributing factor in the structure of vegetation but is not an overriding one.

Finally, abscission of tissue (loss of plant energy ultimately to saprophage herbivores) can range from 20% of GPP in trees to over 90% in annual herbs. Except for the tropics, the annual rain of litter is not entirely consumed each year. Litter turnover time (time for complete decay) is two to three years in the temperate deciduous forest and five to fifteen years in the boreal coniferous forest. Consequently, the litter layer increases from year to year until some episodic disturbance such as fire brings the system back to equilibrium, releasing the store of trapped nutrients. As discussed in Chapter 16, ground fires result in a flush of growth of the remaining vegetation, indicating that some of the released nutrients are quickly cycled back into the biota. The rate of litter consumption depends on soil temperature, soil moisture, and the chemical properties of the litter. Conifer foliage, for example, is acidic and high in lignin, and both properties depress the activity of bacterial populations.

Figure 7-11 summarizes the average drain by herbivores on GPP for a typical plant or plant community. If GPP is given a value of 100 units per square meter per year, then respiration accounts for 50 units, biophage herbivores for 15 units, litter (ultimately saprophage herbivores) for 23 units, and stored plant energy or growth for the remaining 12 units.

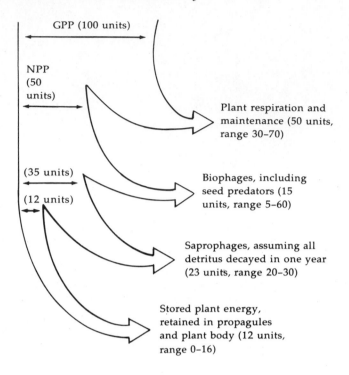

Figure 7-11 The relative consumption of gross primary productivity (GPP) and of net primary productivity (NPP) of a perennial plant by respiration and herbivores. Typical values.

GPP (100 units)

NPP (50 units)

Plant respiration and maintenance (50 units, range 30–70)

(35 units)

(12 units)

Biophages, including seed predators (15 units, range 5–60)

Saprophages, assuming all detritus decayed in one year (23 units, range 20–30)

Stored plant energy, retained in propagules and plant body (12 units, range 0–16)

What Determines the Intensity of Herbivory?

In Chapter 6 we saw that plant chemistry can determine palatability and modify herbivore behavior. Even beyond the presence of toxins, tannins, terpenes, alkaloids, and other specialized feeding deterrents, the basic substance of plant material is quite different from animal tissue (Southwood 1985). Protein content is low, and plant proteins generally differ from animal proteins. Carbohydrate content is high and concentrated in such unique, poorly digestible forms as lignin and cellulose. These differences present a dietary hurdle, in evolutionary terms, for potential herbivores. Southwood believes this is the reason that the fossil record reveals no evidence of vegetative herbivores for 70 million years—from late Devonian to early Permian time.

The above offers a "defensive chemistry hypothesis" as one answer to the question of what determines the intensity of herbivory. An alternative answer is the "nutrient stress hypothesis," which states that either herbivores avoid plant tissue low in such inorganic nutrients as nitrogen or that by feeding they put stress on the nutrient reserves of the plant, which results in nutrient-poor foliage and a subsequent avoidance of the plant by herbivores. Nitrogen content varies over two orders of magnitude, from plant to plant and organ to organ, ranging from 0.03% in wood to 7% in seeds and young tissue, with evergreens generally containing less tissue nitrogen than deciduous species (Mattson 1980). The literature contains many examples showing that diverse herbivores are able to discriminate between tissue containing different amounts of inorganic nutrients and that they prefer to consume the richest tissue, all other factors being equal.

Unfortunately, other factors are seldom equal. Chapter 6 documented a typical correlation between nutrient-rich tissue and the presence of high levels of compounds that deter feeding. Also, the "apparency" of plant tissue—its conspicuousness, predictability in time, or quantitative availability (see Courtney 1985)—is seldom equal from species to species, and this influences herbivores. Finally, the nutrient resources available to the plant determine not only tissue richness but also the kinds of herbivore defense compounds accumulated. A study of Alaskan tundra plants by Bryant et al. (1983) suggested that nitrogen-poor sites may support nitrogen-poor evergreens with carbon-based deterrent compounds (terpenes and phenolics), whereas nitrogen-rich sites may support denser cover by deciduous species competing for light whose tissues are richer in nitrogen and nitrogen-based deterrents (alkaloids and cyanogenic glycosides). All of these factors affected the behavior of vertebrate herbivores in the system. Similarly, a European timberline birch *(Betula pubescens)* growing on nutrient-rich sites responded to herbivory by producing more chemical defenses, while the same species on poor sites responded by lowering its nutrient content. In both cases herbivory declined for very different reasons (Houkioja and Neuvonen 1985).

The Impact of Herbivory: Function

Many ecologists (see reviews by Noy-Meir (1975), Lee and Inman (1975), and Mattson and Addy (1975)) have concluded that the role of herbivores in temperate ecosystems is rather minor, that herbivores do not transfer much energy from one trophic level to another, and that their function is merely to dampen yearly oscillations in plant productivity, thus performing a homeostatic role. But such conclusions may be a bit too conservative—there may be a more pervasive, positive influence of herbivores on plants. Robert Chew (1974) and S. J. McNaughton (1983) have written stimulating reviews showing what this influence might be.

Herbivory is usually thought of as a negative interaction on the part of plants, a one-way flow of benefits to consumers. However, an appropriate intensity of grazing can sometimes result in higher rates of plant photosynthesis and a resulting biomass that surpasses ungrazed, control plots. One reason for this response is that grazing reduces the canopy density to a point where all leaves receive sufficient light for efficient photosynthesis, whereas the ungrazed canopy has many heavily shaded and unproductive leaves. Also, shoot removal affects plant hormone balance, which leads to initiation and vigorous growth of shoots. This positive effect of herbivory has been called compensatory growth.

Compensatory growth can be achieved by mechanical (abiotic) clipping, but there is some evidence that animals provide an additional stimulus. Dyer and Bokhari (1976) found that grasshoppers *(Melanoplus sanguinipes)* feeding on blue grama grass *(Bouteloua gracilis,* a common central prairie plant) stimulated new shoot growth beyond a response from mechanical clipping. They hypothesized that the additional response was due to thiamine in grasshopper saliva being translocated down to the roots and there stimulating metabolic activity. Cattle are thought to have a similar effect (Reardon et al. 1972). (A later study by Detling and Dyer (1981), with a different species of grasshopper *(Brachystola magna),* showed reduced shoot growth of blue grama, as compared to mechanical clipping.) Thiamine is a B vitamin normally produced in leaves and necessary for root growth (Bonner 1940).

Semimutualistic interactions between plants and herbivores should not be surprising. Plants and their herbivores have coevolved over a period of thousands and millions of years. Surely they have come to terms with each other in ways as subtle and pervasive as the ways in which ecotypes have come to terms with their microenvironments (Chapter 3). As an example of plant-herbivore coevolution, Table 7-4 summarizes a series of plant defenses and herbivore adaptations that penetrate the defense. The example involves several closely related species of tropical legumes and associated weevils that feed within their pods.

In the overall view, of course, herbivory has a negative impact on plant growth, reproduction, and survival. The evolution of herbivore defenses, whether chemical, mechanical, or phenological, represents energy costs to the plant. The few cases where herbivores have been shown to have a stimulatory effect on plant growth are exceptional. After all, if one plant is eaten back and another that is not eaten overtops it, then it doesn't matter if new growth is stimulated, for the grazed plant has already lost in the competition for light. A plant has a selective advantage over its neighbor only as long as the cost of avoiding herbivory is greater than the cost of being (partly) eaten.

Summary

It is apparent that any given interaction between two organisms may well have attributes of other interactions. Herbivory, for example, can involve mutualism and amensalism. It is also clear that interactions fall on a continuum, with many intermediate situations among Burkholder's ten types. For example, there are intermediate stages between commensalism (benefit for one organism, no effect on the other, interaction not obligatory) and parasitism (benefit for one organism, negative effect on the other, obligate interaction on the part of the parasite). It is clear that the environment is chemically complex, with plant, animal, and microbial by-products and exudates playing an integral role and possibly mediating such interactions as amensalism, mutualism, and herbivory.

Examples of commensalism are tropical epiphytes and the nurse plant requirement of several desert plants. Little ecological information on epiphytes exists. Lichen epiphytes that contain nitrogen-fixing algal symbionts may contribute significant amounts of nitrogen to their host, so some epiphytic relationships can be mutualistic. Others (for example, the strangler fig) can be damaging.

Root grafts between trees result in protocooperation or parasitism, depending on the relative vigor of each tree. Root grafts between members of the same species are the usual case, but intergeneric grafts are known. Perhaps we should consider the root system beneath some plant communities to be a unit, a mutual benefit society which evens out the distribution of soil resources and photosynthate. Mycorrhizal connections between plant roots may contribute to the same end. It is possible that mycorrhizal relationships are partly responsible for the grassland-forest ecotone in North America.

Table 7-4 Possible plant defenses of some tropical legumes to seed weevil herbivores and corresponding adaptations by the seed weevil species to circumvent these plant defenses. (From "Coevolution of some seed beetles *(Coleoptera: Bruchidae)* and their hosts" by T. D. Center and C. D. Johnson, *Ecology* 55:1096–1103. Copyright © 1974 by the Ecological Society of America. Reprinted by permission.)

Plant defense	Weevil adaptation
Gum production by seed pods following penetration of first larva from egg mass; this may push off remaining eggs or drown or otherwise obstruct young larvae.	A period of quiescence in embryonic development in the egg until seed maturation is completed. Resistance to the gummy fluids. Eggs laid singly instead of in clusters.
Dehiscence, fragmentation, or explosion of pods, scattering the seeds to escape from larvae coming through the pod walls.	Oviposition on seeds only after they have been scattered. Attachment of seeds to one another or to the pod valve.
Production of a pod free of surface cracks, which prevents eggs from adhering. Also, indehiscent pods, excluding weevils, which oviposit only on exposed seeds.	Oviposition into the soft, fleshy exocarp. Chewing holes in the pod walls, in which eggs are then laid. Gluing the eggs to the pod surface.
A layer of material on the seed surface that swells when the pod opens and detaches the attached eggs.	Attachment of the eggs by anchoring strands to allow for substrate expansion. Entry of the larva through the pod into the seed before the pod opens.
Production of allelochemic substances.	Mechanisms for avoidance and detoxification.
Flaking of the seed pod surface, which may remove eggs laid on that surface.	Oviposition beneath the flaking exocarp. Rapid embryonic development or feeding on immature seeds so that entry occurs before flaking begins.
Immature seeds remaining very small throughout the year and then abruptly growing to maturity just before being dispersed.	Entry and feeding upon immature seeds. Delay of embryonic maturation until seeds are mature.
Seeds so small or thin that weevils cannot mature in them.	Utilization of seed contents more completely. Feeding on several seeds.

Herbivores can be subdivided into biophages, seed predators, and saprophages, among others, and the drain that each type places on plant productivity is quite different. Some researchers argue that because biophages remove only 10% of NPP (15% of GPP), they play a minor role in ecosystems, but other recent studies show that they may affect plant distribution and stimulate plant productivity. In addition, of course, many herbivores have long-recognized mutualistic relationships with plants (for example, pollination and seed dispersal).

PART III

THE COMMUNITY AS AN ECOLOGICAL UNIT

In each type of habitat, certain species group together as a community. Fossil records indicate that some of these groups (or very closely related precursors) have lived together for thousands and even millions of years. During that time, it is possible that an intricate balance has been fashioned. Community members share incoming solar radiation, soil water, and nutrients to produce a constant biomass; they recycle nutrients from the soil to living tissue and back again; and they alternate with each other in time and space. Synecologists attempt to determine what is involved in this balance between all the species of a community and their environment.

First, how is the community to be measured and how are the measurements summarized for maximum, useful information? Second, why do some communities change more rapidly over time than others? How accurately can we predict future changes and reconstruct the past? Are there community traits that transcend traits of the individual species making up the community? A synecological view asks such questions, and the answers take us to a level of complexity beyond autecology.

CHAPTER 8

COMMUNITY CONCEPTS AND ATTRIBUTES

T he term **community** is a very general one that can be applied to vegetation types of any size or longevity. It can, for example, be applied to one stratum of plants in a very local area, such as the herbs, woody seedlings, and mosses on the floor of a streambank forest; or to a very widespread, regional vegetation type; or to a transitory plot of vegetation undergoing rapid change in the species that compose it; or to very stable vegetation that has exhibited no significant change for hundreds of years.

An **association** is a particular type of community, which has been described sufficiently and repeatedly in several locations such that we can conclude that it has: (a) a relatively consistent floristic composition, (b) a uniform physiognomy, and (c) a distribution that is characteristic of a particular habitat. In terms of its basic importance to plant ecology and to the classification of vegetation, the association has been compared to the species of taxonomy. Just as a species is an abstract, somewhat artificial synthesis of many individual plants, so an association is a synthesis of many local examples of vegetation called **stands**.

An observant person can measure and describe vegetation in such a way that it can be subdivided neatly into associations (see Chapters 9 and 10). How natural are these units and how much can their existence be attributed to human bias and the need to classify? Some ecologists imply or state that associations are real, closed entities whose component species are interdependent. We think that this organismic view needs more evidence before it can be completely accepted.

Is the Association an Integrated Unit?

The Organismic View

One of the traits of an association, first accepted by the International Botanical Congress of 1910, is a relatively consistent floristic composition. Wherever a particular habitat repeats itself in a given region, the same cluster of associated

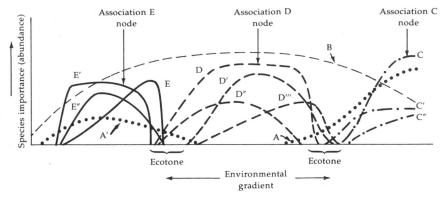

Figure 8-1 Patterns of species importance (abundance) along an environmental gradient as predicted by the discrete view of associations. Clusters of species (C's, D's, and E's) show similar distribution limits and abundance peaks; each cluster of associated species defines an association. Ecotones are narrow. A few species (A and A') have sufficiently broad ranges that they are found in adjacent associations, but in low numbers. A few other species (for example, B) have very broad ranges and are ubiquitous.

taxa is found. This does not mean that every species repeats itself, nor even that the majority of taxa are repeated. Some species are extremely widespread, with broad tolerance ranges, and these can be found in many habitats and in many associations. Other species may have narrower range limits, but nevertheless a few individuals can be found beyond the normal limits, and they are occasional members of many communities.

According to the procedures of one method of community classification, after accidental and ubiquitous taxa are removed from consideration, certain species will remain that show a large degree of association with each other and with a particular habitat. Associations are defined by the presence of such characteristic or indicator species, and a particular stand is placed in an association if it contains a significant fraction of such species. This method—the *table method* of Braun-Blanquet and others—is described more fully in Chapters 9 and 10.

If clusters of species do repeatedly associate together, that is indirect evidence for either positive or neutral (not negative) interactions between them. Such evidence favors a view that communities are indeed integrated units—that the whole is somehow greater than the sum of its parts, much like an organism is greater than the sum of its cells, tissues, or organs. Some of the interactions described in Chapters 6 and 7, such as mycorrhizae, are also evidence that communities are units. Clements (1916, 1920) metaphorically equated associations with organisms. The pattern of species distribution and abundance predicted by this organismic view is diagrammed in Figure 8-1. The species in an association have similar distribution limits along the horizontal axis, and many of them rise to maximum abundance at the same points (noda). The ecotones between adjacent associations are narrow, with very little overlap of species ranges, except for a few ubiquitous taxa found in many associations.

The Continuum View

In 1926, Henry Gleason published a carefully written paper entitled, "The individualistic concept of the plant association." In a letter written 27 years later, he recounted the furor that his proposal generated and the never-never world into which his ideas were cast for some time. Gradually, his ideas did gain acceptance. The Ecological Society of America cited him as a "Distinguished Ecologist" in 1953 and as an "Eminent Ecologist" in 1959.

Gleason (1953) sampled forest vegetation along a north–south gradient in the Midwest and concluded that changes in species abundance and presence occurred so gradually that it was not practical to divide the vegetation into associations. Even within a relatively homogeneous region, there were subtle but important differences in the vegetation from one plot to another.

> The sole conclusion we can draw from all the foregoing considerations is that the vegetation of an area is merely the resultant of two factors, the fluctuating and fortuitous immigration of plants and an equally fluctuating and variable environment. As a result, there is no inherent reason why any two areas of the earth's surface should bear precisely the same vegetation, nor any reason for adhering to our old ideas of the definiteness and distinctness of plant associations. . . . Again, experience has shown that it is impossible for ecologists to agree on the scope of the plant association or on the method of classifying plant communities. Furthermore, it seems that the vegetation of a region is not capable of complete segregation into definite communities, but that there is a considerable development of vegetational mixtures. . . .

Gleason's conclusions have been supported by more extensive and sophisticated studies in such widely different vegetation types as desert scrub in Nevada (Billings 1949), upland forest in Wisconsin (Curtis and McIntosh 1951), montane forest in the Smokey Mountains in North Carolina and Tennessee (Whittaker 1956), and conifer forests in the Siskiyou Mountains of California (Whittaker 1960). These studies show that neither dominance of single taxa nor presence and abundance of groups of species change abruptly along an environmental gradient; noda do not exist (Figure 8-2).

It is now apparent that the method used to sample vegetation will determine whether associations appear as distinct units or as arbitrary segments along a continuum. If sampling is subjective, searching for stands that show the presence of differential species, then the discrete view will be supported, for only similar stands will be sampled, and intermediate stands will be ignored. It is possible that if only climax stands are sampled, many intermediate stands will also be ignored. If sampling is done less subjectively, ecotonal stands will be included, and all the stands will be seen to lie along a continuum. Exceptions to the continuum view do exist, of course, where there are sudden discontinuities in the environment, such as a change in soil parent material, a change in elevation or slope aspect, the advance of fire, or the presence of landslide rubble.

The individualistic nature of species distribution is indirect evidence that communities are not larger than the sum of their parts. It suggests that the level of interactions and interdependence is relatively low, or at least nonspecific. For

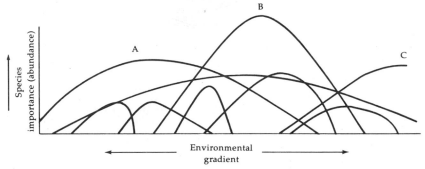

Figure 8-2 Patterns of species importance (abundance) along an environmental gradient as predicted by the continuum view of associations. Noda do not exist. If associations are recognized, based on peaks in abundance of dominant species such as A, B, or C, it can be seen that these associations are merely arbitrary segments along the continuum.

example, forest floor herbs may be well adapted to the phenology, shade cast, and leaf chemistry of an overstory canopy, but the exact species composition of that overstory is not critical. This was shown in a large-scale, natural "experiment" 50 years ago. Between 1906 and 1930, virtually all of the chestnuts (*Castanea dentata*) in the eastern deciduous forest were killed by chestnut blight. Woods and Shanks (1959) and McCormick and Platt (1980) showed that the canopy openings were largely filled by crown expansion or sapling growth of previously associated species (mainly oaks and red maple), and there was no additional major community upheaval. What had been an oak-chestnut community became an oak community, but the understory was little changed. Studies of American elm (*Ulmus americana*) forests struck by Dutch elm disease reveal similar modest adjustments (Parker and Leopold 1983).

Recent studies of understory forest herbs beneath fir forests of Montana (McCune and Antos 1981), deciduous forests of the Lake States (Rogers 1981), and hemlock forests of eastern North America (Rogers 1980) all demonstrate that understory changes are poorly correlated with overstory composition and, further, that understory species change in abundance in an individualistic way. Recall from Chapter 6 that Goldberg and Werner (1983) proposed that plant competition has a non-species-specific nature; these understory studies support their hypothesis. An elegant, 2-year-long field manipulation study of tundra plants in the Brooks Range of Alaska (Chapin 1985) similarly showed that 21 associated species were individualistically affected by such factors as temperature, soil nutrients, shade, competitive interactions, combinations of all these, and evolutionary history.

Associations can be recognized and in some way defined in the field, but we must appreciate their subjective, arbitrary boundaries, the nature of which depends on the bias of the investigator and the sampling methods. Whether associations are real or not, however, stands of vegetation certainly do exist. These stands exhibit collective or emergent attributes beyond those of the indi-

Table 8-1   Some characteristics of plant communities.

Physiognomy Architecture Life forms Cover, leaf area index (LAI) Phenology **Species composition** Characteristic species Accidental and ubiquitous species Relative importance (cover, density, etc.) **Species patterns** Spatial Niche breadth and overlap **Species diversity** Richness Evenness Diversity (within stands and between stands)	**Nutrient cycling** Nutrient demand Storage capacity Rate of nutrient return to the soil Nutrient retention efficiency of the nutrient cycles **Change or development over time** Succession Stability Response to climatic change Evolution (?) **Productivity** Biomass Annual net productivity Efficiency of net productivity Allocation of net production **Creation of, and control over, a microenvironment**

vidual populations that make them up. Some of these attributes (Table 8-1) will be examined in the rest of this chapter.

Some Community Attributes

Physiognomy, Species Composition, and Spatial Patterns

Physiognomy is a combination of the external appearance of vegetation, its vertical structure (its architecture or biomass structure), and the growth forms of its dominant taxa. Chapter 9 discusses methods of sampling, measuring, and numerically describing the physiognomy of communities. Physiognomy is an emergent trait of communities. It could not, for example, be accurately estimated just from a list of all the taxa present within the community.

Life form (see Chapter 1) includes such plant features as size, life-span, degree of woodiness, degree of independence, general morphology, leaf traits, the location of perennating buds, and phenology. Vertical structure refers to the height and canopy coverage of each layer within the community.

Canopy coverage can be expressed as the percentage of ground covered by the canopy, when the edges of the canopy are mentally projected down to the surface. It may also be expressed as **leaf area index (LAI)**. All leaves are collected that project into a column of space that lies above an area of ground (say, 1 m^2), and their cumulative surface area is measured. Then,

$$LAI = \frac{\text{total leaf area, one surface only}}{\text{unit ground area}} \qquad \text{(Equation 8-1)}$$

Many crops, such as corn, have an LAI of about 4, meaning that for every square meter of ground, 4 m^2 of leaves lie above it. Some examples of LAI for natural vegetation types are given in Table 8-4, which appears later in this chapter.

The species composition of a community is also extremely important, because communities are partly defined on a floristic basis. Several communities may have similar physiognomies yet differ in the identity of dominants or other species. The abundance, importance, or dominance of each species can be expressed numerically, so that different communities can be compared on the basis of species similarities and differences (see Chapter 9).

The relative spatial arrangement of species within a community is another community trait. As described in earlier chapters, individuals within a species or individuals of different species may be distributed at random with respect to each other, clumped (positive or neutral interactions), or overdispersed (negative interactions).

The importance of species interactions and interdependence to a discrete view of the community has already been discussed in this chapter, and that assumption will be touched on again in Chapter 11. One theory of community stability discussed in Chapter 11 suggests that stable, long-lasting communities exhibit more species interactions and more species than transient, seral communities. If this is so, then the niches of species in stable communities must be narrower, with less overlap in niche boundaries, than those of less stable communities. Some ecologists have tried to measure niche breadth and overlap to test such hypotheses (Huey and Pianka 1977; Parrish and Bazzaz 1976; Pickett and Bazzaz 1976), but the data are so sparse that generalities cannot yet be made.

Part of a species' niche may include its unique metabolism. For example, flowering plants exhibit three different metabolic pathways of photosynthesis, C_3, C_4, and CAM, each of which is best fitted to different environmental conditions or to different species' phenologies (see Chapter 14). There have been some recent attempts to include such metabolic traits in desert and chaparral community descriptions (Mooney and Dunn 1970; Johnson 1976).

Community descriptions based on physiognomy, life form, niche overlap, and other functional traits (as opposed to taxonomic traits such as the identity of species in the communities) are useful because they permit comparison of widely disjunct stands that have little or no floristic similarity. These comparisons often show a convergence of vegetation types, given a similar macroenvironment. Chaparral stands in southern California and Chile, for example, have little floristic similarity even at the family level, but they share a similar mediterranean-type climate and exhibit some striking similarities in vegetation (see also later pages of this chapter).

Species Richness, Evenness, and Diversity

Species richness is simply the number of species in some area within a community. Each species is not likely, however, to have the same number of individuals. One species may be represented by 1000 plants, another by 200, and a third by only a single plant. The distribution of individuals among the species is called **species evenness**, or species equitability. Evenness is maximum when all species have the same number of individuals. **Species diversity** is a combination of richness and evenness; it is species richness weighted by species evenness, and there are formulae that permit the diversity of a community to be expressed in a single index number.

It is important to emphasize that richness and diversity are quite different. Although richness and diversity are often positively correlated, environmental gradients do exist along which a decrease in richness is accompanied by an increase in diversity (Hurlbert 1971). Community A, with five species but uneven numbers of individuals in each species, has a lower diversity than community B, with four species that have a very similar number of individuals in each. Community A has a higher species richness, however. Biologically, diversity is the measure of population heterogeneity of a community.

Many simplifications are made in every calculation of species diversity (Peet 1974). In the equations, all individuals of any species are equal. This may not be true, especially in regard to animals of different sex or of different developmental stages, or to plants of different phenological stages (dormant, full-leaf, flowering, juvenile, senescent adult), different sizes (seedling, sapling, suppressed adult, overstory adult), or different ecotypes. For this reason, diversity is often calculated for each stratum in a community rather than for the entire community. It is also assumed that all species are equally different, whether the difference is in morphology or niche breadth. This assumption is generally made even when it is unlikely to be valid, simply because we have no way to quantify species differences. Finally, diversity is sometimes expressed as numbers of individuals, biomass, or productivity (see Whittaker 1975), as canopy cover, and sometimes in other ways. Obviously, plant diversity from community to community can be compared only if the same units are used. Some units are more appropriate to plants, and some to animals; possibly there is no one unit suitable for calculation of an ecosystem-wide diversity index (Hurlbert 1971).

A further assumption, and one that is made in any statistical procedure, is that the sample of the community was large enough to represent the community adequately. The data on numbers of species and the relative abundance of each are dependent on sample size. Given a large enough sample, there will be some species with few individuals, some species with many, and many species with an intermediate number. The data will form a bell-shaped curve on a log-normal plot (Figure 8-3). A small sample, however, will probably not include the full spectrum of rare species. Notice, in Figure 8-4, how the relative abundance of lepidopteran insect species shifted and began to approach a bell-shaped curve as the sample collecting periods increased from less than 1 year to 4 years. The truncated appearance of small sample size curves is an artifact, then, and the

Figure 8-3 (a) The log-normal distribution of species, ranked according to the number of individuals in each species. Note that individuals along the horizontal axis are in categories of geometrically increasing numbers, plotted here on a log basis. Most species in this hypothetical example have a moderate number of individuals in the sample, about one to thirty individuals each, and very few species are rare or abundant. (b) A large sample of British birds, which approaches Preston's log-normal curve. [(a) From "The commonness and rarity of species" by F. W. Preston, *Ecology* 29:254–283. Copyright © 1948 by the Ecological Society of America. Reprinted by permission. (b) From Williams 1964 (*Patterns in the Balance of Nature*, by permission of Academic Press) and Krebs 1972 (*Ecology: The Experimental Analysis of Distribution and Abundance*, Harper and Row, Publishers, Inc., New York).]

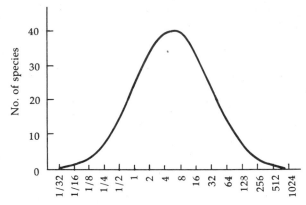

(a)

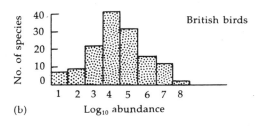

(b)

Figure 8-4 Log-normal distributions of lepidopteran species abundances as determined from sampling at Rothamsted Station, England, for periods of (a) 1/8 year, (b) 1 year, and (c) 4 years. As sample size increased, from 492 to 16,065 individuals trapped, the distribution approached Preston's log-normal curve. [From Williams 1964 (*Patterns in the Balance of Nature*, by permission of Academic Press) and Krebs 1972 (*Ecology: The Experimental Analysis of Distribution and Abundance*, Harper and Row, Publishers, Inc., New York).]

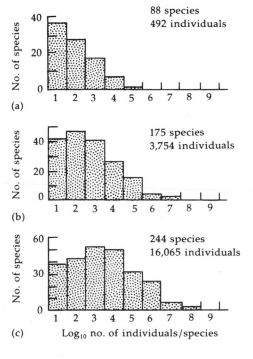

truncated end is a "veil line" that hides additional categories of rarity (Preston 1948).

Diversity Indices   Several indices have been proposed over the past six decades (Auclair and Goff 1971, Hill 1973, Peet 1974, DeJong 1975; see Table 8-2), but we shall emphasize only two here: **Simpson's index** (Simpson 1949) and **Shannon-Wiener's index** (Shannon and Weaver 1949; sometimes called the *Shannon-Weaver index*).

Simpson's index reflects dominance because it weights (is more sensitive to) the most abundant species more heavily than the rare species. The advantage of this attribute is that index values are unlikely to vary much from sample to sample, because it is the rare species that would vary from place to place more than the common ones. Its formula is:

$$C = \sum_{i=1}^{s} (p_i)^2 \qquad \text{(Equation 8-2)}$$

where C is the index number, s is the total number of species in the sample, and p_i is the proportion of all individuals in the sample that belong to species i. Common variations, which measure diversity rather than dominance, include:

$$D = 1 - C \qquad \text{(Equation 8-3)}$$
$$D' = \frac{1}{C} \qquad \text{(Equation 8-4)}$$

The Shannon-Wiener index formula is:

$$H' = -\sum_{i=1}^{s} (p_i)(\ln p_i) \qquad \text{(Equation 8-5)}$$

and a common variation is:

$$H'' = 2^{H'} \qquad \text{(Equation 8-6)}$$

If proportions are expressed on base 10 instead of 2 (the abbreviation ln above), one can simply convert the summation value to base 2 by multiplying it by 3.32.

H' is thought to represent the "uncertainty" or "information" of a community. The more variable its composition, the more variable (the more uncertain or unpredictable) each sample of it would be. The index units have been called *bits, bels, decits,* or *digits* per individual, referring to the binomial units of information used in computers. H' varies from 0, for a community of one species only, to values of 7 or more in rich forests such as those of the Siskiyou Mountains of Oregon and California (DeJong 1975). The deciduous forest of the eastern United States is of intermediate diversity, with H' values for tree species ranging only from >3.0 for the mixed mesophytic forest in the Cumberland and Allegheny Mountains to values <2.0 at the western and northern edges (Monk 1966).

Because the formulae are different, the absolute values of Simpson's and Shannon's indices will differ for the same community. Table 8-3 and Figure 8-5 compare variations of the indices for two herb communities.

Table 8-2 Some of the dominance and diversity indices proposed and used in the twentieth century. For additional details, see Auclair and Goff (1971). N_o = total number of species in the sample, X_o = total number of individuals in the sample, X_i = number of individuals in species i, ln = natural logarithm to base 2.

Index name or description	Formula
Species per unit area	N_o
Species/(individuals/species)	$N_o/(X_o/N_o)$
Species/individuals	N_o/X_o
Species/ln individuals	$N_o/\ln X_o$
ln species/ln individuals	$\ln N_o/\ln X_o$
Species/sq. rt. individuals	$N_o/\sqrt{X_o}$
Simpson dominance index	$\Sigma(X_i/X_o)^2$
Shannon information diversity	$-\Sigma (X_i/X_o) \ln (X_i/X_o)$
McIntosh density diversity	$1 - \sqrt{\Sigma(X_i/X_o)^2}$

Table 8-3 Diversity indices calculated for community samples A and B in Figure 8-5. Note that only the Simpson index (C) shows community A to be the more diverse, reflecting that index's high sensitivity to dominance by one species and insensitivity to low numbers of other species.

Index	A	B
Simpson (C)	0.38	0.20
$D = 1 - C$	0.62	0.80
$D' = 1/C$	2.66	4.92
Shannon (H')	1.78	2.31
$H'' = 2^{H'}$	3.43	4.96

What Does Diversity Signify? A lot of attention has been paid to species diversity in the ecological literature: first, in the development and comparison of several formulae; second, in searching for general trends in diversity along environmental gradients; and third, in trying to attribute functional explanations for diversity gradients and some ecological value that diversity might convey about communities. Despite all this attention, we think it best to treat an index of diversity as simply one descriptive attribute of a community, on a par (for example) with a list of all species found in the community, or an estimate of tons of biomass above the ground, or the leaf area index.

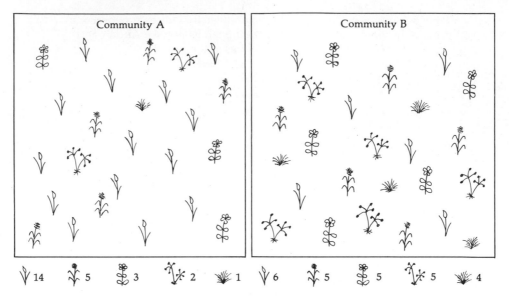

Figure 8-5 Diagrammatic representation of sample plots from two different herbaceous communities, (a) and (b), each with the same five species but in different abundances. Community (a) is heavily dominated by one species, whereas community (b) has a more even distribution of species abundances. The number of individuals in each species is summarized below the plots. Diversity indices are shown in Table 8-3. (Courtesy of M. Rejmanek.)

Generally, there is a gradient of increasing species diversity (and richness) from the poles to the equator, and from high elevations to low elevations. These gradients follow complex environmental gradients of increasing warmth, among other factors. It has been stated that diversity increases as any particular stress lessens (Krebs 1972), but this is not true for all stress gradients. An aridity stress gradient may show the opposite trend, with greater diversity in semiarid grassland and desert than in savanna, woodland, or forest (Hurlbert 1971). Diversity has been equated with productivity and stability, but some very diverse semiarid grasslands and deserts are low in terms of productivity and stability, and there are many other exceptions (van Dobben and Lowe-McConnell 1975). Diversity has been taken as a reflection of the many interactions that supposedly characterize complex communities. However, one model of community structure that assumes no species interactions shows that species diversity "may be maintained in spite of, rather than because of, such interactions" (Caswell 1976).

Maintenance of high diversity appears to require episodic, random (**stochastic**) disturbance. Very stable, regionally extensive, and homogeneous communities exhibit lower species diversity than communities composed of a mosaic of patches disturbed at various times in the past by wind throw, fire, disease, etc. Following disturbance, diversity increases with time up to a point where dominance by a few, long-lived, large-sized species reverses the trend, and diver-

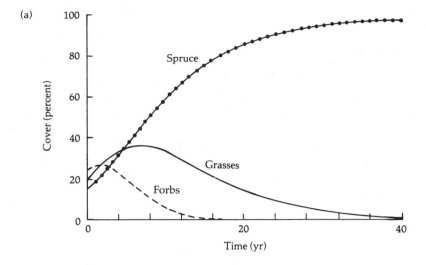

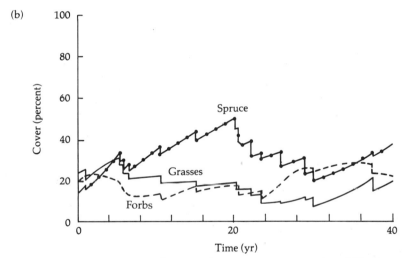

Figure 8-6 Computer simulations of canopy cover over time following (a) a single large disturbance at time 0 and a subsequent period of 40 years without any disturbance and (b) stochastic, small-scale avalanche disturbance every fifth year over the same period. The coexistence pattern of forbs, grasses, and spruce shown in (b) will continue indefinitely. (Redrawn from Rejmanek 1984.)

sity falls thereafter. We can see from a computer simulation (Figure 8-6) of spruce forest in the Krkonoše Mountains of Czechoslovakia that episodic avalanche disturbance maintains three diverse life forms and groups of species indefinitely, whereas absence of disturbance leads to loss of two-thirds of the groups within 40 years.

It is fair to say that the significance of species diversity is not well understood at this time. This deficiency may be due to our lack of a good model of what a community is in a theoretical sense. Whittaker (1975) has shown that any of several theoretical models of species abundance and diversity can be supported with data from actual communities, and that there is as yet no single model that fits all communities. It is also possible that an appropriate formula to express species diversity has not yet been devised.

Nutrient Cycles and Allocation Patterns

Sixteen elements are known to be required for normal growth and development of all higher plants: carbon, hydrogen, oxygen, phosphorus, potassium, nitrogen, sulfur, calcium, magnesium, iron, boron, manganese, copper, zinc, chlorine, and molybdenum. A few other elements are required by certain plant groups, but not by most plants. Sodium, for example, is required in trace amounts by plants with the C_4 pathway of photosynthesis; silicon is required by horsetails (*Equisetum* sp.) (Epstein 1972). Some elements—heavy metals such as lead or gold, and sometimes sodium—are accumulated by certain plants to the point where their foliage is poisonous to livestock, but these elements are not required for normal plant metabolism.

Communities differ in their utilization of certain essential nutrients (that is, how much of each element is required for normal growth, or at least how much is absorbed from the soil solution and translocated to leaves and growing points). They also differ in the rate at which the nutrients are returned to the soil in litter fall and the efficiency of the plant-soil-plant cycle. Early successional communities, for example, may require little soil nitrogen, accumulate very little of any nutrient in their tissues, and return nutrients rapidly to the soil, but erosion removes a large fraction of the returned nutrients because of their low cover or seasonal absence. Climax communities may require greater quantities of some nutrients, store great quantities of nutrients in wood, and return only a small fraction to the soil in leaf litter, but prevent erosive losses by shielding the soil with a permanent, closed canopy (see Chapter 11). Thus, climax communities have fewer leaks in nutrient cycles and more efficiently hold the nutrients in the plant–soil–plant cycle.

A few remarks about the carbon cycle will illustrate major community differences. If vegetation types are ranked according to the amount of **standing** (above ground) **biomass** per hectare, they range from desert communities of only 100 kg ha^{-1} to tropical rain forests of 500,000 kg ha^{-1} (Figure 8-7). Generally, greater biomass indicates greater leaf area, which means that more radiant energy can be trapped each year and a faster growth rate will result, with greater net productivity. **Net productivity** ranges from 100 kg ha^{-1} yr^{-1} for desert communities to 40,000 kg ha^{-1} yr^{-1} for tidal zone, mangrove, marsh, and swamp communities (Figure 8-7). **Efficiency**, the fraction of radiant energy converted into kilocalories of tissue, ranges from 0.04% in deserts to 1.5% in tropical rain forests (Table 8-4).

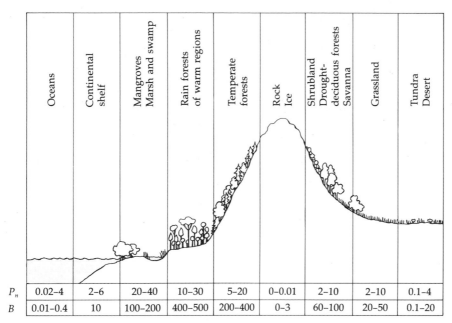

	Oceans	Continental shelf	Mangroves Marsh and swamp	Rain forests of warm regions	Temperate forests	Rock Ice	Shrubland Drought-deciduous forests Savanna	Grassland	Tundra Desert
P_n	0.02–4	2–6	20–40	10–30	5–20	0–0.01	2–10	2–10	0.1–4
B	0.01–0.4	10	100–200	400–500	200–400	0–3	60–100	20–50	0.1–20

Figure 8-7 Differences in net primary productivity (P_n) and in standing biomass (B) of various vegetation types. Data are in metric tons per hectare for B and metric tons per hectare per year for P_n. (From W. Larcher 1975. *Physiological Plant Ecology*. By permission of Springer-Verlag.)

Table 8-4 The leaf area index (LAI, m² leaf surface per m² ground) and efficiency of net productivity for major vegetation types. Efficiency is calculated by converting grams of net production per unit area to kilocalories per unit area (there are approximately 4.2 kcal in each gram dry weight of plant tissue), then dividing by the total radiation received in a year. Only radiation between the wavelengths of 400 and 700 nm is entered in the calculation, because these are the only wavelengths usable for photosynthesis. (From W. Larcher 1975. *Physiological Plant Ecology*. By permission of Springer-Verlag. See also pp. 247, 255, 450.)

Vegetation type	LAI	Efficiency (%)
Tropical rain forest	10–11	1.50
Deciduous forest	5–8	1.00
Boreal conifer forest	9–11 (7–38)	0.75
Grassland	5–8	0.50
Tundra	1–2	0.25
Semiarid desert	1	0.04
Agricultural	3–5	0.60

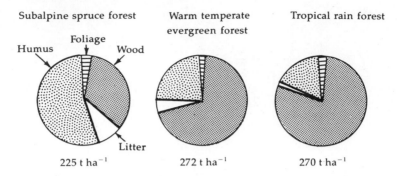

Figure 8-8   Content of organic carbon in the foliage, wood, litter, and humus for three forest types. (From W. Larcher 1975. *Physiological Plant Ecology*. By permission of Springer-Verlag.)

The low efficiency and net productivity of desert communities does not mean that the component species are themselves inefficient or capable of only very low photosynthetic rates. Net productivity is a community trait and to some extent a climatic one, for it is affected by the leaf area index, temperatures during the growing season, distribution of rainfall, soil moisture storage, and the length of the growing season. If net productivity or efficiency is expressed on a 12-month basis, it is clear that the tropical rain forest, with a leaf area index of 10–11 and warm temperatures and adequate moisture all year, should indeed have a greater net productivity and efficiency than a desert, with a leaf area index of 1 or less and a growing season (based on water supply) of only a few months. However, the photosynthetic rates of individual desert shrubs and shrubs of mesic forests are very similar (Barbour 1973*b*).

The **allocation** of net productivity to different organs also differs from community to community. Grassland communities channel much of their energy to below ground biomass, scrub communities less so, and forests even less so. If we compare the distribution of carbon in a subalpine conifer forest, a temperate broadleaf forest, and a tropical rain forest (Figure 8-8), there are significant differences. Most carbon in the subalpine forest is in the form of humus atop or in the surface soil. This is because the acidic, cold soil is not favorable to decomposers. Litter half-life is 10 or more years (Whittaker 1975). Most carbon in the tropical rain forest, in contrast, is locked up as inert wood. This means that leaf litter has a major function in the tropical rain forest, for it represents the only mobile, cycleable part of the nutrient bank. This is one reason why tropical soils cleared of forest vegetation soon become infertile. The annual litter rain has been stopped, and the limited soil reserves are soon depleted by crops or leached from the soil. The litter decay rate is very rapid beneath the rain forest canopy; litter half-life is a fraction of a year. Nutrient and productivity relationships of plant communities are discussed further in Chapters 12 and 13.

Change Over Time

1–500 Years: Succession   All communities are dynamic, changing entities. Change, however, is a relative term, and the time framework must be stated.

Plant communities that exhibit no directional change for several centuries are considered to be in equilibrium with their environment, and are called **climax** communities (see Chapter 11). Other communities may exhibit significant changes in such a time period. Some species decline in abundance and may disappear from the site; invasive species may increase in abundance; and the vegetation type itself may change, for example, from a meadow to a forest, or from a pine forest to a hardwood forest. Such transient communities are called **successional**, or **seral**, communities. It may be possible to recognize and describe an entire sequence of successional communities that replace each other on one site, finally culminating in a climax community. This sequence of communities is called a **succession**, or **sere**.

Stability   Communities differ in their response to disturbance or stress. Intuitively, one can use terms such as *stable* and *fragile* to describe how easily communities are perturbed by stress. More technically, however, *stability* is a complex term that includes several distinctly different qualities (Leps et al. 1982, Pimm 1984).

One component of stability is **resistance**, which is the ability of a community to remain unchanged during a period of stress. Resistance appears to be characteristic of vegetation dominated by long-lived perennials with moderately high species diversity and many paths of interdependence (links or connections) between the component species. Climax communities generally fit this definition. Such communities have considerable inertia, changing slowly even in the face of regional climatic change. Evidence from the fossil record, for example, indicates that vegetation change lagged 1000–3000 years behind a major warming at the Pleistocene–Holocene boundary 13,000–16,000 years ago (Cole 1985).

Resilience is a second component of stability; this is the ability of a community to return to normal, or the rate at which this occurs, following a period of stress or disturbance. Resilience appears to be characteristic of vegetation dominated by short-lived, rapidly maturing species of low diversity and with few interdependence links among them. Early seral communities fit this definition. Climax communities require considerable time to recover from destructive disturbance—thus they have low resilience—but they have great resistance to less destructive stress.

A third component is **variance**, which is the ability to exhibit patches of variable abundance in some of the component species. Many climax communities appear homogeneous on a large scale, but at a very local scale they show considerable variance. A fourth component is **persistence**, which is the ability to remain relatively unchanged over time. Some persistent communities are neither resistant nor resilient but owe their continued existence to a protected, buffered environment. The redwood forest along the northern California coast is a good example; it exists in a very narrow, foggy, temperate belt.

Thousands of Years: Climatic Change   Succession is thought to be driven by biological interactions, such as competition, which occur in the microenvironment created by the plants themselves; it is not driven by macroclimatic change. Climate is assumed to be constant when a succession is investigated and described.

Climate, however, has not been a constant throughout time and it has always undergone significant fluctuations. Weather records, such as those taken by the United States Weather Bureau, only exist for the past 100 or so years. Nevertheless, even these records show statistically significant (though minor) changes. For example, rainfall has been declining and temperatures rising in the southwestern United States (Hastings and Turner 1965).

The climate before the nineteenth century can be inferred by several methods (see, for example, Flint 1957 and Strahler and Strahler 1974). One method, applicable to the southwestern United States, is the analysis of tree rings (**dendrochronology**). Climate of the past 8200 years has been estimated by examining sequences of growth ring widths in the wood of living and dead trunks of bristlecone pine (*Pinus longaeva* = *P. aristata*). The data indicate cycles of warm, arid periods followed by cool, wet periods (Figure 8-9) (LaMarche 1974; Ferguson 1968). The period of time shown in Figure 8-10 begins with an exceptionally arid period called the **Xerothermic period**, which started approximately 8000 years ago and ended approximately 4000 years ago. This period is thought to be responsible for the present distribution limits of many southwestern vegetation types and communities (Axelrod 1977). See Figure 8-10 for average temperatures during geologic time.

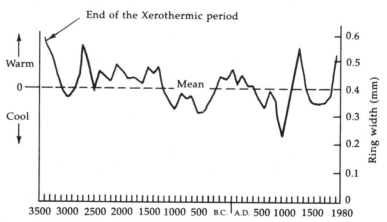

Figure 8-9   Average ring widths in bristlecone pine (*Pinus longaeva*) trunks dated back to 3500 B.C., near the end of the Xerothermic period. Positive departures from mean ring width (0.4 mm) indicate temperatures warmer than average during April–October; negative departures indicate cooler conditions. (From V. C. LaMarche, Jr. *Science* 183:1043–1048. Copyright © 1974 by the American Association for the Advancement of Science. By permission.)

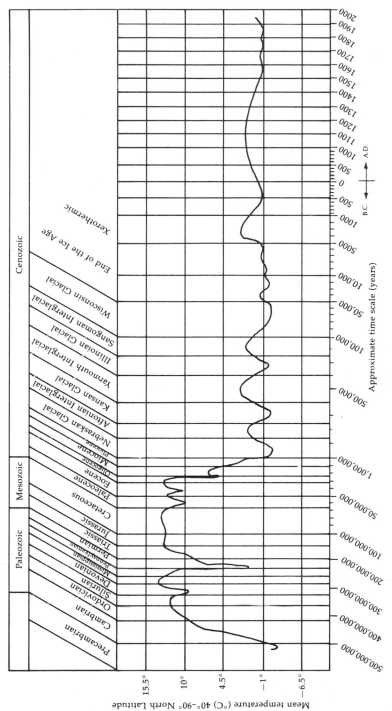

Figure 8-10 Average temperatures for north temperate and polar latitudes during geologic time. The time scale is distorted to show more detail for the last 1 million years. Notice the four glacial advances separated by warm interglacial retreats. The Xerothermic period (a very arid period) is shown here to peak at about 3000 B.C., which is equivalent to 5000 years before the present. (From Dorf 1960. Reprinted by permission of *American Scientist*, journal of Sigma Xi, The Scientific Research Society.)

173

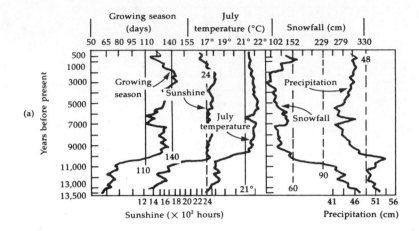

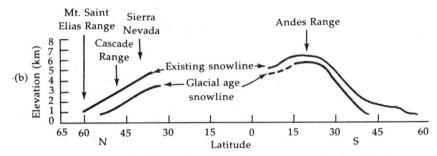

Figure 8-11   (a) A 13,500-year record of growing season length, hours of bright sunshine annually, mean July temperature, annual snowfall, and precipitation during the growing season at Kirchner Marsh, Minnesota. (b) The effect of climatic change over the past 13,000 years on the elevation of permanent snow (the upper limit of alpine vegetation) in western mountain ranges of North and South America. [(a) From R. A. Bryson. *Science* 184:753–759. Copyright © by the American Association for the Advancement of Science. By permission. (b) From Flint 1957 (*Glacial and Pleistocene Geology*, copyright 1957 John Wiley and Sons, Inc., reprinted by permission) and Strahler and Strahler 1974 (*Introduction to Environmental Science*, Hamilton Publishing Co., Santa Barbara, CA).]

Relatively small changes in global temperatures have created striking changes in climate and vegetation. Most careful analyses suggest an average global surface temperature difference between full glacial eras and the present of only 4–6°C (Bryson 1974). This temperature difference, however, has been correlated with changes in cloudiness, rainfall, length of the frost-free season, and other factors that have a large impact on vegetation (Figure 8-11).

The nature of vegetational change since the Ice Age can be documented in several other ways. In the arid southwestern United States, the dried remains of plants cached in underground middens by wood rats (*Neotoma*) have in some cases remained intact and identifiable for thousands of years. Since the foraging activity of wood rats is restricted to a rather limited radius around the nest, the composition of the plant material gives some indication of the nearby vegetation at the time the midden was formed. The plant material can be carbon dated. This kind of evidence permitted Wells and Berger (1967), among others, to determine the changing elevation of vegetation zones following the retreat of the last glaciation in what is now the Mojave Desert.

A more widely used method of documentation utilizes pollen grains, which accumulate at the bottom of slowly filling lakes or ponds. As pollen is shed and transported by wind, some falls onto a lake surface, sinks to the bottom, and becomes incorporated with silt and organic matter into sediment. Pollen of many species is resistant to decay in the anaerobic, cold sediment and may remain intact for thousands or millions of years. The family, genus, or even species of the pollen can be determined under the microscope. A core of sediment, then, reveals a chronological sequence of surrounding vegetation; the deeper the pollen occurs in the sediments, the older it is.

Within limits, ecologists assume that the abundance of pollen in the core is related to the abundance of species in the surrounding vegetation. This assumption, of course, is applied only to wind-pollinated plants such as sedges, grasses, most trees, many shrubs, and some forbs (herbaceous plants other than grasses). The assumption is upheld by measurements of modern pollen "rain." Griffin (1975), for example, found that modern pollen rain in Minnesota plant communities correlated very well with the nearest community type. Pollen is not carried in significant quantities further than 50 km in forested regions (Livingstone 1968). If we consider a pond surrounded by vegetation in a radius of 50 km, then the pond can receive pollen from a total area of 7850 km^2. Most vegetation is not uniform over such an area, but it is likely that the pollen profile will give a good general picture of regional vegetation.

A **pollen profile** constructed from sediment beneath a Nova Scotia lake is shown in Figure 8-12. The present vegetation consists of a deciduous forest on

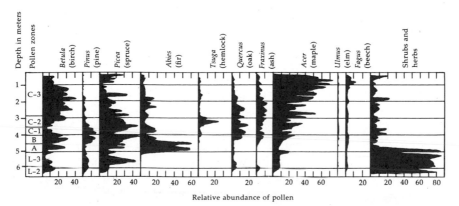

Relative abundance of pollen

Figure 8-12   Pollen profile of sediment from a lake in Nova Scotia. The numbers along the horizontal axis refer to relative abundance of each type of pollen. Depth corresponds to age; layer L-2 has been carbon dated to 9000 years B.C., and the top of layer C-3 represents the present. (From "Some interstadial and postglacial pollen diagrams from eastern Canada" by D. A. Livingstone, *Ecological Monographs* 38:87–125. Copyright © 1968 by the Ecological Society of America. Reprinted by permission.)

the hilltops with sugar maple (*Acer saccharum*), beech (*Fagus grandifolia*), and yellow birch (*Betula lutea*), and a coniferous forest in the valleys, with balsam fir (*Abies balsamea*) and white spruce (*Picea glauca*). Figure 8-12 shows that this pattern of vegetation is relatively recent. Sediments at a depth of 5–6 m, corresponding to an age of about 9000 years before the present, reveal pollen mostly from herbs and shrubs characteristic of tundra vegetation near permanent ice. The nearest such vegetation today occurs 500 km to the north of the lake. Shallower depths, corresponding to an age of about 6000 years before the present, show peaks in spruce and fir, indicating the dominance of typical northern taiga forest, widespread today farther to the north. In yet shallower depths, the pollen of deciduous species (birch and maple) increases in abundance and the pollen of coniferous species declines, indicating a continual warming trend in climate.

Millions of Years: Evolutionary Change   **Microfossils**, such as pollen grains, are used to document vegetational changes over the course of thousands of years. **Macrofossils**, such as leaf impressions, are more useful to document changes over millions of years. As with pollen, it is assumed that the abundance of fossils represents the abundance of species in past vegetation. Care must be taken in interpreting site geology to determine whether plant material was deposited in place or carried by water for some distance and then deposited.

By examining general leaf shape and the pattern of leaf venation, paleo-ecologists have been able to identify fossil species and relate them to their nearest living species. The fossil species are usually extinct, but in many cases they are so close to living species that they can be written as, for example, *Pseudotsuga (menziesii)*, which means a fossil very closely related to the modern species of Douglas fir, *P. menziesii*. It is likely that fossil species were physiologically different from modern taxa, much as modern ecotypes of the same species differ from each other (Axelrod 1977). Nevertheless, the assumption is made that the present is the key to the past, and that modern relatives of fossil plants are in a climate similar to the climate that existed at the time and place the fossil material was deposited. In this way, past climates as well as past vegetation may be reconstructed.

One of the most complete records of fossil plants for western North America during the Cenozoic era (the past 65 million years) is in the John Day Basin of eastern Oregon (Chaney 1948). The record shows a cooling and drying trend. About 60 million years ago the prevailing community contained cinnamon, palms, figs, cycads, avocados, and tropical ferns, which are now found in cool mountain forests of Central America with an annual rainfall of 1500$^+$ mm and no frost. The leaves of these plants were large, with entire margins. About 40 million years ago there was a change to a mixed conifer-hardwood forest with birch, alder, oak, dawn redwood, elm, sycamore, beech, maple, chestnut, sweet gum, and others. The leaves of these plants were smaller, with dentate or convoluted margins, indicating a drier climate, and some trees were deciduous. This exact mixture does not appear anywhere today, but close approximations exist along the cool, wet California coast and on the Cumberland Plateau of Tennessee. Annual rainfall was still high, about 1250 mm, but the climate had grown cooler. About 25 million years ago there was a strong shift to winter-deciduous trees, such as oak, hickory, and maple. This indicates a climate like modern Indiana, with 1000 mm of rainfall annually and prolonged freezing temperatures in winter. Today, the John Day Basin is dominated by sagebrush. Trees are absent except along waterways, and precipitation is about 250 mm per year, including some snow in a cold winter period.

Fossil assemblages for other localities in North America have permitted paleoecologists to reconstruct past climate and vegetation zones on a continental basis (Figure 8-13). In general, vegetation zones have been shifted south and compressed over the past 40 million years; that is, environmental and vegetational gradients from pole to equator have become steeper.

Since the fossil record reveals that communities similar to those existing today have a history extending back millions of years, it is reasonable to ask whether communities evolve. That question has not yet been resolved. Whittaker and Woodwell (1972) have suggested that evolution does occur at the community level, developing their argument as follows: (a) All species evolve in communities rather than in isolation; thus, (b) the evolution of a community occurs as a process of coevolution of the associated species, making the whole community an interactive assemblage; therefore, (c) communities must change in structure and function from ancestral, simpler communities to modern, complex communities, as the component species become more interdependent; and finally, (d) communities have emergent characteristics corresponding to those of organisms, such as growth and maturity, structural differentiation, energy flow, material turnover, homeostasis (the tendency to remain stable or regain stability), adaptive optimization, and organization.

This view is supported by data on the convergence of widely separated communities that share a similar environment, such as chaparral vegetation in California, Chile, southern Australia, South Africa, and the Mediterranean region. The climate of the five regions is similar, and so is the physiognomy of the vegetation. "Similar," of course, is a very subjective term. An extensive IBP study of Chilean and southern Californian chaparral did illustrate remarkable parallelisms, as mentioned earlier in this chapter. But comparative studies of all five chaparral areas have revealed significant differences in physiognomy, growth forms, biomass, phenology, species richness, and response to fire (Table 8-5). Environmental differences in climate, land use history, incidence of fire, and soil nutrient levels have also been identified, and some nonconvergent chaparral differences may be related to those factors.

Examples of convergence in the morphology and/or behavior of widely separated and unrelated species, such as cacti in the southwestern United States and similar-looking euphorbs in Africa, have long served as examples of natural selection at work. Can we accept community convergence in the same light? Is there an optimal solution for community structure, given a certain climate, and does natural selection drive community development towards that solution? We are tempted to answer "yes," but Ricklefs (1973) and others answer "no." They reason that if communities were integrated units and a product of evolution, they would be closed systems like organisms, with sharp boundaries. In fact, however, as we have seen earlier in this chapter, groups of species do not parallel

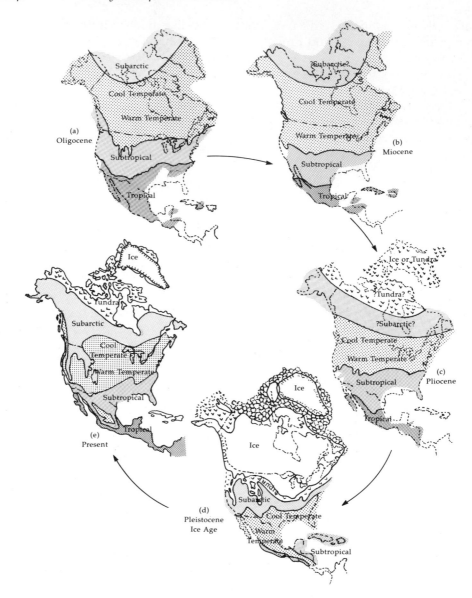

Figure 8-13   Climatic zones in North America during the past 40 million years. (a) Oligocene, about 35 million years ago; (b) Miocene, about 20 million years ago; (c) Pliocene, about 5 million years ago; (d) at the peak of the Pleistocene Ice Age, about 60,000 years ago; and (e) the present. (From Dorf 1960. Reprinted by permission of *American Scientist*, journal of Sigma Xi, The Scientific Research Society.)

Table 8-5 Some important environmental, floristic, and vegetational differences among five regions with chaparral vegetation: Mediterranean rim (Med.), southern California (Cal.), central Chile (Chi.), southwestern Australia (Aus.), and the Cape region of South Africa (Cap.). Relative differences are shown, as high (+), intermediate (0), and low (—), and the ranges of values high to low are shown under Notes. (From many sources, primarily Cody and Mooney 1978, Cowling and Campbell 1980, Milewski 1983, Mooney 1977*b*, Naveh and Whittaker 1979, Parsons 1976, Parsons and Moldenke 1975, Shmida and Barbour 1982, and Specht 1979.)

Trait	Med.	Cal.	Chi.	Aus.	Cap.	Notes
Environment						
Soil nutrients	+	0	0	—	—	N: 0.5–0.0001% P: 0.06–0.0001% Med. rich in Ca
Soil pH	+	—	—	—	—	Med. soils basic
Summer ppt (% of total)	0	—	—	+	+	3–17%
Summer temp.	+	+	—	—	—	24–32°C, mean max.
Winter temp.	+	—	+	+	+	3–8°C, mean min.
Yearly amplitude of temp.	+	+	—	—	—	17–27°C, highest mo. mean max. minus lowest mo. mean min.
Flora						
Species /1000 m^2	0	—	+	0	+	35–120
% woody	0	—	0	+	0	33–57%
% perennial herb	0	—	0	0	+	19–51%
% annual herb	0	+	0	—	—	2–48%
% spinescent	+	0	0	0	—	0–51%
Vegetation						
Growth peak	Spr.	Spr.	Spr.+Su.	Spr.+Su.	Spr.+Su.	Spring or summer
Woody cover	0	+	—	0	0	40–100% absolute
Herb cover	0	—	+	0	0	1–60% absolute
Leptophyll cover	—	0	—	0	+	Relative cover
Biomass/annual productivity	+	+	—	0	0	750–2300 g m^{-2} / 250–670 g m^{-2} yr^{-1}
Fire and response						
Fire frequency	0	—	—	0	+	Perhaps 10–40 yr periodicity
Fire intensity	0	+	0	—	—	Temperatures reached, relative
Herb flush following fire?	Yes	Yes	No	Yes	Yes	Perennial herbs except for Cal.
Lignotubers?	—	+	—	+	—	Relative abundance of spp with them
Serotiny?	—	S+C	Some S	S+F	—	Seeds or fruits or cones

each other in their distribution curves, and when communities are recognized they are relatively arbitrary units, sharing broad ecotones with adjacent communities. Although pairs of species may have coevolved and become interdependent (pollinators and certain plants, parasitic plants and their hosts, mycorrhizal unions), there is no hard evidence that entire communities are integrated, interdependent units. Nevertheless, there have been several recent attempts to develop a theoretical basis or model of community-wide evolution (Wilson 1980, Aarssen and Turkington 1983, Salthe 1985). We can be sure that the topic has not yet been resolved. It is ironic that such recent models bring us back full circle to Clementsian ideas of the community as some kind of superorganism—ideas which stimulated the very beginnings of American plant ecology. The more things change, the more they stay the same.

Summary

The community concept is of general importance to synecology, just as the ecotype concept is central to autecology and the species concept is central to taxonomy. The precise nature of the community is ambiguous because of the biases of individual ecologists and their sampling methods. This does not make the concept useless, however; we must simply appreciate its subjectivity. For the purposes of classification, stands can be grouped into associations that have a fixed floristic composition, physiognomy, and habitat range.

The discrete view of associations assumes that they are closed systems, with interdependent species that synchronously peak in abundance and with narrow ecotones. The discrete view assumes that the whole is greater than the sum of its parts. The individualistic view assumes that associations are open systems, with independent species that happen to associate together wherever their range limits and the chance arrival of propagules overlap; consequently, associations are at most arbitrary units along a continuum.

Whether or not associations exist in the abstract sense, real stands (communities) do exist, and it is useful to consider attributes exhibited by communities beyond the attributes of the component species. These emergent attributes include physiognomy, species importance, spatial and niche patterns, species richness, species evenness, species diversity, the rate of nutrient cycling, the pattern of nutrient allocation to above and below ground parts, and community change over time.

The significance of species diversity to community stability, productivity, interdependence, and environmental stress is still not clear. This may be because (a) we have not yet developed a reasonable method to measure diversity, or (b) there is more than one model of community dynamics and response to stress. Simpson's index may provide the best general index of diversity, if for no other reason than the fact that it is relatively insensitive to sampling error.

Communities differ in their demand for essential nutrient elements, in the efficiency with which radiant energy is converted into net productivity, in the fraction of the nutrient pool that is stored, in the rate at which nutrients are returned to the soil in litter fall, and in the efficiency of the plant–soil–plant cycle.

Succession is community change in a period of up to 500 years, with climate and plant genomes assumed to be constant. In fact, however, climate has not been constant. There have been changes in temperature, rainfall, length of the growing season, and duration of sunshine that have accompanied glacial retreat in the temperate zone over the past 10,000 years. The resulting vegetation changes can be shown by pollen profiles in lake sediment. Vegetation changes over millions of years are documented in macrofossil deposits. These long-term changes, however, include genetic (evolutionary) change in the floras.

There is some argument and indirect evidence to suggest that communities evolve, directed by natural selection toward optimal solutions of environmental problems. However, the conservative opinion is that the most complex level at which natural selection has been shown to operate is with pairs of species, such as a parasite and a host.

CHAPTER 9

METHODS OF SAMPLING THE PLANT COMMUNITY

S ynecologists would like to understand the degree of species interdependence within communities, how the distribution of communities depends upon past and present environmental factors, and what the role of communities is in such ecosystem activities as energy transfer, nutrient cycling, and succession. However, communities must first be measured and summarized in some effective way before these questions can be addressed. The ongoing attempt to inventory the world's vegetation is based on a small sample of the total vegetation cover because of limitations in people, time, and resources. These samples must be taken very carefully to ensure that the resulting estimates will be accurate and useful.

For the sampling to be done rationally and efficiently, the continuum of vegetation that covers the earth must be divided into discrete, describable community or vegetation types, just as the taxonomic continuum of individual plants has been divided into species. Most of the world's vegetation is known to science, but at a relatively simplistic level. It has been broadly classified (UNESCO 1973; Table 1–1), mapped for large areas (Figure 9-1), and photographed from airplanes or satellites. The dominant species and physiognomy of major vegetation types are generally known, but more detailed information about all the component species, the relative importance of each species, and relationships among the species is frequently lacking.

Even if one wishes to describe a particular plant community in a relatively circumscribed, accessible region, one will not usually make a complete census of the community, but instead will take measurements on perhaps only 1% of the total land on which the community exists. If the samples are chosen carefully, investigators feel confident in extrapolating from their sample data to estimate the true values of the **parameters** for the entire community. If the samples are not chosen carefully, the samples will not be representative of the true community parameters and they are said to be **biased.**

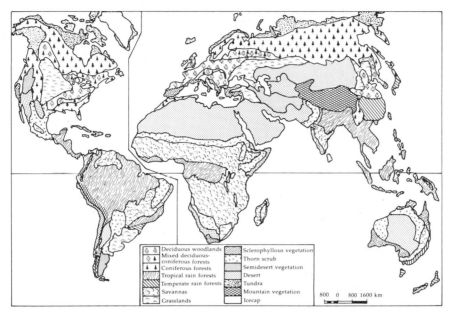

Figure 9-1 Map of major vegetation types of the world. (From *World Vegetation*, by Riley and Young 1974. Reprinted by permission of Cambridge University Press.)

There are two approaches to locating representative samples. One approach is complete subjectivity. A seasoned field worker who has traveled extensively in a region first formulates a concept of a particular community type. Representative stands of that type are found in the field, and one or more sample quadrats are placed so that each quadrat encloses the essence of that stand. This is the **relevé method**. The other approach is to combine subjective selection of stands with an objective (random or regular) placement of sample quadrats within the stands. Theoretically, there is a third approach—random selection of stands and random placement of quadrats within them—but this is so time-consuming that it is seldom done. Completely randomized sampling will inevitably undersample rare, but interesting and ecologically informative, kinds of vegetation.

In this chapter, we will summarize several sampling methods, which is not a very complete review but is enough to get a taste of the diversity among the methods. In Chapter 10, we will discuss data analysis methods that can be used to convey the essence of a vegetation type to others who may be half a world away and have never seen the type, yet wish to compare it to types with which they are familiar.

The Relevé Method

The relevé method was largely codified, if not developed, by Josias Braun-Blanquet, an energetic Swiss ecologist who helped classify much of Europe's vegetation, wrote an impressive text on plant ecology in 1928, founded and directed a center of synecology at Montpellier, France, called SIGMA (Station Internationale de Géobotanique Mediterranéene et Alpine), and was an active editor of the technical journal *Vegetatio* until he was 90. His methods of sampling and classification are sometimes called the **relevé, SIGMA, Braun-Blanquet,** or **Zurich-Montpellier (Z-M) school.**

Our description of the relevé method will be brief. Expanded discussions can be found in van der Maarel et al. (1980), Mueller-Dombois and Ellenberg (1974), Shimwell (1971), Becking (1957), Küchler (1967), and Poore (1955*a,b*). Not many Americans have applied these techniques to North American vegetation; Henry Conard (1935), who helped translate Braun-Blanquet's book into English, was one of the first. Perhaps the most elegant U.S. study was accomplished by Vera Komarkova for Colorado alpine vegetation (1978, 1979). Throughout the rest of the world, the Braun-Blanquet method is extensively used.

An investigator familiar with the vegetation of a region begins to develop concepts about the existence of certain community types that appear to repeat themselves in similar habitats. A number of stands that represent a given community are subjectively chosen. The investigator walks through as much of each stand as possible, compiling a list of all species encountered. Next, an area that best represents the community is located. It is then necessary to determine the **minimal area**— the smallest area within which the species of the community are adequately represented. The minimal area may be determined by a species-area curve. The resulting sample quadrat, based on the concept of minimal area, is called a relevé.

A **species-area curve** is compiled by placing larger and larger quadrats on the ground in such a way that each larger quadrat encompasses all the smaller ones, an arrangement called **nested quadrats** (Figure 9-2a). As each larger quadrat is located, a list is kept of additional species encountered. A point of diminishing return is eventually reached, beyond which increasing the quadrat area results in the addition of only a very few more species. The point on the curve where the slope most rapidly approaches the horizontal is called the *minimal area* (Figure 9-2b–d). Because this definition of minimal area is subjective, some define it instead as that area which contains some standard fraction of the total flora of the stand, for example, 95%. Problems in defining minimal area have been discussed by Rice and Kelting (1955). The most recently proposed solution is to plot the similarity between plots as plot size increases. Beyond some critical area, similarity no longer increases (Dietvorst et al. 1982). Generally, the relevé used is somewhat larger than that which gives the graphical minimal area, for the sake of being conservative. Minimal area is thought by some ecologists to be an important community trait that is just as characteristic of a community type as the species that make it up. Table 9-1 shows how minimal area correlates with vegetation type.

In the relevé method, each species is recorded, and several parameters are estimated within the relevé: cover, sociability, vitality, periodicity, topographic characteristics, and environmental characteristics.

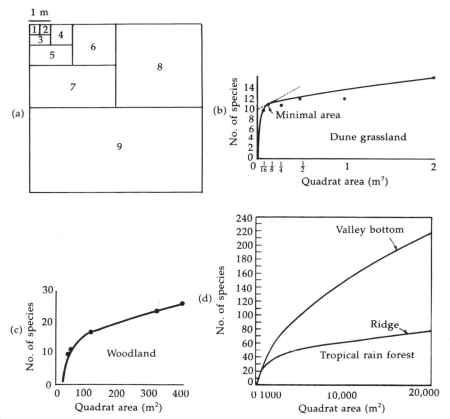

Figure 9-2 The species–area curve. (a) A system of nested plots for determining minimal area. (b) Minimal area for dune grassland in North Carolina is about 0.13 m². (c) Minimal area for an English woodland is about 100 m². (d) Minimal area for two stands of tropical rain forest in Brunei are 1000 m² (a ridge) and 20,000⁺ m² (a valley bottom). [(a) and (d) From *Aims and Methods of Vegetation Ecology.* Mueller-Dombois and Ellenberg. Copyright © 1974 John Wiley and Sons, Inc. Reprinted by permission. (b) From "A discussion of the application of a climatological diagram, the hythergraph, to the distribution of natural vegetation types" by A. D. Smith, *Ecology* 21:184–191. Copyright © 1940 by the Ecological Society of America. Reprinted by permission. (c) From Hopkins 1957. By permission of the British Ecological Society.]

Cover is not measured precisely but is placed in one of seven categories by a visual estimate (Table 9-2). Braun-Blanquet and others, such as Rexford Daubenmire (1968), recognize that plant cover is very heterogeneous from point to point and from time to time even within a small stand. They believe that an exact estimate at one place gives an aura of precision to community description that is not warranted. As another ecologist said, "The ecological world is a sloppy place" (Slobodkin 1974).

Table 9-1   Minimal areas for various vegetation or community types. (From *Aims and Methods of Vegetation Ecology.* Mueller-Dombois and Ellenberg. Copyright 1974 John Wiley and Sons, Inc. Reprinted by permission.)

Type	Minimal area (m²)
Tropical rain forest	1000–50,000
Temperate forest:	
Overstory	200–500
Undergrowth	50–200
Dry temperate grassland	50–100
Heath	10–25
Wet meadow	5–10
Moss and lichen communities	0.1–4

Table 9-2   Cover classes of Braun-Blanquet, Domin-Krajina, and Daubenmire. (From *Aims and Methods of Vegetation Ecology.* Mueller-Dombois and Ellenberg. Copyright 1974 by John Wiley and Sons, Inc. Reprinted by permission.)

Braun-Blanquet			Domin-Krajina			Daubenmire		
Class	Range of cover (%)	Mean	Class	Range of cover (%)	Mean	Class	Range of cover (%)	Mean
5	75–100	87.5	10	100	100.0	6	95–100	97.5
4	50–75	62.5	9	75–99	87.0	5	75–95	85.0
3	25–50	37.5	8	50–75	62.5	4	50–75	62.5
2	5–25	15.0	7	33–50	41.5	3	25–50	37.5
1	1–5	2.5	6	25–33	29.0	2	5–25	15.0
†	<1	0.1	5	10–25	17.5	1	0–5	2.5
r	<<1	*	4	5–10	7.5			
			3	1–5	2.5			
			2	<1	0.5			
			1	<<1	*			
			†	<<<1	*			

*Individuals occurring seldom or only once; cover ignored and assumed to be insignificant.

Table 9-3   Sociability scale of Braun-Blanquet.

Value	Meaning
5	Growing in large, almost pure stands
4	Growing in small colonies or carpets
3	Forming small patches or cushions
2	Forming small but dense clumps
1	Growing singly

Another argument against overly precise cover estimates is that there is a differential bias from one individual to another; it is unlikely that any two estimates would agree closely (Sykes et al. 1983). Schultz et al. (1961) dramatically demonstrated this bias by bringing an artificial quadrat (a 1 m² board with plants represented by discs of different size and color) to a national meeting of professional range management people and asking 100 of them to estimate the total cover of all the "plants" on the board. The resulting range of estimates was impressive: from 6 to 62%. Just as impressive was the fact that the average estimate (27% cover) was 33% in error of the true cover (20%). If each percentage estimate is converted to a class, such as those in Table 9-2, however, more than half of the percentage estimates will fall into the correct class. Seven classes do not provide as much precision as 100 percentage points, but using classes results in greater agreement among investigators. The range of percentage points within each class allows for each observer's deviance from the correct cover percentage.

Sociability is an estimate of the dispersion of members of a species, which does not necessarily have any relationship to cover. Two species may have the same cover, for example, but one could be restricted to a few, dense clumps of individuals while the other may be uniformly scattered throughout the quadrat or stand. Sociability is recorded on a scale of 1 to 5 (Table 9-3) and is written as a decimal addition to the cover value. Thus, species A on a data sheet may be represented by a number such as 2·1, which translates as 5–25% cover, with plants occurring singly.

The investigator also notes the **vitality** (vigor) and **periodicity** (seasonal importance) of species, as well as general topographic and environmental characteristics of the relevé. When all stands have been visited, a summary table of species A stands is prepared; this table will be discussed in Chapter 10.

The summary table reveals **synthetic traits,** which are traits of a community rather than of a single stand. Two synthetic traits are **presence** and **constance.** Presence is the percentage of all stands that contain a given species. If species A occurs in 8 of 10 stands, the species has 80% presence. Presence is calculated from the presence lists that were generated as the investigator walked through the stands. Constance, in contrast, is based on species encountered in relevés.

One relevé, recall, is placed in each stand, and those relevés are all of equal area (though not necessarily of equal shape). Generally, presence is higher than constance. Species A may have been present in 8 stands, but in only 6 of the 10 relevés, thus having 60% constance (sometimes called *constancy*).

Random Quadrats Methods

Most American ecologists select stands subjectively but then sample within them by locating many random quadrats, rather than by subjectively locating a single, large quadrat as in the relevé method. The quadrats can be arranged in a completely random fashion or in a restricted (stratified) random fashion (Avery 1964). The quadrats may also be arranged in a regular, nonrandom fashion, but then statistical conclusions cannot be reached.

Care must be taken in selecting the shape, size, and number of quadrats. A considerable body of literature developed on these subjects in the 1940s and 1950s. Some of that research was done on scale drawings of plots of real vegetation as seen from above, with miniature quadrats of various sizes and shapes placed randomly over the drawings. Some of these maps have been published (e.g., Curtis and Cottam 1962). Other research was done on map-like models of artificial vegetation, generally by placing discs of different size and color in random or other patterns, and then sampling these models with miniature quadrats. Schultz Developments* manufactures very rugged artificial vegetation "maps," suitable for class use.

In any case, whether the maps represent real or artificial vegetation, the point is that one knows the true number of plants and the true cover. Sample estimates of these parameters can then be compared for accuracy. The best sampling method will be both accurate and precise. **Accuracy** is close agreement of sample means with actual parameter means. In Figure 9-3, method A gives values that are very accurate, within 10–20% of the true mean. **Precision** is close agreement of sample means to each other, without reference to the true mean. In Figure 9-3, methods A and B are equally precise because their sample means cluster equally tightly. Method B, however, is much less accurate, sample means being about 30% of the true mean. Method C is neither precise nor accurate.

Precision can be measured without knowing the true mean; it is equal to 1/(variance of sample means). Accuracy, however, can only be measured when the actual parameters are known, and in vegetation sampling this is never the case.

Quadrat Shape, Size, and Number

By sampling a map of forest vegetation in North Carolina, Bourdeau (1953) found that restricted random placement of quadrats yielded greater precision than completely random placement, but the two gave equally accurate estimates. From the same map, Bormann (1953) discovered that precision was best when

---

*Contact Dr. Arnold Schultz, 1200 Capitola Road, No. 3, Santa Cruz, CA. 95062.

Figure 9-3 Accuracy and precision of sampling. The bull's eye represents the true cover for a stand of vegetation, and the radiating circles represent departures in accuracy, from 10 to 40% error. Means estimated by sampling methods A, B, and C are plotted according to their departure from the true cover. Samples taken by methods A and B are both precise (the points are in tight clusters), but A is more accurate. Samples taken by method C show poor precision and poor accuracy.

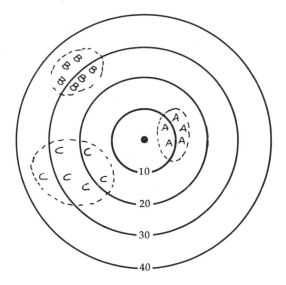

quadrats were long, narrow rectangles, which tended to cross contour lines. Square and round quadrats are often less precise because each one encompasses less heterogeneity within it than a long, narrow plot placed parallel to the major environmental gradient (see also Clapham 1932 and Lindsey et al. 1958).

Accuracy, however, may decline as the plot lengthens because of the **edge effect.** The more perimeter there is to a quadrat, the more often an investigator will have to make subjective decisions as to whether a plant near the edge is "in" or "out," and these decisions are likely to be biased by the taxonomic knowledge of the investigator, how alert the investigator is that day, and how close it is to dinnertime. In this respect, round quadrats are the most accurate because they have the smallest perimeter for a given area. They are also easier to define in the field with a tape measure and center stake, and so a large quadrat need not be carried around. Obviously, compromise choices on quadrat shape are often made.

The best quadrat size to use depends on the items to be measured. If cover alone is important, then size is not a factor. In fact, the quadrat may be shrunk to a line of one dimension or to a point of no dimension and cover can still be measured, as described later in this chapter. But if plant numbers per unit area or pattern of dispersal are to be measured, then quadrat size is critical. One rule of thumb is to use a quadrat at least twice as large as the average canopy spread of the largest species (Greig-Smith 1964); another is to use a quadrat size that permits only one or two species to occur in all quadrats (Daubenmire 1968); another is to use a quadrat size that permits the most common species to occur in no more than 80% of all quadrats (Blackman 1935). There are no fixed rules, however, and the choice is often made by combining intuition and convenience. One person working alone in a desert scrub might choose a quadrat area 2 m long on each side, because the area can all be seen from one point, whereas a team of 2 to 3 people in the same vegetation might choose a quadrat area twice that size.

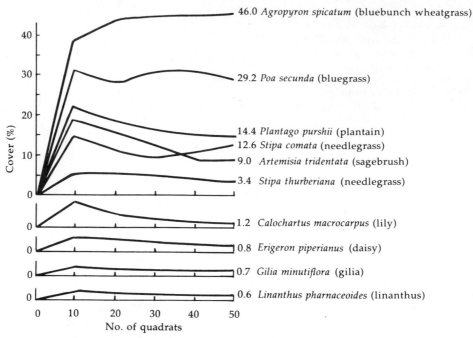

Figure 9-4   Fluctuations in cover are damped out as the number of quadrats increases. In this example from a sagebrush/grassland in the Idaho–Washington area, 40 quadrats might be sufficient to conduct the study, with diminishing reward for added effort beyond that point. (Reprinted by permission from Daubenmire 1959.)

The number of quadrats to use can be determined empirically by plotting the data for any given feature, using different numbers of quadrats, and picking the number of quadrats that corresponds to a point where fluctuations become damped. For example, cover has been tallied in this way in Figure 9-4. Some species, such as bluebunch wheatgrass (*Agropyron spicatum*), were underrepresented by the first 10 quadrats, while others, such as sagebrush (*Artemisia tridentata*), were overrepresented. By 30 to 40 quadrats, values had leveled out, and 50 quadrats did not give additional information, with the exception of one species, needle grass (*Stipa comata*). Therefore, one assumes that 40 quadrats will give a fairly accurate estimate for all additional sampling in this community type.

Alternatively, one can sample until the standard error of the quadrat data is within some previously decided, acceptable bounds. Some field workers suggest that the standard error be ± 15–20% of the mean (that is, two-thirds of all the quadrats supply data that fall within that range about the mean).

Most investigators manipulate quadrat size and number so that 1–20% of a stand is included in the sample. This area may be smaller than the area of a relevé, so there is a possibility that some rare but ecologically significant species will be missed.

Cover, Density, Frequency, Dominance, and Importance

Cover (also called **coverage**) is the percentage of quadrat area beneath the canopy of a given species. The canopy of an overstory species creates a microenvironment that smaller, associated species must contend with. The overstory canopy, therefore, exerts a biotic control over the microclimate of the site. No doubt, the root system of the overstory species extends beneath the ground out to a perimeter corresponding with the canopy edge or even further, so the soil microenvironment is also under the biotic influence of the overstory species. It is assumed that a comparison of cover for each species in a given canopy layer will reveal the relative control or dominance that each species exerts on the community as a whole, such as the relative amount of nutrients or other resources each species commands.

For the practical measurement of cover, holes in the canopy may be viewed as nonexistent, and the canopy edge can be mentally "rounded out," the rationale being that such space is still under the root or shoot influence of the plant in question. The canopy of a plant rooted outside the quadrat is tallied to the extent that the canopy projects into the quadrat space. Thus, in Figure 9-5 and Table

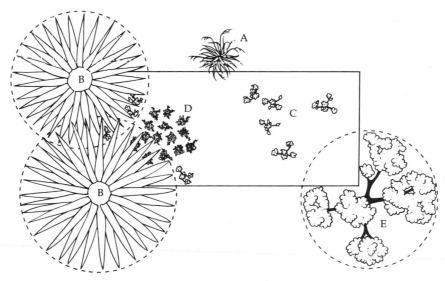

Figure 9-5 Estimation of cover. As seen from above, two members of species B contribute the most cover within the quadrat. Cover for both is estimated, even though only one was rooted in the quadrat. This species could be a plant such as *Yucca*, with radiating basal leaves; the leaf tips describe the perimeter of a circle and cover is estimated as though the circle was completely covered with leaves. Shrub E has a canopy that does not actually penetrate into the quadrat, yet if one fills in the canopy holes by connecting the radiating branches with an imaginary polygon or circle, then there is "cover." Cover values for this 0.1 m² quadrat are given in Table 9-4. (Reprinted by permission from Daubenmire 1959.)

9-4, shrub E does extend into the quadrat space when its canopy is rounded out (dashed lines), and it contributes 7.9% cover. Similarly, the radiating, basal leaves of B that project into the canopy space are tallied by rounding out the edges and estimating each plant separately, but the overlap is not counted twice. The D plants, some small herbs, are mentally grouped together and their cover is estimated separately from the overtopping B plants. In some vegetation, with many overlapping canopies, total cover could exceed 100% and there could still be bare ground. For this reason, bare ground cannot be estimated by subtracting total plant cover from 100%. Some ecologists, however, do not round out the canopies to the extent shown in Figure 9-5; they would award 0% cover for shrub E. Also, some ecologists do count overlapping canopy areas in the same stratum twice.

Relative cover is the cover of a particular species as a percentage of total plant cover. Thus, relative cover will always total 100%, even when total absolute cover is quite low, as in the case of Figure 9-5 and Table 9-4.

Cover of tree canopies can be difficult to estimate, as well as painful after a few hours of neck bending. One solution is to use a "moosehorn" crown closure estimator (Garrison 1949). A periscope-like device is attached to a staff so that the eyepiece is easy to use for viewing while standing; the view of the canopy is seen superimposed on a template of dots. The percentage of dots "covered" by canopies is equivalent to percent cover. These readings could be taken at one location in each quadrat. A more elegant method is to take a picture of the canopy with a fish-eye lens from one location in each quadrat, then to analyze the photographs later for percent cover. Typically, however, canopy cover of trees is assumed to correlate with trunk cross-sectional area (**basal area, BA**) or with trunk **diameter at breast height (dbh)**. To obtain the basal area, the tree is usually measured with a special diameter tape that converts circumference to diameter units. Sometimes **relative dominance** is used as a synonym for relative basal area or relative cover. In this book, we prefer not to equate dominance with BA or any single measure, but to use it as a sum of several measures (see below).

Cover of shrubs and herbs is usually estimated to the nearest whole number or put into cover categories, but if greater detail is required the quadrats may be photographed from above or a scale drawing can be made with the help of a pantograph (Figure 9-6). Neither method works very well when the quadrat is larger than 1 m^2.

Density is the number of plants rooted within each quadrat. The average density per quadrat of each species can be extrapolated to any convenient unit area. For example, Figure 9-5 and Table 9-4 show that herb D had a density of 14 plants per 0.1 m^2 quadrat, which converts to 1.4 million plants per hectare. **Relative density** is the density of one species as a percent of total plant density. **Mean area** is plot area/density; it is the area per plant. Density is independent of cover. For example, many young, slender trees may have a higher density but a lower cover than a few older, branching trees. **Abundance** is a rather nebulous term, but often it is used as a synonym for density.

Frequency is the percentage of total quadrats that contains at least one rooted individual of a given species. It is partly a measure of the same thing the relevé investigators call *sociability*. Rarely, frequency is expressed on a cover basis: any

Table 9-4 Absolute and relative cover and density, based on the 0.1 m² quadrat shown in Figure 9-5.

Species	Absolute cover (%)	Relative cover (%)	Absolute density		Relative density (%)
			Per quadrat	Per ha	
A	0.2	0.4	0	<1	<1
B	33.2	63.6	1	100,000	4.2
C	4.7	9.0	9	900,000	37.5
D	6.2	11.9	14	1,400,000	58.3
E	7.9	15.1	0	<1	<1
Total	52.2	100.0	24	2,400,000	100.0
Overlap	5.3	—	—	—	—
Bare ground	53.1	—	—	—	—

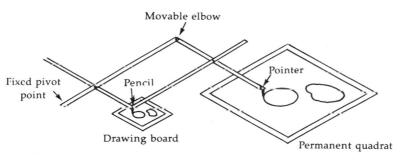

Figure 9-6 A pantograph, which is used to translate actual plant cover and plant location from a quadrat to a scale drawing.

plant, whether rooted in the quadrat or not, that contributes cover for species A is tallied as "present," and frequency becomes the percent of all quadrats in which the canopy of A was "present." **Relative frequency** is the frequency of one species as a percentage of total plant frequency. As mentioned in previous chapters, frequency is an artifact of quadrat size, and in this respect it is a more artificial statistic than cover or density.

Usually, plants are not distributed randomly; then frequency and density are independent of each other. Thus, a clumped species may have a high density but a low frequency, whereas a much less abundant species distributed singly and regularly throughout a stand will have a low density but a high frequency.

It is unfortunate that such an important ecological term as **dominance** is still ambiguously defined by many ecologists. Generally, the dominant species of a community is that overstory species that contributes the most cover or basal area to the community, compared to other overstory species. This definition is based on physiognomy. If oak has the highest relative cover in an eastern deciduous forest with oak, hickory, and elm in the overstory, then oak is said to be the *dominant* species. If all three species contribute about the same amount of cover, or if the balance shifts from one to the other depending on the stand, then the three species are *codominants*. In a savanna or a semidesert woodland, where tree canopies may contribute only 10–30% cover and understory plants such as grasses or shrubs contribute more, density enters the definition, and some grass or shrub species will usually be considered the dominant species.

Another view of dominance is **sociologic dominance** (Kershaw 1973). Sociologic dominants control the reproduction and continued existence of a community, and they may be understory species. For example, the regeneration of ponderosa pine saplings in some ponderosa pine forests is inhibited by root competition for moisture by the understory grass *Festuca arizonica*, but not by another grass *Muhlenbergia montana* (Pearson 1942). The grasses are the sociologic dominants of the ponderosa pine community, even though ponderosa pine is the physiognomic dominant. For this reason, some methods of community description name communities by both overstory and understory species and separate the two species by a slash, for example, the *Pinus ponderosa/Muhlenbergia montana* community, or the *P. ponderosa/Purshia tridentata* community (Daubenmire 1952).

Foresters call any individual tree whose canopy is more than half exposed to full sun a dominant, even though it may not be a member of a species which is a physiognomic or sociologic dominant of that community. In this sense, *dominant* is a synonym for **emergent,** and the latter term should be used. In some other forest studies, dominance is equivalent to trunk basal area. The species with the most basal area per hectare is called the dominant. This use of the term *dominant* corresponds with our definition.

Finally, the term **aspect dominance** is applied to species that are very noticeable and at first glance appear to dominate a community by cover. Careful sampling would reveal, however, that other, less conspicuous species in the same canopy layer contribute more cover and are the actual dominants. Aspect dominance is most common in herbaceous communities, such as grasslands or meadows, where all members of one species will flower synchronously and in this way stand out from the rest of the vegetation.

Throughout this text, we will use the term *dominant* in the physiognomic sense.

Importance refers to the relative contribution of a species to the entire community. It can be used in a very nebulous, almost intuitive, informal sense, or it can be calculated in a precise way. At the investigator's pleasure, importance may be synonymous with any one measure—for example, density. Originally, however, importance was defined as the sum of relative cover, relative density, and relative frequency (Curtis and McIntosh 1951). In the latter case, the **importance value (IV)** of any species in a community ranges between 0 and 300. Table 9-5 illustrates the calculation of IVs for all overstory trees in a Hawaiian rain forest. Notice that two species with similar IVs could have entirely different values for relative cover, density, and frequency; any differences are submerged in the addition process, and the one number that results is a synthetic index of impor-

Table 9-5 Calculation of importance value (IV) for a rather open tropical rainforest at 450 m elevation near Honolulu, Hawaii. The four most abundant overstory trees are summarized below; common understory plants such as tree ferns (*Cibotium splendens*) are not included. In this study, cover is actually the basal area of all stems greater than 3 cm dbh. (From *Aims and Methods of Vegetation Ecology*. Mueller-Dombois and Ellenberg. Copyright 1974 John Wiley and Sons, Inc. Reprinted by permission.)

Species	Relative density	Relative cover	Relative frequency	IV	IV rank
Koa tree (*Acacia koa*)	30.0	78.4	30.8	139.2	1
Ohia lehua (*Metrosideros collina*)	20.0	13.9	23.1	57.0	3
Ohia (*M. tremuloides*)	5.0	5.8	7.7	18.5	4
Guava (*Psidium guajava*)	45.0	1.9	38.5	85.4	2

tance. Other formulae for IV calculation have been developed; they may sum only two relative values rather than three (Bray and Curtis 1957; Ayyad and Dix 1964), or sum more than three values (Lindsey 1956). To avoid confusion, all IVs should be made relative to a 0–100 unit scale.

Importance may also be graphically presented in the form of a four-sided phytograph (Figure 9-7). As originally proposed for tree species by Lutz (1930) (Figure 9-7a), one axis shows relative density of trees over 25 cm dbh, another shows frequency of trees over 25 cm dbh, a third shows the number of individuals in predetermined size (dbh) classes,* and a fourth shows relative basal area. The points on the axis are then connected, and species in a stand can be visually compared by the shape and area of the resulting quadrangles. McCormick and Harcombe (1968) proposed several modifications: (a) that the axes be on a square root scale, rather than an arithmetic scale, which in effect accentuates differences between 0 and 60 (Figure 9-7c); (b) that only relative values be used, rather than mixing relative and absolute scales; and (c) that the axes be relative density, basal area, cover, and frequency, or other traits such as biomass.

Biomass and Productivity

Biomass is the weight of vegetation per unit area; synonyms are *standing crop* and *phytomass*. The dominance or importance of any species can be expressed as the percentage of total biomass. For small quadrats in herbaceous vegetation, biomass may be measured by clipping all above ground matter, drying it in an

---

*Lutz's size classes are: 1 = up to 30 cm tall; 2 = 30 cm–4 m; 3 = saplings 2–8 cm dbh; 4 = poles 10–25 cm dbh; 5 = 25 cm dbh and over.

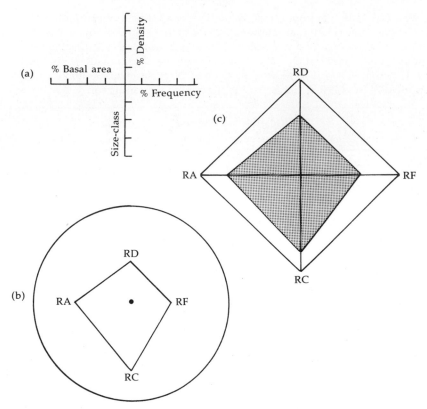

Figure 9-7   The phytograph. (a) Coordinates as originally designed by Lutz. (b) A representation of sugar maple from an Ohio forest, using relative density (RD), relative frequency (RF), relative cover (RC), and relative basal area of stems > 4 cm dbh (RA). (c) The same sugar maple data put on square root axes, with diagonal reference lines added. [(a) Reprinted from *The Description and Classification of Vegetation* by D. W. Shimwell. Copyright 1971 by University of Washington Press, Seattle. (b) and (c) From "Phytograph: Useful tool or decorative doodle?" by J. McCormick and P. A. Harcombe, *Ecology* 49:13–20. Copyright © 1968 by the Ecological Society of America. Reprinted by permission.]

oven, and weighing it. Ideally, roots are also excavated, but they are often ignored; consequently, most biomass data represent only above ground plant matter. Quadrat size and shape are just as important for obtaining accurate estimates of biomass as they are for other measures. Relatively large, circular quadrats may be the most efficient and give the most precise estimates (Van Dyne et al. 1963).

The clearing of large plots in woody vegetation is not practical. Instead, relatively few individuals of different age or size classes are harvested, and a regression line is developed between size and biomass. Sampling for biomass over larger areas can then proceed by measuring plant size, and the data can be converted to biomass.

Productivity is the rate of change in biomass per unit area over the course of a growing season or a year. Productivity and biomass may not be related. A mature forest has a large biomass but may exhibit a small productivity; a grassland has a smaller biomass but may exhibit a larger productivity. Productivity and biomass data may serve to characterize a particular vegetation type, as summarized by Rodin and Bazilevich (1967) and Whittaker (1975). Methods of biomass and productivity sampling have been thoroughly reviewed by Chapman (1976) and Whittaker and Marks (1975).

Species Richness and Diversity

Avi Shmida (1984) has proposed that a quadrat technique developed and widely used by Robert Whittaker until his death in 1980 be adopted worldwide as a standard method to take data for species richness and the calculation of species diversity. We support this attempt at standardization. The "Whittaker method" is described below.

A 20 × 50 m plot (0.1 ha) of homogeneous vegetation is subjectively chosen as representative of a community by the investigator. A 50 m tape is stretched across the plot to serve as a central axis (Figure 9-8). A central 10 m portion is selected, and ten contiguous 1 m² quadrats are marked along one side of the tape. Species presence is noted for each of these ten quadrats. Two 1 × 5 m quadrats on the other side of the tape are then searched for species not already noted in the ten small quadrats, and these are added separately to the list. A single 10 × 10 m quadrat is then searched for new species, and so, finally, is the entire 20 × 50 m plot. Canopy cover for each species can be noted, and additional measures or notes on growth forms, vertical foliage profile, tree and shrub density, and distribution patterns can be taken. Two individuals can complete one plot in as little as 1 hour (desert vegetation) or as many as 4 hours (tropical vegetation).

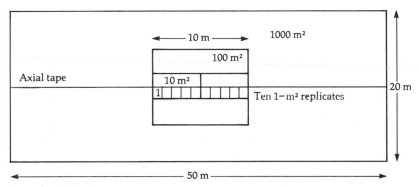

Figure 9-8 Layout of tape and quadrats within a 0.1 ha plot according to the Whittaker method. (From A. Shmida. *Israel Journal of Botany* 33:41–46 published 1984 by the Weizmann Science Press of Israel.)

Two important comparative values can then be calculated. If cumulative number of species is plotted against increasing area on semi-log paper, a best-fit linear relationship can be formulated. The slope of that line is a convenient measure of habitat and biotic diversity and can readily be compared to other habitats and communities. Species diversity can also be calculated for the 0.1 ha plot, weighted by cover rather than density.

Plotless Methods: Line Intercept, Strip Transect, and Bisect

H. L. Bauer (1943) developed the **line intercept** method for dense, shrub-dominated vegetation. He has found it to be as accurate as traditional quadrat methods, but less time-consuming. If a quadrat is reduced to a single dimension, it becomes a line. The line may be thought of as representing one edge of a vertical plane that is perpendicular to the ground; all plant canopies projecting through that plane, over the line, are tallied. The total decimal fraction of the line covered by each species, multiplied by 100, is equal to its percent cover. Just as with quadrats, total cover can be more than 100%. Disadvantages of the method are the loss of density and frequency measures, because there is no area involved (although frequency can be expressed on a cover basis if the line is broken up into segments).

Often, a lengthy line intercept is combined with quadrats which run alongside it. Cover is measured along the line and density or frequency is noted in the quadrats. If the quadrats run continuously along the line, the method is called the **belt transect, strip transect,** or **line strip method.** These methods have been most often applied to forest vegetation (Lindsey 1955).

Bisects are scale drawings of the vegetation within line strips. The idea was originally applied to tropical forests (Davis and Richards 1933; Richards 1936; Beard 1946); Figure 9-9 is an example from the British West Indies. All plants in a strip approximately 60 m long and 8 m wide are shown, drawn as accurately as possible. For those who are not good artists, bisects can be drawn in highly diagrammatic fashion using symbols (Figure 9-10).

These three methods can record cover as a function of height above the ground, if the sampling is done carefully enough. When the data are summarized in bar graphs such as the one in Figure 9-11, striking differences between vegetation types become apparent. The eastern deciduous forest is seen to be composed of four canopy layers, with the most cover being contributed by the overstory tree layer. In contrast, the boreal forest has three canopy layers, with the trees and ground (herbaceous) layers providing nearly continuous cover.

The Point Method

If a quadrat is reduced to no dimension, it becomes an infinitely small point. In practice, metal pins with sharp tips serve as the points, and cover is equal to

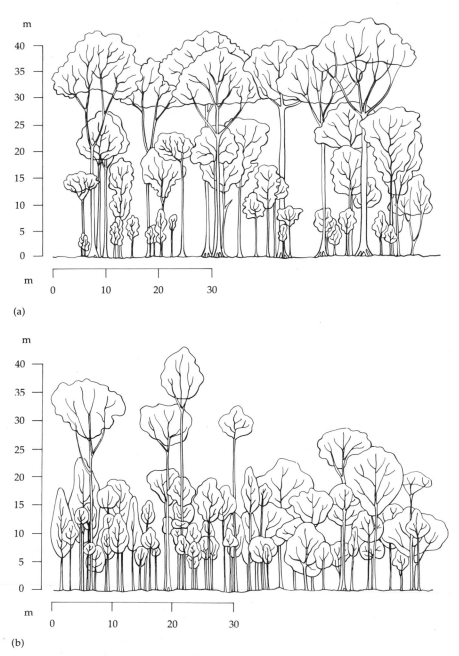

Figure 9-9 Bisects of tropical rainforest. (a) Trinidad, British West Indies. (b) Borneo. Both bisects represent all vegetation within a strip 61 m long and 7.6 m wide. [(a) From Beard 1946. By permission of the British Ecological Society; (b) From Richards 1936. By permission of the British Ecological Society.]

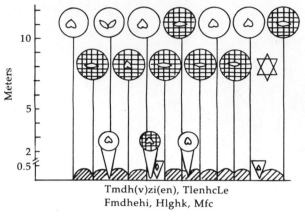

Tmdh(v)zi(en), TlenhcLe
Fmdhehi, Hlghk, Mfc

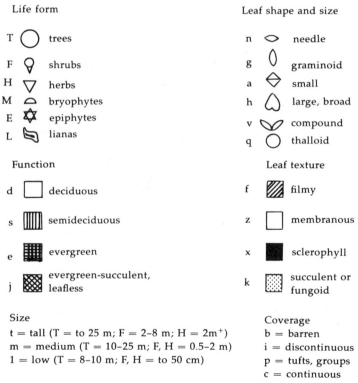

Life form

T ◯ trees
F ♀ shrubs
H ▽ herbs
M ◠ bryophytes
E ✡ epiphytes
L ◞ lianas

Leaf shape and size

n ◇ needle
g ◊ graminoid
a ◇ small
h ♡ large, broad
v ◡ compound
q ◯ thalloid

Function

d ☐ deciduous
s ▥ semideciduous
e ▦ evergreen
j ▧ evergreen-succulent, leafless

Leaf texture

f ▨ filmy
z ☐ membranous
x ■ sclerophyll
k ▦ succulent or fungoid

Size
t = tall (T = to 25 m; F = 2–8 m; H = 2m⁺)
m = medium (T = 10–25 m; F, H = 0.5–2 m)
1 = low (T = 8–10 m; F, H = to 50 cm)

Coverage
b = barren
i = discontinuous
p = tufts, groups
c = continuous

Figure 9-10   Profile diagram of a woodland as represented by some of Dansereau's symbols. (From Dansereau 1951. By permission of the Ecological Society of America.)

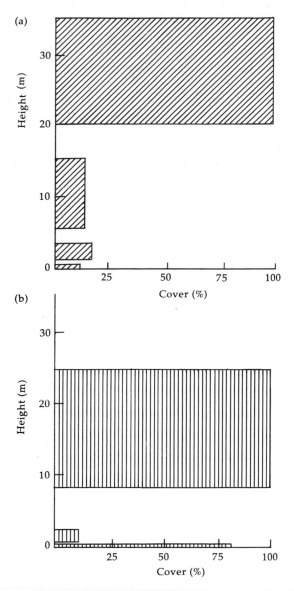

Figure 9-11   Canopy profiles. The height of the horizontal bars represents the average span of canopy height; the length of each bar represents the total cover by all species in that height range. (a) Typical eastern deciduous forest. (b) Typical conifer forest in the taiga across Canada.

the fraction of total pins that touch any plant part as the pin is lowered. Typically the pins are arranged in frames that rigidly limit the pin to a vertical path perpendicular to the ground (Figure 9-12). The frame may be located at several random places in a stand, and one to ten pins can be lowered at each place. For the best precision of estimated cover, lowering only one pin at each place is better than lowering several (Goodall 1957). To measure cover up to 1.5 m tall, recent modifications in frame construction permit rapid placement of as many as 100 pins and projection of pins upward as well as downward (Baker and Thomas 1983, Taha et al. 1983).

As the pin is lowered, the first plant it touches is recorded, then the pin is lowered more until it touches another leaf (of the same or a different plant than the first touch), and so on until bare ground is reached. If no plant is hit, then the point is tallied as bare ground. These data permit two calculations. One is percent cover:

$$\% \text{ cover} = \frac{\text{no. of pins which hit species A at least once}}{\text{total no. of pins}} \times 100 \quad \text{(Equation 9-1)}$$

The other calculation is percent of sward, which weights each species by its canopy thickness, or cover repetition, at each point:

$$\% \text{ sward} = \frac{\text{no. of contacts with species A}}{\text{total no. of contacts}} \times 100 \quad \text{(Equation 9-2)}$$

Disadvantages of the point method are that density and frequency cannot be measured (although cover frequency can), and it is limited to low vegetation, such as grassland, for obvious reasons. But for measuring cover of low vegetation, it may be the most trustworthy and objective method available (Goodall 1957).

Figure 9-12   A point frame suitable for one to ten pins. (From *Aims and Methods of Vegetation Ecology.* Mueller-Dombois and Ellenberg. Copyright 1974 John Wiley and Sons, Inc. Reprinted by permission.)

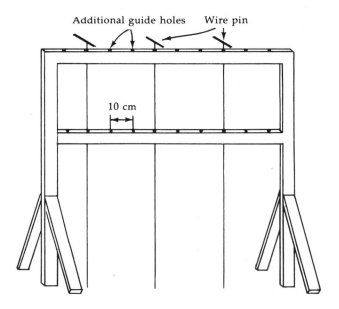

Distance Methods

Distance methods do not use quadrats, lines, or point frames. Only distances (from a random point to the nearest plant, or from plant to plant) are tallied. Average distance, multiplied by an empirically determined correction factor, becomes density. The basic distance methods were developed by Grant Cottam and John Curtis at the University of Wisconsin in the 1950s and were tested and refined on maps of real and artificial forest vegetation. The five methods briefly described here have best been summarized and compared by Cottam and Curtis (1956), Lindsey et al. (1958), and Mueller-Dombois and Ellenberg (1974). Four methods are illustrated in Figure 9-13. These methods have been used with many different types of plants, but most often with trees.

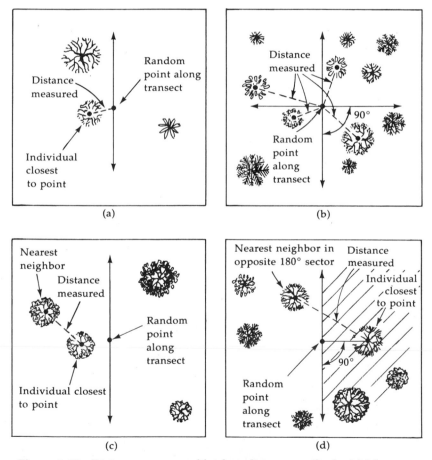

Figure 9-13 Distances measured by four distance methods. (a) Nearest individual method; (b) point-centered quarter method; (c) nearest neighbor method; and (d) random pairs method. The shaded area of (d) is an excluded area which contains the first plant, but cannot contain the second.

Nearest Individual Method

Random points are located in a stand. At each point, the distance to the nearest tree of any species is recorded, the species is identified, and its basal area is measured. Only one measurement is made from each random point. All distances for all species are summed and divided to yield one average distance. Density per hectare (10,000 m^2) for all trees is then:

$$\text{density} = \frac{10{,}000}{2\,(\text{average distance, in meters})^2} \qquad \text{(Equation 9-3)}$$

If distance is measured in feet, then the numerator is 43,560, the number of square feet in an acre. The 2 in the denominator is a constant correction factor. Relative density of each species is:

$$\frac{\text{relative density}}{\text{of species A}} = \frac{\begin{array}{c}\text{no. of trees of}\\ \text{species A encountered}\end{array}}{\begin{array}{c}\text{no. of all}\\ \text{trees encountered}\end{array}} \times \frac{\text{density for}}{\text{all trees}} \qquad \text{(Equation 9-4)}$$

Cover or dominance of each species can then be calculated as its relative density times its average basal area.

Point-Centered Quarter Method

Again, random points are located. The area around each point is divided into four 90° quarters of the compass, and the nearest tree in each quarter is sought. Each tree is identified, its basal area is measured, and its distance from the random point is measured. Again, average distance for all trees taken together is computed, and this is converted to total density by the formula given for the nearest individual method, except that the correction factor vanishes. (The correction factor is 1, not 2.) Since more information is gained at each point than in the nearest individual method, the point-centered quarter method is more efficient, requiring one-fourth the number of points to achieve the same level of accuracy and precision. This method has also been used in grasslands (Dix 1961, Risser and Zedler 1968).

Nearest Neighbor Method

Random points are located in a stand, and the nearest plant is located. The distance measured, in this case, is from that plant to its nearest neighbor (of any species). Total density is calculated as for the nearest individual method, except that the correction factor is 1.67. This nearest neighbor method has also been used to determine whether trees of the same species are distributed at random, are clumped, or are regular (Clark and Evans 1954; Pielou 1961).

Random Pairs Method

In the random pairs method, the nearest plant to a point is located. A line from point to point is imagined. Perpendicular to it and passing through the point is an exclusion line. In Figure 9-13d, the exclusion line happens to correspond with the transect. A nearest neighbor is now searched for, but it cannot be on the same side of this exclusion line as the first tree. The conversion factor in the density formula is 0.8.

Bitterlich Variable Plot Method

A final distance method can be used to calculate basal area only, but basal area is important in calculating board feet of lumber, and the method is extremely fast and has been widely adopted by foresters. It yields more reliable data for less field time than quadrat methods or other distance methods (Lindsey et al. 1958). The method is named after its German inventor,* who originally used a sighting stick 100 cm long with a crosspiece at one end, 1.4 cm across (Figure 9-14a). The stick was held horizontally, with the plain end at one eye, and the viewer would slowly turn in a complete circle. Every tree whose trunk was seen in the line of sight on the circuit was tallied and identified as to species if its trunk appeared to exceed the width of the crosspiece; all other trees were ignored. Using a stick of these dimensions, and based on geometric principles, the total basal area in $m^2 ha^{-1}$ for any species is equivalent to the number of trees of that species tallied, divided by 2. If English units are preferred, then a stick 33 inches long with a crosspiece 1 inch across will give basal area in $ft^2 acre^{-1}$ if the trees tallied are multiplied by 10.

As shown in Figure 9-14b, an angle is being projected—an angle whose size depends on the relative lengths of the stick and the crosspiece. For example, the 33 inches × 1 inch arrangement produces an angle of 1°45' and an English units basal area factor (BAF) of 10. BAF is the number that is multiplied by the number of tallies to obtain basal area (in $ft^2 acre^{-1}$). If the angle becomes smaller, more trees will be tallied and the BAF will become smaller; for example, an angle of 0°33' has an English units BAF of 1. If the angle becomes larger, then fewer trees are tallied and the BAF will become larger; this is useful in a dense forest to avoid miscounting. In this country, angles that give BAFs of 5–20 are commonly used.

More recently, small, hand-held prisms or sophisticated viewing scopes have replaced sighting sticks (Figure 9-14c). Looking both through and over the top of a prism, the lower trunk will appear to be offset partially or completely from the upper trunk; if it is not completely offset, the tree is tallied.

Note that plot size is variable in this method. Plot radius is not fixed but extends as far as the largest tree with an apparent diameter big enough to be

---

*For an English article describing the method and the geometry behind it, see Grosenbaugh 1952.

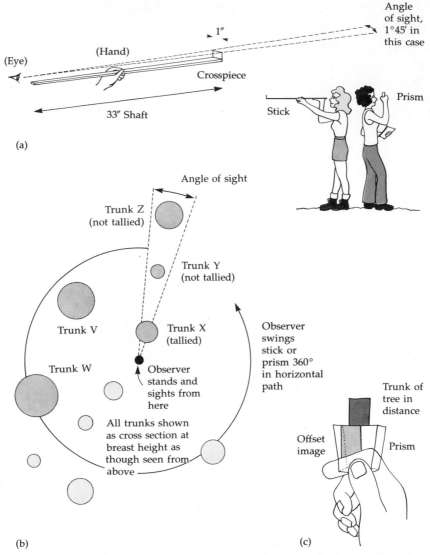

Figure 9-14 The Bitterlich variable plot method for calculation of basal area. (a) The original Bitterlich stick sighting device. This particular one will give total basal area in ft^2 acre^{-1} if the number of trees tallied is multiplied by 10. (b) A bird's eye view of trees that would be tallied as an observer at the center point turns 360°. Tree X is counted because its trunk diameter exceeds the angle projected by the sighting device, but trees Y and Z are not tallied. (c) The prism type of sighting device. In this case, the lower trunk is not completely displaced from the upper trunk, and the tree would be counted.

tallied. A rather complete analysis of the geometric rationale at the base of this method has been compiled by Dilworth and Bell (1978). The method has also been applied to shrubland (Cooper 1957).

Summary

A major objective of synecologists is to complete an inventory of the earth's plant resources—not an inventory of individual species, but an inventory of communities and vegetation types. The results of such an inventory have application to applied and basic science and to autecology as well, for the environment of a plant includes adjacent organisms as well as the physical factors of climate and soil. The inventory is far from complete, because of limitations in researchers, scientific interest, and accessibility of some areas. Conclusions will have to be based on samples representing 1% or less of the earth's surface vegetation. If the estimates are to be accurate and precise, the samples must be chosen carefully. This chapter described procedures for choosing and measuring those samples.

The relevé method uses a more subjective choice of sample locations than any other mentioned, but the process of recording data is relatively rapid and nonmathematical, and its widespread use in the non–English-speaking world makes it an attractive sampling method. Each stand is represented by one large quadrat, whose size must meet minimal area requirements. Data recorded include cover, sociability, vitality, and periodicity. Important synthetic values are presence and constance.

Random quadrats methods involve fewer subjective decisions than the relevé method. Questions that have to do with the size, shape, number, and placement of quadrats are time-consuming to answer, and there is no one best solution to them. A statistical level of confidence can only be applied to data from random quadrats. Such data can include absolute and relative cover, density, frequency, and biomass. Generally, these four measurements are unrelated. (If plants are distributed randomly, density and frequency are related, but plants are usually not distributed randomly.) Cover may be the most meaningful of the four. It can be estimated as actual crown cover, but for trees it is assumed to correlate with trunk basal area. Cover and basal area are both aspects of size. Frequency is the most artificial of the four, being highly dependent on quadrat size.

Two synthetic values from quadrat data are dominance and importance. A single definition of dominance has not yet been accepted. Most often, however, dominance is related to the cover of overstory species, but in some cases density is also considered. Typically, importance is a summation of relative cover or basal area, density, and frequency, but the summation may involve one more or one less component, and it can be expressed as a phytograph.

The line intercept sampling method reduces quadrats to a single dimension; consequently, only cover can be measured. Line intercepts may be combined with quadrats, as in strip transect and bisect methods. The point method reduces quadrats to no dimension; it is most useful for estimating cover in short vegetation types.

Distance methods measure distances from random points to nearest plants or distances between plants. The data can be converted to plant density, and in some cases the data can reveal whether plants are randomly or nonrandomly distributed. The most efficient distance method is the point-centered quarter method. The Bitterlich variable plot method was treated as a distance method, even though it only permits the calculation of basal area, because the geometric assumptions behind it relate to distance from observer to tree and to tree diameter.

CHAPTER 10

CLASSIFICATION AND ORDINATION OF PLANT COMMUNITIES

No matter how sampling is done—by quadrats, line intercept, points, or distances—the raw data are too voluminous and disorganized to mean much to the investigator, let alone to someone half a world away with whom the investigator would like to communicate. This chapter describes the commonly used methods of summarizing sampling data in order to relate one stand to another.

Sampling and data analysis, of course, are not independent of each other. Some methods of analysis require certain methods of sampling. For example, association analysis requires the use of random quadrats, and the Braun-Blanquet table method requires relevé information.

The biases and objectives of the investigator also influence the choice of a data analysis method. If the aim is to draw lines and describe discrete entities, then the table method and vegetation mapping are good choices. If the aim is to let the reader draw the lines between entities and to emphasize the continuum of vegetation, then gradient analysis and ordination are good choices.

A word of caution: The statistical tests and the mathematical rigor behind them are only superficially presented in Chapter 9 and in this chapter. The old saying, "A little knowledge can be a dangerous thing," fits the area of sampling and analysis very well. Although we will continue to cite the original literature in this chapter, these citations may not be helpful unless you possess some mathematical background, for their subject matter is technical and the reading sometimes difficult. Our objective is to survey some commonly used procedures, so that you are reasonably familiar with the options available. Many professional ecologists find the mathematics and computer requirements behind certain methods formidable and require the help of specialists in order to use them. Other techniques, such as ordination, are still being amended and refined, since even mathematically oriented ecologists are not satisfied with them.

Classification

Classification methods attempt to place similar stands together in discrete entities (associations or other units), cleanly separated from all other stands and units. In North America, particularly in the West, classification has emphasized dominant species. In Europe and some other parts of the world, classification has emphasized indicator or character species that seldom are dominants. They are often understory species of low cover but exhibit strong fidelity to certain units—typically present in one unit and absent in all others.

Classification Based on Dominance

Clements classified North American vegetation on the basis of only one or two dominant species in each unit. Sometimes the dominant taxon was listed only to genus, implying that the species might vary locally within the vegetation unit. This is an extreme emphasis on overstory dominants, and the results may mask important variations. In some regions, the overstory dominant may remain constant while ecologically important understory changes occur. Daubenmire (1952) modified Clements' simple scheme by listing two species per unit: an overstory and an understory species, separated by a slash. In this way, a *Pinus ponderosa/Purshia tridentata* tree–shrub mosaic could be classified separately from a *Pinus ponderosa/Agropyron spicatum* woodland–grass mosaic. In a simpler scheme, they would be lumped together as a *Pinus ponderosa* unit. Recently, the USDA Forest Service has been classifying national forest vegetation on a dominance basis, but with even finer scales of differentiation. The classification is being done in such a way that any field biologist would be able to walk through a vegetation unit and use a key to identify it accurately with a standardized name in published lists (see, for example, Paysen et al. 1982).

In order to classify vegetation on the basis of dominance, any quadrat, transect, point, or distance sampling method can be used if it yields data that can be relativized. Dominance (as described in earlier chapters) can be based on relative canopy cover, density, basal area, biomass, or on sums of any of these (i.e., importance values).

Classification Based on the Entire Flora

The Braun-Blanquet relevé technique (Chapter 9) lends itself to floristic classification of vegetation. However, any sampling scheme can be used, so long as all species encountered are recorded. Once sufficient stands have been sampled to represent variation within an area, the results are summarized in a **primary data table** (matrix) such as Table 10-1, which gives data for English pastures. In such a primary matrix, the stands and species are listed in the order in which they were sampled and encountered. The objective now becomes to generate a second, **differentiated table,** where similar stands lie near each other and species with similar distributions also lie near each other. As the rows (species) and columns are shifted about with scissors and tape, it soon becomes apparent that

Table 10-1 The raw data table, listing species (down) chronologically or phylogenetically, and listing relevés (across) in numerical sequence. The units digit represents cover, and the tenths digit represents sociability. The cross (†) indicates <1% cover and then sociability is understood to be 1 unless noted. (Reprinted from *The Description and Classification of Vegetation* by D. W. Shimwell. Copyright 1971 by University of Washington Press, Seattle.)

Relevé number	1	2	3	4	5	6	7	8	9	10	11	12	13	14	15	16	17	18	19	20
Seed plants																				
Hippophae rhamnoides	5·1	5·1	5·1	5·1	5·1	3·2	2·2	4·3	2·2	5·1	3·2	5·1	3·3	4·3	3·2	5·1	5·1	5·1	5·1	5·1
Senecio jacobaea	1·1	+	1·1	+	+	+	+		+		+		+	+	1·1	+	+	+	1·1	+
Solanum dulcamara	2·1	2·1	+	+	+	+	+			1·1	+	+		+		1·1		+	+	1·1
Rubus fruticosus s.l	+	1·1		+	1·1					+	+	+		+		+	+	+	+	1·1
Urtica dioica	3·3	1·3	1·3																+·3	2·3
Rumex crispus	+	+	+			+		+	+		+		+	+		+		+	1·1	+
Montia perfoliata				3·4	4·4					4·5		2·3				4·5	2·3	1·3		
Stellaria media				1·2	+							3·4				+	2·3	1·2		
Festuca rubra						1·1	+	1·1	+		2·3		3·3	1·3	1·3					
Agropyron repens							2·3	+	3·3		+		+	2·3	†2					
Ammophila arenaria						2·3	4·3	2·3	+		3·3		1·3	†3	4·3					
Sonchus arvensis						+		+	+	+	1·1			+	+				+	
Ononis repens						1·1														
Galium verum						+								+						
Calystegia soldanella							+	1·1	+	+	+	+		1·1	+		+	+		
Poa pratensis								+	+		1·1		+	+	+					
Agrostis stolonifera								+	†2					+	†2		+			+
Ranunculus bulbosus									+		+		+	+						
Plantago lanceolata									+		+		1·1	+	1·1					
Veronica chamaedrys									+	+				+					+	
Chamaenerion angustifolium										+	1·1	1·1		+		+			2·3	
Cerastium vulgatum										+		+				+	1·1	1·1		+

(continued on next page)

Table 10-1 The raw data table, listing species (down) chronologically or phylogenetically, and listing relevés (across) in numerical sequence. The units digit represents cover, and the tenths digit represents sociability. The cross (†) indicates <1% cover and then sociability is understood to be 1 unless noted. (Reprinted from *The Description and Classification of Vegetation* by D. W. Shimwell. Copyright 1971 by University of Washington Press, Seattle.) *(continued)*

Relevé number	1	2	3	4	5	6	7	8	9	10	11	12	13	14	15	16	17	18	19	20
Seed plants *(continued)*																				
Sambucus nigra																+				†2
Cirsium vulgare																+	+			
Heracleum sphondylium																	+			
Inula conyza																	+	1·1		+
Cardamine hirsuta																	+	+		
Hypochaeris radicata																			+	
Arrhenatherum elatius																			1·1	
Sonchus asper																				+
Seedless plants																				
Eurynchium praelongum	1·3		1·3																1·3	1·3
Hypnum cupressiforme	+	†2	†2		+															
Brachythecium rutabulum		+	1·3	+						†3		+		+			+	†3		
Geastrum fornicatum				+	+						†3	+								
Brachythecium albicans						†3			1·3					+						
Bryum inclinatum						†3								+						
Tortula ruraliformis								†3						†3						
Cladonia rangiformis													+	+						
Bovista nigrescens																				
Lophocolea heterophylla																				+
Number of Species	8	9	8	8	9	11	7	10	14	12	12	10	10	22	10	11	12	12	12	13

some species are of little use in differentiating groups of stands. Some species occur so rarely that only one or two stands show them; others occur so often that nearly every stand shows them; still others do not form species groups. Such species will be eliminated from the differentiated table. In the case of Table 10-1, 28 of the total 39 taxa were in this unusable category, and they are not included in Table 10-2.

The remaining species are called *differential* or *characteristic* species. Their **fidelity** (faithfulness) to a given association can be expressed on the basis of how few stands outside the association contain them. The exact level of fidelity demanded varies from investigator to investigator, but a general rule is that a species that helps to define an association cannot occur in more than 20% of the stands outside that association.

Fidelity is not related to constancy. A species may be restricted to association X, but it may occur rarely even there and have a constancy of only 10%. Useful species must have both moderately high fidelity and moderately high constancy. Again, the required level of constancy varies, but in general it must be above 50%.

As characteristic species are searched for, a reciprocal problem must be solved simultaneously: How many characteristic species must be shared by any two stands before they are considered part of the same association? A common answer is that more than 50% of the total list of characteristic species for association X must be present in any of the stands belonging to it.

All of these criteria have been met in the differentiated table shown in Table 10-2. Three associations (Groups A, B, C) are outlined in boxes. A computer program has been developed to accomplish the preliminary task of rearranging columns and rows to obtain a differentiated table (Ceska and Roemer 1971).

Once there has been sufficient sampling of associations in a region, it may be possible to group similar associations together into a higher level of classification, analogous to the way many species can be grouped into a genus. This higher level is called an **alliance.** Alliances can then be grouped into **orders,** and orders into **classes.** In this way, a hierarchical classification of all vegetation in a region is possible, based largely on floristic similarities. At these higher levels, however, dominance by particular growth forms can also become an important criterion (Rejmanek 1977).

Cluster Analysis

The objective of **cluster analysis** is to simplify the data and to present them in graphical, rather than table, form. The resulting figure is a **dendrogram of stands** (Figure 10-1), similar to the dendrogram of species that a numerical taxonomist constructs.

The first step is to express the similarity between any two stands in a single number, called the **community coefficient (CC)**. There are several ways to calculate CC, as Table 10-3 shows (see also Goodall 1973 and Mueller-Dombois and Ellenberg 1974). Basically all the formulae indicate the relative number of species

Table 10-2 The differentiated table. Species and relevés shown in Table 10-1 have been rearranged so that similar groups (associations or groups A, B, C, boxed) stand together. A large group of species that did not contribute to defining associations has been left out of the table. Also, relevé 14 has been omitted because it was a disturbed site. (Reprinted from *The Description and Classification of Vegetation* by D. W. Shimwell. Copyright 1971 by University of Washington Press, Seattle.)

Revised relevé order	1	2	3	19	20	4	5	10	12	16	17	18	7	6	8	9	11	13	15
Group A																			
Urtica dioica	3·3	1·3	3·3	+3	2·3														
Eurynchium praelongum	1·3	+3	1·3	1·3	1·3														
Group B																			
Montia perfoliata						3·4	4·4	4·5	2·3	4·5	2·3	1·3							
Stellaria media						1·2	+	+	3·4	+	2·3	1·2							
Geastrum fornicatum						+	+		+	+	+	+							
Cerastium vulgatum								+	+	+	1·1	1·1							
Group C																			
Festuca rubra							+						+	1·1	1·1	+	2·3	3·3	1·3
Agropyron repens													2·3	+	+	3·3	+	+	+2
Ammophila arenaria													4·3	2·3	2·3	+	3·3	1·3	4·3
Poa pratensis														1·1	+	+	1·1	+	+
Plantago lanceolata																+	+	1·1	1·1

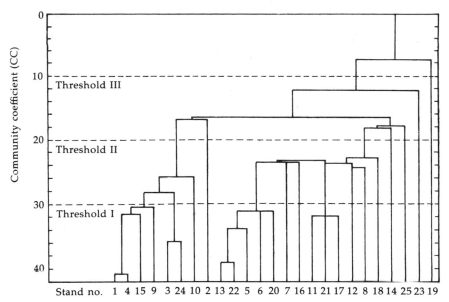

Figure 10-1 A dendrogram resulting from cluster analysis of 25 stands. If one decides that members of an association must exhibit a community coefficient (CC) of 30 or more, then this dendrogram indicates that 15 associations exist, because 15 lines project up through the dashed line called Threshold I. If a CC of 10 were instead selected, there would be only two associations (two lines project through the dashed line called Threshold III). (Simplified from *Aims and Methods of Vegetation Ecology.* Mueller-Dombois and Ellenberg. Copyright © 1974 by John Wiley and Sons, Inc. Reprinted by permission.)

shared by any two stands or quadrats. A CC of 100 represents identity, while a CC of 0 represents complete difference. A rule of thumb is that any two plots with a CC of more than 50 represent the same association.

After CC values are computed for every pair of stands, the two stands with the highest CC (closest similarity) are plotted on a graph as vertical lines, which are joined by a horizontal line at that CC value. In Figure 10-1, stands 1 and 4 have the highest CC, 41, and they are joined at that level. Now these two stands are lumped into a new, artificial, second-level stand, and CC values are computed all over again. In this case, it turned out that the highest CC value in the second calculation was for stands 13 and 22. They are joined, CC values are recomputed, and so on until all stands are joined at some, generally low, CC value (a CC of 8 in Figure 10-1).

A cluster analysis dendrogram can be used for classification if the investigator selects some threshold value at which to define associations. As already mentioned, stands of one association are often expected to share a CC of above 50. Using that criterion, every one of the 25 stands in Figure 10-1 represents a

Table 10-3 Four methods of calculating the community coefficient (CC) from presence and % cover data.

Species	Presence and % cover data	
	Stand (quadrat) A	**Stand B**
No. 1	10	20
No. 2	4	12
No. 3	—	7
No. 4	—	15
No. 5	32	15
No. 6	15	—
No. 7	2	—
No. 8	1	1
Total % cover	64	70

Calculation methods

Formula	Calculation	Community coefficient
Jaccard, presence only	$\dfrac{C}{A + B - C} \times 100 = \dfrac{4}{6 + 6 - 4} \times 100 =$	50
Jaccard, weighted by cover	$\dfrac{MC}{MA + MB} \times 100 = \dfrac{(10 + 4 + 15 + 1)}{64 + 70} \times 100 =$	22
Sorensen, presence only	$\dfrac{2C}{A + B} \times 100 = \dfrac{2(4)}{6 + 6} \times 100 =$	75
Sorensen, weighted by cover	$\dfrac{2MC}{MA + MB} \times 100 = \dfrac{2(10 + 4 + 15 + 1)}{64 + 70} \times 100 =$	45

A = total number of species in stand A
B = total number of species in stand B
C = total number of species in both stand A and stand B
MA = total % cover of species in stand A
MB = total % cover of species in stand B
MC = total % cover of species in both stand A and stand B, using the lower % cover figure
 for each species

different association. If a threshold value of 30 is chosen instead of 50, then 15 associations would result, as represented by the 15 vertical lines projecting up through the dashed Threshold I line in Figure 10-1. If a threshold of 20 is selected instead, then seven associations would result.

The advantage of cluster analysis over the table method is that the classification process is quantified, because some threshold value is chosen as the lower limit to an association. Also, the relatedness of different associations can be quantified. Various computer programs have been developed to produce cluster diagrams. TWINSPAN is one frequently used program that yields results very comparable to Braun-Blanquet analyses for the same data sets. This clustering technique and four others were described and compared by Gauch and Whittaker (1981).

Association Analysis

Association analysis builds up a dendrogram from the top down, rather than from the bottom up as in clustering. Associations are divided on the basis of differential species just as in the table method, but the selection of differential species is based on probabilistic, statistical equations rather than on fidelity and constancy.

Recall that a positive or negative association between two species can be revealed with a contingency table and a chi-square calculation, and the higher the chi-square value, the stronger the positive or negative association. Use of the chi-square formula requires the placement of random quadrats, so this method of analysis cannot be applied to relevé data. The first step is to compute chi-square values for every pair of species. The species having the highest sum of chi-square values with all other species is selected as the first differential species. In the salt marsh example shown in Figure 10-2, with 77 species, *Puccinellia maritima* (species no. 32) is the primary differential species, with a chi-square sum of 48. The 77 species encountered in the 70 quadrats are represented as a vertical line that splits at a chi-square sum of 48; at the left end of the horizontal line are those quadrats that contain *Puccinellia*, and at the right end are those quadrats without this species.

In those quadrats with *Puccinellia*, the chi-square procedure is repeated to find a second differential species, which is *Spergularia media* (species no. 38) in this example. A horizontal line at a chi-square value of 10 separates the quadrats once again. At the left end are quadrats with both *Puccinellia* and *Spergularia*, and at the right end are quadrats with *Puccinellia* but not *Spergularia*. Each of these groups is analyzed by the chi-square procedure, but in this example no species with a significantly high value was found (a 99% level of significance was used, which is represented by the dashed line in Figure 10-2). The chi-square procedure is repeated for those quadrats without *Puccinellia*, resulting in a total of eight groups of quadrats. Each group is relatively homogeneous and differs from other groups by the inclusion or exclusion of one of the seven differential species. Each group may be considered an association.

This method was first described in a slightly different form by Goodall (1953) in the days before computers. It was elaborated and given the name *association analysis* by Williams and Lambert (1959); their approach requires computer facilities. Still later (1961), they referred to this method as **normal association analysis,** and they presented a complementary procedure called **inverse association analysis.** Inverse association analysis results in groupings of species rather than quadrats (see also Ivimey-Cook and Proctor 1966, and Goldsmith and Harrison 1976). Together, normal association analysis and inverse association analysis do the same thing that the table method does—that is, species and stands, or quad-

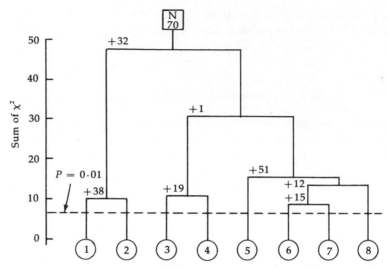

Figure 10-2   A dendrogram resulting from normal association analysis of 77 species in 70 quadrats in a salt marsh. The +32, in the upper left, represents a line of quadrats which all contain species no. 32, *Puccinellia maritima*. The +38 represents *Spergularia media*, and so on. Each of the seven species that are numbered (*Puccinellia, Spergularia*, etc.) serves to separate quadrat groups, and each species exhibited positive and/or negative associations with other species, as indicated by the high chi-square values along the left axis. A chi-square value below 7 was considered nonsignificant in this example. The result was a division of the quadrats into eight groups, each group representing an association. (From Ivimey-Cook and Proctor 1966. By permission of the British Ecological Society.)

rats, are rearranged so that similar ones lie near each other in the summary picture—but these methods do at least give a numerical value to the degree of similarity and they invite the reader to decide what an association is.

Vegetation Mapping

The most generally useful vegetation maps show both physiognomic and floristic information. In such maps, both the life form and the identity of the dominants of each type are indicated. Needleleaf forests are separated from broadleaf forests, and pine-dominated needleleaf forests are separated from fir-dominated needleleaf forests. If physiognomy alone is mapped, then considerable information is lost.

The most sophisticated maps, such as those of Henri Gaussen, use symbols and colors to provide environmental as well as vegetational information. A temperate zone deciduous forest with moderately high humidity and moderately warm summer temperatures, for example, would be colored light blue (for moderately high humidity) applied in a flat tint (for a forest), plus yellow (for moderate temperatures), giving an overall effect of light green. An overlay symbol of ♀ indicates that the dominant species are broadleaf, deciduous trees. A warm desert scrub would be colored in red lines (red for very hot, lines for scrub), plus orange (very dry), with an overlay symbol of ⸕Ѵ̵ to indicate xeromorphic shrubs. UNESCO adopted this method and published a booklet that describes the symbols in some detail (UNESCO 1973). Most vegetation maps, however, use color only to help the reader separate one vegetation type from another; Küchler's maps of the conterminous United States (1964), of Kansas (1974), and of California (1977) are good examples.

One problem with vegetation mapping is that vegetation types are generally described in paragraphs of prose, rather than by tables, numbers, or graphs, and the results reflect the biases of what the mapper regards as important differences.

Mapping generally begins by examining aerial photographs, such as that shown in Figure 10-3. These pictures are taken from directly above with black-and-white, color, or infrared-sensitive film, and they are generally timed so that there is considerable overlap from one frame to the next. When prints from adjacent frames are placed together correctly and examined with a stereoscopic viewer, the vegetation and topography take on a three-dimensional quality that makes mapping easier. From the photographs, the investigator tries to distinguish as many different communities or vegetation types as possible, outlining the boundaries of each on acetate sheets placed on top of the prints.

The next step is to go back to the field and examine representatives of each tentatively mapped type in order to identify the dominants and check the reliability of the interpretation of the photographs. This phase is often called *ground truth*. A short description of each mapped type is prepared, the boundary lines

Figure 10-3 Aerial photograph, showing several contiguous vegetation types or communities. A synecologist has marked the boundaries of each type with ink lines in preparation for publishing a vegetation map. (From *Vegetation Mapping* by A. W. Küchler 1967. By permission of Ronald Press, New York.)

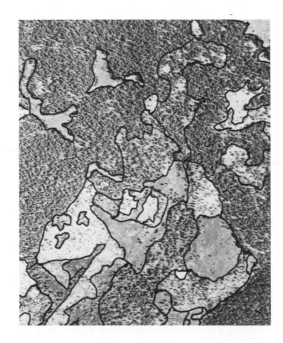

may be amended, and then the acetate overlays are transferred to some political, topographic, or other kind of base map.

Limitations Imposed by Scale and Objectives   The level of detail that can be shown on a map is limited by its scale. The smallest area that can be easily seen by a reader is a circle 1 mm in diameter. At a scale of 1:1,000,000, this corresponds to a real area of 247 acres (100 ha). Obviously, such a scale is inappropriate to show the location of associations that might have a mean area of 1 acre or less; for that level of detail, a larger scale would have to be used, perhaps 1:100,000.

Many of the boundary lines drawn on vegetation maps do not exist in nature. One community or vegetation type will typically change gradually into an adjacent one through a broad ecotone. The line on the map may represent the midpoint of that ecotone. If the ecotone is unusually large, or if several types coexist in some complex mosaic, then the ecotone or mosaic may be mapped as a separate unit in its own right.

A final word of caution: Some vegetation maps represent **actual vegetation,** existing at the time the map is made; other maps represent **virgin, prehuman vegetation;** and still others represent the **potential vegetation,** which could return if human activities ceased in the area. Depending on the objective, the results may be quite different. Küchler's 1977 map of California vegetation is a blend of all three types of maps. For example, the central California valley is mapped as bunchgrass prairie, which is a primeval, virgin type. Today the area is either cultivated land or an annual grassland dominated by aggressive, introduced species; it is highly unlikely that the original grassland could return even if man and his domesticated animals were to vanish. Sierran montane brush fields that today cover thousands of hectares are mapped instead as forest. Fire and logging have removed the original forest cover, but young conifers are slowly growing up through the brush, and forest is the potential vegetation if given enough time. In contrast, many other montane areas are mapped according to existing vegetation. It is likely that most vegetation maps show a bit of all three approaches and are no more inconsistent than this California map. Küchler and McCormick (1965) have compiled a valuable multivolume reference to all published vegetation maps for the world, and Küchler (1967) has written a book dealing with the procedures of vegetation mapping.

Remote Sensing

Remote sensing literally means to observe and measure an object without touching it. In environmental science, it refers to the use of sensors detecting electromagnetic radiation reflected (visible, near infrared, shortwave infrared, and microwave) or emitted (thermal infrared) from vegetation and soil surfaces (Short 1982; Curran 1985). Because satellite sensors acquire data in infrared and microwave spectral bands in addition to visible bands, they provide a new source of information for research. Since the data are recorded as numerical brightness values for each band, statistical analysis and data transformation are possible. Photograph-like pictures are routinely made for visual presentation by assigning shades of gray to digital values, and false-color composites are formed by combining the data from several spectral bands.

The field of environmental remote sensing really developed with the space program and the Landsat satellite series, beginning in 1972. Today, many second-generation satellites are in operational use, such as the Landsat 5 and the French Earth Observation Test System, SPOT (Barrett and Curtis 1982; Curran 1985). These systems have better resolving capability than earlier sensors because they detect smaller ground units, with more and narrower spectral bands capable of greater sensitivity. Reflectance bands for the Landsat 5 sensors, the multispectral scanner (MSS), and thematic mapper (TM) are shown in Figure 10-4. Third-generation sensors, such as the imaging spectrometers, will be capable of pro-

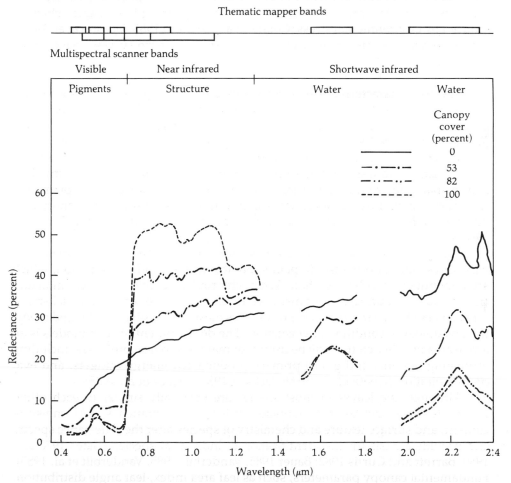

Figure 10-4 Reflectance properties of bare soil (0 canopy cover) and of different amounts of vegetation cover (53, 82, and 100% cover by alfalfa). Reflectance for two wavelength regions, about 1.4 and 1.9 μm, are omitted from the figure because they contain "noise" from nonvegetation sources. (Redrawn from Short 1982.)

ducing high-resolution spectra (bandwidths < 10 nm) at spatial resolutions of 10 to 15 m from space and will be operational in the 1990s (Goetz et al. 1985*a* and *b*). We may soon be able to identify vegetation features on the Earth's surface as accurately as substances are analyzed in laboratories today. Global coverage by the Landsat 5 satellite is repeated every 18 days, and the data are stored by the National Oceanic and Space Administration.

Photo interpretation for environmental classification and mapping remains the most widely used application of this information (Colwell 1983). Other uses include determining the spatial distribution of vegetation types (Carneggie et al. 1983; Meyers 1983), patterns associated with habitat boundaries and land use (Short 1982), local environmental conditions such as location of heavy metals or hydrocarbon seeps (Goetz et al. 1983), short-term vegetation changes associated with canopy moisture content (Jackson and Ezra 1985), phenological changes associated with the growing cycle (Justice et al. 1985), physiological changes induced by environmental stress (Collins et al. 1981 and 1983; Jackson et al. 1983), habitat disturbance (Heller and Ulliman 1983), and long-term successional changes (Carneggie et al. 1983).

Spectral Characteristics of Leaves and Canopies   Reflectance spectra for leaves are different from those for bare soil, and they differ from species to species. Even within a species, the spectra differ with LAI or biomass, as shown for alfalfa in Figure 10-4. Relatively little energy is reflected by leaves in the visible portion of the spectrum (0.4–0.7 μm), most is reflected in the near infrared (0.7–1.3 μm), and less is reflected in the shortwave infrared (1.5–2.5 μm). This is principally due to the absorption of visible light by photosynthetic pigments, reflection of near infrared by cell wall–air interfaces, and absorption of shortwave infrared by leaf water (Gates et al. 1956; Gausman et al. 1977 and 1978; Gates 1980).

These spectral features are the basis for obtaining information about two basic physiological processes in plants, photosynthesis and transpiration, which are dependent on solar radiation. When chlorophyll content and leaf moisture change, such as during development, senescence, or under environmental stress, reflectances change in predictable ways. This makes it possible to infer and model the physiological condition of vegetation. The development of such models is an active area of research that will be useful for many ecological purposes, including estimating regional and global primary productivity, energy budgets, and eva- potranspiration (Goetz et al. 1985*b*; Jackson 1985; Tucker et al. 1985).

Although the leaves of most species are spectrally similar to each other, canopy features such as the orientation of leaves and branches, LAI, moisture content, and surface texture and chemistry of species alter the reflectance spectra and provide a basis for the identification of individual species (Gates 1970 and 1980; Barrett and Curtis 1982; Bauer 1985; Vanderbilt 1985; Vanderbilt et al. 1985). Fundamental canopy parameters, such as leaf area index, leaf angle distribution, and green biomass, have been measured for a number of crop types (Bauer 1985; Goel 1985). These estimates have led to the routine prediction of the worldwide regional wheat yield 1–2 months before harvest with better than 90% accuracy (Barrett and Curtis 1982) and a procedure for the yearly estimation of crop inventory and acreage for the state of California (Thomas et al. 1984). Although natural vegetation is inherently more difficult to interpret due to greater environmental heterogeneity, ecologists and resource managers are optimistic about applying analytical techniques derived from agricultural systems.

Ordination

Ordination attempts to summarize sampling data in a simpler, less space-consuming fashion than the table method. Even a rather small differentiated table such as Table 10-2 contains 285 cells or bits of data (19 stands × 15 differential species). An ordination of the same data could be one small graph showing 19 points spread out in space. Each point represents a stand, and the distance between points represents their degree of similarity or difference. At a glance, one can see if there are any patterns of relatedness. Are some points (stands) clustered together? Do others seem to form a continuous progression from one extreme to another? The objective of ordination is not to draw lines around similar stands and label them part of an association; rather, it is to show a pattern of continuous relationships. Obviously, much of the information contained in the original data is lost in the ordination diagram, but this loss is a consequence of any kind of data reduction, not just of ordination.

Polar Ordination

The simplest, earliest methods of ordination involve the least complicated computations and the most involvement by the investigator. The first step is to create a matrix of CC values between all possible pairs of stands. For n stands, there will be $(n)(n - 1)/2$ different CC calculations. Figure 10-5a shows hypothetical data for seven stands, A–G; the CC values range from 90 (stands A and G) to 20 (stands A and B).

The second step is calculation of a matrix of dissimilarity rather than similarity. Each pair of stands has an **index of difference (ID)**, which is equal to $100 - CC$ (Figure 10-5b).

The third step is a transfer of the ID values to a graph. There are several ways to make the transfer, and some require the use of computers. In the simplest method, called **polar ordination,** two highly dissimilar stands are selected as endpoints (poles) on a horizontal axis. Figure 10-5c shows a hypothetical example, where stands A and B have an ID of 80, which is the largest of all pairings for stands A–G. The ID of 80 becomes 80 graph units on axis 1. All other stands are now placed on the same axis by plotting their IDs with reference to A and B. For example, stand C has an ID of 67 with A and an ID of 63 with B. A compass with a radius of 67 units is swung from A, and a compass with a radius of 63 units is swung from B, forming two arcs. The intersections of the two arcs define a line perpendicular to the axis. Where that line crosses the axis is the position of stand C on that axis. The length of the line from where the arcs cross to the axis (e) is a measure of the poorness of fit for stand C to this one-dimensional

	A	B	C	D	E	F	G
A	—	20	33	65	30	86	90
B		—	37	60	70	52	50
C			—	55	68	49	45
D				—	70	70	75
E					—	45	47
F						—	40

CC values

(a)

	A	B	C	D	E	F	G
A		80	67	35	70	14	10
B			63	40	30	48	50
C				45	32	51	55
D					30	30	25
E						55	53
F							60

ID values

(b)

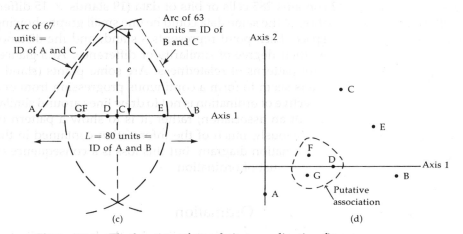

(c)

(d)

Figure 10-5   The location of stands in an ordination figure.
(a) Community similarity matrix for seven hypothetical stands, A–G,
showing community coefficients (CC). (b) Community dissimilarity
matrix for the seven stands, showing indices of difference (ID). (c) After
the endpoint stands A and B have been located on axis 1, all other
stands are located between them by swinging arcs that correspond to
their ID values with stands A and B. Stand C, for example, has an ID
of 63 with B and an ID of 67 with A. The crossing arcs define a line
perpendicular to the axis; where that line crosses the axis is the location
of stand C. The poorness of fit is the distance e, from the crossing arcs
to the axis. (d) Creation of a second axis serves to pull the stands apart
much better than did the first axis alone. Stands F, G, and D (within
the dashed line) appear to cluster together and they represent an
approximately homogeneous unit, perhaps an association.

graph. The location of stands on the axis can also be calculated instead of plotted.
The location of stand C, for example, from the left end of the axis (point A) is:

$$x = \frac{L^2 + d\text{AC}^2 - d\text{BC}^2}{2L}$$

(Equation 10–1)

where x is the distance along the axis from the left end, L is the ID of the endpoint
stands (A and B), dAC is the ID between stands A and C, and dBC is the ID
between stands B and C. The poorness of fit, e, for stand C is:

$$e = \sqrt{dAC^2 - x^2} \qquad\qquad \text{(Equation 10–2)}$$

e varies for each point on the line, which means that the quality of fit for each point varies. Two points may be next to each other on the x axis because they have similar IDs with the endpoints. However, if one point has a high e value, it does not fit the line well and therefore does not really belong next to the other point; the proximity of the two points is an artifact created by this method. A two-dimensional graph would provide a more accurate analysis of the data.

To spread the points in two dimensions, another axis must be generated, with its own reference stands. One reference stand may be that stand with the largest e value, as calculated above. In the hypothetical example of Figure 10-5, stand C fills that criterion. The other reference stand is now the stand that has a large ID value with stand C. However, the objective of creating a second axis is to spread out stands that are close together on axis 1 (the x axis), so by agreement the search for the other reference stand is limited to $\pm 0.1L$ units away from C on axis 1. In this case, L is 80 units long, so stands ± 8 units from C are examined. D then becomes the other reference point of axis 2. Now all other stands, R, can be placed along axis 2 using a formula similar to that used for axis 1:

$$y = \frac{L'^2 + dCR^2 - dDR^2}{2L'} \qquad\qquad \text{(Equation 10–3)}$$

where y is the number of units along the second axis from stand D in our example, dCR is the ID between the new stand, R, and the C stand at one end, dDR is the ID between the new stand, R, and the D stand at the other end, and L' is the ID between the reference stands, C and D. The values of x and y then become the coordinates for locating the stands on a two-dimensional graph (Figure 10-5d).

A third axis may also be constructed, generally at right angles to the first two axes, but often two dimensions serve to spread the stands out well enough and to account for most interstand variability. Step-by-step procedures for this method are well illustrated in a laboratory manual by Cox (1985).

Some Limitations This method of ordination is called *polar ordination* to emphasize that the reference points of each axis, the poles, are chosen by the investigator. It was the first ordination method to be widely applied, and it was developed by Bray and Curtis (1957). There are some important limitations to the method.

First, the ordination figure that results does not explain the "why" of vegetation any better than the table method. If the axes do lead to a good separation of stands, then the next level of questioning concerns the meaning of the axes. Do they correlate with certain environmental factors? Do they represent environmental gradients of major importance to plant and community distribution? To answer that, one has to go back to the stands in nature and start looking at the microenvironment of each. Further, it will be difficult to relate environmental factors in any exact way to the axes, because species and stands will probably not respond to any factor in a linear fashion.

Second, since the objective of ordination is not the separation of stands into associations, the results do not lend themselves to the aims of community classification. If some stands appear to cluster together, as stands D, F, and G do in Figure 10-5d, we may be tempted to put a dashed line around them and call them representatives of a relatively homogeneous association, but there are no guides to draw such a dashed line with any precision. (For example, should stand A also be included?)

Third, the original data of stands and species are lost to the reader. Most research papers that include ordinations do not include the raw data, so the reader cannot get much more from the ordination graph than what the author concluded from it. There is no way for someone with a different point of view to reach an independent conclusion, and this is not good science. After all, no one yet has the final vision of "truth"; all we do is improve our approximation of it. The truncated data of ordination may not permit that improvement process. In fairness, however, it must be added that this criticism applies to some extent to all the data analysis methods presented in this chapter.

Fourth, the array of stands in an ordination may be changed by selecting a different CC formula or by choosing the reference stands differently. If some few stands are quite unlike any others, the polar ordination method just described will not yield much separation of stands. The endpoints will be far apart, but all other stands will cluster between them. Some subjectivity has to go into the selection of reference stands. Ordination works best when the extremes are not too far apart. A partial solution to this problem of subjectivity is the use of different ordination methods.

Other Ordination Techniques

All ordination procedures attempt graphic display of stands (or species) in two or more dimensions, so that patterns of relationship can more easily be seen. When ordinated, stands of similar composition will be located closer together in the ordination figure than they will to other, dissimilar, stands. Second- and third-generation techniques have become less subjective by placing the "rules" in the realm of computer subroutines rather than in the hands of the investigator.

The development of reciprocal averaging (RA) and principal components analysis (PCA) ordinations was an improvement over polar ordination, because the computer is directed to select the single "best" separation of stands in ordination space. All possible combinations are internally calculated and continually adjusted ("weighted") until stand positions stabilize and the final figure is produced. One weakness was that the second axis was sometimes dependent on the first, rather than being mathematically independent. A second weakness was that stands on the extreme ends of an axis sometimes came to be positioned closer than expected, resulting in a horseshoe array or an arched set of points instead of a linear or uniform spread of points. This distortion was sometimes traced to inclusion of too much inter-stand difference or of some stands with unusually few species.

A third-generation technique, *detrended correspondence analysis* (DCA or DECORANA), avoids those weaknesses. Mathematically, it ensures that similar floristic or ecological differences between stands will be expressed by similar distances on the ordination figure. The process and the computer program are best described by Hill (1979) and Hill and Gauch (1980). More technical comparisons of ordination techniques appear in Pielou (1984), Gauch (1982), Orloci and Kenkel (1985), and Minchin (1986). We may expect that the current popularity of DCA will shift in the future to techniques not yet developed that will prove to be even less ambiguous. Whatever the method, ordination figures are only diagrams that reveal patterns. The ecological explanation for the patterns is the ultimate goal, and these explanations must eventually be experimentally determined by ecologists who are biologists as well as mathematicians.

Gradient Analysis Ordination techniques so far described are examples of **indirect gradient analysis:** Only after the mathematical procedures have spread the stands in ordination space do we attempt to look for environmental gradients that will explain the figure axes. **Direct gradient analysis** is a type of ordination, but the stands are not spread out along mathematically determined axes. At the very beginning of the sampling, the investigator determines that the axes shall represent some obvious environmental gradient. The quadrats are located along that gradient, rather than randomly. The environmental gradients are generally complex. For example, quadrats located along an elevational cline from a shaded, moist stream bank to an exposed ridge with dry, shallow soil lie on a gradient that involves changes in temperature, moisture, winter snow cover, light intensity, and probably soil nutrient status. Each quadrat along that gradient is given a relative, but numerical, gradient position value.

The importance value or other traits of major species are then plotted on a graph, in relation to quadrat position along the gradient. Figure 10-6 shows how

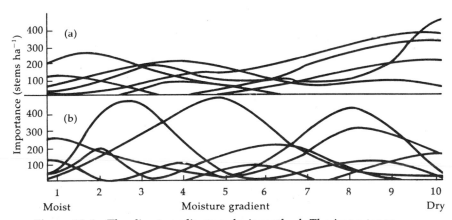

Figure 10-6 The direct gradient analysis method. The importance (measured as density, stems ha^{-1}) of several species is plotted as a function of stand position along a complex environmental gradient in the Siskiyou Mountains of Oregon (a) and the Santa Catalina Mountains of Arizona (b). Each line represents a different species. The gradient (horizontal axis) is labeled as a moisture gradient, but it actually relates to several other factors as well. (Reprinted with permission of Macmillan Publishing Co., Inc. from *Communities and Ecosystems* by R. H. Whittaker. Copyright 1975 by Robert H. Whittaker.)

the density of many tree species changes along moisture gradients in the Siskiyou Mountains of Oregon and the Santa Catalina Mountains of Arizona. This method of sampling and data analysis cannot generally be used for classification because few species appear to share the same peaks and endpoints in their curves on the graph.

Direct gradient analysis was first used by Paczowski at the turn of the century, but it was not described in English and applied to American vegetation until the 1950s (Curtis and McIntosh 1951). It has been widely used by Robert Whittaker (1967, 1973, 1975, 1978), among many others. The results of direct gradient analysis have contributed to the individualistic concept of the community as promoted by Gleason (Chapter 8).

Summary

The investigator's biases and objectives affect the choice of method for data analysis. Further, some methods of analysis require certain sampling procedures. In the table method, associations are defined on the basis of differential or characteristic species that have high fidelity and constance values. Associations are presented in a large differentiated table, which manages to preserve most of the original sampling data of species and stands.

In contrast, ordination reduces the sampling data to one or two graphs that show stands as points in space. The distance between stands on a graph represents their degree of similarity, and the graph axes may correspond to gradients of environmental factors. Some limitations to the simplest form of ordination are partially correctable, at the cost of increased calculation, and sometimes the result is difficult to interpret ecologically.

Direct gradient analysis is a form of ordination that requires the investigator to sample along complex environmental gradients, assigning each quadrat or stand a numerical position along that gradient. The direct gradient graph plots the importance of species as a function of each stand's gradient position. Generally, nonsynchronous curves for all the species result, so the graph is not useful for classification.

Cluster analysis uses community coefficient (CC) values of stand pairs to construct dendrograms that show the relatedness of stands. In contrast to the table method, the result invites readers to choose their own criteria for defining associations, and it quantifies the amount of relatedness for stands of the same association and between associations. However, the table method and the cluster analysis method are equally subjective.

Association analysis also builds a dendrogram of stand-to-stand relationships, but its construction is based on differential species rather than on CC values. The differential species chosen are those that show the greatest degree of nonrandom association with other species; they are not chosen on the basis of fidelity and constancy, as in the table method. Normal and inverse association analyses, however, accomplish the same effect as the table method.

Sampling results can also be summarized in the form of vegetation maps. Some mapping schemes permit environmental data to be shown in addition to vegetational data. The sources of mapping data are aerial photographs and ground truth. The level of detail is dependent upon map scale and the skill and biases of the cartographer. Some vegetation maps show only existing vegetation, but others may include potential vegetation and virgin (prehuman, climax) vegetation. Recent developments in remote sensing techniques not only improve vegetation mapping accuracy and extent but can describe the growth and vigor of plant biomass over large areas.

CHAPTER 11

SUCCESSION

P **lant succession** is a directional, cumulative change in the species that occupy a given area, through time. This definition must immediately be qualified by putting limits on the time involved. Many communities undergo significant, spectacular change through the seasons of one year. A grassland, for example, may be dominated by annual or perennial dicotyledonous herbs in the spring, when the grasses are just beginning their growth; then, in mid- to late-summer, the taller grasses dominate (Figure 11-1). In the hot deserts, annual herbs may form a relatively dense ground cover in spring or summer, providing sufficient rain has fallen some months earlier to trigger germination. The density of desert annuals is notoriously variable from year to year, depending on rainfall, and an additional source of variation is the taxonomic difference between annuals that flower in spring and those that flower in summer (Mulroy and Rundel 1977). Such seasonal changes are not usually included in a definition of succession.

If we look at the other extreme of time—thousands or millions of years—we introduce the factors of climatic change and evolutionary change. There is evidence, of course, that climatic change and genetic change (the formation of new ecotypes, for example) have occurred over shorter periods of time, such as hundreds of years, but these changes are minor compared to climatic changes in the temperate zones over the past 10,000 years or to floristic changes everywhere over the past 10,000,000 years. Such long-term changes are usually not included in a definition of succession.

Most of the thousands of words that have been written about succession have discussed changes occurring in a time span of 1–500 years. If significant changes in species composition for a given area do not occur within such a period, the community is said to be a mature or **climax community**. Climax communities are not static. Changes do occur, but they are not cumulative in their effect. Instead, the random, small changes in plant numbers or even in the flora merely result in fluctuations about some long-term mean. This is a state of **dynamic equilibrium**, similar to a chemical balance in a solution.

If a community does exhibit some directional, cumulative, nonrandom change in a period of 1–500 years, it is said to be a **successional** or **seral community**. It

(a)   (b)

Figure 11-1   Seasonal changes of dominance in a California grassland.
(a) Spring aspect, dominance by several annual forbs; (b) summer aspect,
dominance by grasses.

is often possible to estimate a community's future composition by extrapolating
from changes measured in a short time, by comparing other communities that
have plants of different ages, or by noting differences between overstory plants
and understory seedlings. Seral communities or species will replace one another
until a climax community is achieved. The entire progression of seral stages,
from the first one to occupy bare ground (the *pioneer* community) to the climax
community, is called a **succession** or a **sere** (Figure 11-2).

The phenomenon of succession has been investigated and documented for
hundreds of years. Clements devoted an entire book to the subject in 1916 when
succession and its prevalence in nature came to be widely recognized. Clements
described succession in such a dramatic, authoritarian manner that a fresh, unbiased examination of the subject did not begin until recently, more than a half
century after his book was published (see, for example, reviews in Golley 1977,
MacMahon 1980, and McIntosh 1980).

Clements equated formations (regional climax communities) with organisms. He compared the seral stages that led to formations, and the disturbances
(fire, etc.) that destroyed them, to the developmental stages of an organism:

> . . . the unit or climax formation is an organic entity. . . . As an organism the for-
> mation arises, grows, matures, and dies. . . . Furthermore, each climax formation
> is able to reproduce itself, repeating with essential fidelity the stages of its devel-
> opment. The life-history of a formation is a complex but definite process, comparable
> in its chief features with the life-history of an individual plant.

The parallel between a community and an organism is not, however, as close as
Clements proposed, and the metaphor has muddied some ecological theories
about succession.

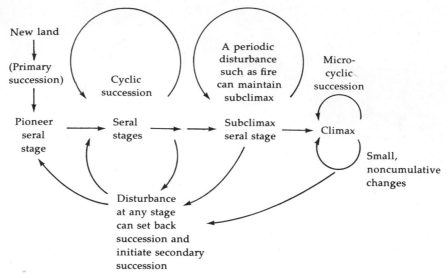

Figure 11-2 Diagrammatic pathway of different types of succession: primary, secondary, and cyclic. The climax stage is in a state of dynamic equilibrium.

Types of Succession

Primary Versus Secondary

New land is continuously being exposed to colonization by plants, such as when volcanic explosions create new islands or cover a pre-existing landscape with a new surface of ash or lava, when a landslide buries a swath of vegetation beneath coarse rubble, or when an advancing sand dune smothers a coastal forest. New land is also exposed by mountain building, by the filling of ponds, by the changing of river courses, and by the expansion of tropical deltas under the influence of silt-trapping mangroves.

The establishment of plants on land not previously vegetated is called **primary succession**. If the pioneer community becomes established on a wet substrate, such as the edge of a gradually filling bog, then the invasion is a **hydrarch** primary succession. If the pioneer community becomes established on a dry substrate, such as exposed granite, then the invasion is a **xerarch** primary succession. In either case, the direction of succession is usually towards the most mesic site and the most mesophytic community possible, given the limitations of regional climate, topography, and soil parent material.

Secondary succession is the invasion of land that has been previously vegetated, the pre-existing vegetation having been destroyed by natural or human disturbances such as wind-throw, fire, logging, or cultivation. However, the barren surface is not as severe as the surface in primary succession because much of the soil remains (although it is sometimes depleted of nitrogen or other nutrients), and many plant propagules (seeds, bits of rhizomes, etc.) are already present in the soil. Consequently, secondary successions often proceed five to ten times faster than primary successions (Mueller-Dombois and Ellenberg 1974; Major 1974.) In some cases, secondary succession can require nearly as much time as primary succession. The example of primary succession detailed later in this chapter for hemlock-spruce forest in coastal Alaska requires 200 years for completion; secondary succession following fire or logging in the same forest requires 130–180 years (Alabock 1982). Secondary succession on abandoned cropland is called **old-field succession**.

Autogenic Versus Allogenic

One of the driving forces behind succession is the effect plants may have on their habitat. Plants cast shade, add to the litter, dampen temperature oscillations, and increase the humidity, and their roots change the soil structure and chemistry. Some of these modifications put the seedlings of overstory species that are less well-adapted to shade at a competitive disadvantage with the seedlings of other species better adapted to the moderated conditions. With time, then, a different association of species will come to dominate the site. Tansley (1935) called succession that is driven in this way **autogenic (biotic) succession**. Both the environment and the community change, and this metamorphosis is due to the activities of the organisms themselves.

Allogenic succession, on the other hand, is due to major environmental changes beyond the control of the indigenous organisms. A slight drying trend over the past 100 years in the southwestern United States may, for example, be *partly* responsible for succession from desert grassland to desert scrub. During the Dust Bowl, accelerated cycles of erosion and deposition of loess (wind blown material) changed the pattern of vegetation. River meanders that fluctuate in time may serve to move the succession process back to a meadow or tundra stage. Changes in sea level or topography may give some coastal sites additional protection from wind and salt spray, permitting the invasion of more inland species. The introduction of exotic, weedy species has changed western grasslands over the past 200 years. These are all examples of allogenic factors, which initiate allogenic succession.

In this chapter, we will emphasize examples of autogenic succession.

Progressive Versus Retrogressive

Succession often leads to communities with greater and greater complexity and biomass and to habitats that are progressively more and more mesic (moist). This type of succession is called **progressive succession**.

Retrogressive succession leads in the opposite direction, toward simpler, more depauperate communities (with fewer species) and toward either a more hydric (wet) or a more xeric (dry) habitat. Some retrogressive successions are allogenic. For example, the introduction of cattle, weedy annuals, and fire to

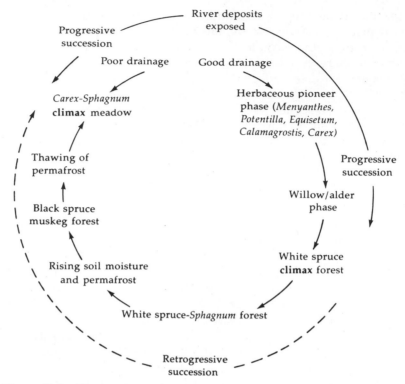

Figure 11-3   Progressive and retrogressive succession on Alaskan flood plains. (From Drury, W. H., Jr. 1956. "Bog flats and physiographic processes in the upper Kuskowin River region, Alaska." *Contributions from the Gray Herbarium* 178:1–30. By permission of the Gray Herbarium.)

Great Basin sagebrush steppe vegetation has resulted in degenerated rangeland (Young et al. 1975). Other retrogressive successions are autogenic. Drury (1956) noted that succession in Alaskan flood plains may at first be progressive, leading from a sedge meadow to a white spruce forest with low shrubs of cranberry and blueberry. The dense shade, however, encourages the growth of a dense moss carpet and the encroachment of a shallow permafrost (frozen soil and water) layer. As the soil moisture rises, sphagnum moss invades, white spruce is replaced by black spruce, and ultimately retrogression to a sedge meadow can result (Figure 11-3).

Another example of retrogressive succession comes from the Mendocino coast of California. There is a series of coastal terraces, elevated and exposed to plant invasion over a period of 500,000 years, which shows degeneration in vegetation and soil environment over time, leading from a rich, mesic forest of redwood, spruce, fir, and hemlock to an open, dwarf forest of pygmy pine and cypress. The dwarf forest is underlain by a very acidic, leached podzolic soil with a shallow hardpan that creates flooding in winter and drought in summer (Figure 11-4) (Jenny et al. 1969).

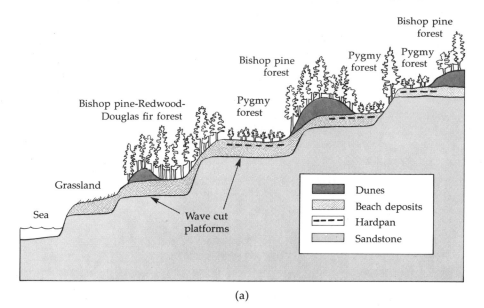

Figure 11-4 (a) The sequence of older terraces along part of the Mendocino coast of California. (b) On the oldest (highest) terraces, soil development has led to the formation of a shallow hardpan; above such soil is the climax community—an open, dwarf forest of pine and cypress, much more depauperate than earlier seral stages of spruce, fir, hemlock, and redwood. [(a) Reprinted by permission from Jenny et al. 1969; (b) Courtesy of H. Jenny.]

(b)

Cyclic Versus Directional

We have, until now, been discussing **directional succession**, which is characterized by an accumulation of changes that leads to community-wide changes. Even in a climax community, however, there continue to be **cyclic** successional changes on a very local scale. These changes occur because the life-span of overstory plants is finite, and their disappearance from the canopy may open the site to an invasion by new species. In some climax communities, the juvenile forms of overstory plants are well adapted to life beneath the parent; when the parent dies, they will replace it in the overstory. In such cases, there will be no local, cyclic succession. In other communities, however, the overstory may inhibit the growth of juveniles beneath it—juveniles of its own kind or those of any species—and in such cases, local, cyclic succession will occur when the overstory plant dies.

Shrub-dominated communities often exhibit cyclic succession. Open areas within desert scrub in Texas, for example, appear to go through a short cycle of invasion by creosote bush (*Larrea tridentata)* followed by invasion by a cactus,

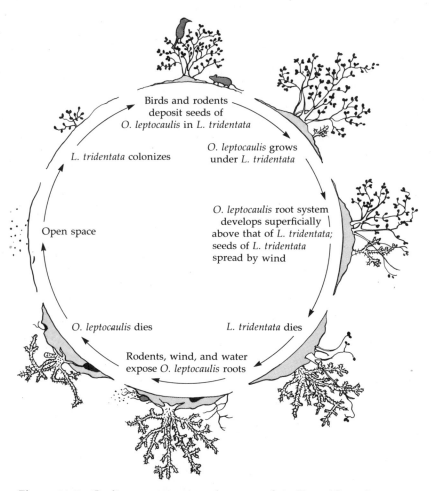

Figure 11-5 Cyclic succession in a desert scrub in Texas (*Opuntia leptocaulis* and *Larrea tridentata*). (From Yeaton 1978. By permission of the British Ecological Society.)

Christmas tree cholla (*Opuntia leptocaulis*), followed by a reversion back to bare ground (Figure 11-5) (Yeaton 1978). Bare sites may be invaded by *Larrea* seedlings because the small seeds and fruits are abundant and may be widely wind-dispersed. Once a shrub is established, it may attract birds and rodents that scatter the seeds and fruits of *Opuntia*. As the cactus grows, its roots may compete for soil moisture with *Larrea*, leading to *Larrea* mortality. Now removed from the protective influence of a shrub canopy, the shallow root system of the cactus may be subject to erosive forces. Large *Opuntia* plants also attract burrowing rodents that further weaken the root system, and the plant dies. Now open space is available for *Larrea* seedlings to invade.

A northern hardwoods climax forest in New Hampshire offers another example of local, cyclic succession. Seedlings of the sugar maple (*Acer saccharum*), American beech (*Fagus grandifola*), and yellow birch (*Betula alleghaniensis)* are not positively associated with overstory parents (Forcier 1975). Beech seedlings and saplings, for example, are positively associated with overstory sugar maple but are negatively associated with overstory beech. This means that when a beech tree dies, its space in the canopy will not immediately be filled by another beech. Forcier concluded that the following cyclic microsuccession would occur (Figure 11-6). Yellow birch, whose seedlings were the most widespread of all three species, would most likely be the first species to fill the gap. It would grow rapidly and a sugar maple understory would develop beneath it (sugar maple seedlings are positively associated with birch overstory). When the relatively short-lived birch died, it would be supplanted by a sugar maple tree. However, beech seedlings are positively associated with maple overstory trees, and beech would ultimately succeed to the canopy when the maple later died. A similar small-scale successional pattern was reported for climax stands of more complex hemlock-hardwood forest throughout the Lake States (Woods 1984). Hemlock trees

Figure 11-6 Cyclic microsuccession in a climax hardwood forest in New Hampshire. (From "Reproductive strategies and the co-occurrence of climax tree species," by L. K. Forcier, *Science* Vol. 189, pp. 808–810, 5 September 1975. Copyright 1975 by the American Association for the Advancement of Science.)

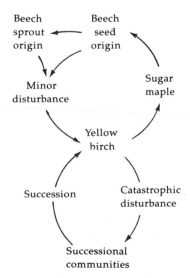

discriminated against beech and maple saplings beneath them, basswood trees discriminated against all other species (except its own sucker sprouts), sugar maple favored beech saplings, and yellow birch saplings were rarely seen under any canopy.

Chronosequence Versus Toposequence

Typically, many plant communities coexist in a complex mosaic pattern. That is, one climax community does not cover an entire region. Sometimes the mosaic reflects a periodic, local disturbance, such as fire, or it might reflect the progressive exposure of new land, as behind a retreating glacier. In the boreal forest of Canada, for example, a spruce-fir forest forms a matrix within which localized patches of meadow, aspen, and aspen with an understory of spruce-fir occur. These patches represent different stages of recovery (seral stages) from fire, wind-throw, or other disturbances to the matrix type. The mosaic expresses a successional relationship and is called a **chronosequence**.

In other cases, the mosaic reflects topographic differences, such as south-facing versus north-facing slopes, basins with poor drainage and fine-textured soil versus upland slopes with good drainage and coarser soil, or different distances from a stress such as salt spray. In such cases, the communities within the mosaic do not bear a successional relationship to one another; they constitute a **toposequence**.

Each community in a toposequence may, in fact, be a climax community. A climax community on level mesic ground that reflects the regional climate is called a **climatic climax community**. Another community may be on atypical parent material or in a poorly drained depression so that the soil will not support the climatic climax; this is an **edaphic climax community**. Still other climax communities occur on north- or south-facing slopes where a unique microclimate will not support the regional (climatic) climax. Each region, then, is characterized by a mosaic of climax types; each region has a **polyclimax** landscape (see reviews by Wittaker 1953, 1973a).

Some ecologists have attempted to fit toposequences into successional schemes, but the results stretch the definition of succession too far and can be misleading. For example, Cooper (1919) and McBride and Stone (1976) summarized succession on dunes in the Monterey Bay region of California as shown in Figure 11-7. In actuality, however, the diagram summarizes a spatial zonation of communities inland from the beach. Open beach and dune pioneer communities characterize areas nearest to shore, with highest winds, sand blast, and salt spray. Dune scrub occurs further inland, and "climax" pine or oak forest occurs yet further inland. "Climax" forest will never invade the scrub or the beach, no matter how much time goes by, because the microenvironment near the beach is too severe. Succession would occur only if the sea level were to drop and the tide line were to regress some distance out to sea, giving more protection from the wind.

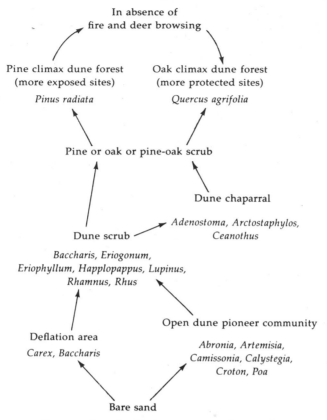

Figure 11-7   Presumed path of primary succession on dunes near Monterey Bay, California, according to Cooper (1919) and McBride and Stone (1976). It might be better to consider the communities as part of a toposequence, rather than successionally related in a chronosequence, for succession may not occur close to shore and oak or pine forest may only be possible some distance inland. (Reprinted from Cooper 1919, courtesy of the Carnegie Institution of Washington. McBride and Stone 1976 reprinted by permission of *American Midland Naturalist*.)

Indeed, sites with a particularly severe microenvironment, such as beaches, salt pans, outcrops of hard, nutritionally poor rock, or deserts, may not support succession. The few, scattered pioneer species able to colonize the habitat constitute the climax species as well, although cyclic succession on a very local scale may occur (Yeaton 1978). Forest Shreve (1942) stated that succession does not occur in deserts, but evidence accumulated since, suggests that both primary succession (Vasek and Lund 1980) and secondary succession (Wells 1961; Vasek et al. 1975*a* and *b*; Vasek 1979; Prose and Metzger 1985; Carpenter et al. 1986) can occur in Mojave Desert habitats. Succession there proceeds through a seral stage of short-lived perennials, and climax is achieved in 40–500 years.

Methods of Documenting Succession

Repeated Measures on One Plot

The most direct, unambiguous way to document succession is to make repeated observations of the same area over time. Permanent quadrats or exclosures can be established, and measurements of plant cover, biomass, density, diversity, demography, or the like can be taken every year, every decade, or over longer periods of time. The establishment and initial sampling of such plots takes a large measure of unselfish foresight, for it is likely that only sampling done by some future investigator 50–100 years later will reveal a pattern of late succession in a long sere. A great deal of demographic information, of course, can be obtained in just a few years (see, for example, Werner 1976). If the area to be followed is relatively large and homogeneous, one may periodically sample a collection of random quadrats within it, a different collection of quadrats serving as the sample each time.

The notes and photographs of nineteenth-century boundary and land surveyors have been used as general summaries of past vegetation, and then compared to modern vegetation for evidence of a succession (Hastings and Turner 1965; Potzger et al. 1956; Whitney 1986). Surveyors' notes, which list the identities of "witness" trees marked in the field at intervals, cannot be taken as random samples without bias, however. Certain species may have been favored because of their large size, long life, commercial importance, prominent position, or simply because the surveyor could identify those and not others (Lorimer 1977). Paired photographs, taken decades apart in time, avoid these biases (see, for example, Hastings and Turner 1965).

Succession may take place rapidly enough to occur within the memory span of residents still living. If so, their passive observations can be accumulated by an investigator as additional documentation of succession. Kerner (1863) used this method in his pioneering description of succession in central Europe.

In forested areas, a comparison of sapling density to overstory tree density by species may reveal successional patterns. Figure 11-8 shows that an oak-beech forest near Washington, D.C., is unlikely to remain stable, for there are no saplings of the most common overstory oak, and very few saplings of the other oaks; instead, beech and maples dominate the sapling layer. One could extrapolate from this information, collected at one moment in time, and predict that the oak-beech forest is seral to a beech-maple forest, but it is important to realize that two assumptions are being made.

One assumption is that the sapling population will not significantly change from year to year or decade to decade. It is not always safe to assume that the pattern of reproduction for all tree species through time is the same. It is possible that some years will yield a large seed crop for one species and many seedlings and saplings for several years thereafter, while another species has a complementary cycle.

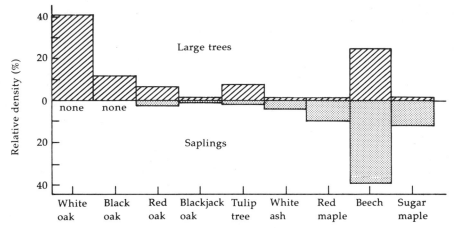

Figure 11-8   Relative density of overstory trees and saplings, by species, for an oak forest near Washington, D.C. The sapling stratum is dominated by beech and maple, possibly indicating that the oak forest (large trees) is seral to a beech-maple forest. (From "Sugar maple in forest succession at Washington, D.C." by R. L. Dix, *Ecology* 38:663–665. Copyright © 1957 by the Ecological Society of America. Reprinted by permission.)

Another assumption is that the mortality rate of saplings is uniform from species to species. This assumption is not always appropriate. For example, the spruce-fir forests of the Appalachians are dominated by spruce in the overstory but by fir in the understory. Nevertheless, the overstory balance does not shift to fir, because fir seedlings are more susceptible to damping-off disease and grow less well in dense shade than spruce (Oosting and Billings 1951).

Observations on Nearby Plots of Different Successional Ages

Most successional schemes have been intuitively determined from observations on nearby plots of different successional ages, rather than from observations on one plot over time. The objective is to find a series of plots that have been exposed to primary succession or disturbed and opened to secondary succession at different, known times. One must assume or demonstrate that all other factors, such as slope, aspect, parent material, and macroclimate, are uniform.

Sometimes the plots of different ages are adjacent to each other, and zoned in obvious bands. Such is the case behind a retreating glacier, and Crocker and Major (1955) used these zones to deduce a path of primary xerarch succession leading from open tundra to spruce forest, as described in the next section. The assumption was made that the spatial sequence of communities back from the glacier's front was repeated in time—or would be repeated in time—for any one location. Thus, where a spruce forest now occurs as a climax community, sometime earlier there had been an alder thicket, and before that an open meadow, and before that ground covered by the glacier itself.

Similar concentric zones of vegetation occur around the edges of filling bogs in the northeastern United States. Open water is followed by (a) a zone of floating or emergent aquatics (the pioneer seral stage), (b) a sedge or sphagnum mat, (c) a zone of shrubs with scattered trees tolerant of waterlogged soil, such as black spruce (*Picea mariana*) or tamarack (*Larix laricina*), and finally (d) the matrix of more mesophytic, climax vegetation of spruce-fir or maple and other hardwoods, depending on the latitude (Figure 11-9).

Figure 11-9   (a) The edge of concentric circles of vegetation around a filling lake. (b) Diagrammatic summary. Presumably, the communities from left to right represent seral stages in a primary hydrarch succession, the floating aquatics being the pioneer stage and the surrounding forest matrix being the climax. (Reprinted with modifications from *Plant Communities* by Rexford Daubenmire 1968*a*. Copyright 1968 by Harper and Row Publishing Co.)

(a)

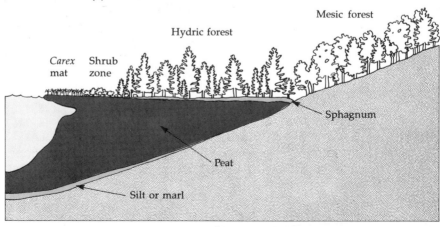

(b)

In many other cases, however, the sites of different ages are randomly scattered throughout a region, and their age sequence must be determined from land use records. Oosting (1942) used this method in documenting secondary succession, as described later in this chapter. He located sites that had been abandoned as cropland 1–100$^+$ years before. The assumption was made that the older plots of vegetation had all progressed through seral stages still present on younger plots. In many western vegetation types, fire, rather than cultivation, is the disturbance that must be dated. Large, recent fires can be dated with assurance from Forest or Park Service records, but older or more local ones must be dated indirectly by noting fire scars on nearby, surviving trees or by carbon dating beds of charcoal that remain in the soil. A series of charcoal and silt deposits beneath a coast redwood grove (*Sequoia sempervirens*) in northern California has permitted a history of fire and flood disturbance to be traced back 1000 years (Zinke 1977).

An Example of Primary Xerarch Succession   Crocker and Major (1955) conducted an elegant, thorough study of succession behind a retreating glacier in Glacier Bay, Alaska, describing soil and vegetation changes that take place in a period of 200 years, from pioneer mosses on bare rock glacial debris to a spruce-hemlock climax forest. Their work built on an earlier description of the area by Cooper (e.g., 1923 and 1931).

A period of 200 years is exceptionally short for primary succession; the short time no doubt reflects (a) the fact that glacial till decomposes easily, (b) the wet, moderate climate, and (c) the presence of nitrogen-fixing plants (*Alnus, Dryas*) in the sere. Reference to primary succession rates elsewhere helps to put 200 years into perspective:

1. Rainforest can develop on fresh lava in Hawaii in 400 years (Atkinson 1970).
2. A pine scrub can develop on bare granite outcrops in Georgia in 700 years (Burbanck and Platt 1964; Burbanck and Phillips 1983).
3. A spruce-hemlock forest can develop on river terrace sediment in the Hoh Valley of Washington in 750 years (Fonda 1974).
4. A deciduous forest can develop on dune sand around Lake Michigan in 1000 years (Olson 1958).
5. A sagebrush-bitterbrush desert scrub can develop on inland dunes in Idaho in 1000–4000 years (Chadwick and Dalke 1965).
6. A moss-birch-tussock grass tundra can develop on glacial debris in Alaska in 5000 years (Viereck 1966).

The retreat of several glaciers in Glacier Bay has been well documented since 1760 by human records, according to Crocker and Major (1955). John Muir, for example, built a cabin near the ice front in the 1890s and that site is now 24 km from the ice front. Retreat prior to then had been even more rapid; altogether, the glaciers have retreated nearly 100 km in 200 years.

The zonation of communities back from the present position of the glacier suggests the sere and the accompanying soil changes summarized in Table 11-1.

Table 11-1   Primary xerarch succession at Glacier Bay, Alaska. The table includes dominant species, canopy cover, soil bulk density, soil pH, and biomass of litter on the soil surface. (From Crocker and Major 1955. By permission of the British Ecological Society.)

Year	Community name and dominant species	Cover (%)	Bulk density (g cm^{-3})	Soil pH	Litter (kg m^{-2})
0–15	Pioneer: moss (*Rhacomitrium*), fireweed (*Epilobium*), horsetail (*Equisetum*), *Dryas drummondii*	<50	1.5	8.0	0
15–35	Willow: willow shrubs (*Salix*, 3 species), alder (*Alnus*), *Dryas*, cottonwood (*Populus*)	85	1.3	7.5	1.0
35–80	Alder thicket: alder (*Alnus crispa* ssp. *sinuata*)	100	1.2	6.0–5.0	3.0
80–115	Transition to forest: alder (*Alnus*), Sitka spruce (*Picea sitchensis*)	100	0.9	5.0	6.0
115–200	Spruce forest: Sitka spruce (*Picea sitchensis*)	100	0.8	4.8	9.0
>200	Hemlock-spruce forest: hemlock (*Tsuga heterophylla, T. mertensiana*), spruce (*Picea*), *Vaccinium* shrubs	100	?	?	?

Shortly after the parent rock is exposed, it is colonized by pioneer mosses and vascular plants, the most important of which is *Dryas drummondii* (Figure 11-10), which forms extensive mats and harbors nitrogen-fixing actinomycetes in root nodules. About 15 years after the substrate has been exposed, these pioneers are joined by patches of willow and alder shrubs and an occasional cottonwood tree. Gradually, the alder becomes dominant and the canopy closes to give an alder thicket, about 35 years after the onset of succession. Soil changes occur most rapidly in this seral stage because alder root nodules also contain nitrogen-fixing actinomycetes, and a great amount of litter is added to the soil. Sitka spruce (*Picea sitchensis*) saplings become established and grow up through the alder canopy; as they form a closed overstory, the alders senesce. This long transition period from alder thicket to spruce forest takes place 40–115 years after the start of succession. About 200 years after the pioneer stage, the spruce forest is invaded by two species of hemlock, which ultimately become dominant. An understory

Figure 11-10 (a) *Dryas drummondii* (× ½). (b) scattered *Dryas drummondii* discs and small thickets of *Alnus crispa sinuata* on a 15-year-old surface near Glacier Bay, Alaska. *Dryas* is a dominant of the pioneer community and a plant which contains symbiotic nitrogen-fixing actinomycetes in root nodules. (Photo courtesy of D. B. Lawrence, Professor Emeritus, Dept. of Botany, University of Minnesota, St. Paul, MN.)

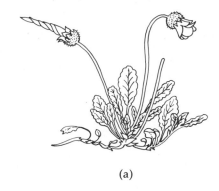

(a)

(b)

of mosses and low ericaceous shrubs completes the physiognomy (Figure 11-11).

Soil factors that Crocker and Major (1955) examined along the sere include pH, nitrogen content, and bulk density. Bulk density declines as the porosity of the soil increases. In this sere, bulk density was cut in half, revealing an improvement in soil properties favorable for root growth. Litter biomass and soil nitrogen also increased, though nitrogen reached a peak prior to the climax (Figure 11-12). Soil pH shifted from a very basic reaction (the marble till has a high calcium content) at the start of succession to a very acidic reaction (caused by 100–200 years of soil leaching and the acidic reaction of decomposing conifer needles). Successional studies elsewhere by other researchers have shown that additional soil features, such as texture, depth, and water holding capacity, also change during succession. The soil microflora also undergoes succession (Webley et al. 1952).

Figure 11-11   Sitka spruce forest on the terminal moraine from which ice receded about 200 years earlier. Younger tiers of trees are on an emerging shore that is rising 2.5 cm per year due to release from ice loading. Glacier Bay, Alaska. (Photo courtesy of D. B. Lawrence, Professor Emeritus, Dept. of Botany, University of Minnesota, St. Paul, MN.)

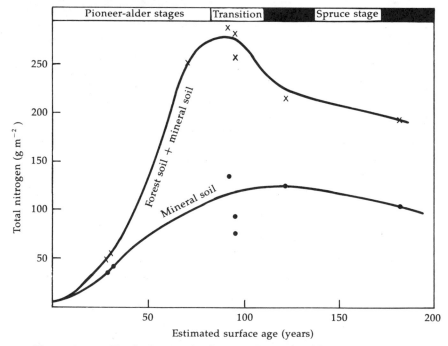

Figure 11-12   Total nitrogen in the top 45 cm of soil beneath various seral communities at Glacier Bay, Alaska. (From Crocker and Major 1955. By permission of the British Ecological Society.)

An Example of Secondary Succession   The Piedmont (literally, foothill) region to the east of the Appalachians is a gently rolling, still heavily forested area with a mosaic of hardwood forest, pine forest, recently abandoned fields, and many cultivated fields. Unlike the large farms of the Midwest and West, farms in the Piedmont region may extend for only a few hectares. Most farms are planted with corn and cash crops such as tobacco. Secondary succession, leading from freshly abandoned cropland to climax forest, has been examined in the Piedmont region by several ecologists (Bard 1952; Billings 1938; Keever 1950 and 1983; Oosting 1942) and it is reasonably well documented.

Oosting (1942) sampled abandoned fields of varying age in North Carolina, trying to replicate each age with fields as similar as possible in soil, slope, aspect, and other physical features. His conclusions as to the vegetation on one-year-old fields, for example, were based on five fields. On older, forested plots, he compared sapling density with tree density as an additional source of indirect evidence for succession. The sere is summarized in Table 11-2.

One-year-old fields, sampled in the early summer one year after abandonment, show a total of 35 species, all annual or perennial herbs. Not every field exhibits all 35 species, but two species have the highest consistent density and frequency on all fields: crabgrass (*Digitaria sanguinalis*) and horseweed (*Conyza canadensis*). These are the pioneer dominants. Two-year-old fields still contain these two species, as well as virtually all the first-year species, but the dominants are aster (*Aster ericoides*, not in first-year fields) and ragweed (*Ambrosia artemisiifolia*, present but not dominant in first-year fields). Despite the addition of 26 new species in second-year fields, the Sorensen community coefficient calculated by presence for first- and second-year fields is high, 0.63.

Three-year-old fields decline in species richness because of the almost complete dominance of large clumps of the perennial grass broomsedge (*Andropogon virginicus*); (Figure 11-13). *Andropogon* continues its dominance for several years,

Figure 11-13   Old fields dominated by broomsedge (*Andropogon virginicus*), typical of secondary succession three to six years after abandonment and prior to dominance by pine. (*Introduction to Ecology* by P.A. Colinvaux. Copyright 1973 John Wiley and Sons, Inc. Reprinted by permission.)

Table 11-2   Secondary (old-field) succession in the Piedmont region. Only the common names of the dominant species are listed; many other taxa are associated with each seral stage. [From Oosting 1942 (reprinted by permission of *American Midland Naturalist*), Bard 1952 (by permission of the Ecological Society of America), and Richardson 1977 (*Dimensions of Ecology*, by permission of Williams and Wilkins, Baltimore).]

Years after abandonment	North Carolina Piedmont	
0	Cropland	
1	Crabgrass, horseweed	
2	White aster, ragweed	
3	Broomsedge	
5	Broomsedge, pine seedlings	
10	Young pines, Broomsedge	
	Shortleaf pine (drier sites)	Loblolly pine (moister sites)
20		
30		
40		
60	Shortleaf pine, hardwood understory	Loblolly pine, hardwood understory
100		
150	White oak, post oak, hickory, dogwood, etc.	White oak, many hickories, dogwood, sourwood, etc.

but pine saplings gradually overtop the grass and form a closed canopy 10 years after abandonment. The pine seedlings first appear 3–5 years after abandonment, scattered between the clumps of broomsedge, and in any given field they are generally all of one species: either loblolly pine (*Pinus taeda*), shortleaf pine (*P. echinata*), or Virginia pine (*P. virginiana*). Upland sites usually come to be dominated by pure stands of loblolly, with all the trees the same age. Pines reproduce poorly in their own shade, however (possibly because of an obligate requirement for relatively high light intensity or because they are poor competitors for soil moisture; see Shirley 1945), and so a 20-year-old stand will have a well-developed understory of hardwoods. Many of the hardwoods, such as dogwood (*Cornus florida*) and redbud (*Cercis canadensis*), are genetically limited to being understory trees throughout their life-span, but others, such as hickory (*Carya* spp.) and oak (*Quercus* spp.) are future overstory trees, able to replace senile pines.

Pine stands 50–75 years old are over 25 m tall on good sites; beneath them are the oaks and hickories, perhaps 10 m tall, and beneath them are scattered understory hardwoods (Figure 11-14). As shown in Figure 11-15, stands 100 years of age show as many hardwoods in the overstory as pines, and the balance shifts more and more towards hardwoods with time. Climax hardwood stands 200$^+$ years old still have scattered pines in the overstory and a few seedlings and saplings of pine below, but the great majority of overstory trees, understory trees, and saplings are hardwoods.

Secondary succession can stop at the pine seral stage if ground fires sweep the area every several years (and there is evidence that natural fires of this frequency did occur along the southeastern coastal plain centuries ago; see Chapter 16). Hardwood saplings are susceptible to relatively "cool" surface fires, but pines 10 years old or older have a thick enough bark and a tall enough canopy to withstand such fires without injury.

Figure 11-14   A loblolly pine stand with a hardwoods understory, typical of secondary succession in the Piedmont 50 years after abandonment.

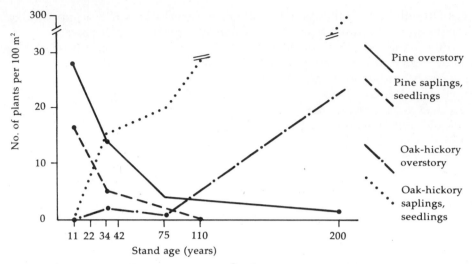

Figure 11-15   Density (per 100 m²) of overstory and understory pines and oaks and hickories throughout secondary succession in the Piedmont of North Carolina. (From Oosting 1942. Reprinted by permission of *American Midland Naturalist*.)

Much of the story of secondary succession in the Piedmont was originally elaborated by Henry Oosting and his student, Catherine Keever. In a retrospective look at research since that time, Keever (1983) concluded that their choice of North Carolina was fortuitous. The patterns of successional change are more distinct, dramatic, and rapid there than in other regions. Piedmont succession may not be typical of succession elsewhere. Now, she would place more emphasis on the role of chance in dispersing propagules to early successional sites, and she would temper broad conclusions made on the basis of any one sample of seral communities.

General Trends During Succession

For half a century, ecologists have been intent on documenting (or inferring) successional pathways in many types of vegetation. Only recently has there been an attempt to synthesize this information and to formulate general theories on the process of succession. Odum (1969) was among the first to attempt to make the transition from taxonomic specifics to functional generalities. He presented an impressive table of 24 ecosystem traits which he thought changed significantly during succession. He was careful to explain that documentation for these trends was very scanty, in some cases coming from laboratory studies of microsuccessions in aquaria. He presented his table in the spirit of a hypothesis that invited testing, rather than as established fact.

Odum's table has been a great stimulus to further research and thought on succession (see, for example, Drury and Nisbet 1973). Some of his hypotheses

Table 11-3 Some vegetation and ecosystem traits that often change during progressive succession. The status of each trait is shown for early and late stages of succession (*not* for pioneer and climax stages necessarily, for some trends peak at intermediate seral stages). Each trend is briefly discussed in the text.

Trait	Early stages	Late stages
Biomass	Small	Large
Physiognomy	Simple	Complex
Leaf orientation	Multilayered	Monolayered
Major site of nutrient storage	Soil	Biomass
Role of detritus	Minor	Important
Mineral cycles	Open (leaky), rapid transfer	Closed (tight), slow transfer
Net primary production	High	Low
Site quality	Extreme	Mesic
Importance of the macroenvironment	Great	Moderated and dampened; less
Stability (absence or slowness of change)	Low	High
Plant species diversity	Low	High
Species life-history character	*r*	*K*
Propagule dispersal vector	Wind	Animals
Propagule longevity	Long	Short

have not been supported by this later work, others have been amended, some still appear to be valid, and a few have been added. We have chosen 13 community and ecosystem attributes to illustrate successional trends (Table 11-3). Some items are paraphrased from Odum's table. Each trait will be discussed briefly in the text following. It is important to remember that the table is only concerned with progressive, directional succession, that early seral stages are being compared to later stages (not pioneer to climax, for some trends peak in mid-to-late succession rather than in the climax stage), that there are exceptions to every item, and that the rate of change in any factor is probably not uniform throughout succession (Major 1974).

Vegetation and Site Quality

Biomass increases during succession. Plant cover, the density of foliage above the ground (the leaf area index; see Chapter 8), and the height of the plants also increase. If succession leads from herbs or shrubs to trees, the percentage of total biomass that is below the ground may decline, but the total biomass still increases.

Physiognomy increases in complexity because the variety of growth forms increases as succession proceeds. If successional stages involve trees, the structure and leaf orientation of those trees may change (Horn 1971 and 1975). Tree species characteristic of early successional stages, such as aspens or pines, tend to be tall, thin, conical, and capable of rapid growth; their leaves are small, numerous, and randomly oriented and are borne in such a way that many leaves are shaded by others above them (multilayered) (Figure 11-16a). Tree species characteristic of the climax community may have a similar profile but grow more slowly, have fewer and larger leaves, and bear those leaves in a planar fashion such that self-shading is minimized (monolayered) (Figure 11-16b).

The major site of nutrient storage in the ecosystem shifts from soil to plant biomass (perennial roots, storage organs, and tree trunks). The role of detritus (leaves, twigs, and other litter) in nutrient cycles thus becomes more important during succession, because soil nutrients have been depleted and are being stored for long periods of time in plant biomass. Detritus carries at least a fraction of the nutrient pool back to the soil, where it can be tapped to satisfy new plant growth demands the next season.

The speed with which nutrients cycle from soil to plant and back again slows down during succession because many nutrients are stored in long-lived but inert parts of plants. Such storage does, however, prevent erosional losses of nutrients out of the ecosystem, so we can hypothesize that the cycling of nutrients is more efficient later in succession, even though it occurs very slowly.

Net primary production (photosynthesis minus respiration) may decline with succession, for a variety of reasons. (a) There is a great deal of supporting but nonphotosynthetic tissue whose maintenance (respiration) reduces net production. (b) Nutrients may be limiting because of their storage in inert tissue. (c) Many plants in the overstory may be senescent, with lower photosynthetic rates than young plants. (d) The leaf orientation of climax trees may be inappropriate for high photosynthetic rates (Horn 1974).

The environment becomes more mesic during succession, and the effect of the macroclimate is dampened. As the canopy closes, diurnal fluctuations in temperature and humidity are moderated. Increased humus, increased soil depth, and a finer soil texture in later successional stages lead to greater soil moisture retention and a buffering of seasonal changes in precipitation.

Stability and Diversity

Stability has many definitions. Depending on the definition, stability may increase or decrease with succession (Colinvaux 1973; Horn 1974; Ricklefs 1973; Woodwell and Smith 1969). If we use the most straightforward definition—that stability is equivalent to lack of change—then stability increases with succession: The climax species are longer-lived, and any changes occurring in the undisturbed climax community are in the form of random, minor fluctuations around some long-term mean. Table 11-3 uses this definition. If stability is defined as resistance to minor changes in the macroenvironment, then stability also increases with succession because the plant cover dampens fluctuations and extremes in

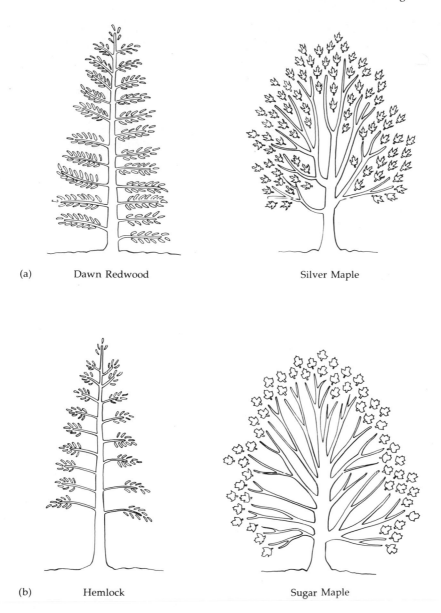

(a)   Dawn Redwood   Silver Maple

(b)   Hemlock   Sugar Maple

Figure 11-16 Distribution of leaves in multilayered trees (inner leaves shaded) and monolayered trees (self-shading minimized). (a) Dawn redwood and silver maple, which are multilayered; and (b) hemlock and sugar maple, which are monolayered. (Reprinted by permission from "Forest succession" by H. S. Horn. Copyright © 1975 by *Scientific American, Inc.* All rights reserved.)

the macroenvironment. If stability is defined as the ability to return rapidly to a balance point **(homeostasis)** following a major recurring disturbance such as a fire or windstorm, then preclimax communities are the most stable and climax communities are the least stable, requiring hundreds of years to return. If we adopt such a definition, then we have to agree with Henry Horn (1974) that climax communities are fragile:

> Conservationists . . . have often cited the conventional generalization that diversity conveys stability, arguing that diverse [climax] natural communities should be conserved for their stabilizing influence. . . . One could equally argue that if complex [climax] systems were inherently stable, they should need no protection. The opposite view, that . . . climax communities are inherently fragile, is a much more powerful reason for their requiring protection. . . .

Plant species diversity (see Chapter 8) increases throughout early succession, but it decreases in the temperate zone in late succession as the canopy closes and a few species become major dominants. Thus, in the temperate zone, a periodic local disturbance that sets succession back to earlier seral stages is necessary to maintain maximum diversity (Loucks 1970). There are exceptions to this trend of increasing diversity, just as there are exceptions to all the trends listed in Table 11-3. Old-field succession in Oklahoma, for example, shows a steady decline in species diversity from pioneer weeds through climax prairie (Perino and Risser 1972). Diversity is nearly halved, even though species richness (the total number of species) rises throughout the sere.

Autecology

Pioneer species are typically r-selected, with rapid growth rates, short life-spans, relatively large energy allocations to sexual reproduction, and high photosynthetic rates. Climax species are typically K-selected, with slow growth rates, long life-spans, small energy allocation to reproduction in any one year, lower photosynthetic rates, and heavy, animal-dispersed seeds. However, r-selected species can still exist in climax communities. They may occupy sites beneath openings in the overstory, and their seeds can lie dormant in the soil for long periods of time until such openings occur. Or they may have light, abundant seeds that are imprecisely dispersed by wind.

It has been hypothesized that, in general, more plant–animal, plant–plant, and plant–microbe interactions (what Margalef (1968) calls *information*) occur in later successional stages than in early ones. There is no conclusive evidence to show that this hypothesis is correct. In fact, some pioneer communities composed of lichens or nitrogen-fixing plants may show a more complete dependence on interactions than climax communities. Certainly, many of the examples of allelochemics and other types of interactions discussed in Chapters 6 and 7 come from seral rather than climax communities. This could be just coincidental, however, reflecting greater research interest in seral communities.

Driving Forces of Succession

Clements and Relay Floristics

Clements wrote of succession as a six-step process:

1. **Nudation**, the exposure of a new surface in primary succession or the clearing away of previous vegetation in secondary succession.
2. **Migration** of seeds, spores, or vegetative propagules from adjacent areas, though in secondary succession many of these are already present in the soil.
3. **Ecesis**, the germination, early growth, and establishment of plants.
4. **Competition** among the established plants.
5. **Reaction**, the autogenic effects of plants on the habitat.
6. **Stabilization**, the climax.

Clements visualized succession as composed of several discrete seral communities. The associated species of each community undergo steps 2–5 together, changing the habitat in such a way that new species (associates of the next seral community) are at a competitive advantage and thus supplant the previous community. The process is repeated until a climax community develops that can propagate itself continuously and prevent the establishment of additional taxa.

This view of succession is called **relay floristics** (Egler 1954; McCormick 1968); it is diagrammed in Figure 11-17. Each seral community relays the site to the next community. The driving force behind succession, then, is the reaction of the site to the plants living on it.

An example of relay floristics in primary succession comes from a study of rock outcrops in the foothills of the southern Appalachians by Sharitz and McCormick (1973). Small islands of vegetation, 3–5 m in diameter, are scattered over the rock surface (Figure 11-18), and each is made up of concentric zones of seral communities. The outermost zone, on bare rock, consists of mosses and lichens; the next zones on 1–4 cm of soil are dominated by annuals; more interior zones on deeper soils up to 15 cm thick are dominated by herbaceous and woody perennials. The spatial distribution of some species can be quite distinct. The pioneer species modify the habitat by contributing to the formation of soil and thereby aid in the establishment of plants in the following seral stage. Sharitz and McCormick showed that competition for soil moisture is responsible for the abruptness of early seral stages. The stonecrop *Sedum smallii*, for example, which dominates the community on the shallowest soil, has a high tolerance for low soil moisture and soils less than 4 cm deep. Stonecrop is competitively inferior to *Minuartia uniflora*, however, which dominates the adjacent interior community on soils 4–10 cm deep.

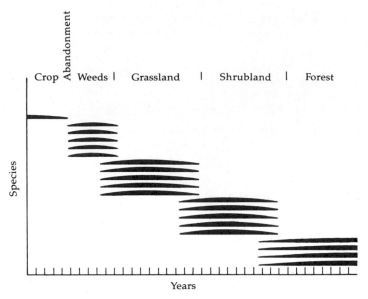

Figure 11-17 Diagrammatic summary of the relay floristics model of succession. The thicker the line, the more important the species at a given time. (From "Vegetational science concepts. I. Initial floristic composition, a factor in old-field vegetation development" by F. E. Egler. *Vegetatio* 4:412–417. Reproduced by permission of Dr. W. Junk BV.)

Figure 11-18 The edge of an island of vegetation on a rock outcrop, showing several seral stages. The outermost community dominated by lichens and mosses (0) lies beyond one dominated by *Sedum smallii* (1), which is followed by a community dominated by *Minuartia uniflora* (2). More interior communities (3) are dominated by perennial plants. (From "Population dynamics of two competing annual plant species" by R. R. Sharitz and J. F. McCormick, *Ecology* 54:723–739. Copyright © 1973 by the Ecological Society of America. Reprinted by permission.)

Chance and Initial Floristic Composition (IFC)

If the relay floristics model of succession is entirely correct, then we should expect to find groups of species appearing and disappearing together during the course of succession. We would also have to agree with Clements that the course of succession is predictable, for each seral stage relies upon the previous one. Some recent work on primary and secondary succession shows that these assumptions are not always correct.

In England, Walker (1970) took corings of sediment from the edges of 66 lakes that had been undergoing hydrarch succession for hundreds of years. Sediment had been laid down in the order of the succession that had actually occurred, with the deepest sediments harboring the oldest plants. Because of the slow rate of decomposition, plant remains could be identified throughout the cores. Walker examined the sequence of sediments in each core to see if the sediments corresponded to the traditional sequence of succession. His results showed that there was considerable variability in the progression of seral communities. In fact, the "traditional" sere was never followed.

Frank Egler (1954) initiated plots of secondary succession on his forested estate in Connecticut, and similarly concluded that the progression of seral communities was neither fixed nor predictable. He concluded that the path of succession (at least in the early stages) is driven by chance and the differential longevity of plants. Chance determines which propagules are in the soil or soon reach the site at the time succession begins. He referred to this collection of starting propagules as the **initial floristic composition (IFC)** of the site. All of the pioneer species, many of the seral species, and some of the climax species are present in this initial floristic composition. Some of them germinate and become established quickly, others germinate quickly but grow more slowly and for a longer period of time, and still others may become established even later. The larger, longer-lived, slower-growing species ultimately outcompete the smaller pioneer species, and dominance shifts. The site is not relayed from one collection of species to another in this view of secondary succession. Rather, there is a sorting out of species one at a time, based mainly on their longevity (Figure 11-19).

Egler did not present any data to document his IFC theory of succession, but others subsequently did. McCormick (1968) manipulated 36 plots of abandoned cropland at an experiment station in Pennsylvania and probably accumulated more data on old-field succession than exist anywhere else. Unfortunately, many of his findings have not yet been published. He found a great deal of variability in the composition of early seral communities. Some first-year plots were dominated by young woody vines such as honeysuckle (*Lonicera japonica*) and blackberry (*Rubus* species), rather than by annual herbs. Fourth-year plots "should" have been dominated by *Solidago* and *Aster* species, according to traditional views, but very few plots were so dominated. To test the theory that early stages prepare a site for later stages, McCormick weeded out annual herbs from some plots as soon as the seedlings could be identified. Then he compared the vigor of perennial herbs that dominated subsequent seral communities in both weeded and nonweeded plots. He found that the perennials grew more

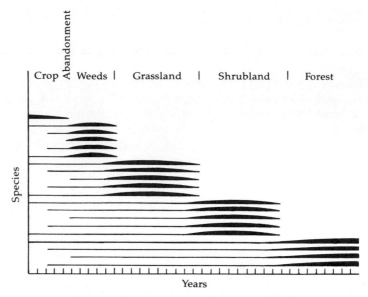

Figure 11-19 Diagrammatic summary of the initial floristic composition model of succession. The thicker the line, the more important the species at a given time. (From "Vegetational science concepts. I. Initial floristic composition, a factor in old-field vegetation development" by F. E. Egler. *Vegetatio* 4:412–417. Reproduced by permission of Dr. W. Junk BV.)

vigorously in the weeded plots, not less vigorously as the relay floristics model would predict.

Similarly, Hils and Vankat (1982) manipulated first-year old fields in Ohio by removing annuals and biennials, or perennials, or particular species (*Erigeron* annual species or *Aster* and *Solidago* perennial species). They then observed the growth and succession of vegetation for several years after the manipulations. Removal of annuals stimulated growth of perennials, and vice versa. The Clementsian model did not apply, because the later perennial-dominated stage did better if the earlier annual-dominated stage was eliminated. Subsequent data (Vankat, personal communication) show a virtual convergence of all plots by the fifth year, despite 100% difference in composition of first-year plots.

A close examination of species lists for old fields that Oosting (1942) compiled shows some agreement with the IFC model. Dominants of later stages were sometimes present in earlier stages, and many species were shared among the first several seral communities, giving a high community coefficient. However, pine seedlings did not appear until broomsedge became dominant, and hardwoods did not appear until a pine overstory became established. Keever (1983) retrospectively pointed out that her first-year plots contained "some individuals of most future dominants . . . [including] asters, broomsedge, and a few pines . . . [but that] oaks and hickories with big heavy seeds are the last to invade."

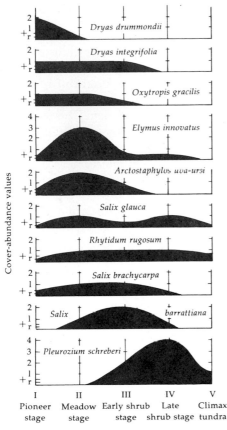

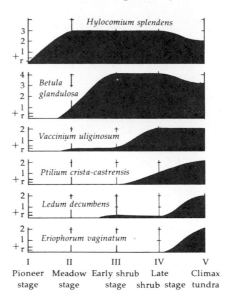

Figure 11-20   The importance (according to the cover-abundance scale of the Braun-Blanquet system) of representative species in a primary sere from glacial outwash to tundra in Alaska. r = rare; + = less than 5% cover and found only occasionally; 1 = less than 5% cover, but commonly found; 2 = 5–25% cover; 3 = 25–50% cover; 4 = 50–75% cover; and 5 = 75–100% cover. (From "Plant succession and soil development on gravel outwash of the Muldrow Glacier, Alaska" by L. A. Viereck, *Ecological Monographs* 36:181–199. Copyright © 1966 by the Ecological Society of America. Reprinted by permission.)

Another example of an overlapping species replacement pattern, rather than an abrupt relay pattern, comes from a primary sere on glacial outwash in Alaska (Figure 11-20). Most of the dominants for each of the five seral stages were present in more than one stage. The moss *Hylocomium*, as an extreme example, was rare but present in the pioneer stage, was very abundant in the next three seral stages, and still provided 5–25% cover in the climax tundra. Only *Dryas* and *Eriophorum* were restricted to one stage.

Modern Syntheses

There has been increased interest in succession over the past decade, and new models of succession have been proposed. In general, these draw upon and validate both Clements' relay floristics ideas and Egler's IFC ideas. An important stimulus to this renewed interest was a very readable, thoughtful review by Connell and Slatyer (1977). They concluded that there could be three succession models, which they termed *facilitation, tolerance,* and *inhibition.* The facilitation model is a modern paraphrasing of Clements' relay floristics, without the super-organism aspect, and they criticized it for a lack of evidence on sufficient feedback mechanisms to maintain the climax steady state. The tolerance model is a modern paraphrasing of the IFC approach: Climax species are merely better competitors, longer-lived, and bigger than seral species, and all may be present together early in succession. They criticized this model as also lacking in sufficient evidence for application beyond forest vegetation. The inhibition model is a reverse of relay floristics: Early species inhibit succession by allelopathy and are replaced only when they die or are damaged. This model seems to hold for marine intertidal communities and for some shrublands, but there is little evidence to show that it might be a universal model.

MacMahon (1980 and 1981) reviewed successional processes in many biomes and concluded that Clements' six-step view of succession can be essentially correct, if we realize that the driving forces he described work on the associated species in an individualistic way instead of in some holistic, superorganism manner. In his integrated model (Figure 11-21), succession at one locale can proceed to many endpoints. Convergence to some single "true" climax is not predicted, and cases where this has been reported are "partially an illusion."

Most recently, Tilman (1982 and 1985) has proposed a model of succession based on resource use. Essentially, it states that changes in resource availability with time, and competition by plants for those limited resources, drive succession. In this view, plant species are specialized in terms of their tolerance limits for ratios of various resources, and the composition of a community will change as the availability of those resources changes. This model incorporates the essence of Clements' ideas, but it elaborates the underlying mechanisms in more detailed, physiological terms. Tilman challenges ecologists to test his model by experimentally identifying the limiting resources in a given sere and determining if changes in them can explain the successional patterns observed.

It is likely that there is some value in both the relay and the initial floristic composition models of succession, but more research is needed to formulate a realistic, comprehensive model of succession. Two other driving forces that have been proposed are species interactions (especially allelopathy) and natural selection for maximum efficiency in the cycling of energy and nutrients.

Is Allelopathy a Driving Force?

There is evidence that allelopathy plays a role in some seres (Chapter 6), but that evidence is so scattered and sometimes contradictory that it is not possible

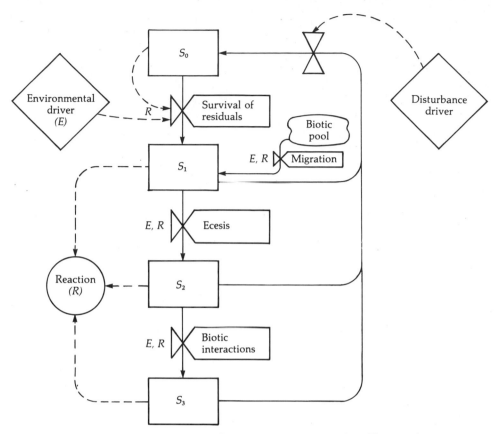

Figure 11-21   A flow model of succession, incorporating Clementsian ideas in a modern, modified framework. The boxes are seral stages (S) in a plot of ground over time (time 0, 1, 2, and 3). The diamonds are driving forces, the circle is an intermediate variable, and dashed lines show information flow. The letters E and R next to control gates replace other dashed lines that would go from the gates to environmental driver (E) or reaction (R); these lines have been omitted for simplicity in this diagram. (From MacMahon 1980.)

to reach a general conclusion. Catherine Keever (1950), for example, initiated a doctoral dissertation to determine whether allelopathy might be of major importance to old-field succession in the Piedmont. Of the many species and seral stages examined, she found allelopathy to be a factor in only one case: The decaying roots of horseweed (*Conyza canadensis*) inhibited the growth of its own seedlings. Consequently, the horseweed seral stage is self-limiting. She found all other species replacements in succession to be due to competition for light or soil moisture or to differences in longevity, time of germination, and pattern of seed dispersal. *Aster*, for example, is a poor competitor for soil moisture with *Andropogon*; biennials that are insignificant the first year as rosettes outcompete and overtop annuals a second year; pine seeds are not as widely distributed as seeds of ragweed or rhizome fragments of other herbs.

Other manipulative experiments in the Piedmont region corroborate Keever's general conclusion of the limited importance of allelopathy. Pinder (1975), for example, removed all *Andropogon* clumps from old-field plots in Georgia early in July, then harvested the subdominant forbs in September. These forbs were remnants of early seral stages. Removal of *Andropogon* resulted in a tripling of forb growth, giving a yield comparable to that for earlier seral stages. Pinder concluded that forb suppression during the *Andropogon* stage was due to competition, not allelopathy. A more complex experiment in New Jersey by Allen and Forman (1976) involved the selective removal of many different old-field species. The effect of the removals on growth of remaining species was not uniform, but the authors concluded that the magnitude of the effect depended on the growth form of the species removed, not on allelopathic interactions.

Is Natural Selection a Driving Force?

Clearly, succession occurs much too quickly for evolution to play a direct role. What some ecologists believe, however, is that climax communities have been subjected to natural selection in much the same way that populations have. They believe that community-level processes such as nutrient cycling and energy flow may act as selective forces at the community level, and that populations replace one another through time, thereby optimizing community functions. In other words, during evolutionary time, species and niches have shifted in response to selective pressure so that solutions to problems of time and energy allocation have been maximized or optimized (Cody 1974). Succession might also be viewed as a sequence of communities, each one within the chain being a more efficient system than the previous one.

It is ironic that this recent view of succession and the nature of climax communities—a view that treats communities as integrated entities affected by natural selection, much like populations or organisms—brings us back full circle to the organismic view of Clements. As discussed in Chapter 8, there is no direct evidence that natural selection operates at the community or ecosystem level.

The simplest concept of succession is that succession is a population phenomenon, involving the gradual, inevitable replacement of opportunist species (*r*-selected) with equilibrium species (*K*-selected). The rate at which the transformation occurs, and the exact pathway it takes, simply reflects the rate at which propagules arrive and the nature of the macroenvironment. In the absence of any disturbance, equilibrium species are always at a competitive advantage, as dominants, over opportunists. This advantage is due to, or enhanced by, such features as longevity, large biomass, moderated microenvironment, and symbiotic or antagonistic interactions with other organisms (Drury and Nisbet 1973). The temporal and spatial frequency of disturbances, however, has always been great enough to maintain opportunist species as well as equilibrium species.

Summary

Succession is a directional, cumulative change in the species composition of vegetation at one location over the course of 1–500 years. Seral (successional) communities exhibit significant, cumulative change during such a period, whereas climax communities do not.

Primary succession begins with plant colonization of new land; secondary succession occurs on land that still supports some residual soil and plant propagules. A secondary sere may be completed in 5–300 years, but a primary sere requires $200-1000^+$ years because of the greater severity of the pioneer conditions. Progressive succession leads to more mesic sites and communities of greater complexity and biomass. However, there are many well-documented cases of retrogressive succession (both allogenic and autogenic), which result in a more extreme habitat and simpler communities. There is also a type of local, nondirectional succession called *cyclic succession*. And in some severe habitats there is no succession, the pioneer community being the climax community as well. Some seres do not reach a climax because the periodicity of some disturbance, such as fire, is shorter than the time required for the progression of a complete sere. Most regions are characterized by a mosaic of climax types and seral stages leading to them.

Succession may be documented by repeated measures over time on a single plot or by reference to historical records for that plot, but most seres have been inferred from indirect evidence. One indirect method is to sample vegetation on many separated plots that differ in age. Also, the species composition in seedling and sapling strata can be compared to the overstory stratum.

Many changes in soil properties, microclimate, and vegetation occur during progressive succession. Some trends peak in the climax and others peak in mid-to-late succession; most do not have a constant rate of change. The following attributes increase during succession: biomass, complexity of the physiognomy, nutrient storage in biomass, the importance of detritus in mineral cycles, the mesic and equable nature of the site, stability (in the sense of longer-lived species and slower organism turnover times), species diversity, and the proportion of species that are *K*-selected. Net productivity may decline.

Several factors have been hypothesized to be the driving force behind succession. Clements theorized that successional species modify the site in such a way that they are at a competitive disadvantage with invading species of the next seral stage. The Clementsian model of succession is a series of discrete communities that relay the site to the climax community. Others view chance and differential longevity as the driving forces of succession. Species, rather than discrete communities, replace each other during succession, and seral communities are not discrete. Biotic interactions may also direct some successions, but there is not yet sufficient evidence to reach a general conclusion about the importance of these interactions. Recent theories on succession integrate aspects of both Clementsian and Eglerian ideas.

Finally, some ecologists hypothesize that climax communities have been subjected to natural selection for optimum solutions to problems of time, space, or energy allocation, and that successional trends may also be determined by selective forces operating above the population level. It is not known, however, whether natural selection can operate at the community or ecosystem level, nor what forces might drive succession towards levels of greater efficiency.

Possibly the best model of succession is a very simple one. Succession involves the gradual, inevitable replacement of opportunist (r-selected) species with equilibrium (K-selected) species. The rate and exact pathway of the replacement depends on the rate at which propagules arrive and the nature of the macroenvironment, indicating that succession is probably a population phenomenon, regulated by the invasive capacity of new species.

CHAPTER 12

PRODUCTIVITY

I n this chapter, we will discuss the interaction between plants and the global carbon cycle, consider the functioning of plant communities in the context of energy flow, and contrast the abilities of different communities to convert solar radiation into plant biomass. Plant biomass represents the energy store for the entire heterotrophic community. Therefore, the measurement and understanding of the factors controlling rates of biomass accumulation are basic not only to our interpretation of plant community dynamics but also to our understanding of the factors that control the energy dynamics of the biosphere.

Terrestrial Vegetation and the Global Carbon Cycle

The most active reservoir for carbon is the atmosphere (Figure 12-1), where carbon dioxide occurs at concentrations of 300–400 ppm. Carbon dioxide is removed from the atmosphere by photosynthesis and released by respiration and combustion of fossil fuels. Photosynthetic uptake and respiratory release are essentially in equilibrium in prevailing climates, except where large-scale clearing of forests has resulted in increased transfer of biological carbon to the atmosphere. The clearing of a forest for agriculture results in an estimated 90% decrease in carbon held in living tissues and a 20–50% decrease in soil carbon. Because of he widespread use of fossil fuels and clearing of forests, the concentration of carbon dioxide is increasing at an annual rate now exceeding 1 ppm (Figure 12-2), even though the oceans buffer changes in ambient concentrations to a large extent. Local concentrations may exceed 400 ppm on a daily basis in areas with a high level of fossil fuel combustion and topographic conditions that reduce atmospheric mixing. Such increases will continue while widespread use of fossil fuels and forest clearing are prevalent; there will be a dramatic decrease when fossil fuel supplies and forests are depleted (Woodwell et al. 1983).

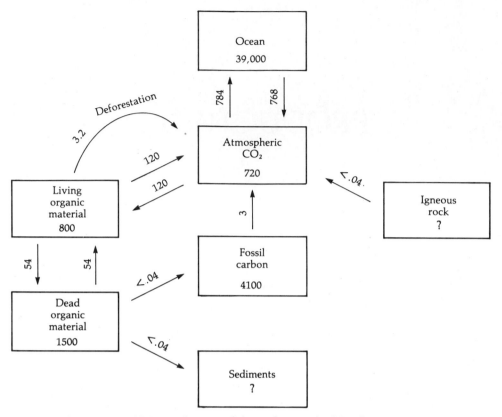

Figure 12-1   Major pathways of the carbon cycle. Numbers are annual carbon storage and transfer rates in gigatons.

The importance of vegetation in the global patterns of carbon dioxide are illustrated in Figure 12-2, where the yearly oscillations in CO_2 can be attributed almost exclusively to the annual cycle of net productivity of temperate zone forests. The highest levels of CO_2 at Mauna Loa are found in winter, when the temperate forests are largely dormant, and lowest in summer, when photosynthetic rates are highest. Similar but less dramatic shifts are found in the appropriate season in the southern hemisphere (Machta 1983).

Increasing CO_2 levels support higher photosynthetic fixation rates by C_3 plants (most green plants). It is possible that the competitive interactions of some species will change, due to CO_2-induced changes in their photosynthetic rates or to temperature-related responses caused by fluctuations in atmospheric CO_2, which traps solar energy much like the glass of a greenhouse does. We might expect subtle changes in physiological response, vegetational structure, and vegetational composition as the carbon cycle fluctuates and finally reaches equilibrium. The implications of increasing atmospheric CO_2 for global climate are not entirely clear. For interesting and conflicting views, see Idso (1984) and Dickinson (1982).

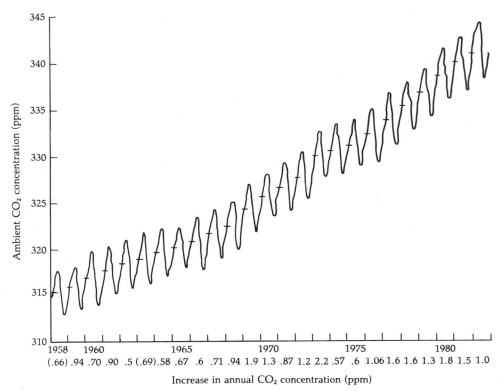

Figure 12-2 Mean monthly concentrations of atmospheric CO_2 at Mauna Loa, Hawaii. The crosshatches along the line are the annual mean. (Data from Geophysical Monitoring for Climatic Change, National Oceanic and Atmospheric Administration.)

Energy Flow Model

Before we begin our discussion of terrestrial plant productivity, it is necessary to define some terms applicable to community energetics. Energy enters the ecosystem as light and is fixed into chemical energy by photosynthesis. The total amount of energy fixed by photosynthesis per unit area, per unit time is referred to as **gross primary productivity** or GPP (see Figure 12-3). Not all energy fixed by photosynthesis is converted to biomass; a significant part is released by respiration to supply energy for plant metabolic activities. Gross photosynthesis minus respiration is equal to **net primary productivity** (NPP)—the rate at which energy is stored in plant tissues. Note that productivity values always represent the rate of carbon or energy flow and are expressed as grams of biomass (or calories) per unit leaf area (or gram biomass) per unit time. The dry weight of plant material present at any point in time is referred to as **biomass** (or **standing crop**, or sometimes **phytomass**). Existing biomass is not a measure of community productivity, since biomass turnover rate is frequently not related to the total

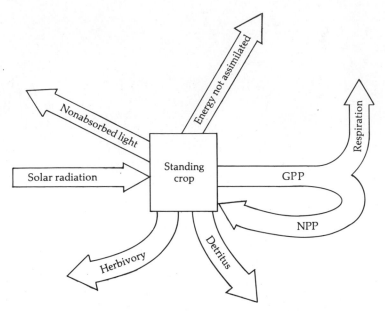

Figure 12-3   Potential pathways of energy partitioning at the primary trophic level.

standing crop. For example, productivity per kilogram of biomass in a temperate grassland may be 375 g yr^{-1}, but only 40 g yr^{-1} in a temperate deciduous forest; the larger amount of nonphotosynthetic biomass in a deciduous forest is responsible for most of the difference. Net primary productivity is a measure of the rate at which energy is stored or incorporated into living tissues. We can measure NPP as change in biomass through time or as net community photosynthetic rate.

Part of NPP is the food source of herbivores and decomposers and the remainder accumulates within the community as standing crop biomass or is consumed by fire. The distribution and rate of turnover of biomass is closely related to community physiognomy, the type of herbivores present, and the relative importance of decomposers in the system. These phenomena are discussed in more detail when we consider individual communities later in this chapter.

It is common practice to compare the efficiency with which a species or a stand transmits energy from one form or physical state to another. Efficiency values are the ratio of output to input of energy at various points along the pathways of energy flow within a plant or a community (Odum 1971). Efficiency values can be calculated for any energy transfer. Here, we mention three that are important in understanding differences in community productivity. **Exploitation efficiency** (Table 12-1) is related to the ability of plants to intercept light. Characteristics important in this regard include latitude, topographic location, leaf area index (LAI) (see Chapter 7), and leaf orientation. **Assimilation efficiency**

Table 12-1 Formulae defining common efficiency values as applied to primary producers. (Modified from Ricklefs 1979, *Ecology*, 2nd edition, by permission of Chiron Press, Inc., New York, N.Y.)

Exploitation Efficiency (%)	=	$\dfrac{\text{Gross primary productivity}}{\text{Solar radiation}} \times 100$
Assimilation efficiency (%)	=	$\dfrac{\text{Gross primary productivity}}{\text{Absorbed radiation}} \times 100$
Net production efficiency (%)	=	$\dfrac{\text{Net primary productivity}}{\text{Gross primary productivity}} \times 100$

(or **quantum yield**) refers to the ability of plants to convert absorbed radiation into photosynthate. Factors modifying assimilation efficiency are those governing the photosynthetic process (see Chapter 15 for details), such as resistance to CO_2 assimilation, water and light availability, evaporative demands of the atmosphere, and temperature. **Net production efficiency** is a measure of the capacity to convert photosynthate into growth and reproductive biomass rather than utilizing it for maintenance respiration. The amount of energy used for maintenance depends on such factors as temperature (because of the direct effect on rates of chemical reactions) and the amount of nonphotosynthetic biomass that must be supported. For a more complete evaluation of the concepts and terminology of efficiencies, see Kozlovsky (1968). The general trends and variations to be expected for ecological efficiency values are discussed later in this chapter.

Methods of Measuring Productivity

The most accurate means of measuring net primary productivity would be to measure the net photosynthetic rates of photosynthetic tissues, then subtract the respiration rates of nonphotosynthetic tissues, and finally extrapolate to the community level, using the net production per gram of biomass of each species in the community. This assessment of NPP is not possible on a large scale, because we do not have photosynthesis and respiration measurements for all the species in any community, nor are we prepared to predict physiological responses to the wide range of conditions experienced by plants in the natural environment. Consequently, we usually use methods that depend on the accumulation and disappearance of biomass through time.

Net primary productivity (NPP) is frequently measured by calculating the change in biomass through time:

$$\text{NPP} = (W_{t+1} - W_t) + D + H \qquad\qquad \text{(Equation 12-1)}$$

where $W_{t+1} - W_t$ is the difference in standing crop biomass between two harvest times, D is the biomass lost to decomposition, and H is the biomass consumed by herbivores during the period between harvests. Productivity may be expressed as $g\,m^{-2}yr^{-1}$ or, if the caloric content of the material is known, as $cal\,m^{-2}yr^{-1}$. The latter units are more meaningful when efficiency of light conversion or respiration is important, since these are measured in calories rather than grams.

Aboveground biomass can be measured with little error in herbaceous vegetation by replicate samples harvested randomly from a grid. This technique is most effective with annual vegetation, where little biomass is lost to decomposition during the growing season. If herbivore activity is significant, comparisons between replicate samples taken inside and outside herbivore exclosures are often employed. If the herbivores are primarily insects that cannot be excluded, the area of tissue or the number of reproductive parts consumed must be monitored by direct observation. See Singh et al. (1975) for a review of harvest techniques applied to grasslands.

Dimension analysis is an alternate way of estimating productivity in those situations where the volume of individual plants is very large or regrowth is so slow that extensive damage may be caused by complete harvest of sample plots. The technique is based upon the assumption that some easily measured parameter, such as plant height, diameter at breast height (dbh), or plant volume, can be correlated with standing crop. Whittaker and Woodwell (1968) devised a sophisticated dimension analysis scheme in which branch or plant age, as determined by growth ring or bud scale scar analysis, is related to biomass, production, and surface relations of woody plants. A few individuals must be harvested to determine the slope of a regression line, which may then be used to predict plant biomass from the easily measured parameter. For example, satisfactory predictability of standing crop has been obtained from basal diameter, dbh, or diameter^2 × height (Madgwick and Satoo 1975; Whittaker and Woodwell 1968), using the regression model proposed by Kittredge (1944):

$$\log_e Y = a + b \log_e X \qquad \text{(Equation 12-2)}$$

where Y is the estimate of standing crop, X is the measured parameter, a is the point at which the regression line crosses the Y axis, and b is the slope of the regression line. The values of the constants are determined in a preliminary harvest. Errors in making estimates of biomass from log-log regression lines may be avoided by using the methods outlined by Beauchamp and Olson (1973).

Production studies must also account for loss of leaves, branches, etc., during the period between harvests. Several types of **litter traps** have been devised to quantify the litter production of shrubs and trees (Figure 12-4). The most effective trap design will differ, depending on local conditions and the nature of the material being sampled.

Root production is generally estimated to exceed aboveground production, but technical limitations make it difficult to assess the accuracy of the estimates. The usual approach is comparable to the methods described for measuring aboveground productivity—that is, to estimate the relative biomass of roots extracted from soil cores taken periodically over a growing season (Kelly 1975; Bohm 1979; Kummerow et al. 1978; Vogt et al. 1982). Accuracy is problematic

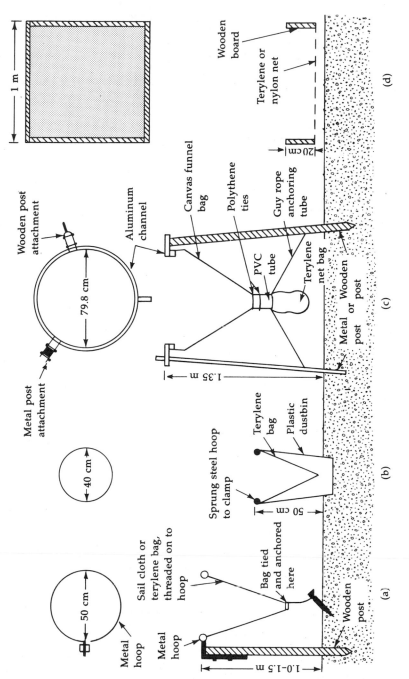

Figure 12-4 Examples of litter traps. (Modified from Newbould 1967. Reprinted by permission of Blackwell Scientific Publications, Ltd., from IBP Handbook No. 2.)

because we cannot know when to sample to obtain estimates that represent minimum and maximum root biomass levels during the growing season and because there is no way to gauge the loss of roots to herbivory and decomposition. The problem is further complicated by a high degree of variability in biomass estimates from soil cores (Singh et al. 1984). The problem of determining maximum and minimum biomass levels can be reduced by frequent sampling, and variability can be decreased by taking as many core samples as can be processed. The values obtained by harvest techniques may be conservative, since production balanced by death and herbivory occurring between sampling dates is not accounted for. For example, estimates taken at consecutive harvests may show no increase in root biomass when, in reality, a large portion of the roots may have been shed and replaced during the period.

Others (e.g., Lieth 1968; Newbould 1968) have monitored root growth by direct observation in glass-sided root pits. This technique has all of the problems of sequential harvests for estimating productivity but is very effective in determining the temporal aspects of root growth. Recently, Carman (1982) has described a nondestructive staining technique to trace the temporal aspects of root growth.

Belowground productivity has been estimated to account for 10–75% of total productivity, and much of this production is in the form of fine (<3.0 mm diameter) roots. Most older production studies in the literature either do not account for fine root production or make gross estimates from biomass samples or from data on the ratio of root and shoot growth as determined in controlled environments. There has been a considerable increase in research activity concerning fine-root dynamics in recent years (e.g., McClaugherty and Aber 1982; Nadelhoffer et al. 1985). These studies generally use intense, frequent sampling programs to minimize the problems of phenology and variability mentioned above. However, Nadelhoffer et al. (1985) have devised an indirect method of measuring turnover in fine roots that may solve some of the problems associated with sequential biomass sampling. Figure 12-5 shows their model of nitrogen flux for temperate forest ecosystems. Since we can measure the flux of nitrogen represented by each arrow in the model, except that associated with belowground litter (turnover of fine roots), the values for all other fluxes can be determined, and the value for belowground litter can be calculated. Techniques that indirectly measure turnover in roots may be a significant step toward solving the technical problems associated with measuring belowground productivity.

Patterns of Productivity and the Distribution of Biomass

Production and Biomass in Shoots and Roots

The relative productivity and biomass of shoots and roots has important adaptive significance in plants. For example, colonizing species with annual life histories should show greater aboveground productivity, because the following

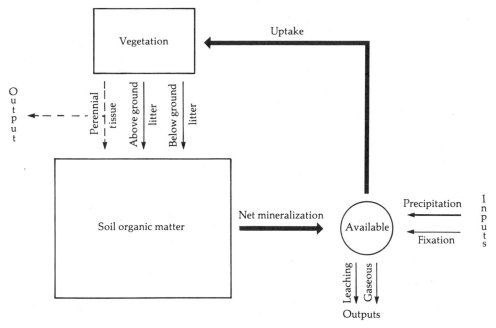

Figure 12-5 Nitrogen (N) model for a temperate forest ecosystem. Boxes represent ecosystem N pools. The net amount of N made available annually for uptake by vegetation is represented by a circle. Solid arrows are annual transfers between pools. Arrow widths are approximately proportional to flux sizes. The dotted arrow represents N transferred irregularly from perennial tissues to soil or out of the system. Annual N uptake is the sum of net N mineralization and precipitation N measurements minus any leaching losses. Transfer of N from vegetation to soil organic matter via belowground litter (*) is estimated at each site using measures of other fluxes. (From "Fine roots, net primary production, and soil nitrogen availability: A new hypothesis" by K. J. Nadelhoffer et al., *Ecology* 66:1377-1390. Copyright © 1985 by the Ecological Society of America. Reprinted by permission.)

generation depends on seed output. On the other hand, a perennial plant growing in a nutrient or water-limiting environment might have a greater belowground productivity and root biomass, because uptake of limiting nutrients and water is increased by having a large and rapidly growing root system (Grime 1977; Chapin 1980; Waring 1983). Root production is costly, since it requires transport of carbohydrate from the shoot. Therefore, plants growing in xeric or nutrient poor habitats should maintain a larger active root system and have a lower rate of root tissue turnover than plants of resource-abundant habitats (Orians and Solbrig 1977; Chapin 1980). This explains why we find larger root/shoot ratios in plants of resource-stressed habitats. Do roots of plants in resource-limiting environments therefore have higher productivity? The answer is apparently "no," because of a rapid fine-root turnover in plants of resource-abundant habitats. For example, Chapin and Van Cleve (1981) found that the fine roots of boreal forest trees growing in resource-limiting situations were maintained for a full year, whereas those from less limiting habitats have few fine roots that survive the winter. It appears, then, that plants of resource-limiting situations have larger root/shoot ratios but lower root productivity than plants of resource-abundant habitats. Productivity of fine roots and associated mycorrhizal fungi represents as much as 75% of the total net primary productivity in mature Pacific silver fir stands in western Washington (Vogt et al. 1982).

Litter Production and Decomposition

Biomass may remain living, serve a support function, be consumed directly by herbivores, or become detritus. The latter possibility is most likely; more than half of annual net productivity is deposited as litter. Litter is the food source for decomposers and detritivores and the means by which nutrients are returned to the cycling pool. It is thus important to consider the rate of accumulation and decomposition of dead plant parts. Olson (1963) calculated the ratio of litter production to litter accumulation as an expression of the rate of decomposition. The ratio is high in tropical environments, reflecting the high rate of leaf production and rapid decomposition, and decreases on a gradient toward the poles. Litterfall is generally related to LAI, causing the ratio to decrease with latitude in a fashion similar to that of NPP. Within forests, litterfall decreases with latitude and/or altitude. However, Jordan (1971) pointed out that litterfall does not show the same relationship along a moisture gradient. No differences in litter production are found when grassland, old-field, and tundra values are compared with forests at similar latitudes (Table 12-2).

Plant decomposition rates have been compiled by Singh and Gupta (1977), showing that decomposition varies with vegetation type and environment (Table 12-3). Broadleaf temperate forests require about 1 year for complete litter decomposition, whereas coniferous forests typically require 3–5 years. The extremes reported by Singh and Gupta are represented by high mountain pine forests of California, which need over 30 years for complete degradation, and tropical forests, where complete breakdown may occur in less than 2 months. These turnover rates depend primarily on the chemical makeup of the litter, the temperature, and the moisture conditions of the habitat.

Moisture and temperature together exert a significant influence on rates of decomposition (e.g., Wiegert and Evans 1964; Rochow 1974). It is not possible to separate the effects of temperature and moisture because they are not environmentally independent, and extreme levels of either will exceed the tolerance of decomposers. Douglas and Tegrow (1959) found decomposition rates of only about 1.0 t ha^{-1}yr^{-1} in arctic tundra communities. Such low rates are probably a result of the very short growing season. Contrast the eight times greater rate in tropical rainforests with continuous growing seasons (Wanner 1970). Bleak (1970) measured the disappearance of litter under winter snow in central Utah and found losses of up to 50% despite temperatures of $+1.2°$ to $-2.5°C$. Fungi and bacteria were active at these temperatures and were the primary agents for breakdown. Winn (1977) examined decomposition rates in chaparral stands in

Table 12-2   Rate of litter production in various forest, perennial herb, and grass ecosystems. (From data compiled by Jordan 1971. Reprinted by permission of *American Scientist*, journal of Sigma Xi, The Scientific Research Society.)

Community	Location	Litter fall ($g\ m^{-2}\ yr^{-1}$)	Source
Tropical rainforest	Thailand	2322	Kira et al. 1967
Tropical rainforest	Average of several	1600	Rodin and Basilevic 1968
Subtropical forests	Average of several	1200	Rodin and Basilevic 1968
Dry savanna	Russia	290	Rodin and Basilevic 1968
Oak forest	Russia	350	Mina, cited in Rodin and Basilevic 1968
Fir-taiga	Russia	250–300	Rodin and Basilevic 1968
Oak-pine forest	New York	406	Whittaker and Woodwell 1969
Pine forest	Virginia	490	Madgwick 1968
Tropical seasonal forest	Ivory Coast	440	Muller and Nielsen, cited in Kira et al. 1967
10 Angiosperm forests	Europe	280	Bray and Gorhman 1964
10 Angiosperm forests	Tennessee	320	Whittaker 1966
13 Gymnosperm forests	Tennessee	267	Whittaker 1966
Old-field upland	Michigan	312	Weigert and Evans 1964
Old-field swale	Michigan	1003	Weigert and Evans 1964
Perennial herbs	Japan	1484	Iwaki et al. 1966
Tallgrass prairie	Missouri	520	Kucera et al. 1967; Dahlman and Kucera 1965
Mesic alpine tundra	Wyoming	162	Scott and Billings 1964

California where temperatures rarely limit decomposer activity. Microbes responded quickly to small fluctuations in moisture conditions, and decomposition is apparently controlled by the level of hydration of litter.

Numerous other factors, such as secondary metabolites leached into the litter, soil characteristics, the kinds of organisms in the detritus food chain, and herbivore activity influence decomposition on a local basis. Many of these factors can be related to water, temperature, and the chemical nature of the litter. Meentemeyer (1978) studied actual evapotranspiration (AET), as a measure of energy and water in the environment, and lignin content, as a measure of litter quality, to develop a model predicting decomposition in temperate and boreal forests. More than half (52%) of the variation in the data could be explained by AET. Litter quality added little predictive power to the model. Whitford et al. (1981)

Table 12-3   Representative rates of plant litter decomposition.

Climate	Biome and location	Decomposition rate (% per day)	Source
Tropical			
	Rainforest:		
	Trinidad	0.45	Cornforth 1970
	Grassland:		
	India	0.30	Gupta and Singh 1977
	Other	0.17–1.5	
Temperate			
	Oak forest:		
	Minnesota	0.018	Reiners and Reiners 1970
	Missouri	0.095	Rochow 1974
	New Jersey	0.018	Lang 1974
	England	0.30	Edwards and Heath 1963
	Pine forest:		
	California	0.0027–0.0082	Jenny et al. 1949
	Missouri	0.036	Crosby 1961
	Southeastern United States	0.07	Olson 1963
	Other	0.0027–0.12	
	Deciduous forest: Eastern United States	0.057	Shanks and Olson 1961
	England	0.043–0.06	Anderson 1973
	Australia	0.04–0.15	Ashton 1975
	Grassland: North Dakota	0.082–0.11	Redmann 1975
	Missouri	0.14	Koelling and Kucera 1965
	Utah	0.082–0.14	Bleak 1970

tested the Meentemeyer model in desert habitats and in severely disturbed temperate mesic forests. They found that the ability of the AET model to predict decomposition rates was dramatically reduced in these systems. In clear-cut forests, the tolerance limits of the detritivores were surpassed, reducing decomposition and causing the AET model to overestimate decomposition. Desert organisms, and most likely organisms from any harsh environment, have behavioral and physiological mechanisms to overcome water and temperature restraints on activity, and decomposition is more rapid than would be predicted by the AET model.

Global Patterns of Productivity and Biomass

Terrestrial vegetation occupies approximately 30% of the globe's surface and provides 62% of the total world primary productivity. Also, most of the world's biomass consists of terrestrial vegetation (Lieth 1973). Whittaker and Likens (1975) estimated world net primary productivity to be about $170 \times 10^9 \, t \, yr^{-1}$, and of this they estimated that $90–120 \times 10^9 \, t \, yr^{-1}$, or 53–71%, is produced in terrestrial systems. Russian estimates are somewhat higher, but the estimates quoted here are based on average values from actual productivity measurements, and they are conservative. Table 12-4 reveals the dramatic differences in productivity and biomass values reported for the major terrestrial ecosystems. Productivity estimates are lowest in deserts and highest in tropical rainforests, spanning the entire range between 0 and approximately $3000 \, g \, m^{-2} yr^{-1}$. Whittaker suggested that 3000 to $3500 \, g \, m^{-2} yr^{-1}$ is a maximum productivity value for terrestrial systems. Biomass estimates range from $1.0 \, kg \, m^{-2}$ in desert and tundra ecosystems to $200 \, kg \, m^{-2}$ in some temperate rainforests. The usual range, however, is between 1.0 and $60 \, kg \, m^{-2}$. A careful evaluation and comparison of these values later in this section will further refine our understanding of the distribution of productivity and biomass.

The relationship of biomass to productivity is often expressed as the **biomass accumulation ratio** (BAR) (Whittaker 1975). BAR is the ratio of dry weight biomass to annual net primary productivity. BAR values are a measure of the accumulation of primarily woody material, a characteristic related to environmental harshness and the potential age of dominant species. The BAR represents the average residence time in years of organic matter in the community. There is a broad overlap in BAR values between ecosystems. Representative values are 1 for communities of annuals, 2–10 for deserts, 1.3–5 for grasslands, 3–12 for shrublands, 10–30 in woodlands, and 20–50 for mature forests. Whittaker and Niering (1975) calculated BAR values for an elevational gradient from desert to subalpine forest in the Santa Catalina Mountains of Arizona and found that in forest and woodland zones, BAR decreased as a function of biomass. However, once the low-elevation desert shrublands were encountered, BAR values and biomass did not change significantly with further drops in elevation.

Jordan (1971) also recognized the relationship between major environmental gradients and the biomass allocation patterns of the dominant plants. He calculated the ratio of wood production to litter production and found high correlations between the ratio and total light energy available during the growing season (Figure 12-6) and between the ratio and annual precipitation. This suggested that rapid wood production is an advantage in areas of low light such as the boreal forests. Perhaps larger plants have more energy reserves and can thus resist greater environmental stress. Greater energy storage capacity of larger plants increases their chances of survival in stressful environments, where they may frequently suffer defoliation or very short growing seasons.

In general, trends in world productivity are loosely related to biomass, in that *K*-selected plants in less harsh or more predictable environments are able to accumulate extensive root and branch systems. These form the bases for efficient

Table 12-4 Net primary productivity and related characteristics of terrestrial biomes. (From R. H. Whittaker and G. E. Likens, 1975. The Biosphere and Man. In *Primary Productivity of the Biosphere* edited by Lieth and Whittaker. By permission of Springer-Verlag, New York.)

Ecosystem type	Area (10^6 km^2)	Net primary productivity (dry matter)			Biomass (dry matter)			Leaf surface area	
		Normal Range (g m^-2 yr^-1)	Mean (g m^-2 yr^-1)	Total (10^9 t yr^-1)	Normal range (kg m^-2)	Mean (kg m^-2)	Total (10^9 t)	Mean (m^2 m^-2)	Total (10^6 km^2)
Tropical rainforest	17.0	1000–3500	2200	37.4	6–80	45	765	8	136
Tropical seasonal forest	7.5	1000–2500	1600	12.0	6–60	35	260	5	38
Temperate forest:									
Evergreen	5.0	600–2500	1300	6.5	6–200	35	175	12	60
Deciduous	7.0	600–2500	1200	8.4	6–60	30	210	5	35
Boreal forest	12.0	400–2000	800	9.6	6–40	20	240	12	144
Woodland and shrubland	8.5	250–1200	700	6.0	2–20	6	50	4	34
Savanna	15.0	200–2000	900	13.5	0.2–15	4	60	4	60
Temperate grassland	9.0	200–1500	600	5.4	0.2–5	1.6	14	3.6	32
Tundra and alpine	8.0	10– 400	140	1.1	0.1–3	0.6	5	2	16
Desert and semidesert scrub	18.0	10– 250	90	1.6	0.1–4	0.7	13	1	18
Extreme desert: rock, sand, ice	24.0	0–10	3	0.07	0–0.2	0.02	0.5	0.05	1.2
Cultivated land	14.0	100–4000	650	9.1	0.4–12	1	14	4	56
Swamp and marsh	2.0	800–6000	3000	6.0	3–50	15	30	7	14
Lake and stream	2.0	100–1500	400	0.8	0–0.1	0.02	0.05	—	—
Total	149		782	117.5		12.2	1837	4.3	644

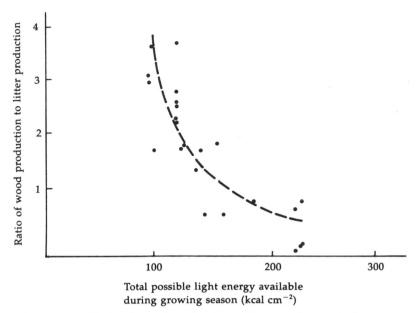

Figure 12-6 Ratio of wood production to litter production in forest communities as a function of light energy available during the growing season. (From Jordan 1971. Reprinted by permission of *American Scientist*, journal of Sigma Xi, The Scientific Research Society.)

light interception, nutrient uptake, and, in turn, productivity. It should be remembered, however, that much of the accumulation of biomass serves only a support function, which explains the different rates of change in productivity and biomass values observed from the poles toward the tropics.

When we consider the information on forested ecosystems in Table 12-4, several generalizations are apparent. Even though the productivity of tropical rainforests has been estimated to exceed 3000 g m^{-2}yr^{-1}, the range of productivity values overlaps all other forest and woodland types and may even include values for some temperate grassland systems. There is some evidence suggesting that temperate evergreen forests are more productive than their deciduous counterparts (e.g., Post 1970; Reiners 1972). At first thought, this seems improbable because evergreen leaves have 30–60% lower maximum photosynthetic rates per unit surface area than deciduous leaves (Mooney 1972); however, this is offset by high leaf area index (LAI) for evergreen forests. Kira (1975) explained that deciduous plants have very low LAI during periods of leaf development and may miss periods of favorable moisture and temperature following photoperiodically induced leaf drop. Kira also suggests that the high net productivity of tropical evergreen forests is related to increased height and greater solar radiation, which allows increased LAI and therefore higher exploitation efficiency. Needleleaved evergreens frequently have nearly twice the LAI of deciduous trees of similar biomass and height, thus allowing plants with a lower photosynthetic capacity per unit leaf area to have higher potential stand production. This very

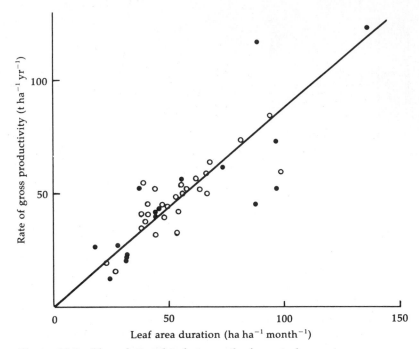

Figure 12-7   The relationship between leaf area index, evergreenness and gross productivity. Leaf area duration values are calculated as LAI times length of growing season in months. Closed circles are for broadleaved forests and open circles for needleleaved forests. (From Kira 1975 in *Photosynthesis and Productivity in Different Environments* IBP 3 edited by J. D. Cooper. Reprinted by permission of Cambridge University Press.)

close relationship between LAI, evergreenness, and gross productivity is plotted in Figure 12-7. A high LAI is made possible by large size, necessitating larger amounts of nonproductive structural tissue, thereby reducing the net production efficiency of forest trees.

Waring (1983) has developed a conceptual approach of expressing growth efficiency in trees as the ratio of wood production to leaf area. Since the area of sapwood involved in water conduction is directly related to canopy leaf area, the growth efficiency and potential aboveground productivity of trees can be calculated simply by regression. The product of growth efficiency and canopy leaf area gives an estimate of production and can thus be used to assess the degree to which water stress, nutrient stress, overcrowding, etc., influence a particular forest stand. Such general indices of vigor may be useful in predicting the susceptibility of a forest stand to disease or insect attack. (See "Herbivores," p. 290.)

Grassland productivity is typically lower than in communities with trees because of the LAI relationships considered above. Ovington et al. (1963) compared prairie productivity with the productivity of adjacent savannah and oak-wood communities and found grassland productivity to be 10–20% of the productivity of communities with woody plants. The absence of woody tissues results in low BAR values (1.3–5) and short-lived aboveground parts in grasslands. Kucera et al. (1967), Penfound (1964), and Hadley (1970) have observed dramatic increases in productivity when aboveground parts are removed mechanically or by fire. Rice and Parenti (1978) suggested that such increases may be in response to increased soil temperatures following biomass removal.

Arctic and alpine communities include life forms that range from shrubs to cryptogams and have LAIs comparable to other communities in severe climates. Net primary productivity is typically 100–150 g m^{-2}yr^{-1}, with biomass as low or lower than desert communities. Miller and Tieszen (1972) modeled production processes and noted the same increases in production in response to removal of dead material in arctic and alpine communities as in grasslands. Lemmings remove large quantities of biomass in the natural system (Dennis and Johnson 1970) and may cause increases in production similar to the increases caused by fire in grasslands. The main restrictions on production in the tundra are low LAI, low temperatures, and low angles of incident radiation.

Primary production in arid lands varies from 10 to over 200 g m^{-2}yr^{-1}, depending on the amount and pattern of rainfall. LAI is severely limited by stomatal closure during dry periods, and variations occur seasonally in response to water availability (Chew and Chew 1965; Burk and Dick-Peddie 1973).

Successional Patterns of Productivity and Biomass

Productivity follows a pattern of change during succession similar to the pattern of change noted in Chapter 10 for species diversity. Figure 12-8 depicts this trend of gradually increasing productivity during the pioneer and early tree

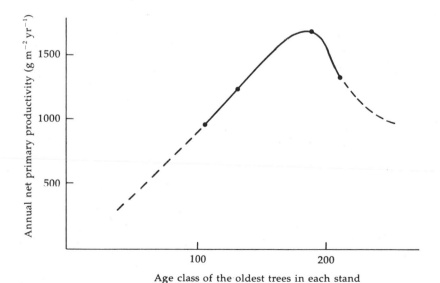

Figure 12-8 Annual net primary productivity, as estimated by tree stem measurements, plotted by age class of the oldest trees in each stand. (Recalculated from Loucks 1970, Evolution of diversity, efficiency, and community stability. *American Zoologist* 10:17–25.)

stages, followed by decreasing productivity as the self-perpetuating climax trees reach maturity (Loucks 1970). Botkin et al. (1972) constructed a model of forest growth predicting that standing crop reaches a peak within about 200 years, then drops 30–40% during the following 200 years. This biomass reduction could, in itself, account for some declining productivity. Other trends, such as a decline in the photosynthetic efficiency of overmature individuals (Hellmers 1964), the allocation of a greater proportion of net productivity to nonphotosynthetic structural biomass, limitations imposed upon LAI by canopy form and leaf orientation (Horn 1974), and the binding of nutrients into structural biomass (Connell and Slatyer 1977), also lead to reduced productivity in mature communities. These changes in biomass allocation and physiological response lead to greater susceptibility of many communities to fire, insect attack, and wind-throw. Loucks (1970) suggested that these characteristics are adaptations to repeating patterns of environmental change that stimulate periodic returns to states of high net primary productivity.

The biomass accumulation ratio mentioned earlier is often used in connection with succession studies (Whittaker and Likens 1975). The BAR increases from 1 in the pioneer stage of succession to between 30 and 50 in mature forest communities (Figure 12-9). Attiwill (1979) suggests that the growth of a forest may be considered a sequence of three stages, defined by changing relationships

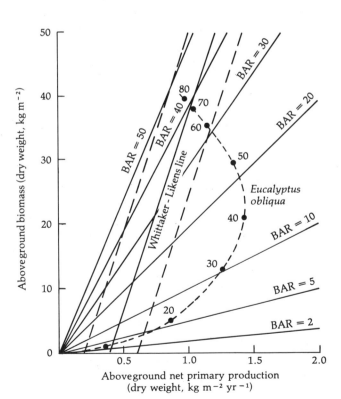

Figure 12-9   Relationship between biomass and NPP for temperate forests of the United States, and the trajectory of the relationship (biomass accumulation ratio, BAR) with age in years for *E. obliqua* (the heavier dashed line). A range of values of BAR in years is shown as a family of curves. (From "Nutrient cycling in a *Eucalyptus obliqua* (L'Herit.) forest III. Growth, biomass, and net primary production" by P. M. Attiwill, *Australian Journal of Botany* 23:439-458, 1979.)

between aboveground net primary production and aboveground biomass. The first stage consists of rapid productivity without significant increases in biomass; thus the BAR remains low. This stage lasts up to about 20 years. After that, continued increases in biomass depend on attaining larger size. The second stage is characterized by a period of rapid heartwood formation and rapidly increasing values of BAR. As the forest trees reach their genetic or environmental size limit, the rate of increase in BAR values tapers off. This final stage is characterized by an average residence time of organic matter in the biomass of 40–50 years.

Environmental Factors and Productivity

Light and Temperature

Global radiation varies with atmospheric conditions, latitude, and altitude. However, the effect of radiation on community productivity is indirect and is seen primarily through differences in growing season (Whittaker 1975) and temperature (Figure 12-10). Productivity increases along the mean annual temperature gradient from the poles toward the equator. Optimum temperatures for productivity coincide with the 15–25°C optimum range of photosynthesis. This close correlation between the photosynthetic temperature optimum and produc-

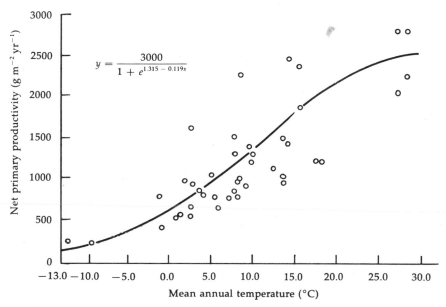

$$y = \frac{3000}{1 + e^{1.315 - 0.119x}}$$

Figure 12-10 Net primary productivity, including both above- and belowground productivity, in relation to mean annual temperature. (From "Primary production: Terrestrial ecosystems" by H. Lieth, *Human Ecology* 1:303-332, 1973. By permission of Plenum Publishing Corporation.)

tivity may be coincidental, since length of growing season is perhaps a more important determinant than temperature itself. The wide spread of points in Figure 12-10 is indicative of the many environmental factors other than temperature that affect productivity. Light and temperature influence water use and availability and, as a result, have few completely independent affects.

Water

The environmental factor most directly correlated with productivity is water. Lieth (1973) (Figure 12-11a) showed the correlation between productivity and mean annual precipitation, and Rosenzweig (1968) (Figure 12-11b) compared productivity with actual evapotranspiration (AET). Evapotranspiration is a measure of the total amount of water lost by transpiration and evaporation. Evapotranspiration has high predictive value in productivity studies because it includes the influence not only of water but also of light and temperature. Precipitation and evapotranspiration approach equality in arid environments, so the relationship between productivity and annual rainfall is nearly linear at values below 500 mm. At higher levels of precipitation, more water is lost as runoff or drains below the root zone, where it no longer influences production. A fourfold increase in precipitation and associated cooler temperatures resulted in a species-specific 100–600% increase in aboveground productivity for Mojave Desert shrubs (Bamberg et al. 1976). Cable (1975) studied the response of perennial grasses near Tucson, Arizona and developed a model in which the product of current precipitation and the past summer's precipitation was 203 times more accurate as a predictor of growth than the current precipitation alone. The reason we see such close correlations between annual means and overall productivity (annual means are usually poor predictors of plant response) may be this delayed response, which tends to reduce the impact of unusually wet or dry years.

Direct evidence for productivity trends can be obtained by studying adjacent communities along a gradient of temperature, moisture, and evaporation. Whittaker and Niering (1975) conducted such a study in the Santa Catalina Mountains near Tucson, Arizona, along an uninterrupted vegetational gradient from subalpine forest through woodlands and grasslands to desert. Figure 12-12 compares the Santa Catalina Mountain results with those predicted by Rosenzweig (1968) and Lieth (1973) for the relationship between AET or mean annual precipitation and net primary productivity. The actual measurements show a more complex relationship than predicted. There is an abrupt change of slope (line c in Figure 12-12) at 400–500 g m^{-2}yr^{-1}. If one were to guess at what point such a change might occur, the margin of the desert would seem a reasonable place because of the dramatic change in life form and the nature of precipitation extremes in that ecotone. Interestingly, however, the transition does not occur at the desert margin, but at the transition from open to dense woodland, where trees become the dominant producing life form. Net productivity of about 1300 g m^{-2}yr^{-1} represents the upper level measured. This must indicate another abrupt change in slope, since estimates of maximum productivity for climax temperate forests are approximately 1500 g m^{-2}yr^{-1}. (Note that the dashed portion of line (c) is hypothetical.)

(a)

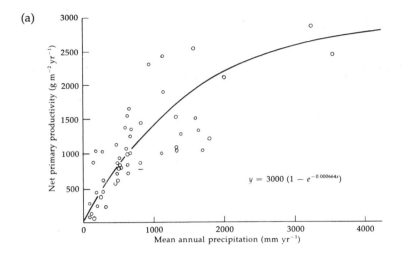

$$y = 3000\,(1 - e^{-0.000664x})$$

Figure 12-11 Patterns of terrestrial net primary productivity as predicted (a) by Lieth (1973), using mean annual precipitation, and (b) by Rosenzweig (1968), using actual evapotranspiration. [(a) from "Primary production: Terrestrial ecosystems" by H. Lieth, *Human Ecology* 1:303-332, 1973. By permission of Plenum Publishing Corporation. (b) from Rosenzweig, M. L. 1968. Net primary productivity of terrestrial communities: Prediction from climatological data. *American Naturalist* 102:67–74. Copyright © 1968 by The University of Chicago.]

(b)

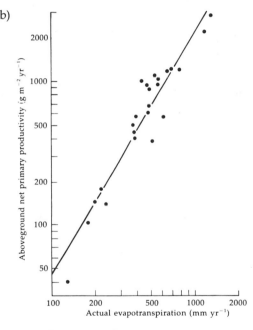

Assuming that this sort of nonlinear relationship is typical, Whittaker and Niering offer some hypotheses. They suggest that the concave lower part of curve (c) resembles the theoretical curve (a) of Rosenzweig because, in these arid environments, evapotranspiration is essentially the same as precipitation. Productivity falls lower because a leaf area index small enough to minimize water loss necessarily limits productivity. The steep central portion of curve (c) represents evergreen forests in environments humid enough to support a less limiting leaf area index, thereby allowing a more rapid increase in productivity per unit of available moisture than in more arid habitats. Further increases in available mois-

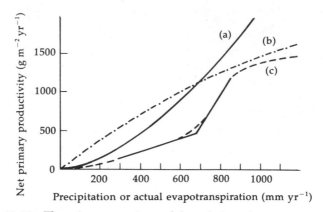

Figure 12-12   Three interpretations of the relation of net primary productivity (dry g $m^{-2}yr^{-1}$) to precipitation and actual evapotranspiration. (a) The curve fitted by Rosenzweig (1968) for aboveground net primary productivity of forests and shrublands in relation to actual evapotranspiration (mm yr^{-1}) using the formula NPP $= 1.66 \log_{10}AE - 1.66$. (b) The curve fitted by Lieth (1973) for total net primary productivity in relation to mean annual precipitation (mm yr^{-1}) using the formula NPP $= 3000/(1 - e^{1.315 - 0.119 \text{ MAP}})$. (c) Curve relating the two slopes of the Santa Catalina total net primary productivity estimates to probable mean annual precipitation, and adding to these a third, upper slope for limitation of climax temperate forest productivity at around 1500 g $m^{-2}yr^{-1}$. The dashed portion of the curve is hypothetical. [(a) From Rosenzweig, M. L. 1968. Net primary productivity of terrestrial communities: Prediction from climatological data. *American Naturalist* 102:67–74. Copyright © 1968 by The University of Chicago. (b) From H. Lieth, Primary production: Terrestrial ecosystems, *Human Ecology* 1:303–332, 1973. (c) Modified from Whittaker and Niering 1975. Copyright 1975 by the Ecological Society of America.]

ture (above 800–900 mm) are associated with distinctly less rapid increases in net productivity. Here, factors such as nutrient turnover, balance of supporting and photosynthetic tissue, and light absorption may severely limit increases in productivity in these temperate climax forests. Webb et al. (1978; 1983) suggest that the lack of continued increase in productivity at high levels of water availability is typical of ecosystems where water is not the most critical limiting factor. The lower slope for desert and open woodland communities may be due to the fact that, because of unpredictable conditions or more restricted water availability, the plants are not capable of high levels of productivity even when water is available. Those communities at intermediate elevations in the Whittaker and Niering (1975) data have sufficient aboveground biomass and metabolic adaptations to respond more dramatically to increases in precipitation. The forest communities represented by the dashed portion of curve (c) may represent communities that are not water-stressed. In non-water-stressed forests, Webb and his colleagues suggest that such factors as light, temperature, leaf biomass, and aboveground biomass are better correlated with productivity.

Productivity responses along environmental gradients are complex and non-linear. It will be necessary to accumulate more data on these relationships before the interactions of climate, life form, and productivity are understood and become predictable.

Carbon Dioxide

Earlier in this chapter, we mentioned that the level of CO_2 in the atmosphere is gradually increasing, but we did not consider the consequences for plant productivity. Carbon dioxide concentration is known to be a limiting factor in photosynthesis of C_3 plants (see Chapter 15). The predominance of C_3 photosynthesis in the world flora would lead one to speculate that dramatic increases in production should occur as atmospheric levels of CO_2 increase. However, the problem is not that well defined, because CO_2 is not evenly distributed within the plant canopy (Saeki 1973), and its concentration depends on other factors such as light and wind. CO_2 concentrations (Figure 12-13) in windless daytime condition are minimum near the top or middle of the canopy in the vicinity of the greatest leaf area. It is during windless, high-light periods that CO_2 concentrations are most limiting (Lemon 1960). Maximum levels occur at the soil surface because of the respiratory activity of roots, soil-inhabiting heterotrophs, and the low diffusivity of CO_2 near the surface. A gradual gradient from soil surface to ambient concentrations is rapidly reestablished even by low rates of air flow.

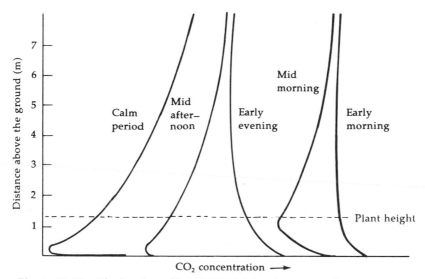

Figure 12-13   Idealized profiles of CO_2 in the air surrounding a photosynthesizing corn field as a function of height at different times during the day. (E. R. Lemon 1960. Reproduced from *Agronomy Journal*, Volume 52, pages 697–703, 1960 by permission of the American Society of Agronomy.)

Another factor that complicates our predictions of future productivity in the face of increasing atmospheric carbon dioxide is that most of the increase in productivity will probably be observed in agricultural plants. Plants in natural communities are subjected to competition from neighboring individuals, and limitations of light, nutrients, and water will more severely limit their growth than that of agricultural plants, where resources and competition can be controlled. (See M. C. Anderson (1973) for a review of the problems of understanding the impact of changes in CO_2 concentrations on productivity.)

There have been experiments on the photosynthetic capacity of some montane plants of southern California that are exposed to periodic high concentrations of CO_2 (Wright 1974). As expected, all C_3 plants showed a dramatic increase in photosynthetic rate when concentrations of CO_2 were increased. However, the response was more pronounced in the angiosperms tested than in the conifers. If these photosynthetic responses are translated into higher productivity, and therefore greater competitive ability, we might expect both greater production as a result of increases in CO_2 and a change in species composition of our mixed forest types. The latter effect is compounded by the greater susceptibility of conifers to air pollutants such as ozone (Miller 1969), which are often associated with locally high CO_2 concentrations.

Increasing the amount of carbon dioxide in the atmosphere may increase productivity by increasing water use efficiency (g CO_2 fixed/kg H_2O transpired) even in C_4 species that have little or no direct photosynthetic response to ambient CO_2. Plants that experience water stress during midday may, through increased water use efficiency, avoid stress and fix a higher amount of CO_2 during the day than they could at lower CO_2 levels (Rogers et al. 1983). Thus plants growing in water-stressed habitats may show greater response to CO_2 enhancement than plants of wet habitats because they have the photosynthetic advantage of higher CO_2 and/or more conservative water use.

It is clear, then, that a consideration of the influence of carbon dioxide on productivity must also consider the synergistic effects of light, temperature, and water.

Edaphic Factors

Soil characteristics such as texture, nutrient status, and depth are well documented as important factors determining the competitive relationships and growth rates of plants in a wide variety of environments (Bradshaw 1969). However, the translation of edaphic influences to productivity studies is sometimes difficult, because not all species have the same nutrient use efficiency (amount of nutrient used to produce a given amount of biomass). Edaphic differences may be masked by equal productivity when plants growing in poor sites have a higher nutrient use efficiency (Pastor and Bockheim 1984). In nutrient limiting sites, both the species composition and productivity can be influenced by modifying the nutrient regime. For example, Willis (1963) studied the nitrogen- and phosphorus-limited calcareous dune grasslands of England, where fertilizer applications increased density, productivity, and the importance of grasses at the expense of broadleaved species.

A dramatic productivity response to edaphic factors is documented for the pygmy forests on the coastal terraces of Mendocino County, California. Pygmy cypress (*Cupressus pygmaea*) and Bolander pine (*Pinus contorta* var. *bolanderi*) form an open, stunted forest, frequently less than 2.0 m tall, on highly acidic (to pH 3.7), shallow (to 2 dm), podzolic soils (Chapter 17). (See Figure 11-4 and Vogl et al. 1977.) Pygmy individuals 50–125 years old grow 10–30 m from redwoods (*Sequoia sempervirens*) of similar age that are up to 30 times their size. The Bolander pine is endemic to the pygmy barrens. However, where pygmy cypress grows in less limiting habitats, it may reach 50 m in height. The restrictions on growth are clearly soil-related and probably due to the combined effect of low pH, potential aluminum toxicity, poor drainage, and shallow soils (Westman 1975). Westman and Whittaker (1975) compared production samples from a pygmy forest, a bishop pine (*Pinus muricata*) forest, and a coastal redwood forest (Table 12-5). Productivity and biomass differences are impressive when one considers that these vegetation types occur in the same humid, maritime climate, on the same parent material, and that such great differences are due to edaphic factors. It is interesting to compare productivity per unit leaf area; the pygmy forest has a high production efficiency for an evergreen forest (most fall between 60 and 120 g m^{-2}). Not all pygmy forests are the result of edaphic conditions. For example, stunted forests of the New Jersey Piedmont were once thought to be a result of soil nutrient influences, but studies by Good and Good (1975) show that physiological or genetic influences reduce the growth rates of individuals and that the overall growth form is a complex response to frequent fire (see also McCormick and Buell 1968).

Edaphic factors, then, are most important in determining the species composition of certain communities; only in isolated situations is there clearly an edaphic influence on community productivity. One must be aware, however, that it is difficult to isolate the influence of soil factors in most habitats. Calcareous dunes of the British Isles and the pygmy forests of California are examples of extreme edaphic circumstances. More subtle soil-related controls of productivity remain to be identified.

Table 12-5 Comparisons of productivity and biomass of a *Sequoia sempervirens* forest, a *Pinus muricata* dominated forest, and a pygmy cypress (*Cupressus pygmaea*) forest, all from Mendocino County, California. (From Westman and Whittaker 1975. By permission of the British Ecological Society.)

	Sequoia sempervirens forest	*Pinus muricata* forest	Pygmy cypress forest
Aboveground biomass (t ha^{-1})	3200	415	27
Aboveground net productivity (g m^{-2} yr^{-1})	1401	1089	307
Leaf area index (m^{2}m^{-2})	20	16	2.1
Productivity / leaf area (g m^{-2})	72	115	147

Herbivores

The role of herbivores in ecosystem productivity is extremely important, largely because of the consumption of photosynthetic tissues (Petrusewicz and Grodzinski 1975). Activities such as pollination, seed dispersal, and soil aeration are no less important but have less direct impact on productivity. Consumption of photosynthetic tissues does not always result in a predictable impact on primary productivity. For example, a reduction in tuber yield of only 1 to 2% was noted when 20% of potato plant leaves were consumed by the Colorado potato beetle. In contrast, Varley (1967) reported that the loss of wood production far surpassed the relative consumption of oak leaves by caterpillars.

The impact of herbivores also depends on the kind of plant community in question. In grasslands, a larger proportion of the biomass is readily available for consumption than in forest communities, but only 12–20% of grassland biomass is consumed (Wiegert and Owen 1971). Therefore, the kinds of plants present and the quality of standing crop biomass are important determinants of the importance of herbivores in regulating primary productivity. Morrow and LaMarche (1978) conducted an interesting experiment on *Eucalyptus* in Australia, where insects commonly consume 20–50% or more of the foliage. They removed insects from experimental trees and found dramatic increases in the width of growth rings. This continuous heavy consumption of leaves severely reduces growth and may be a factor in competitive interactions between species of *Eucalyptus*. Variations in the width of annual rings may be a reflection of changes in the numbers of herbivores rather than of weather patterns, thus complicating things for the dendrochronologists.

Herbivore activity influences primary productivity directly, by reducing photosynthetic area, and indirectly, by modifying other environmental factors. For example, nutrients may leach more readily from the damaged foliage, and litter may have higher nutrient content when removed by herbivore activity. This is because natural abscission is preceded by remobilization and reabsorption of certain nutrients in the expendable part. Herbivore feces and urine make nutrients more readily available for decomposition, thus increasing the turnover rate. Availability of partially broken-down plant material may stimulate microbial activity, thereby increasing the turnover of nutrients. Removal of living biomass may change species composition, increase light penetration into the canopy, or reduce competition for light, water, and nutrients, any of which will affect primary production.

Mattson and Addy (1975) considered the importance of plant-eating insects in regulating primary productivity of forest communities. They concluded that insects help maintain consistently high primary productivity in natural systems by consuming biomass from less vigorous plants, thus opening the canopy for younger, more vigorous plants to establish themselves. Such insect activity may occur on a large scale, as in the case of the Douglas fir tussock moth, the gypsy moth, the spruce budworm, and the southern pine beetle, which periodically cause widespread defoliation and destruction of trees (Figure 12-14). Insect outbreaks may be related to increased nutritional value of the insect food and lower

Figure 12-14 Spruce forest in southern Colorado killed by insect damage. (Photo compliments of Harold Bradford.)

host resistance brought about by tree age, marginally adequate habitat conditions such as drought, or low levels of soil nutrients. Pine beetles (*Dendroctonus ponderosae*) may also be associated with reduced resistance of trees in stands where density is high and carbohydrates are allocated to meet metabolic needs rather than to defend against herbivores (Waring and Pitman 1985). Thinning of the forest by insects increases the availability of soil nutrients, reduces competition for light and water, and increases the resistance of trees to insect attack. The more vigorous growth of the surviving trees is probably the reason that epidemic populations begin to subside in 3–5 years. Carlson et al. (1985) found that a stand of Douglas fir (*Pseudotsuga menziesii*) thinned before a western spruce bud worm outbreak suffered much less severe defoliation than adjacent unthinned stands. Similar changes in grasslands and chaparral are stimulated by periodic fire and in forests by fire, wind-throw, or cutting. Loucks (1970) suggested that such drastic perturbations are essential to stimulate new waves of high productivity and species diversity in preclimax forests. Reduction of fire frequency in this century and the resulting high-density forests may have created a situation in which many forests are now much more susceptible to insect outbreaks.

Even a modest level of defoliation may stimulate vegetative and reproductive growth of plants. Reichle et al. (1973) analyzed insect consumption in a tulip poplar (*Liriodendron tulipifera*) forest and found that, on the average, 2.6% of the net primary production of foliage was consumed yearly. This nutrient return may be important in maintaining vigorous growth of the forest. Churchill et al. (1964)

support this conclusion, showing that net production was greater following partial defoliation in aspen (*Populus tremuloides*).

An accumulation of data indicates that the revival of productivity by herbivores may be due not only to the passive changes in abiotic factors stimulated by foliage removal or increased nutrients but also to direct chemical stimulation from the animals (see chapter 7 and Reardon et al. 1972, 1974; Dyer 1975; Harris 1974; Dyer and Bokhari 1976).

There is abundant evidence (e.g., Louda 1983; Rockwood 1974, 1976; Janzen 1973; Whittaker and Feeny 1971; Ehrlich and Raven 1964) that the activity of herbivores is very sensitive to the chemical makeup of the host. Herbivores have a direct influence on the active and potential productivity of plants by regulating the amount of photosynthetic tissue. Also, the chemicals that control the quality and palatability of the food source are modified as a result of herbivore–plant interactions. These relationships are the result of a long coevolutionary history between plants and herbivores. Is it possible, as Mattson and Addy (1975) suggest, that relations between foliage-eating insects and plants are, in the long term, mutualistic? The data are, in an evolutionary sense, inconclusive because the presence of a broad range of herbivore defense mechanisms implies that the plant's benefits from the relationship do not offset the energetic costs of herbivory.

Summary

The total energy fixed by photosynthesis per unit time (gross primary productivity) is used, in part, as maintenance energy. The remainder (net primary productivity) is used for new plant biomass or reproduction and is the food source of herbivores. The amount, distribution, and turnover rate of biomass determines community physiognomy, types of herbivores present, and the relative importance of decomposers in a community. The efficiency with which plants absorb light (exploitation efficiency), incorporate absorbed energy into photosynthate (assimilation efficiency), and convert photosynthate into structural components (net production efficiency) are important determinants of population and community characteristics.

Productivity is measured by determining photosynthetic and respiratory rates, by determining changes in biomass through time, or by dimension analysis, which estimates standing crop by determining the relationship between an easily measured parameter and biomass. Dimension analysis and harvest techniques are the most commonly and easily applied methods of studying productivity.

Terrestrial vegetation supports about 62% of the total world primary productivity, totaling about $100 \times 10^9 \ t \ yr^{-1}$. Productivity varies from a theoretical maximum of 3000–3500 g m^{-2}yr^{-1} in tropical rainforests to near zero in some deserts. These trends in productivity are related to severity of habitat and are paralleled by changes in biomass. High levels of productivity occur in communities with maximum leaf area indices (LAI). LAI is maximized in trees, but the large amounts of energy necessary to maintain nonproductive support biomass reduces the potential advantage of the tree life form.

Litter production and decomposition rates have an important influence on mineral cycling and the composition and abundance of detritivores. Rates of decomposition vary with species, the chemical composition of litter, the availability of nitrogen and water, and temperature. Litter production and decomposition rates are highest in the tropics and generally decrease toward the poles.

Carbon dioxide, soil, light, temperature, water, consumers, and level of community development are important factors influencing productivity. Habitats with environmental factors in the optimal range for photosynthesis for the longest periods of time are most productive. Local imbalances of soil nutrients may severely restrain productivity, as in the pygmy forests of California. The successional stage of the community also affects productivity; productivity increases during the pioneer and early tree stages and decreases as the community reaches maturity.

CHAPTER 13

MINERAL CYCLES

Nutrients do not move through living systems in smooth, even-flowing transition, but in pulses, jerks, and floods. The cycling of matter is inherent in the functioning of ecosystems and is integral to their structure. Both essential and nonessential materials move in cyclic fashion. In contrast, energy flows one way and is noncyclic: It is continually replaced by the sun and is lost from the system as heat or exported as energy-rich plant and animal parts and wholes (Figure 13-1).

The goal of nutrient cycling research is to quantify (a) the sum total of nutrients and nonessential elements present within a system, (b) the cycling times, turnover rates and residence times of nutrients within the system, (c) the biotic and abiotic factors that govern nutrient cycles and cycling times, (d) nutrient use efficiency, and (e) ecosystem regulation of nutrient loss following disturbance. Our approach in this chapter will be to define the roles of various physical and biotic forces with respect to nutrient cycling, to compare and contrast nutrient cycling in three biomes (coastal dune, grassland, and temperate forest), and finally to examine nutrient use efficiency and ecosystem mechanisms that promote intraecosystem cycling and thereby prevent loss of nutrients.

Introduction

Biogeochemical Cycles

We refer to the movement of chemical elements within the environment as **biogeochemical cycles.** *Bio* refers to living systems, and *geo* to the rocks, water and air of the earth. Geochemistry deals with the exchange of elements among the physical components of the earth. Thus, *biogeochemistry* refers to the transfer or flux of materials back and forth between the living and nonliving components of the biosphere.

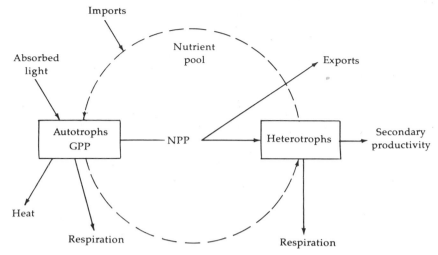

Figure 13-1   A biogeochemical cycle (dashed circle) superimposed on a simplified energy flow diagram contrasting the cycling of material and the one-way flow of energy. GPP = gross primary productivity and NPP = net primary productivity, which may be consumed within the system by heterotrophs or exported from the system. (Adapted from "Limits of remote ecosystems containing man," by H. T. Odum, *The American Biology Teacher*, National Assoc. of Biology Teachers, 25:429-443, 1963.)

Nutrient cycling is the movement of those materials that are essential to organisms. **Intrasystem cycling** is a homeostatic mechanism whereby plants build up an available (or exchangeable) pool of nutrients, incorporated as living or dead biomass. The movement of nutrients within an ecosystem is often an order of magnitude more rapid than the movement into and out of that same system (interecosystem cycling, which connects the system to the global ecosystem). The retention of nutrients, through time, accompanies succession (see Chapter 11 for a complete discussion). Conversely, a drastic reduction in the nutrient pool accompanies the degradation of an ecosystem.

Gaseous and Sedimentary Cycles

The movement of materials is largely by one of two basic avenues. (1) Those elements that have a major gaseous phase are involved in gaseous cycles and participate in regional and/or global cycles. (2) Those elements which lack a major gaseous phase move in sedimentary cycles. The gaseous nutrients are part of what Odum (1971) terms perfect cycles, in that local perturbation is quickly compensated for and equilibrium rapidly reestablished. Some nutrients, such as S, may have both a gaseous and a sedimentary phase; however, they may be placed in a specific category depending on what phase is incorporated into living tissues.

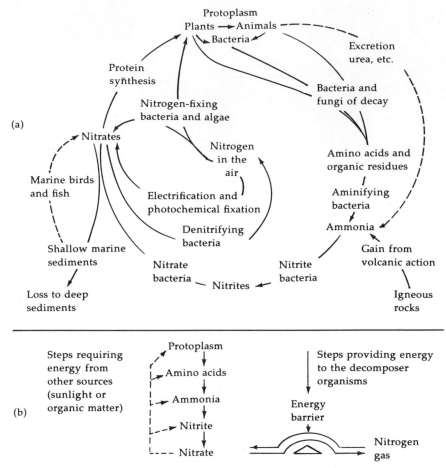

Figure 13-2 Two ways of picturing the nitrogen biogeochemical cycle, which is an example of a relatively perfect, self-regulating cycle with a large gaseous reservoir. In (a) the circulation of nitrogen between organisms and the environment is depicted along with microorganisms that are responsible for key steps. In (b) the same basic steps are arranged in an ascending-descending energy series, with the high-energy forms on top to distinguish steps that require energy from those that release energy. (From *Fundamentals of Ecology*, 3rd edition, by E. P. Odum. Copyright © 1971 by W. B. Saunders Company. Copyright 1953 and 1959 by W. B. Saunders Company. Reprinted by permission of CBS College Publishing.)

An example of a gaseous cycle, the nitrogen cycle, is diagrammed in Figure 13-2. Contrast this "perfect cycle" with the phosphorus cycle, shown in Figure 13-3. Note that the reservoir for nitrogen is the atmosphere. The shallow and deep sediments and deposits of the earth's crust are the reservoir for phosphorus. Further, phosphorus is rare, compared to nitrogen. We will have more to say about both these nutrient elements later.

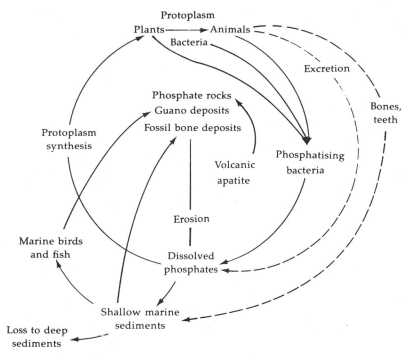

Figure 13-3 The phosphorus cycle. Phosphorus is a rare element compared with nitrogen. Its ratio to nitrogen in natural waters is about 1 to 23. Chemical erosion in the United States has been estimated at 34 t km^{-2} yr^{-1}. Fifty-year cultivation of virgin soils of the Middle West reduced the P_2O_5 content by 36%. As shown in the diagram, the evidence indicates that return of phosphorus to the land has not been keeping up with the loss to the ocean. (From *Fundamentals of Ecology*, 3rd edition, by E. P. Odum. Copyright © 1971 by W. B. Saunders Company. Copyright 1953 and 1959 by W. B. Saunders Company. Reprinted by permission of CBS College Publishing.)

Those elements that do not have a prominent gaseous phase, such as calcium, phosphorus and iron, share the earth's crust as their reservoir. All of these elements ultimately tend to be deposited in the sea, in some pond, or in a lake, and only returned to the nutrient pool by geological activity—mountain building or some other tectonic phenomenon. Exceptions to this limitation include the short-term recycling that occurs when predatory birds return to land after feeding on ocean fishes. Their rookeries become covered with gleaming white deposits of guano which is rich in phosphorus and other materials. Anadromous fishes (which have matured in the sea) return to spawn and die inland. This migration also recycles nutrients quickly, but the contribution of such transfer is minimal.

Early research on nutrient cycling focused upon cycling of calcium, magnesium, and potassium (Bormann and Likens 1967; Likens et al. 1970, 1977; Siccama et al. 1970; Likens and Bormann 1972; Bormann et al. 1977). More recent literature (Vitousek et al. 1979 and 1982; Johnson and Cole 1980) and this chapter reflect an increased focus upon nitrogen cycling. This emphasis is defensible for several reasons (modified from Vitousek et al. 1982, but see also Vitousek et al. 1979 and Johnson and Cole 1980):

1. Nitrogen is often the primary limiting nutrient in many communities, including grassland and forest biomes. Recovery following disturbance could be similarly limited.

2. Nitrogen losses in a disturbed ecosystem may be substantially higher than losses of other nutrients.

3. Because cation leaching is dependent upon the supply of mobile anions, nitrate loss can lead to increased cation loss.

Nutrient Storage and Flux

Figure 13-4 gives a diagrammatical view of nutrient storage and cycling. The nutrients are viewed here as compartmentalized, occurring as primary (soil) or secondary (rock) minerals (see Chapter 17), as available nutrients in the atmosphere, or in living or dead organic matter (Bormann et al. 1974). Odum (1971) points out that the flux of nutrients is more significant for the functioning of ecosystems than is the absolute amount present in a given compartment at a specified time. Intrasystem cycling may occur through plant uptake and assimilation of nutrients, leaching of nutrients from plants and parts, and biological decomposition.

The vehicles for intersystem cycling, which transport nutrients to and from the ecosystem, include: (a) meteorological factors, for example, rain bringing dissolved sulfates, carbonates, and particulates, and wind bringing aerosols*; (b) geologic forces, such as surface or subsurface water, which move nutrients such as phosphates and calcium into or out of a system and which facilitate weathering of primary and secondary minerals; (c) biologic flux, which is generally caused by animal transport.

Does a nutrient storage compartment ever become a vault, locked against internal cycling? Large, old trees become repositories of enormous quantities of organic matter. Yet wood is notoriously low in nutrient concentration, so that the lockup, although not trivial, is not disastrous for the ecosystem. In the boreal[†] jack pine (*Pinus banksiana*) forest, for example, the vault is in the litter and humus of the forest floor.

A jack pine stand up to about 20 years old is very productive, and organic matter accumulates on the forest floor at about 800 kg ha^{-1} yr^{-1}. After that, the annual increments of organic matter added to the litter decrease to about half as the growth of the stand slows. Especially apparent is the decline in productivity of stands between 30 and 65 years old.

Foster and Morrison (1976) point out that very early in the life of the stand, phosphorus concentration stabilizes at about 12-15 kg ha^{-1} in the aboveground biomass of jack pine. From age 20–30, N and P are immobilized to an increasing

---

*An aerosol is a suspension of fine particles, solid or liquid, in a gas.
[†]Boreal refers to the forest areas of the north temperate zone and Arctic region.

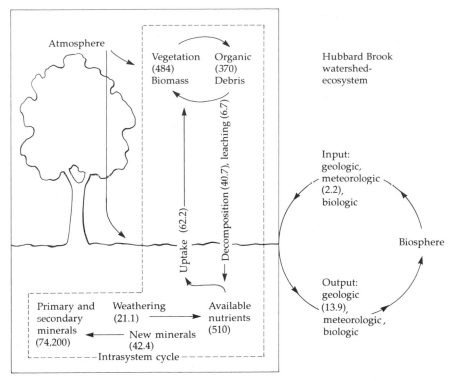

Figure 13-4 A nutrient cycling model. A nutrient element, such as calcium, may be found within the ecosystem in the organic compartment, where it is a part of living and dead biomass, or in the available nutrients compartment, where it occurs as a cation on exchange surfaces or is dissolved in the soil solution. Shown also are extrasystem compartments, the atmosphere and primary and secondary minerals (see Chapter 17). Calcium may be added to the system by weathering (geological input) or by atmospheric deposition (meteorological input). Nutrients are moved within the system (intrasystem cycling) along such pathways as foliar leaching and nutrient uptake, plant assimilation and use, and biological decomposition. Nutrients depart by the formation of new minerals or by leaching from the system. The ecosystem is thus connected to the larger biogeochemical cycles by meteorologic, geologic, and biologic vectors that move nutrients across ecosystem boundaries. Data are from the Hubbard Brook watershed ecosystem, where input of nutrients with sedimentary cycles is almost wholly meteorologic, while output is almost wholly geologic. All data are in kg ha^{-1} and kg ha^{-1} yr^{-1}. (Modified from Likens et al. 1977. *Biogeochemistry of a Forested Ecosystem*. By permission of Springer-Verlag.)

degree in microbial biomass. Using incorporation of elements into the forest stand as a measure of growth, we find little to no incorporation in stands 30–65 years old. The flow of minerals from soil to roots to vegetation and back to soil is maintained, but the percentage of nutrients retained in plant biomass is very small. The lack of decomposition locks up N, P, and Mg during this stage in the life of the stand, and so plant growth falters. Return to high productivity may be accomplished by wildfire (see also chapter 16).

The time frame for nutrient cycling depends on the compartment. Metabolic processes may turn over elements in minutes. For example, in photorespiration,* carbon is moved from the atmosphere into the plant and back in a matter of minutes. Nutrients in the leaf litter in a forest may turn over within an ecosystem annually or require several years or decades to turn around. Some biogeochemical processes (e.g., deposition, tectonic movement that exposes sediments, and subsequent retrieval by some root system) may require hundreds of millions of years and are therefore of little ecological importance.

Plant Nutrients

Essential elements and inorganic compounds are categorized by plant physiologists according to relative quantities required by plants for adequate nutrition. Major nutrients (macronutrients) are those required in rather large amounts. Those required in only small or trace amounts are minor nutrients (micronutrients). An example of differences in amounts of macro- and micronutrients for corn plants is shown in Table 13-1.

The nutritional state of a plant in regard to elements present in the growth medium may be viewed as (a) deficient, (b) adequate, or (c) injuriously excessive. Deficiencies may result in stunted growth or premature sexual maturity and

Table 13-1 The percentage of dry weight concentrations of essential elements in higher plants. (Modified from *Plants, Man, and the Ecosystem*, 2nd Ed. by W. D. Billings. © 1970 by Wadsworth Publishing Co., Inc. Used by permission.)

Carbon	45%	Sulfur	0.1%
Oxygen	45	Chlorine	0.01
Hydrogen	6	Iron	0.01
Nitrogen	1.5	Manganese	0.005
Potassium	1.0	Boron	0.002
Calcium	0.5	Zinc	0.002
Magnesium	0.2	Copper	0.0001
Phosphorus	0.2	Molybdenum	0.0001

---

*Photorespiration is a phenomenon involving the light-dependent production of glycolic acid (in chloroplasts) and its subsequent oxidation in peroxisomes (see Chapter 15).

senescence. Specific deficiencies, caused by lack of only one or a few of the vital elements, are often revealed by characteristic symptoms. If nutrient concentrations exceed the limits of tolerance, even those nutrients essential to proper nutrition may be toxic. Examples of excess include high salinity in estuarine habitats, which limits the distribution of glycophytes and, at extremes, even halophytes such as cord grass (*Spartina alterniflora*) (see Chapter 6 for more details).

Although carbon, oxygen, and hydrogen are in the strictest sense plant nutrients, the assimilation and use of these is covered elsewhere, and we will not consider them here. Nitrogen is a constituent of proteins and nucleic acids and is structurally involved in most catalytic molecules. It accumulates in young tissues, seeds, and storage organs. Phosphorus is involved in the structure of many vital molecules, such as nucleic acids and phospholipids. The energy that is used in the cell is released largely by hydrolysis of various phosphate bonds. Sulfur forms part of some amino acids and is crucial to the stability of the tertiary structure of enzymes and other proteins. Potassium is very soluble and is easily leached from light or sandy soils. Although it has no apparent structural role, potassium plays a catalytic role: It is used primarily in meristematic zones in young tissues and is important in regulating guard cell turgor. Magnesium is a structural part of the chlorophyll molecule and serves to activate many of the enzymatic reactions that transfer phosphates. It tends to accumulate in young leaves and will be translocated out before leaf abscission. Calcium is essential for the formation and metabolism of the mitochondria and nucleus. Cell membranes lose their integrity when there is a calcium deficiency. Apparently Ca^{++} also alters the permeability of cell membranes to other minerals. Calcium is virtually immobile in the plant, due to its structural role, and is therefore usually released only by decomposition.

Iron is essential for the synthesis of chlorophyll, though it is not part of the structure of the molecule. It is the center of the porphyrin ring of the cytochromes and so is involved both in the transformation of radiant energy and in the utilization of energy within the cell.

Deficiencies of copper and chlorine in the soil are rare, but deficiencies of boron, manganese and molybdenum occur often. Molybdenum, a minor nutrient, is essential for nitrogen fixation. Manganese and copper are enzyme catalysts, and boron may exert influence on the activity of various enzymes. Zinc is essential to the synthesis of the important plant hormone indole acetic acid (IAA) and may be involved in protein synthesis. Chloride, which probably is absorbed in ionic form and remains so, plays a vital role in photosynthesis.

Certain elements behave as analogs to nutrients and will replace them to some extent. The uptake of strontium, which mimics calcium and will relieve deficiency symptoms for a time, has been of great interest to nuclear scientists and nutritionists in recent years. Radioactive strontium was released by atmospheric testing of nuclear devices and tended to be concentrated in the Northern Hemisphere. Lichens in the tundra took up the strontium as if it were calcium. Reindeer fed upon the lichens, and Laplanders, who were nourished by caribou (reindeer), showed higher concentrations of the harmful isotope than peoples of other lands.

Factors in Nutrient Cycling

Every physical force and organism impedes, hastens, or otherwise affects nutrient cycling. Our knowledge of general pathways of nutrient movement has preceeded our understanding of how much is moved and in what time frame the transfer occurs. However, quantitative studies have appeared with increasing frequency (e.g., Woodwell et al. 1975; Schlesinger and Hasey 1981; Melillo et al. 1982; Vitousek 1982; Gray 1983; Vitousek and Matson 1984; McClaugherty et al. 1985; Lindberg et al. 1986). A look at some of the factors involved in nutrient flux will be of value.

The Hydrologic Cycle

The following mnemonic device

$$P = E + T + R + I$$

reminds us of the balance that must exist in the hydrologic cycle, such that P (precipitation) is equal to E (evaporation) plus T (transpiration) plus R (runoff) plus I (infiltration, the downward entry of water into the soil). Both intrasystem and intersystem cycling are associated with the hydrologic cycle. Precipitation carries nutrients in solution, runoff and infiltration remove nutrients from a system or move them down the soil column, and evapotranspiration of water concentrates and conserves nutrients. In Europe, the contribution of nutrients from the atmosphere (primarily Cl, Na, Ca, S, K, Mg and N) may range as high as 25-75 kg ha^{-1} yr^{-1}, compared to 19-40 kg ha^{-1} yr^{-1} for a deciduous forest in the eastern United States (Larcher 1975; Likens and Bormann 1972). Groundwater and water rising by capillary action transport nutrients into the root zone. Although transfer by runoff is often only to a region of lower elevation, streamflow, with its load of nutrients, must be viewed as loss on a local scale. This section will give several illustrations of the role of water in nutrient cycling.

Deposition and Canopy Interaction   There is a significant nutrient contribution from the atmosphere, both in rainfall and intercepted by the canopy. Precipitation may provide the largest nutrient source to many forest systems (Parker 1983). The contribution of rainfall to nutrient input can be quantified by the use of precipitation gauges and subsequent chemical analysis (Likens et al. 1970; Cole et al. 1967). Precipitation brings sulfate and hydrogen ions, nitrate, and significant amounts of ammonium, chloride, sodium and calcium to the eastern deciduous forest (Table 13-2). Rainfall is often acidic, frequently with a pH of less than 4.0 (Likens and Bormann 1972). This is attributable, at least in part, to human activities. Acid precipitation may contribute significant nutrients, but may also enhance foliar leaching (Evans 1984). Acid deposition in boreal forest is discussed in Chapter 20.

When acid precipitation infiltrates the soil, cations such as K and Ca may be displaced by the hydrogen ions of the acid and will be leached away by seepage

Table 13-2 Weighted average concentrations of various dissolved substances in bulk precipitation and stream water for undisturbed watersheds 1–6, Hubbard Brook Experimental Forest, 1963–1969. (After Likens, G. and F. H. Bormann, "Nutrient Cycling in Ecosystems," in *Ecosystem Structure and Function*, copyright 1972 Oregon State University Press.)

	Precipitation	Stream water
	(mg 1^{-1})	(mg 1^{-1})
Calcium	0.21	1.58
Magnesium	0.06	0.39
Potassium	0.09	0.23
Sodium	0.12	0.92
Aluminum	*	0.24
Ammonium	0.22	0.05
Sulfate	3.10	6.40
Nitrate	1.31	1.14
Chloride	0.42	0.64
Bicarbonate	*	1.9[a]
Dissolved silica	*	4.61

*Not determined but very low.
[a]Watershed 4 only.

and surface runoff. Under these circumstances, there is an apparent increase in export of nutrients in the eastern deciduous forest (Table 13-2). Is the nutrient flux the result of increased weathering because of the acid, or does it represent an actual loss of available nutrient from the system? We will consider this again when we discuss runoff.

Schlesinger and Reiners (1974) measured atmospheric deposition of water, calcium, magnesium, sodium, potassium, and lead, both in bulk precipitation and on artificial foliar collectors. Their data suggested that *both* bulk precipitation and interception by the canopy should be considered when assessing atmospheric nutrient input, especially in montane areas subject to high winds and clouds of fog (see also Schlesinger et al. 1982). Art et al. (1974) investigated the dune forest system on Fire Island, New York, and reached a similar conclusion. The canopy network intercepts the airstream, minerals are deposited on the foliage, and subsequent leaching by rainfall carries the nutrients to the forest floor. Uptake by plants increases productivity, biomass, and leaf area, which in turn traps more minerals, generating a positive feedback loop until the mineral-laden winds that provide nutrition become intense enough to kill back the twigs of the upper canopy.

More recent studies of a mixed hardwood forest demonstrated again the role of canopy interception in forest nutrition. Lindberg et al. (1986) report that, in a "forest typical of the ridge and valley physiographic province of the eastern United States," the major mechanism of atmospheric input was dry deposition, supplying sulfur, nitrogen, and free acidity by vapor uptake and calcium and potassium by particle deposition. In fact, they estimated that such deposition supplied calcium and nitrogen for 40% of the annual needs of the woody increment, and sulfur (in this sulfur-rich ecosystem) to more than 100% of the annual total forest requirement. It is noteworthy that standard bulk deposition collectors underestimated these contributions by a significant amount.

Leaching   **Leaching** refers to the removal of soluble constituents from the soil or litter by percolating water. As a factor in intersystem cycling, leaching is a primary cause of nutrient loss from a system. The most easily leached nutrients are K and Ca (Tappeiner and Alm 1975). Hunt (1977) has calculated that 45–65% of the labile component (the component that readily undergoes chemical change) may be leached from grassland soils and litter in as short a time as 4 hours. This process is positively correlated with increasing temperature.

Leaching from attached leaves, on the other hand, adds nutrients to the soil (see Chapter 6). As rain runs over the foliage, it passively absorbs nutrients. The nutrients are then carried to the ground in **throughfall**[*] and **stemflow**.[†] These are major factors in intraecosystem cycling. Potassium, sodium, and sulfur are leachable and therefore cycled predominantly in this fashion, rather than in litterfall. Throughfall nutrient flux may be as much as 11 times as high as precipitation concentration for the same nutrient (potassium, for example; Parker 1983). The nutrient content of waters that move onto and through the canopy is altered by that contact. Foliar leaching is the major process controlling nutrient enhancement in throughfall and stemflow, according to Parker (1983), but canopy interception, mentioned earlier, plays a major role in certain forests. Mosses gain nutrients primarily from throughfall and stemflow. In some vegetation types with considerable moss cover on the ground, nutrients derived from throughfall must pass through mosses before they are available to trees (Foster and Morrison 1976).

Evapotranspiration, Runoff, and Infiltration   In a region with an impermeable geologic substrate, water loss by evapotranspiration may be inferred from the difference between precipitation and stream flow. Chemical concentrations in precipitation are a measure of nutrient input, and concentrations in stream water are a direct measure of nutrient output in such a region. At the Hubbard Brook Experimental Forest, stream-gauging stations were anchored to bedrock to meter nutrients in stream water flowing from a watershed underlain by an impermeable substrate. The losses shown in Table 13-2 appear enormous, but water lost by evapotranspiration accounts for much of the increased nutrient concentration (Likens and Bormann 1972). Nutrient and evapotranspirational loss is relatively stable from year to year in an undisturbed forest. The biotic

---

[*]Throughfall is rainwater which falls through a canopy.
[†]Stemflow is the fraction of precipitation that flows down the stem or trunk of a plant and then enters the soil.

contribution to this stability will be referred to again in a later section of this chapter. Rather than promoting nutrient loss, transpiration actually conserves mineral elements because excess water, which would promote loss through leaching, is removed without harm to the system, leaving the nutrients behind.

In a region where the substrate is permeable, evapotranspirational losses and water movement through the soil may be estimated by the use of a **lysimeter** (Figure 13-5). In a tropical rainforest, which has a permeable substrate, lysimeters

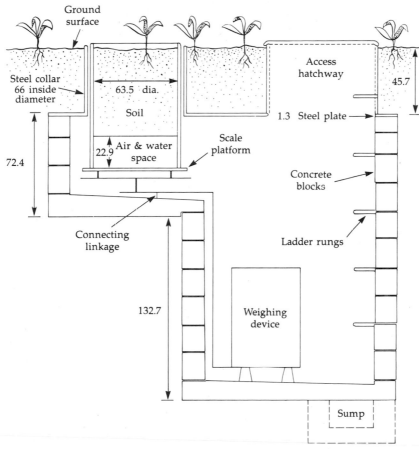

Figure 13-5 Diagram showing the principle of a weighing lysimeter (measurements are in cm). A lysimeter consists of a container holding a mass of soil mounted flush with the soil surface and arranged so it can move up and down on a weighing device (modified from England and Lesesne 1962). The lysimeter pictured here uses a scale to follow changes in weight, but electronic weighing devices have advantages. For the results to be applicable to specific crops or stands of vegetation, the lysimeter must be surrounded by similar vegetation. Some problems of lysimetry are discussed by Hagan et al. (1967). (Reprinted by permission of McGraw-Hill Book Co. from *Plant and Soil Water Relationships: A Modern Synthesis* by Kramer, 1969.)

are employed to determine the volume of water running through the soil. Water thus collected can be used to determine the concentrations of nutrients leached. The product of the volume and the concentration of each nutrient is the loss rate for that nutrient. Evapotranspirational losses are inferred from the difference between precipitation input and percolation output. One drawback of lysimetry is that the hydrologic regime is disturbed, sometimes greatly so, by manipulation of the soil to put a lysimeter in place.

Nutrient Budgets   In general, the successional status of community can be deduced from nutritional budgets derived from studies such as those just described, which measure nutrient input and output. If elements are being stored, as indicated by a positive input/output balance, new net biomass is being accumulated, and the system is aggrading (i.e., it is a seral stage). Loss by leaching from an Amazonian rainforest (which is not aggrading) during each year from 1975 to 1980 was less than or equal to atmospheric input, suggesting that this mature forest is maintained on intrasystem cycling and nutrient input from the atmosphere (Jordan 1982). This situation is a long-term but fluctuating steady state, in which tree growth balances tree falls. A small negative balance may represent weathering of parent material and suggests that the system is a mature or climax community.

Though no ecosystem is ever precisely in steady state (Jordan 1982), the watersheds of Hubbard Brook reflect a "time-invariant mode" or steady state *condition*, in which additions and deletions will occur at the perimeter of the system. The system is therefore an open or continuous system, rather than closed. It is important to distinguish steady state from equilibrium, even though the former may be similar to or even approach the latter (Johnson 1971).

The chemical concentrations in stream water at the Hubbard Brook Experimental Forest are amazingly constant (concentration of nutrients being independent of volume of water flow), especially those of magnesium, sulfate, chloride and calcium (Johnson 1971; Likens and Bormann 1972). This constancy may be due to the high exchange capacity of the soils there. As ions are removed from the soil water by roots or runoff, they are replaced from the soil. There is a near-saturation of exchange capacity, and adequate nutrition for whatever grows in that soil. Rich, fertile systems may show a greater loss of nutrients than do impoverished soils as precipitation increases, because nutrient loss is proportional to the volume of runoff. The acidity of rainwater also influences nutrient loss, as was stated earlier. Nutrient-unsaturated soils tend to lose a lower proportion of nutrients even with increased runoff (Jordan and Kline 1972).

To illustrate the concept of nutrient constancy in stream flow, consider calcium both in a Puerto Rican tropical rainforest and in the temperate deciduous forest. The concentration of calcium in stream flow is relatively insensitive to volume fluctuations in both ecosystems. In the temperate deciduous forest, stream flow varies by more than four orders of magnitude, and yet the calcium concentration in the water remains unchanged (Likens and Bormann 1972). In the rain forest, increases of 100 times or more in the volume of soil water reduced calcium concentration by less than 25% (Jordan and Kline 1972).

In contrast, the concentration of potassium is a very sensitive index of biological activity, for it may be reduced dramatically in stream flow during periods of plant growth and then rebound during dormant times.

Salt Spray   The nutrients that are received in a dune community from salt spray are essential to the continued functioning of that system. Sea water contains all the necessary mineral ions, except nitrates and phosphates (Boyce 1954; Salisbury 1952). Young dune soils may be as high as 99% sand, with very poor capacity for cation storage because they lack clay and organic matter. Dune soils are known to have a cation exchange capacity (CEC) of only about 10–15 meq 100 g^{-1} of soil. (A millequivalent (meq) is the amount of material that will combine with or replace one milligram of hydrogen.) For comparison, one may expect a CEC of 40–50 meq 100 g^{-1} of clay, and a CEC of 160–250 meq 100 g^{-1} of organic matter (Ranwell 1972).

The dune system at Cape Hatteras, North Carolina, receives the following weights ($\text{kg ha}^{-1} \text{ yr}^{-1}$) from salt spray: Na, 250–1300; Mg, 37–120; Ca, 19–120; and K, 13–77. By contrast, rainfall contributes only 1.0–1.3, 21–26, 2.3–3.4 and 3.0–3.9 $\text{kg ha}^{-1} \text{ yr}^{-1}$ of each nutrient, respectively (van der Valk 1974). In spite of the very heavy input, the habitat is not saline, due to the extreme leachability of these porous soils.

Fire   Wildfire plays an important role in the maintenance and functioning of many communities. The reproduction of giant sequoia (*Sequoiadendron giganteum*), jack pine (*Pinus banksiana*), Bishop pine (*P. muricata*), and various species of California lilac (*Ceanothus*) is enhanced by fire of a certain intensity. Fire is also important to cycling of nutrients in certain regions. In the far north, low temperatures prevent decomposition, and the result is acid soil covered by raw humus. In Norway, for example, burning actually improves forest site quality by raising the pH, thus favoring the growth of populations of nitrogen-fixing bacteria (*Azotobacter*, *Rhizobium*). In chaparral regions, fire may serve to: (a) break down sclerophyllous litter, which is resistant to biotic decomposition; (b) remove inhibitors of microbial decomposition; and (c) alter wettability of the soil.

Nutrients are released in the form of soluble mineral ash by slash and burn agriculture. Thus, in a tropical rainforest, slash and burn temporarily enhances the nutrient regime. Crop yields are good at first but diminish quickly. Reasons for this vary, but at least three should be mentioned here:

1. The fire may destroy such recycling mechanisms as ectomycorrhizae within the soil.
2. The nutrients are released in a single large pulse, and their availability probably exceeds the exchange capacity of the soil.
3. Thus, nutrients are quickly leached out of the root zone (Jordan and Kline 1972).

Earth Movement   Mountain building, movement along a lateral fault, volcanism, and other types of tectonic activity exert major influence on mineral cycling. The impact may be sudden or require eons of time. For example, the volcano Paracutin in Michoacan, Mexico, poured an estimated one billion tons of ash, cinders, and bombs* on the land in its first year.

Such events, rare in our time frame, are common in geological time and even in ecological time. In addition, the magnitude of the event counterbalances its infrequent occurrence. Geophysical phenomena rank as primary physical agents of biogeochemical cycling.

Biotic Factors

Biotic factors function primarily in intraecosystem cycling. The transfer of biomass, with its nutrient content, from producer to herbivore to carnivore to decomposer, etc., is well known. Most of us recognize that respiration, excretion, defecation (in animals), leaf fall (in plants), and death are integral components of nutrient cycling (see Figure 13-2).

Microbial Decomposition   The action of microorganisms in the cycling of nutrients has been well studied. In the nitrogen cycle, microbial action can **immobilize**[†] nitrogen or can **mineralize**[‡] it. The ratio of carbon to nitrogen in the substrate to be decomposed largely determines which of these processes will occur. At ratios of C:N greater than 32:1, some immobilization will occur, as microorganisms will then use soil nitrogen (as NO_3 or NH_4) to build their own proteins. At narrower ratios, some mineralization will occur, and decomposition proceeds. We will refer to this process again as we discuss decomposition and nutrient use.

Microbial decomposition has been studied by Brinson (1977) (see also Triska and Sedell 1976). There are three stages of decomposition:

1. Formation of particulate detritus.
2. Production by saprophytes of immobile humus and the concomitant release of immobile, though soluble, organic compounds (in a relatively short time).
3. Mineralization (mobilization) of humus (a slower process and one that needs further study).

Although the products of stages one and two are not available as nutrients, humic substances and other organics play a vital role in nutrient storage, forming complexes with minerals that enhance uptake by plants. Immobilization through decomposition serves to retain nutrients beyond the period of dormancy to the time of growth.

Ingestion and Digestion   Perhaps less widely recognized than microbial decomposition is the nutrient regeneration in soils due to ingestion and digestion

---

*Volcanic bombs are molten material thrown from volcanoes which solidifies and falls as an igneous rock with a bomb-like shape.
[†]Immobilization in this context refers to the conversion of inorganic ions to organic forms.
[‡]Mineralization refers to the portion of the nitrogen cycle in which organic matter is decomposed and inorganic ions are released.

of bacteria and fungi by other organisms. Within the soil, metazoan animals account for only 10% of the total metabolism, which leaves the bulk of it (90%) for microorganisms. Thus, we may infer that protozoan digestion and mineralization of bacteria play a major, vital role in the intrasystem cycling of nutrients. Evidence supports this inference (Pomeroy 1970). In addition, protozoan predation removes cells from overcrowded bacterial populations, which stimulates active bacterial growth and enhances bacterial nitrogen fixation.

Litter   The role of litter in nutrient cycling must be viewed in an ecosystem framework (see also Chapter 12). Two factors determine the amount of litter in an ecosystem: the total litter produced in a unit of time and the rate at which it is decomposed. Litter production is governed by the type of vegetation. For example, in the jack pine forest system in northern Ontario, Canada, organic matter can accumulate at the rate of 800 kg ha^{-1} yr^{-1}. Organic material decomposes very slowly, requiring as much as 16 years for some elements to recycle (Foster and Morrison 1976). The acid soils that are produced contain a well-defined surface layer of organic material in which nutrients are taken up and immobilized in the biomass of microbes.

In contrast to the slowly turning mineral cycle of the boreal forest, litter in the tropical rainforest decomposes quickly, and there is nearly complete turnover each year. Litter production in a tropical rainforest can be staggeringly high; leaf fall alone is 45-126 kg ha^{-1} every day (Walter 1979). As Figure 8-8 shows, the bulk of the carbon in a subalpine conifer forest is in organic litter and humus on or in the soil, but it mainly resides in the wood of the tropical rain forest.

The deciduous forest of the temperate zone stands midway between these two extremes. It exhibits moderate turnover times, such that a year's litter production is decomposed in 2–3 years (Larcher 1975).

Rates of decomposition vary, then, from one forest type to another. These rates are governed by many factors, notably climate (moisture and heat availability) and soil microflora and fauna. Does litter quality also influence the rate of decomposition and therefore nutrient release? In recent years, this question has been addressed by many researchers (McClaugherty et al. 1985; Schlesinger 1985; Melillo et al. 1982; Schlesinger and Hasey 1981), who have studied the chemical composition of the decomposing substrate to determine its role in the rate of decomposition. Although calcium, nitrogen, organic matter and phosphorus all move largely via litterfall, it is the roles of lignin, nitrogen, and phosphorus that have been most widely reported. We will discuss here decomposition dynamics in southern California chaparral and the Hubbard Brook forest in New Hampshire.

The chaparral of southern California characteristically contains both evergreen and deciduous species, such as evergreen *Ceanothus megacarpus* and drought-deciduous *Salvia mellifera* (sage). Schlesinger and Hasey (1981) found that, during the first year of decomposition of leaf litter of these two species, more than 70% of the original contents of soluble carbohydrate, phenolics, and potassium were lost. During that same year, *Ceanothus* leaf litter lost a lesser amount of the initial ash-free dry mass than did *Salvia*. As we might predict, the lignin concentration of abscised leaves of *Ceanothus* was higher initially than that of *Salvia*. Even so, mean residence time for litter of *Ceanothus* was only 4.6 years.

There was no apparent mineralization of either nitrogen or phosphorus during the study. A later study (Schlesinger 1985) of the same two species provided data showing that neither nitrogen nor phosphorus was released from *Ceanothus* during a three-year period. Only nitrogen was immobilized in *Salvia*. There was an apparent increase of lignin substances during the study. Availability of phosphorus may be limiting to *Ceanothus* in senescent stands. In this fire-maintained vegetation, the slow decomposition of leaf litter probably provides a continuing and significant source of plant nutrients between fires.

The suggestion in the chaparral study of an interference role of lignin in decomposition of leaf litter is strengthened by a study at the Hubbard Brook eastern deciduous forest. A high initial lignin content of litter may slow decomposition rates, through interference in the enzymatic degradation of carbohydrates, including cellulose, and proteins. Initial lignin content controls by the amount of nitrogen immobilized per unit of carbon respired by microbes of decay. Mellilo et al. (1982) examined the decomposition of six species of hardwood leaves with regard to lignin and nitrogen contents of the litter.

The New Hampshire data were used to test a simple inverse linear regression model of leaf litter mass loss (see Cromack 1973). The model relates the decomposition constant (k) to the ratio of initial lignin:initial nitrogen (see Figure 13-6a). Melillo and his colleagues did find an inverse linear relationship between rates of decay and the initial lignin:initial nitrogen ratios, but only for a narrow range. If initial ratios are much greater than 29, for example, note that the model would predict that there would be no decomposition. They then proposed a more general model of decomposition, one in which a curvilinear relationship exists between the parameter k and the initial lignin:initial nitrogen ratios. Data from needle decomposition studies (Daubenmire and Prusso, 1963) were used to test the model (Figure 13-6b). The model fits for 11 of the 13 types studied (but see Melillo et al. for more detail).

The development of accurate mathematical models allows for *prediction* in ecosystem analysis. However, a given model may not fit every situation. What generalizations can we draw from the work at Hubbard Brook and elsewhere? First, remaining biomass is most highly correlated with the initial lignin:nitrogen ratio ($r^2 = .91$). Second, initial nitrogen exerts this significant influence only when the exogenous (soil) nitrogen is low. If exogenous nitrogen is high, then initial nitrogen in the substrate has less importance, as the microbes of decay are not relying upon the original nitrogen of decomposing tissues for their needs. In that case, lignin content may exert more control in rate of decomposition, as we saw in the first model presented (Figure 13-6a).

In an analysis of data drawn from studies of nutrient cycling in forests at least 20 years old, Vitousek (1982) found, on a very broad scale, litterfall mass to be coupled with nitrogen circulation to a greater degree than to either phosphorus or calcium.

And finally, the sequestering of nitrogen in "lignin-derivative" or humus substances is significant to long-term cycling of nitrogen in forest systems. Humic complexes are resistant to decay and release nutrients over time, not in a large, leachable pulse.

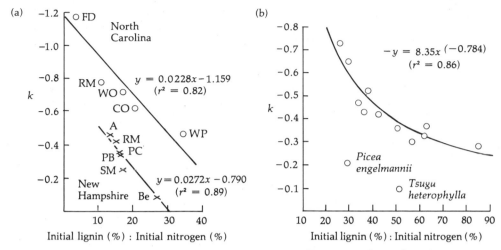

Figure 13-6 (a) The decomposition constant (k) expressed as a function of the ratio of initial lignin concentration to initial nitrogen concentration. (b) The decomposition constant (k) expressed as a function of the ratio of initial lignin concentration to initial nitrogen concentration. (From "Nitrogen and lignin control of hardwood leaf litter decomposition dynamics" by J. M. Melillo et al., *Ecology* 63:621-626. Copyright © 1982 by the Ecological Society of America. Reprinted by permission.)

The decomposition of litter is a major and significant factor in intrasystem cycling. Soluble substances tend to disappear first, along with litter rich in solubles. Nitrogen concentrations often increase linearly, with cumulative mass loss during the early periods. Finally, the disappearance of more decay-resistant substances dominates the loss of litter mass (but see also McClaugherty et al. 1985).

Evergreenness and Deciduousness Evergreen vegetation occurs on acid, infertile soil in the Sierra Madre Occidental of northwestern Mexico. The more favorable, fertile, less acid soils are dominated by deciduous species, with very little overlap between the two types of vegetation. Goldberg (1982) proposed that the evergreens would be successful on fertile soils, growing as well or better than on poorer soils, but they do not establish in the better sites because of competition with dense, highly productive deciduous plants.

In California, Gray and Schlesinger (1983) and Gray (1983) have tested the hypothesis that the growth of *Salvia leucophylla*, a drought-deciduous shrub of the coastal sage scrub, is more closely dependent upon soil nitrogen than is the growth of the evergreen chaparral shrub, *Ceanothus megacarpus*. Their data can be summarized as follows. *Ceanothus* stands exhibit greater nutrient pools, annual accumulation, and annual return, as well as greater efficiency of redistribution of nutrients from abscising leaves. *Salvia* leaves are more susceptible to loss by leaching, and the foliar nutrient concentration is both greater and more variable than that of *Ceanothus*. Further, following nitrogen stress, *Salvia* exhibited a greater response to increased nitrogen supplies, in terms of nitrogen concentration, root absorption rates and growth rates. The deciduous shrub also displayed greater relative leaf area, growth rate, dry matter yield, and higher nitrogen concentration than did *Ceanothus*. Their data are consistent with the hypothesis: Growth in *Ceanothus* is less dependent upon soil nitrogen than is growth in *Salvia*.

There are real advantages to each leaf type. In deciduous species, nutrient retrieval/reabsorption from abscising, expendable leaves provides significant recycling of nutrients. In an aspen-mixed-hardwood-spodosol ecosystem in northern Wisconsin, Pastor and Bockheim (1984) found that nutrient retrieval by translocation was significant in both aspen and sugar maple for nitrogen, phosphorus and potassium, of lesser importance for sulfur, magnesium and iron, and of no value for calcium and zinc. On the other hand, evergreens may use, in the production of stem wood, many times more of a nutrient than is lost in litterfall each year (Vitousek 1982). Also, the capacity for survival with low growth rates may be advantageous in depauperate situations.

Some Roles of Forest Trees

In general, trees have not only a very extensive root network but sometimes have longer and deeper roots than do shrubs and herbs. Therefore, the portion of the soil column confronted by tree roots is greater, making possible a unique contribution to the ecosystem. Deep-lying minerals are extracted by tree roots, raised into the plant and incorporated into plant biomass. Nutrients such as potassium may be leached from the leaves or bark, and other nutrients, such as calcium, may be returned as litter. Then, decomposition and mineralization add these nutrients to upper soil layers; now these formerly deep-lying nutrients are accessible to the herbaceous and shrub strata of the forest. An analogous situation exists in the salt marsh ecosystem. The transfer of phosphorus and other nutrients from sediments more than a meter deep to the water is accomplished in several steps: (a) cordgrass root systems take in P; (b) bacteria degrade the cordgrass; (c) detritus feeders ingest bacteria and return P to the marsh waters. This "pump" role of cordgrass is analogous to the action of deep roots of forest trees, bringing nutrients up from the depths and depositing them in surface layers.

Substances released into the soil from roots may provide significant amounts of carbohydrates and the like to the **rhizosphere**, the zone in the immediate vicinity of the plant root. As discussed in Chapter 6, Smith (1976) analyzed root exudates from sugar maple (*Acer saccharum*), yellow birch (*Betula alleghaniensis*), and beech (*Fagus grandifolia*), all of a northern hardwood forest. The beech released the most amino acids and organic acids per hectare, and the birch released the most carbohydrates. The cations released were mainly Na, K, and Ca; the anions were chiefly sulfate and chloride.

Smith pointed out that the significance of root exudation is twofold. First, root exudates enhance the growth of the microbial saprophytes and parasites in the rhizosphere; and, second, although root exudates are beneficial to microbial saprophytes and parasites, they may be either beneficial or harmful to individual plants and to the ecosystem.

The atmospheric contribution of inorganic nitrogen to the watershed of Lake Tahoe, California–Nevada, ranges from 1–2 kg ha^{-1} yr^{-1}. Very little of this drains into the lake, for it is removed efficiently by well-developed stands of conifers. However, perturbation (construction or logging, for instance) reduces the efficiency of nitrate uptake, resulting in an increased export of growth-stimulating nutrients to the lake itself. Thus, there is increased algal growth in the lake, reducing the clarity of the water. Alders are associated with both nitrogen fixation and nitrification, and so significant amounts of nitrate nitrogen are also contributed to the soil water by alder stands in the Lake Tahoe watershed, especially in the fall and early winter (Coats et al. 1976).

Nutrient Cycling in Different Vegetation Types

Maritime Ecosystems

How does nutrient cycling differ from one maritime ecosystem to another? Have conservative mechanisms for retention and internal recycling evolved as a means of obtaining elements in short supply? A coastal salt marsh* ecosystem exhibits a very tight, conservative mineral nutrient regime, in contrast to the loose, open regime of the dune system at Cape Hatteras, North Carolina, for instance.

In general, the annual output of nutrients from a terrestrial ecosystem exceeds the meteorological input, and that input is small compared to the nutrient reservoir of the system. This is true of the salt marsh. The marsh ecosystem is very stable, although it has little biological structure, is not very diverse, and is often subjected to a stressful environment. There is efficient internal cycling. The contributions of the large storage compartment in the clay sediments and the internal biological processes that hold and recycle nutrients provide a base for a highly productive system. However, in dune ecosystems the atmospheric contribution of cations slightly exceeds the calculated output through leaching and greatly exceeds the nutrients in storage in the system.

The dune systems at Cape Hatteras National Seashore, North Carolina (which includes the dunes of Bodie, Ocracoke, and Hatteras Islands), were studied by van der Valk (1974), with special emphasis on their nutrient cycles. Bodie Island, in the northern part of the National Seashore, is oriented such that it receives the full fury of winter storm winds, which blow mostly from the northeast. The coastline of Ocracoke Island to the south runs in an east-northeast direction and so is parallel to the winds' path. Also, it is partly protected by Hatteras Island. Bodie Island receives greater cation input than Ocracoke Island, in salt spray and rainfall, probably due to its orientation. The foredune (the most seaward dune) at Bodie is higher than that at Ocracoke Island, and this may also influence salt spray deposition.

The vegetation of the dunes, primarily American beachgrass (*Ammophila breviligulata*) and sea oats (*Uniola paniculata*), forms a sparse cover; coverage is

---

*A coastal salt marsh is an herb-dominated ecosystem in which the rooting medium is inundated by tidal water for long periods, if not continuously. The substrate is chiefly mineral, typically with high humus content.

3–4% for each of the grasses. There are few or no forbs* on the front of the foredune, and perhaps 20 forb species on the back of the foredune. Some contribute as much or more cover than the grasses. The principal forb species are horseweed (*Conyza canadensis*), cudweed (*Gnaphalium obtusifolium*) and goldenrod (*Solidago sempervirens*).

Nutrient Supply   The primary nutrient source in this community is salt spray; there is almost no internal reservoir of nutrients. At Bodie Island, the salt spray contribution of mineral ions exceeds that of rainfall as much as eightfold (Table 13-3). Notice, for instance, that the atmospheric input of cations at Bodie Island in bulk precipitation (rainfall plus salt spray) ranges from 14 kg ha^{-1} yr^{-1} for K, to 433 kg ha^{-1} yr^{-1} for Na, with annual inputs of 35 kg ha^{-1} for Ca and 64 kg ha^{-1} for Mg. By comparison, Ca input from bulk precipitation for an inland Tennessee *Liriodendron* forest was estimated at 10.5 kg ha^{-1} yr^{-1}.

What happens to the large supply of nutrients that arrive on the wind? Recall that the foredune is not a truly saline habitat, and that the exchange capacity in soils with a very high sand to clay ratio is poor. Excess cations are quickly leached away.

Output of each cation for a certain depth of soil may be calculated, since

$$S_2 = S_1 + IN - OUT$$

Table 13-3   Average concentration of cations (ppm) and annual input of cations (kg ha^{-1}yr^1) in the bulk precipitation (rainfall plus salt spray) and in the rainwater at Bodie Island and Ocracoke Island. (From "Mineral cycling in coastal foredune plant communities Cape Hatteras National Seashore" by A. G. van der Valk, *Ecology* 55:1349–1358. Copyright © 1974 by the Ecological Society of America. Reprinted by permission.)

	K	Na	Ca	Mg
Bodie Island:				
Bulk precipitation (ppm)	1.30	26.00	2.30	3.90
Rainfall (ppm)[a]	0.15	7.16	1.02	1.30
Bulk precipitation (kg ha^{-1} yr^{-1})[b]	14.40	432.60	34.80	64.20
Rainfall (kg ha^{-1} yr^{-1})	2.30	118.00	16.00	21.00
Ocracoke Island:				
Bulk precipitation (ppm)	1.00	21.00	3.40	3.00
Rainfall (ppm)[a]	0.15	4.36	0.82	0.59
Bulk precipitation (kg ha^{-1} yr^{-1})[b]	15.60	319.80	51.60	46.20
Rainfall (kg ha^{-1} yr^{-1})	2.20	65.00	12.00	8.90

[a]Data from Gambell and Fisher 1966.
[b]Estimated from data for May 1971 to January 1973.

*Herbaceous plants other than grasses.

where S_2 is the quantity of the cation present at a certain depth at sampling time T_2, S_1 is the quantity of the cation present at the same depth at the previous sampling time T_1, IN is the input of the cation in salt spray or in salt spray leachate during the time period from T_1 to T_2, and OUT is the output of the cation due to leaching by rainwater during the time period from T_1 to T_2. About 160 kg ha^{-1} yr^{-1} of Ca is exported from the 60 cm depth, which is significantly more than the amount of Ca that entered as salt spray. The difference is probably due to leaching of calcium carbonate from shell fragments in the sand. The turnover times for nutrients in dune systems are very short compared to other terrestrial ecosystems: 11–37 days at Bodie Island for K, Na, and Mg. For Ca, a longer time of 32–206 days reflects input from dissolution of shell fragments. Total annual meterological inputs are always higher than exports for the cations K, Na, Ca and Mg.

Limits to Growth   Many of the dune perennials have extensive root systems, or rhizome networks, which form clones over large areas. These may pull nutrients from the front of the dune, which receives ample supply of K, to the back, where K may be limiting to vegetation growth. As shown in Figure 13-7, the annual input of Na and Mg at Bodie Island greatly exceeds that present in vegetation.

The sparse vegetation provides almost all the nutrient storage of the system. There is no soil reservoir. The dune system shows virtually no conservative mechanisms for retention and internal cycling of nutrients. Thus, mineral cycles in seashore dunes are less stable than in other terrestrial ecosystems (Jordan and Kline 1972).

Grasslands

In North America, the grasslands extend in a broad, sweeping belt from Texas north to Saskatchewan, replaced at the margins by mesquite (*Prosopis*), aspen (*Populus*) or oak (*Quercus*) savannah. Grasslands are characterized by frequent fires, heavy soils and a continental climate with 300–500 mm precipitation annually. Potential evaporation exceeds precipitation. In the **pampa** or grass steppe of South America, precipitation increases to 1000 mm yr^{-1}, but potential evaporation is correspondingly high (Walter 1979), and the precipitation to evaporation ratio is still less than 1. Recall that evapotranspiration is a conservative force, positively correlated with efficient intrasystematic cycling of nutrients; that is, intrasystem cycling is usually enhanced by increased evapotranspiration.

We will not trace the flux of every nutrient in the grassland ecosystem, but will select one: nitrogen. In grasslands, almost without exception, primary productivity is limited by the available nitrogen supply. Although loss is unlikely during normal functioning, both grasses and decomposers use up soil nitrate and ammonium ions so rapidly that their concentration remains low. When live aboveground biomass reaches 300 kg ha^{-1}, nitrogen becomes severely limiting. In California serpentine grassland, both nitrogen and phosphorus have been demonstrated to be limiting to herbaceous growth, with nitrogen probably the more important of the two (Turitzen 1982).

Nitrogen fixation rates, both by symbiotic and by free-living organisms, are

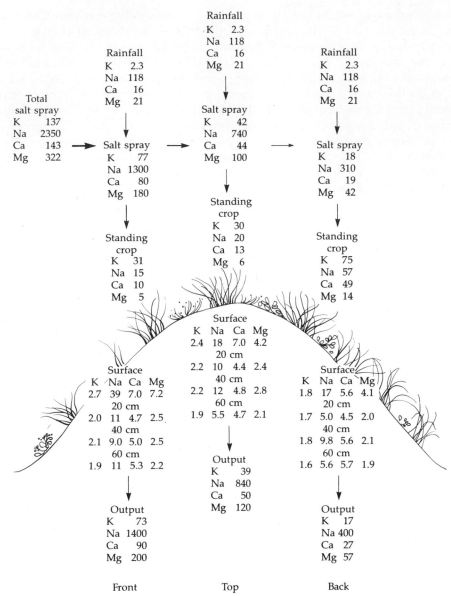

Figure 13-7 Estimated or calculated annual inputs (kg ha^{-1} yr^{-1}) of cations in salt spray and rainwater, and output (kg ha^{-1} yr^{-1}) in drainage water from 60 cm below the surface on the front, top, and back of the foredune at Bodie Island. The cation content (kg ha^{-1}) of vegetation at maximum standing crop and average concentrations (kg ha^{-1}) of cations in soil at the surface and 20, 40, and 60 cm below the surface are given in the boxes. Outputs equal salt spray leachates plus rainfall. (From "Mineral cycling in coastal foredune plant communities in Cape Hattaras National Seashore" by A. G. van der Valk. *Ecology* 55:1349-1358. Copyright © 1974 by the Ecological Society of America. Reprinted by permission.)

low in grasslands (about 1 kg ha^{-1} yr^{-1}, compared to 1.5–8 kg ha^{-1} yr^{-1} for fixation by the lichen component of the tropical rainforest). Contribution in precipitation is about 3 kg ha^{-1} yr^{-1}. Nitrogen loss may occur through leaching, denitrification or volatilization. Estimates of volatilization range to about 1.5 ha^{-1} yr^{-1}; leaching and denitrification require the presence of significant concentrations of nitrate in wet soils, a situation rarely encountered in grasslands.

A generalized biogeochemical cycle for N was shown in Figure 13.2. Let us begin with uptake of nitrogen into plants through living roots as ammonium ion or nitrate ion. The rate of uptake of nitrate nitrogen increases as nitrate concentration increases, but only to a point, leveling off as soil nitrate becomes high. Translocation and distribution within the plant varies seasonally, with the ratio of live shoot N concentration to live root N concentration changing with plant phenology. The ratio of shoot to root N varies from 3.0 to 1.0, remaining high during the first third of growth, then declining. The decline in shoot to root N may be adaptive, pulling N back to living, persisting roots late in the season, before aboveground biomass dies back.

Decomposition and mineralization transfer N back to the various soil compartments. Partitioning of N from decomposing roots and litter into soil organic matter or ammonium is determined by the soil C:N ratio. When this ratio exceeds 50, most of the N in tissue being decomposed will be transformed to stable organic matter. As the C:N ratio in soil declines, more and more N will be mineralized, i.e., converted to ammonium. Mobilization of soil organic N (mineralized to ammonium) provides an annual turnover amounting to about 2% of the substrate. Factors affecting this rate are: (a) the concentration of organic nitrogen, and (b) abiotic factors such as temperature and soil water.

It is apparent that mechanisms have evolved to conserve and recycle vital N within the grassland ecosystem. The natural force that circumvents these mechanisms is fire. It reduces aboveground biomass to mineral ash but allows the escape of volatile elements, including N. However,

> the prairie had two strings to its bow. Fires thinned its grasses, but they thickened its stand of leguminous herbs . . . each carrying its own bacteria housed in nodules. . . . Each nodule pumped nitrogen out of the air into the plant, and then ultimately into the soil. (Leopold 1966)

The Hubbard Brook Experimental Forest

The study of an entire ecosystem requires many workers of related and widely divergent disciplines, many hours, and an excellent integration system. In part, our understanding of ecosystem function has been enhanced through the use of models, which are essentially of two types: (a) a microcosm, utilizing a closed system with quantifiable inputs and outputs of energy and materials— for instance, a balanced aquarium—and (b) systems analysis, using computers to simulate the functioning of a real ecosystem (for example, see Jordan and Kline 1972 or Shugart et al. 1976).

Generalizations and predictions that derive from a microcosm study may not be valid at the ecosystem level. Relationships may not be linear, nor even logarithmic. Extrapolation from exceedingly small, closed systems to, for example, a forest region, may be unwarranted. Simulation also has its drawbacks. Data collection to provide an informational base must be unbiased and painstakingly accurate. Often, simplifying assumptions are made, which of necessity eliminate consideration of short-term fluctuation, of separate categories of producers and consumers, etc.

Ecosystem Analysis   What must be done is analysis of real ecosystems. The small watershed approach of Likens and Bormann (1972; also **Bormann** and Likens 1967; Bormann et al. 1977; Siccama et al. 1970) combines some of the most attractive features of the microcosm approach with large-scale ecosystem-encompassing measurements. Their criteria for a suitable study site are as follows:

1. The watershed ecosystem is part of a larger, homogeneous biotic and geologic unit.
2. It has an impermeable geologic substrate.
3. It is characterized by relatively humid conditions, and the input/output budget for nongaseous nutrients can be easily determined from the difference between the meteorologic input (dissolved substances and particulate matter in rain and snow) and the geologic output (dissolved substances and particulate matter in drainage waters).

The Hubbard Brook Experimental Forest is located north of Plymouth, New Hampshire, in the lower elevations of the White Mountains. The watersheds studied contain a cool-temperate, mesophytic, broadleaf deciduous forest at lower altitudes, with some conifers higher up. Although the area was logged extensively in the early part of this century, there are some older trees, and the stands are believed to represent a climax association. The physical environment, including edaphological, climatological, geological and hydrological factors, is discussed extensively elsewhere (Likens et al. 1977).

The data compiled over the last decade or so form the basis for formulating general principles and concepts about nutrient cycling in this eastern deciduous forest. Experimental manipulation provides additional information, which can be used for prediction about the behavior of such systems in response to perturbation, whether the change is natural or is caused by human impact. In this section, we will discuss the nutrient budgets at Hubbard Brook. We will examine two cycles in some detail: one in which there is net loss (the calcium cycle) and one in which there is net gain (the nitrogen cycle).

Nutrient Budgets: Input, Output, and Balance   Nutrient input—geological, biological and atmospheric—can be estimated. Rainfall, throughfall and dry fallout can all be gauged as the product of the concentration per unit of time and the total precipitation, etc. The Hubbard Brook region is not subject to geological influx, and biological output is assumed to balance biological input, since the area holds no special attraction (e.g., for migratory animals), so that these two sources for nutrients can be disregarded. The climax forest is unlikely to lose many nutrients as wind-borne particulates or aerosols, although an occasional leaf fragment may be lost. Significant losses would most likely be due to geological output, specifically in stream flow. In addition to dissolved nutrients, stream water carries inorganic and organic matter suspended by the turbulence of the stream, and other material which rolls along the bottom of the stream, carried by the current.

The streams at Hubbard Brook are subjected to sensitive hydrologic and chemical measurements, made possible by the geologically impermeable substrate. The assumption is that the streams of the area carry all of the water leaving the system, and that they carry or deposit all of the particulates and dissolved nutrients leaving the watershed. A concrete ponding basin behind a v-notch weir (Figure 13-8) allows for assessment of flow, as well as measurement of chemical concentrations and particulate load.

As the stream flows over the weir, constant monitoring of flow provides baseline data. Periodic sampling measures solution nutrient loss from the system. The chemical output is derived as nutrient concentration (mg l^{-1}) times the volume of water draining from the watershed. The stream flow slows in the ponding basin behind the weir, so that as the suspended matter and the bed load are deposited, they can be measured also.

Nutrient budgets are inextricably connected to the hydrologic cycle. Since stream-flow nutrient concentration is virtually constant, nutrient loss is high in wet years and low in drought years. Nutrient budgets developed from these long-term studies at Hubbard Brook show a net gain for nitrate nitrogen and ammonium nitrate, but a net loss for Si, Ca, S, Na, Mg and Al. Input and output for K and Cl is balanced. Net losses are assumed to be possible because of weathering, and so net loss data is used as an estimate of weathering within the system. Annual living and dead biomass accumulation at one of the watersheds is 3930 kg ha^{-1}, which indicates that the system is aggrading and therefore seral.

Figure 13-8   A weir showing the v-notch recording house and ponding basin. (Courtesy of the Northeastern Forest Experiment Station, Forest Service, U.S. Dept. of Agriculture.)

The Calcium and Nitrogen Cycles   In the Hubbard Brook system, the vegetation plus organic debris contains about 854 kg Ca ha^{-1}. Uptake by vegetation exceeds decomposition by about 9.5 kg ha^{-1} yr^{-1} due to accretion. Sources of Ca are atmospheric inputs (2.2 kg ha^{-1} yr^{-1}) and soil water and exchange surfaces (510 kg ha^{-1}). The estimated 74,200 kg Ca ha^{-1} that is locked up in soil and rocks is unavailable.

The output of Ca (13.9 kg ha^{-1} yr^{-1}) includes 13.7 kg ha^{-1} yr^{-1} carried as solution in stream water. The Ca cycle is effectively closed, with internal cycling of about 1/8 of the Ca pool. Only about 2.3% of the available Ca is lost each year, and this is assumed to come from intersystem input—weathering (11.7 kg ha^{-1} yr^{-1} lost). The system efficiently conserves Ca, recycling it in throughfall, litterfall, decomposition and renewed uptake.

An estimated 20.7 kg N ha^{-1} yr^{-1} enters the system. Of this, 81% is held in long-term storage, partly in litter (46%), with the rest (54%) in living biomass. The bulk of N input (68%) comes from N fixation. About 32% derives from precipitation, and virtually none from weathering (Bormann et al. 1977).

Mineralization contributes inorganic N, of which 1% comes from root exudates. Apparently some 33% of the 119 kg N ha^{-1} used annually by vegetation is drawn from within the plant, used in growth, then withdrawn from leaves and replaced in storage at the end of the growing season. Thus, internal cycling is an important conservation mechanism for the forest N cycle. N tends to be stored primarily in soil organic matter (90%) with 9.5% in the vegetation, and the remaining 0.5% in the soil.

Deforestation   Watershed 2 was deforested in 1965, and regrowth was prevented for three summers by herbicide application. The purpose of this drastic manipulation was to test the resilience of the system, after the severance of uptake transfer of nutrients by vegetation (refer back to Figure 13-4). That is, if one compartment is removed from the system, how will the others function?

The deforestation and herbicide application had a great effect on the hydrologic cycle. Brief, intense runoff from the deforested region carried an increased load of particulates. Average annual particulate export on undisturbed sites was about 2.5 metric tons km^{-2}, but was 38 metric tons km^{-2} yr^{-1} from the disturbed site. The ratio of dissolved substances to particulate averages about 2.3; but following manipulation, this ratio leaped to more than 8.0 the first 2 years.

Immediate cessation of transpiration resulted in increased stream flow as follows: in the first year, 35 cm of stream flow, which is 40% above what would be expected in an undisturbed site; in the second year, 27 cm or 28% above expected; and in the third year after deforestation, 24 cm or 26% above expected. This increased stream flow due to lack of transpiration is not surprising; stream flow regularly increases on undisturbed sites in autumn, before the rains begin, in response to synchronous leaf fall.

Since stream flow is related to nutrient loss, the deforestation and herbicide application also had an effect on nutrient loss. Export of ions increased for all major ions, except for ammonium, sulfate and bicarbonate ions. Nitrate, from decaying vegetation, was one of the nutrients whose concentration in stream flow increased greatly. Reflecting the decrease in sulfate concentration and the increase in nitrate concentration, the stream pH went down (5.1 to 4.3), and its character changed from a dilute sulfuric acid to a stronger nitric acid.

Increased export of dissolved organic substances (14–15 times greater than controls) was due to increased erodability, to increased stream flow, and to accelerated chemical decomposition of inorganic materials in the deforested region.

Stability

Stability is a measure of a system's response to some outside perturbation. A *stable* system returns to steady-state conditions in a short time following some irregularity in either the physical environment or in the flow and cycle of energy and materials. The shorter the recovery time, the higher the stability. Ecosystem stability involves nutrient pools that are small and quickly revolving and rely upon the soil microflora and fauna. Also involved, on an ecological time frame of perhaps many hundreds of years, are pools that are very large and slowly turning: decay-resistant litter or shallow lacustrine or marine sediments, for instance (Coleman et al. 1983).

What factors are involved in the stability of ecosystem nutrient cycles? Efficiency of nutrient use, size of storage compartments, erosion control, and immobilization of nutrients are properties of ecosystems that serve to conserve and recycle nutrients. We will briefly discuss three of these.

Nutrient Use Efficiency

With regard to long-lived perennials, Vitousek (1982) defined nutrient use efficiency as "the grams of organic matter lost from plants or permanently stored within plants per unit of nutrient lost or permanently stored." In other words, nutrient use efficiency is the inverse of nutrient concentration in the organic matter increment and in litterfall and root turnover.

In many forest ecosystems, increasing efficiency of nitrogen use exists (i.e., increasing litterfall dry mass per unit of nitrogen) with decreasing aboveground nitrogen circulation. What are the ecological/evolutionary origins of this efficiency?

Net mineralization in forest floors is low in sites with small amounts of nitrogen circulating annually. That is, there is a positive feedback system wherein litterfall with high C:N ratio would favor nitrogen retention by decomposers and would reduce available soil nitrogen. This condition leads to greater efficiency of nitrogen use, enhancing the C:N ratio further. We could predict, then, that long-term ecological (and ultimately, evolutionary) advantage in nitrogen-poor sites would go to those species with higher nitrogen use efficiency, which provide litter with high C:N ratios (Vitousek 1982).

Storage

The coupling of metabolism and mineral cycling is a stabilizing force within an ecosystem. One bridge between metabolism and mineral cycling is the availability of large storage compartments within the system. Large storage capability tends to buffer nutrient cycles against perturbation. A large standing crop (e.g., trees in tropical rain forest) provides a spacious storage compartment; so do ample exchange sites for cations (e.g., young volcanic soils). The organically bound component in soil also provides secure storage for nutrients. An element's replacing power on soil surfaces is an important feature of nutrient stability. Knowing the magnitude of the replacing power allows prediction of the resistance of that cycle to loss; thus the efficiency of internal cycling can be estimated.

Certain apparently open systems are also stable. Salt marshes seem to have a superabundance of energy supply, nutrients are rarely limiting, and storage compartments are never empty. Although diversity is low, stability remains high. This is probably attributable to a remarkably high metabolic activity level appropriate to the abundance of nutrients available in the system.

Immobilization of Nutrients

The largest *source* of biologically available nitrogen is the ammonium released from decaying organic matter; the *sink* is uptake of ammonium and nitrate by microorganisms and plants. Nitrogen pool sizes are small and turnover is rapid, because biological uptake is effective and rapid (Vitousek and Matson 1985a).

There is, however, a potential for significant nitrogen loss from fertile sites, directly related to the amounts of nitrogen circulating in an aggrading, undisturbed system. In the face of disturbance, the ability of plants to take up nitrogen may be reduced, but uptake by plants is but one of many processes that prevent or delay solution losses of nitrate following disturbance. "Nine other processes can prevent solution losses of mineral nitrogen from disturbed ecosystems even in absence of plant uptake" (Vitousek et al. 1982); these are shown in Figure 13-9. Ammonium immobilization, nitrogen immobilization, lags in nitrification, clay fixation, low net nitrogen mineralization, and/or lack of water for nitrate transport could be involved in retention (see Figure 13-9).

Disturbance could cause elevated soil ammonium concentrations. What would prevent subsequent nitrate production and loss? Vitousek and Matson (1985a) proposed that interaction of labile inhibitors of nitrification and low pH or low initial populations of nitrifying bacteria delayed nitrification (even though soil ammonium was elevated) and thus enhanced nitrogen retention. Microbial activity is the predominant process regulating nitrogen availability following disturbance (Vitousek and Matson 1985b).

Summary

Materials cycle within the biosphere, but energy flows one way. Biogeochemical cycling includes nutrient cycling, which may be one of two types: gaseous (or perfect), which tends to be regional or global (e.g., CO_2); or sedimentary, which operates primarily within one ecosystem (e.g., Ca). Nutrients may be accumulated with successive seral stages. Nutrients are compartmentalized, occurring in soil and rocks, in soil solution, in the atmosphere, and in biomass. Transfer into or out of an ecosystem may be meteorologic, geologic, or biologic. Cycles may require minutes, decades or millenia to revolve.

Essential nutrients include N, P, S, K, Mg, Ca, Fe, Mn, Zn, Cu, Mo, B, and Cl. Factors involved in nutrient cycling include fire and earth movement and the various aspects of the hydrologic cycle: rainfall and leaching, atmospheric deposition and canopy interception, evapotranspiration, runoff, infiltration, and salt

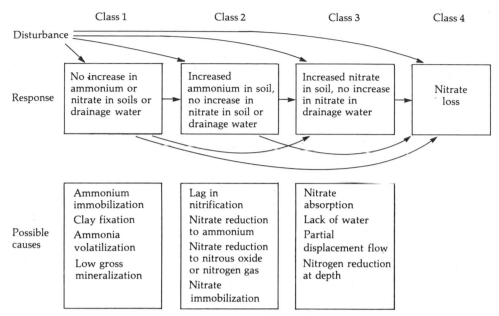

Figure 13-9 The possible responses of soil and solution inorganic nitrogen to disturbance, and the processes that could cause each response. All of the responses could be observed sequentially in a single site, or any of the responses could be absent. (From "A comparative analysis of potential nitrification and nitrate mobility in forest ecosystems" by P. M. Vitousek et al., *Ecological Monographs*. 52:155–177. Copyright © 1982 by the Ecological Society of America. Reprinted by permission.)

spray. Biotic factors include decomposition, root exudation, and retrieval of deep-lying nutrients.

Salt marshes are extremely productive and exhibit high stability. Their efficient nutrient cycling regime is in part attributable to large storage compartments in clay sediments. Coastal dune ecosystems are open, nonconserving systems relying on salt spray for nutrient input, with no soil reservoir for nutrients.

Nitrogen commonly limits productivity in grasslands. N is provided by fixation, both from free-living and symbiotic fixation, and lost through volatilization.

The soil provides a large storage compartment in the broadleaf temperate deciduous forest of the eastern United States. The exchange sites are saturated with cations, and stream-flow discharge maintains a constant chemical concentration for most major nutrients except K. Experimental deforestation of one watershed resulted in greatly increased runoff and enhanced export of particulates and all major nutrient ions except ammonium, sulfate and bicarbonate. The undisturbed system is stable and conservative, with efficient intrasystem cycling.

Stability relies upon resilience. Efficiency of nutrient use, replenishment power and retention of nutrients, resistance to erosion, large storage compartments, and exploitative pioneer species are essential homeostatic mechanisms for ecosystems.

PART IV

ENVIRONMENTAL
FACTORS

We have seen in earlier chapters that the population and community attributes of plant species are determined by the response of individuals to the external environment. It is possible that autecological information may, when enough taxa have been characterized, lead to predictability at the community level. In this part, we will consider the environmental factors of light, temperature, fire, soil, and water, including their effect on physiological and community processes.

It should be kept in mind that separate discussions of various factors is artificial; the interaction of all environmental factors determines the response of the plant. Chapter 20 attempts to show how the distribution of major vegetation types in North America is a response to interacting environmental factors.

CHAPTER 14

LIGHT AND TEMPERATURE

S olar radiation is one of the most important qualities of the environment influencing plants. Light is the ultimate source of energy for most organisms, so the variations of light in time and space and the photosynthetic responses to those variations are basic to our understanding of plant ecology. In this chapter, we will consider the physical nature of light and look at how it varies spatially and temporally. We will also consider the variations in the heat energy (temperature) in the ecosystem.

Heat is the aggregate internal energy of motion of atoms and molecules of a body. It can be transferred by means of radiation, convection, or conduction. **Convection** refers to the heating of air; **conduction** refers to the transference of heat through solids. **Radiation** refers to energy propagated as an electromagnetic wave, which would pass through a vacuum with the speed of light (3×10^{10} cm sec^{-1}) but would also travel through a medium such as air or water. In actual use, the term radiation refers both to the energy transferred and to the process of transferring energy. The solar energy that supports the life of this planet comes to us as radiation, transmitted in a broad spectrum that includes visible light and ultraviolet radiation at one extreme and heat or infrared radiation at the other extreme.

There is a continual exchange of energy between objects and the environment. Direct sunlight is absorbed by plants, rocks, soil, and the air around them. Infrared radiation (heat) is transferred from plants to the air, from soil to plants, and so forth. Unless the absorbed radiation is reradiated or in some way transferred, the body absorbing it would continue to increase its energy content.

Because our eyes do not perceive the longer wavelengths, the radiant energy given off by rocks, trees, and the air itself is invisible to us. On a warm spring day (300°K, or 27°C), the young trees, the mature trees, the various rocks, open soils, herbaceous cover, and the various woodland animals radiate energy across a rather broad spectrum but at a maximum wavelength of about 10 microns (10,000 nm), invisible to human eyes.

Solar Energy Budget

Let us develop a solar energy budget and envision the allocation of energy in equation form:

$$S = R + C + G + Ps + LE \qquad\qquad \text{(Equation 14-1)}$$

where S = solar radiation; R = reflected (shortwave, diurnal), or reradiated (longwave, both day and night); C = convective heating of air (called H for heat by some authors); G = heating of solid objects (soil, for instance); Ps = photosynthesis; and LE = latent heat of evaporation.

The amount of energy reaching any given spot on the earth varies seasonally and diurnally, but on the average there are 1.94 gram-calories (or cal) $\text{cm}^{-2}\,\text{min}^{-1}$ received as direct solar radiation measured perpendicular to the sun's rays just outside the earth's atmosphere at the mean distance of the earth to the sun. (A gram-calorie is that quantity of energy that is required to raise the temperature of one gram of water from 14.5°C to 15.5°C.) This 1.94 cal $\text{cm}^{-2}\,\text{min}^{-1}$ is known as the **solar constant,** and it represents one of the most fundamental and vital components that is known of our environment.

Total solar radiation (S) is usually expressed in langleys min^{-1}, a unit of **radiant flux,** which is the amount of energy received on a unit surface in a certain time. Net radiation (N) is equal to solar radiation minus reflectivity and reradiation: $N = S - R$. This can be diurnally negative, because at night S is zero (therefore less than R), and there is a net loss of heat from earth. During the day, S exceeds R, because reflectivity and reradiation are less than solar radiation, and there will be a positive energy balance. During the long winter of the polar regions, N is seasonally negative; it may be so even in more temperate regions. In Siberia, the annual net radiation is negative (Figure 14-1a). Or there may be virtually no monthly period that shows a negative balance, with a resulting large net positive balance; such is the case at Aswan, Egypt (Figure 14-1b). In the tropics, there is little seasonal change. The angle of the sun does not vary enough to create seasonal deficits. N is positive throughout the year. In fact, diurnal variation exceeds the seasonal variation of daily temperature means (Figure 14-1c).

The distribution of solar energy (Equation 14-1) varies over time and from one community to another. For instance, the maximum S incident upon a wet mountain meadow is $+1.3$ ly min^{-1}. Meadow vegetation has an **albedo** of 0.7. Albedo refers to the reflectivity of a material summed over all wavelengths for a sunlit surface. S times albedo equals R, so in this example $R = (0.7)(1.3\text{ ly min}^{-1})$ $= 0.91$ ly min^{-1}. The rest of the solar radiation is apportioned among G, C, Ps and LE.

Mountain meadows are often densely covered with herbaceous vegetation, and at 100% cover, G will be rather low. For various physical reasons, C is also low. Only 1–2% of incident light is usually converted into chemical energy, and photosynthesis here will be lower than usual due to low temperatures. Although Ps is low, LE will be high, due to the wetness of this mountain meadow system.

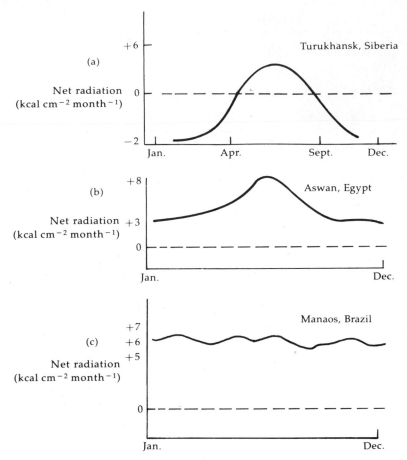

Figure 14-1   (a) Net radiation for a weather station in Siberia. Note that the annual average is negative. (b) Net radiation for a weather station at Aswan, Egypt. Note that there is virtually no period of negative radiation balance. (c) Net radiation for a tropical weather station. Note that there is very little seasonal flux in the radiation balance. These are monthly averages and do not necessarily reflect diurnal changes. (Modified from *Plants, Man, and the Ecosystem,* 2nd Edition by W. D. Billings. © 1970 by Wadsworth Publishing Co., Inc. Reprinted by permission.)

A contrasting situation is seen in an arid area such as the Mojave Desert. Again, S_{max} is 1.3 ly min^{-1}. R will be higher than in the mountain meadow. In very arid regions plants are very widely distributed, and the open expanses of exposed soil absorb a relatively high fraction of S. Thus, G will be much higher than in the wet mountain meadow system and photosynthesis will be very low, less than 0.1% of S. C will be high, as the denser air absorbs radiant energy from sunlight. There is very little liquid water available, and LE will be correspondingly low.

Measurement and Physical Properties of Radiation

Global Radiation

There are several forms of radiant energy that are important influences on an exposed leaf (Figure 14-2). Solar radiation may arrive at the surface of a leaf directly or indirectly, as diffuse radiation scattered by the atmosphere (skylight) (S^{sky}) and by clouds (cloudlight) (S^{cloud}), or as radiation reflected from the soil or other objects in the habitat (rS^{direct}).

The spectral qualities and intensity of radiation (Figure 14-3) vary depending on the distance the radiation travels through the atmosphere and the features of the habitat that absorb, reflect, or transmit the light. The atmosphere effectively removes a portion of direct solar radiation, reducing the intensity from the extra-terrestrial level of about 2 cal cm^{-2} min^{-1} (the solar constant) to about 1.3 cal cm^{-2} min^{-1} at sea level on a clear summer day (Gates 1965b). Most radiation received is direct, with only a small amount that is rich in the blue wavelengths arriving as skylight. The energy of solar radiation can be measured with devices called pyranometers, which will be discussed later in this chapter.

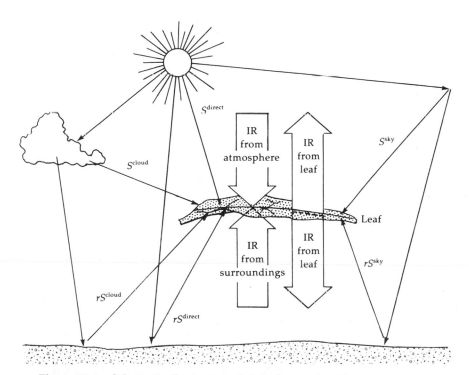

Figure 14-2 Schematic illustration of eight forms of radiant energy incident on an exposed leaf, and the infrared radiation emitted from its two surfaces. (From *Introduction to Biophysical Plant Physiology* by Park S. Nobel. W. H. Freeman and Company. Copyright © 1983.)

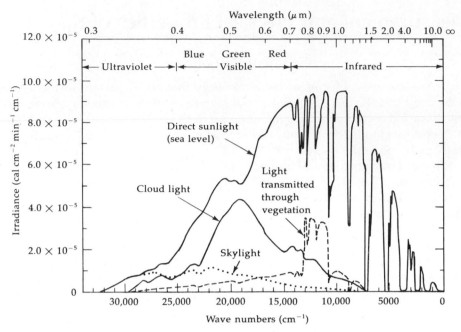

Figure 14-3   Spectral distribution of direct sunlight, skylight, cloudlight, and sunlight penetrating a stand of vegetation. Each curve represents the energy incident on a horizontal surface. (From Gates et al. 1965. By permission of the American Meteorological Society.)

As light passes through the atmosphere, the balance of wavelengths changes. Caldwell and Robberecht (1980) have shown that maximum daily total shortwave irradiance received in the arctic alpine life zone of the Northern Hemisphere varies by only a factor of 1.6 along a latitudinal gradient from polar regions at low elevations to alpine regions at high elevations in low latitudes. In contrast, along the same gradient, the maximum integrated effective UV-B irradiance (280–320 nm) varies by a full order of magnitude—sevenfold for total daily effective radiation.

The light relationships of an idealized green leaf are diagrammed in Figure 14-4. Reflection is measured by placing an appropriate light sensor above the leaf surface so that it reads only light reflected by the leaf. Placing the light sensor in the shadow of the leaf, one can measure the amount of light transmitted through the leaf. Total light minus reflectance and transmission is equal to absorbance. The amount of radiation absorbed is relatively high at wavelengths shorter than 0.7 μm (700 nm), which includes the visible and ultraviolet wavelengths. The very high energy ultraviolet light (wavelengths shorter than 0.4 μm [400 nm]) can be damaging to biological material, but water present in leaf cells is very efficient in absorbing these wavelengths. High rates of absorption in the visible portion of the spectrum are caused by the presence of chlorophylls, carotenes, and xanthophylls—the pigments in greatest abundance in plant cells. Light transmitted through vegetation is dramatically reduced in the visible region (Figure 14-3) because the chlorophylls reflect green light and absorb much of violet,

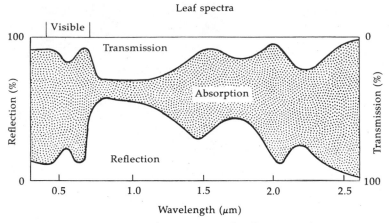

Figure 14-4 Transmission, absorption, and reflection with respect to wavelength for an idealized green leaf. (After Monteith, J. L. *Principles of Environmental Physics*, Edward Arnold, London, 1973.)

blue, and red, while the carotenoids (including xanthophylls) reflect yellow to orange light and absorb heavily in the blue to green range. This leaves little visible light under the canopy of a dense forest.

Radiation

Through a global system of weather stations, we have a reasonably accurate picture of the world's climate. We can also measure radiation, the source of heat, but measurement of energy flux has rarely been done, and only recently has the instrumentation been developed to support global-scale quantification of the energy environment.

Radiation can be measured by devices called **radiometers,** which integrate all incoming wavelengths and measure in energy units. **Pyranometers** are used to measure direct and diffuse sunlight. They are sensitive to 360–2500 nm, which includes some ultraviolet light (<450 nm), all visible light, and considerable infrared light (>700 nm). Filters can be added to a pyranometer, restricting its sensitivity to only the visible light, or all but a portion of the visible spectrum can be filtered out, if that is needed.

An Eppley black-and-white pyranometer (Figure 14-5) utilizes the temperature differential between hot and cold areas to monitor radiation intensity. However, it does not separate direct rays from indirect rays. On a clear day, direct sunlight accounts for approximately 85% of the energy received. On an overcast day, up to 100% of the incoming energy is due to diffuse, indirect sunlight. If we wish to monitor only diffuse sunlight, a shading device called an occulting ring, or shadow band, would be used with the pyranometer. The Eppley Precision Spectral Pyranometer (Figure 14-6) is believed to be the most accurate instrument produced commercially for measurement of global sun and sky radiation. It may also be used with a shading device, to screen out either the sun or the diffuse sky component.

Figure 14-5 The Eppley pyranometer. The sensing element under the glass dome is a thermopile with hot (black) and cold (white) areas.

Figure 14-6 The Eppley Precision Spectral Pyranometer. This instrument utilizes a plated, wirewound thermopile that is temperature compensated to be virtually unaffected by ambient temperature and is able to withstand severe mechanical vibration and shock.

An instrument called a **pyrheliometer** measures only direct solar radiation. It consists of a sort of tube and a sensor to track the sun such that only solar radiation of normal incidence enters the monitoring device, the tube. The output can be calibrated to read in ly min^{-1}. Without the use of the pyrheliometer, the intensity of direct beam (direct sunlight) may be estimated, using data obtained with the pyranometer alone, and with the pyranometer combined with the occulting ring:

$$R_s \text{(direct beam)} = R_s \text{(total global)} - R_s \text{(diffuse)} \qquad \text{(Equation 14-2)}$$

The monitoring of net radiation is of special interest to plant ecologists. Ideally, the net radiometer absorbs all incoming (downward) radiation (I_R), and all outgoing (upward) radiation (O_R). The difference between I_R and O_R gives the net radiation, which is the energy available in the community for work. (For information in meteorological instrumentation, see also Wang 1979.)

In a forested area, estimates of light received, length of sunflecks, and annual variations have relied upon continuously recording instruments, requiring considerable time and effort, and have produced results that are imprecise at best. Patterns of direct beam irradiance may be analyzed by hemispherical canopy photography, using a 180° equidistant fisheye lens. The camera is placed at the desired position beneath the canopy, at the appropriate level, a standard is placed to indicate true north, and a slide or series of slides is taken. Later, the slides are projected onto graphs of solar tracks enlarged from Smithsonian Institute tables for the appropriate latitude (List 1951). The researcher can then measure minutes of potential incident radiation and number and length of potential sunflecks for an entire year (Evans and Coombe 1959; Ustin et al. 1985; Selter et al. 1986). Growth of saplings of species in the understory of a Hawaiian forest was highly correlated with estimates of minutes of sunflecks received by that stratum. Photosynthetic photon flux density sensors and hemispheric fisheye photographs were employed to estimate sunfleck duration and photon flux densities (Pearcy 1983).

Photosynthetically Active Radiation (PAR)

The wavelengths absorbed by chlorophyll, and therefore active in the photosynthetic process, are between 0.4 μm and 0.7 μm. Light in this band has been labeled **photosynthetically active radiation (PAR).** Physicists have shown that radiant energy is transported in discrete bundles called photons. A mole of photons (6.02×10^{23} photons) is referred to as an Einstein (E)—the unit of photosynthetically active radiation. (PAR is also referred to as photosynthetic photon flux density (PPFD), expressed as $\mu mol\ m^{-2}s^{-1}$.) The intensity of PAR is expressed as the photon (quantum) flux per unit area, which, for example, would be about 2000 $\mu E\ m^{-2}\ sec^{-1}$ (or as PPFD, 2000 $\mu mol\ m^{-2}s^{-1}$) near midday on a clear day.

Other units of light measurement were used in plant ecological studies prior to the advent of instrumentation to measure only PAR. Common units and their equivalents are listed in Table 14-1. It is not possible to convert light values reported in conventional units directly to PAR or PPFD, because the broad spectral range of lux or footcandle measurements reports light as perceived by the human eye, which is not identical to PAR. Portable light meters fitted with quantum sensors that measure only PAR are available commercially.

Table 14-1   Common units for the measurement of light energy.

Unit name	Units or equivalents	Value at sea level in full sun
langley (ly)	1 cal cm^{-2}	1.3 ly
foot-candle	10.76 lux	10,000 ft-c
lux	1 lumen m^{-2}	107,600 lux
watt	1 joule sec^{-1} or 10^7 ergs	1,000 watts m^{-2}
Einstein	6.02×10^{23} photons	2,000 $\mu E\ m^{-2}\ sec^{-1}$ = 200 nE cm$^{-2}\ sec^{-1}$

Temperature Measurement

Temperature may be sensed by a thermocouple, a mercury or alcohol thermometer, or various other devices. Integration then may be used, if temperature data over a long period of time are needed. An inexpensive chemical integrator of average temperature over time is sucrose. The rate at which a sucrose solution is hydrolyzed (inverted) to glucose and fructose is a function of temperature. About 10–15 ml of buffered sucrose solution is sealed in a tube and left at a desired height aboveground or belowground at a field site. The amount of sucrose that was inverted during the field exposure period can be determined by measuring the optical rotation of the solution with a polarimeter. The method has been described in detail by Berthet (1960, in French) and by Lee (1969).

Workman (1980) has described another chemical method for the measurement of mean integrated temperature, with an accuracy of ±0.5°C. A spectrophotometer, which is more commonly available than a polarimeter, is employed to trace the course of a chemical reaction. The employment of such chemical devices has economic and logistic advantages over the use of climatological data or of automatic temperature recorders (Workman 1980).

Variations in Light and Temperature

Insolation (exposure to solar radiation) varies temporally and spatially. The primary factors that influence spatial variations in temperature are latitude, altitude, and proximity to water. Topography, cloud cover, vegetation, and slope aspect are contributing factors to temperature variation. The obvious and immediate temporal differences derive from the earth's rotation. During the day, radiant energy from sunlight usually replaces what was lost during the night, so there is an energy balance.

Because the earth moves about the sun in an elliptical orbit, there is some variation in the earth–sun distance. However, it is the tilt of the earth that gives us our seasons. There are periods of equability, when sunlight shines from pole to pole, and day and night are virtually equal in length. These periods, when the earth's polar axis is perpendicular to the radius between the earth and the sun, are called *vernal* and *autumnal* (spring and fall) equinoxes. Alternatively, when the earth's polar axis is in a plane parallel to the earth–sun radius, one pole will be in virtual 24-hour darkness, the other bathed both day and night in sunlight. In the Northern Hemisphere, our longest night, or winter solstice, is in December; our longest day, or summer solstice, is in June.

Latitude

The general distribution of total yearly radiation over the surface of the earth is plotted in Figure 14-7. There is a gradual decrease in global radiation over the land areas from latitudes of about 30°N and S toward the poles. This decline is due primarily to the reduced number of daylight hours during the winter and to

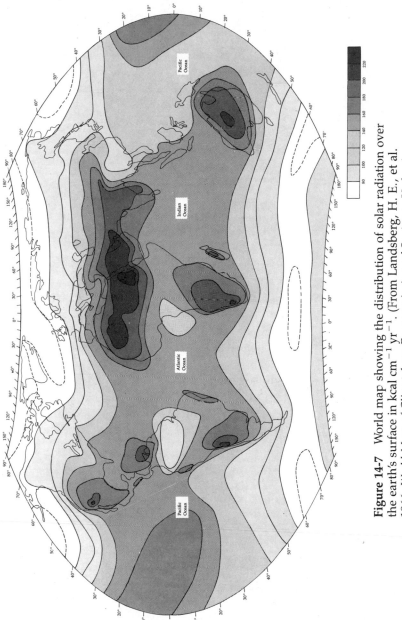

Figure 14-7 World map showing the distribution of solar radiation over the earth's surface in kcal cm^{-1} yr^{-1}. (From Landsberg, H. E., et al. 1966. *World Maps of Climatology*. By permission of Springer-Verlag.)

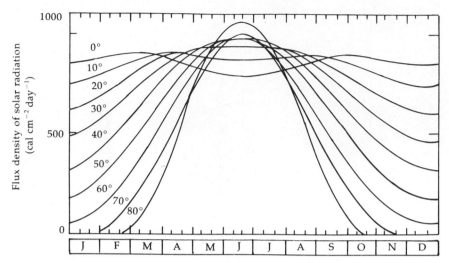

Figure 14-8   Daily total of the undepleted solar radiation received on a horizontal surface as a function of latitude and time of year. (From *Energy Exchange in the Biosphere* (Figure 2, p. 8) by David M. Gates. Copyright © 1962 by Harper and Row, Publishers, Inc. Reprinted by permission.)

the greater distance solar radiation must pass through the atmosphere in temperate regions and poleward. From the equator to 30°N and S, radiation is generally high, with limited areas of very high radiation or very low radiation. Radiation over the Amazon Basin is low because of frequent heavy cloud cover and heavy rainfall. Areas of very high solar radiation occur on all the major continents where high pressure systems prevail and cloud cover is very low.

A comparison of solar radiation received throughout the year at latitudes ranging from 0° to 80° shows the greatest equability at the lowest latitudes, and the greatest range and the highest energy received in a day at the highest latitudes (Figure 14-8).

Altitude

At higher elevations, light rays pass though less atmosphere. As a result, the ultraviolet light is a larger fraction of the total incoming radiation.

In general, temperature will be lower at higher altitudes. This inverse correlation of altitude with temperature creates a gradient or **lapse rate,** usually given as about 10°C per 1000 m of elevation. There are several lapse rates; the one given here is a **dry adiabatic** lapse rate, which means the cooling of air as it rises without condensation and without cloud formation. Lowry (1969) discusses lapse rates and their significance. There is less atmospheric pressure at the higher elevations, so air there expands and loses energy. Conversely, air coming downslope is more compressed and therefore warmer. Thus, in addition to latitudinal differences, there will be, at any given latitude, elevational differences in temperature (see Figure 20-12).

Topography

The decline in air temperature due to increased elevation is not constant, especially in dissected topography. The situation is complicated by various physical phenomena, including temperature inversion. During the night, the soil gives up heat to the air, sometimes so quickly that the soil may become cooled below the temperature of the overlying layers of air. These layers return heat by conduction to the soil and may become cooler than the overlying strata, resulting in a **temperature inversion.** The normal vertical temperature gradient, that is, temperature decreasing with increasing elevation, is inverted. There is thus an upslope thermal belt warmer than the lower-lying regions. The warm air capping the cold eventually is cooled (usually by 300 m in elevation) and the normal lapse rate is resumed.

Another means by which the temperature gradient can become inverted is by cold air drainage. Cold air is denser than warm air and drains into ravines and valleys at night. These low-lying canyon bottoms and river valley floors are thus colder than the slopes above them, augmenting the thermal belt effect. The flow of air may be reversed in the daytime as warmed air moves back upslope.

There are often tiny basins or frost pockets in which cold air accumulates and remains due to feeble air movement. A further complication of temperature variation in mountainous regions is the width of ravines and valleys. Narrow mountain valleys or canyons receive proportionately less sunlight. The walls therefore reflect less heat into the canyon, and it will be colder and damper than the surrounding area. By contrast, broad, open mountain valleys build up high temperatures due to the accumulation of heat reflected from the wide sides, and they are consequently hotter and drier than surrounding regions.

Proximity to Water

The temperature of a body of water, especially a large lake or ocean, is much more stable than that of a comparable land mass. This is due to the high specific heat of water, its ability to lose energy by evaporation, its high reflectivity, and the contribution of vertical mixing to stability of temperature. Therefore, proximity to a large lake or ocean buffers the climate for adjacent land areas. The water gives up energy during the winter, warming the nearby area, whose average winter temperature is therefore higher than a comparable area at the same latitude further inland. The inland region is said to have a *continental climate.* Conversely, summer temperature will also be more moderate in a marine climatic region as water absorbs energy from the surrounding air. Thus, the amplitude of temperature oscillations in an extreme continental area is greater than that of a maritime region. The effect of a maritime climatic regime may be considered with regard to dates of last and first killing frost. These data are especially important to agriculturalists and horticulturalists, as well as plant ecologists. Miller and Thompson (1975) have shown in map form the average dates of the last killing frost in spring and the first killing frost in autumn for the continental United States (Figure 14-9).

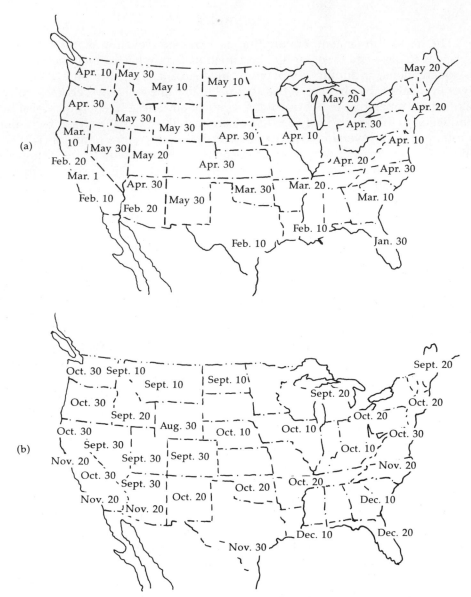

Figure 14-9   (a) Average dates of last killing frost in spring, and (b) average dates of first killing frost in autumn. (From *Elements of Meteorology*, 2nd ed., by Miller and Thompson 1975. Reprinted by permission of Charles Merrill Publishing Co.)

Let us now combine considerations of diurnal variation with the effect of proximity to water. Note the narrow range of temperatures and the minor seasonal differences at a maritime station, San Francisco, California, compared with a more continental station, El Paso, Texas, shown in Figure 14-10.

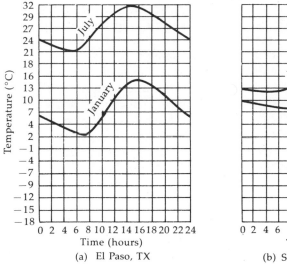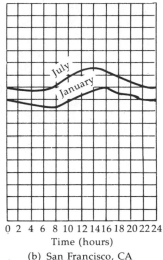

Figure 14-10 Diurnal temperature variation at (a) a continental weather station and (b) a maritime station. (From *Elements of Meteorology*, 2nd ed., by Miller and Thompson 1975. Reprinted by permission of Charles Merrill Publishing Co.)

Cloud Cover

The influence of cloud cover is twofold. Water vapor is opaque to certain wavelengths of solar radiation. A place frequently covered by clouds will not be warmed nearly as much as an area under clear skies. There is, however, some reflection of sunlight from clouds to the ground and some reradiation heat from vegetation, soil surface, and rocks, so that cloud cover may augment direct sunlight and impede heat loss from a land surface. For instance, direct sunlight and sunlight reflected from clouds may simultaneously warm an alpine slope such that radiation values are as high as 2.2 ly min^{-1}, which is in excess of the solar constant. The considerations mentioned regarding the moderating influence of liquid water apply also to water vapor. We cannot overemphasize the important role of moisture in the atmosphere in holding and reradiating energy. Temperature variations in the tropics vary as little as 2°C on cloudy days but can vary as much as 9°C on sunny days. Without the moderating influence of cloud cover, deserts experience great diurnal extremes of temperature. Also, frosts are much more likely on clear nights because of the absence of that moderating influence. In general, moisture in the atmosphere has a stabilizing influence on temperature variations.

Slope Aspect

Due to the inclination of the sun, a south-facing slope in northern temperate latitudes will always experience greater total insolation than will the north-facing slope of the same region. Thus, there will be warmer air and soil, less moisture, and sparser vegetation, in general, on the south-facing slope. North-south slope aspect difference is often manifested in very different plant cover. In central California, for instance, a north-facing slope might be oak woodland (blue oak, *Quercus douglasii*) and the south-facing slope of the same ravine will be covered by chaparral (chamise, *Adenostoma fasciculatum*). See Figure 14-11.

Figure 14-12 illustrates the variation in energy reception on sloping surfaces as a function of the time of day. Figure 14-12a depicts energy received at winter solstice, 22 December. Note that north-facing slopes steeper than 22.5° receive no direct insolation on that date. Figure 14-12b depicts energy reception at the vernal equinox. The north vertical receives no direct radiation, but the south slope at 45° intercepts almost as much as it receives at the summer solstice, which is depicted in Figure 14-12c. As we would predict, the total energy received is the greatest for the community at the summer solstice, 22 June (see also Figure 20-11).

The study of the relationship between climatic factors and seasonal phenomena is referred to as **phenology.** A significant climatic factor is temperature. Phenological events concerned with reproductive activity in plants may be retarded by several days on a north-facing slope with respect to plants of the same species on an adjacent south-facing slope. Jackson (1966) found significant correlation between air temperature sums and flowering dates. The increased insolation received by south-facing slopes, as compared to nearby north-facing slopes, results in warmer temperatures earlier in spring and therefore advanced flowering.

Figure 14-11   Slope aspect difference.

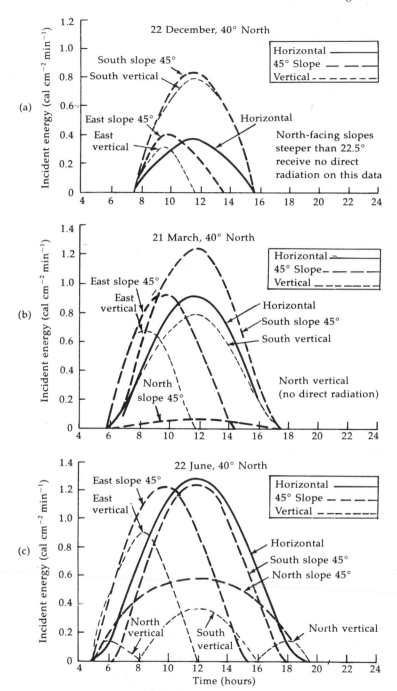

Figure 14-12   The amount of solar radiation incident upon sloping surfaces as a function of time of day at latitude 40°N for (a) the winter solstice, (b) the vernal equinox, and (c) the summer solstice. (From *Man and His Environment: Climate* (p. 36) by David M. Gates. Copyright © 1972 by David M. Gates. Reprinted by permission of Harper and Row, Publishers, Inc.)

We might relate time range of flowering dates to latitude (northward) and to elevation (upward). For instance, one day of retardation or advance of flowering might be equivalent to more than 27 km in distance northward, or more than 30 m in elevation. Jackson found flowering variation of an average of 6.0 days between a north-facing slope of a large gorge and the south-facing slope only 46 m distant. Across a small east-west gorge, plants of the same species blooming less than 8 m apart, but on opposite sides, evidenced a delay of 2.8 days on the north-facing slope as compared to the south-facing one. This could be assigned a difference comparable to 80 km (northward) or 85 m (upward). He found a mean range of flowering dates of 7.2 days for all species studied. This may be interpreted as equivalent to a geographic distance of 201 km northward and 219 m upward (Hopkins 1938), or as 217 km northward and 168 m upward (Jeffree 1960).

Vegetation: Influence on Temperature

The modification of temperature by plant cover is both significant and complex. Shaded ground is, of course, cooler than open area, if there is little exchange of air. Vegetation interrupts the laminar flow of air, impeding heat exchange by convection. The sensual impact of a forest is usually one of moist coolness. This is attributable first to the reduction of soil heating due to shading: The soil absorbs energy from the warmer air above it, cooling the forest air. Secondly, because the forest vegetation transpires as photosynthesis proceeds, the forest microclimate will be more humid than adjacent clearings, and the quantity of energy needed to raise the air temperature is increased. The reradiation of heat is impeded by plant cover, so that the nocturnal forest air and soil are warmer than those of nearby clearings. Temperature oscillations are thus damped both day and night by heavy, continuous plant cover (Table 14-2).

Table 14-2   Comaprison of certain habitat factors in a virgin pine forest and in an adjacent clearing in northern Idaho. Data for the month of August. (From "Effect of removal of the virgin white pine stand upon the physical factors of site" by J. A. Larsen, *Ecology* 3:302–305. Copyright © 1922 by the Ecological Society of America. Reprinted by permission.)

Factor		Forest	Clearing
	Maximum	25.9	30.0
Air temperature (°C)	Minimum	7.4	3.9
	Range	18.5	26.1
Mean relative humidity at 5 P.M. (%)		38.8	35.2
Mean daily evaporation[a] (ml)		14.1	36.1
Mean soil temperature at 15 cm (°C)		12.8	17.0
Mean soil moisture at 15 cm (%)		32.0	43.2

[a]Livingston atmometer mounted 15 cm above the ground.

Vegetation: Influence on the Light Regime

Vegetation modifies the light environment in different ways, depending on the geometry of the plants, the distribution of biomass, and the spectral properties and orientation of leaves. The modifications due to these factors are variable because of seasonal differences in vegetation and in the elevation of the sun.

Hutchinson and Matt (1977) studied the distribution of solar radiation in a tulip poplar forest in Tennessee (Figure 14-13a). The year was divided into **phenoseasons,** each of which represents a period when the solar elevation and phenological state (the state of organism properties that change seasonally, e.g., leaf formation, bud formation, and flowering) of the forest canopy create a unique set of influences on the radiation regime within the forest. The light environment was measured as the amount of total solar radiation received per unit time at different levels within the forest. Variations above the canopy were due to the phenological state of the canopy and fluctuations in solar radiation (Figure 14-13b). The highest radiant flux reaches the forest floor between the spring leafless phenoseason and early summer, when the canopy is in full leaf (Figure 14-13a). It is during this period that sufficient light and adequate temperature allow the spring bloom of herbaceous plants on the forest floor. The gradient in light reduction during the leafless phenoseasons was much less than during the fully leafed phenoseasons at equivalent solar angle (Figure 14-13b). Other less obvious variations in leaf orientation and canopy structure modify the extinction rate of light as it passes through the forest canopy.

Mahall and Bormann (1978), in a detailed study of forest herb phenology in New Hampshire, found that the plants could be divided into four groups, depending on phenoseason activity. The vernal photosynthetic species (Ea in Figure 14-13c) developed and died synchronously, taking advantage of the spring leafless and leafing phenoseasons. Summer green species (Us in Figure 14-13c) developed together in early spring along with the vernal species but remained active through most of the summer. Late-summer species (Aa in Figure 14-13c) develop gradually in summer and die as a group after taking advantage of additional light on the forest floor in late autumn. The last group, which is green from early spring through autumn (Om in Figure 14-13c), can photosynthesize on the low light of the forest floor in summer, enhanced by the periods of more light in spring and autumn. Clearly, light is a significant limiting factor for these forest floor herbs.

The extinction of light as it travels through a vegetation canopy depends on the total leaf area index—the area of leaves projected on a unit area of ground surface. Campbell (1977) considered the theoretical leaf area that would receive direct sunlight as a function of leaf area index (LAI) and leaf orientation (Figure 14-14a). Very little additional sunlight penetrates a canopy of horizontal leaves with a leaf area index >3, but when more leaves have a vertical orientation, greater light penetration occurs in canopies of higher LAI. Not all leaves within a canopy or even on an individual plant are at the same inclination. Forest communities have LAI of 8 or more. Campbell has simulated gross photosynthesis in model canopies assuming that light was the only limiting factor. Figure 14-14b shows the results for LAI values of 1, 3, and 5. Notice that high photosynthesis

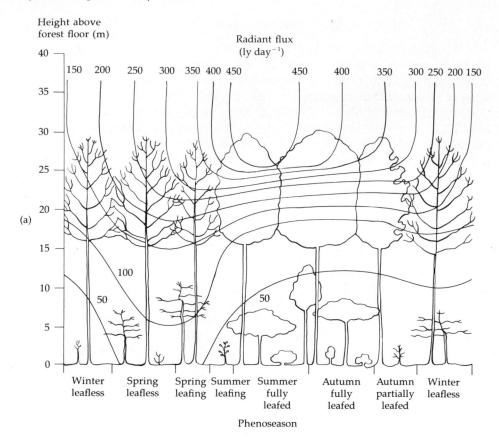

Height above
forest floor (m)

Radiant flux
(ly day^{-1})

(a)

Phenoseason

Winter leafless | Spring leafless | Spring leafing | Summer leafing | Summer fully leafed | Autumn fully leafed | Autumn partially leafed | Winter leafless

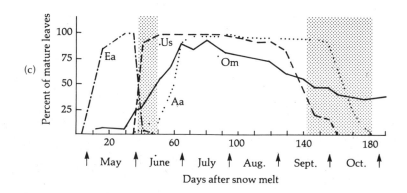

Full sunlight radiant flux (ly day^{-1})

| 400 | 475 | 610 | 800 | 905 | 975 | 980 | 940 | 820 | 675 | 530 |

(b)

Jan.   Mar.   May   July   Sept.   Nov.

(c)

Percent of mature leaves

Days after snow melt

◀ **Figure 14-13** Distribution of solar radiation. (a) Light and phenology in deciduous forests: synthesized annual course of average daily total solar radiation received within and above a tulip poplar forest. (b) Approximate daily total of the undepleted solar radiation received on a horizontal surface at 35°N. (c) A comparison of the percentage of leaves in the mature leaf stages of four taxa at various times of the year. Stippled bands indicate expanding or falling leaves in the canopy. [(a) From "The distribution of solar radiation within a deciduous forest" by B. A. Hutchison and D. R. Matt, *Ecological Monographs* 47:185–207. Copyright © 1977 by the Ecological Society of America. Reprinted by permission. (c) From B. Mahall and F. H. Bormann. *Botanical Gazette* 139: 467–481. Copyright © 1978 by University of Chicago Press. By permission.]

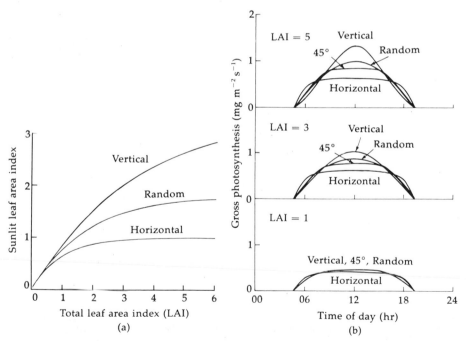

Figure 14-14 (a) Light interception versus leaf area for vertically, randomly, and horizontally oriented leaves. Sunlit leaf area index is found by subtracting the sunlit area below the canopy from 1. (b) Gross photosynthetic rate in model canopies having various leaf areas and leaf distributions as a function of times of day (48°N on 1 June). (From G. S. Campbell, 1977. *An Introduction to Environmental Biophysics.* By permission of Springer-Verlag.)

is possible with large LAI values when leaf orientations are random, but at low LAI, leaf orientation has little effect.

Light- and Temperature-Mediated Plant Responses

We have seen how energy is allocated within the community. We have recognized that a balance is needed if a plant, or even a leaf, is to live. Outgoing energy must equal incoming energy, except for that radiant energy which is transformed to chemical energy and can thus be stored. How is the energy that reaches the plant dealt with so that balance is achieved? We will discuss the concept of energy allocation by the plant in broad terms and then examine some of the phenological events that are governed in part by changes in temperature: events associated with thermoperiodism, dormancy, stratification, vernalization, and heat pretreatment or **sumorization.**

The activity of a plant should be adaptive to the plant or it may perish, even before reproduction can occur. The timing of various phenologic events must be in response to environmental cues: light duration, light quality, light intensity, moisture availability, various chemicals in solution, gravity, touch, and temperature. Plant responses to many of these various stimuli are discussed elsewhere in this book and in general botany and plant physiology books.

Allocation of *S* Within the Plant

How is the right side of the solar energy budget in Equation 14-1 dealt with by plants? Recall that energy is continually being exchanged between a plant and its environment (Figure 14-4). Incident radiation from a variety of sources will be absorbed by a leaf: direct sunlight, scattered skylight, reflected sunlight, and infrared wavelengths. If the air is cooler than a leaf, energy moves away from the leaf. If the air is warmer, the leaf will take on energy.

Cooling results from the evaporation of water (*LE*). The vaporization of a gram of water at 28° C requires about 580 cal of energy, a process that cools the leaf much as an evaporative cooler cools a room. There may be a limited water supply, and desert plants, for example, cannot depend upon unlimited transpiration for balancing the energy budget. Laminar flow of air removes heat from the leaf (*C*) if its temperature is warmer than that of the air. The transfer of heat is directly proportional to the square root of the wind velocity and inversely proportional to the square root of the leaf width. In other words, the larger the leaf, the less effective the cooling. At the same air temperature and wind velocity, a small leaf will lose heat more effectively than a larger leaf. Small leaves, dissected leaves, and lobed leaves are characteristic of plants of arid regions.

Either a very simple system or a stressed one, such as a desert plant community, can provide a model system of heat transfer. Daytime temperatures of desert soils and the air immediately above them are often a great deal higher than that of the air at the standard height above the ground for measuring temperature (1.5 m). For instance, the highest standard air temperature ever measured was 57.8°C (Azizia, Tunisia, on 13 September 1922, and San Luis, Mexico, on 11 August 1933). However, air next to soil surfaces (dark soil, in full summer sun) may exceed 70°C (Gates 1972). As the spring season gives way to early summer, some small desert annuals assume a basket shape, as the rosetted leaves are lifted into the cooler air just a few cm above the soil surface. This adaptation of leaves of desert plants increases chances for heat loss and reduces absorption of heat from the soil (G). Raising leaves may also change the angle of the leaf relative to the sun and thus reduce S.

Desert plants must deal with strong insolation, often in still air. Insolation (S) may raise leaf temperature above that of the surrounding air. If convection C) and evaporation (LE) cannot transfer energy as rapidly as it is absorbed, then the leaf will store heat and may then have a temperature 10°C, or even 15–20°C, above the ambient temperature. Heat exchange is enhanced by winds, which may remove air to within a few mm of the leaf surface. Conversely, heat exchange is hampered in still air, although the layers of warm air above the leaf will tend to rise above cooler air, generating a turbulence that also hastens cooling. Loss of leaf area, and therefore reduction of leaf exposure to insolation (S), is achieved in a variety of ways. Most members of the Cactaceae, for instance, have greatly reduced leaves. A different method for avoiding heat from the ground is seen in members of the Fouquieriaceae, a family with only one genus and with several species found in Mexico and the southwestern United States. Ocotillo (*Fouquieria splendens*) (Figure 20-41) loses its leaves seasonally in response to drought stress, replacing its leaves as many as five times in a single year. Vertical leaf orientation, an adaptation often exhibited by desert shrubs, is another way to avoid excessive S.

Both winter and summer desert ephemerals exhibit heliotropic leaf movements (Ehleringer and Forseth 1980). These annuals also have the capacity to utilize high levels of solar radiation (S) in photosynthesis (P). There are some species that utilize diaheliotropic leaf movements (orientation of the leaf lamina perpendicular to the sun's rays) to enhance reception of solar radiation. Paraheliotropic leaf movements (orientation of leaf lamina parallel to the sun's rays) are used to reduce transpiration rates and leaf temperature, both adaptive strategies to deal with drought stress (Ehleringer and Forseth 1980).

Work with two species of winter annuals in Imperial County and Death Valley, California, *Malvastrum rotundifolium* and *Lupinus arizonicus*, showed contrasting patterns of heliotropic movements that may relate to the origins of these taxa (Forseth and Ehleringer 1982). *M. rotundifolium* maintained tracking movements over a wider range of leaf water potential, namely to the wilting point, −4MPa, than did *L. arizonicus*. The latter utilized paraheliotropic leaf movements at −1.8 MPa and exhibited complete stomatal closure at a higher water potential than did *M. rotundifolium*. The behavior of *M. rotundifolium*, which included alteration of leaf osmotic potential to maintain turgor, is characterized as *drought tolerance*. This winter annual probably invaded the California flora from desert ancestors to the south. *L. arizonicus* exhibits *drought avoidance* and may be derived from the northern, Arcto-tertiary flora (see also Raven and Axelrod 1978; Mooney and Ehleringer 1978).

Figure 14-15 Photosynthetically activeradiation incident on three leaf types over the course of a midsummer day: a diaheliotropic leaf (cosine of incidence = 1.0): a fixed leaf angle of 0°, the horizontal leaf, and a paraheliotropic leaf (cosine of incidence = 0.1.) (Modified from J. Ehleringer and I. Forseth, *Science* 210:1094–1098. Copyright © 1980 by the American Association for the Advancement of Science.)

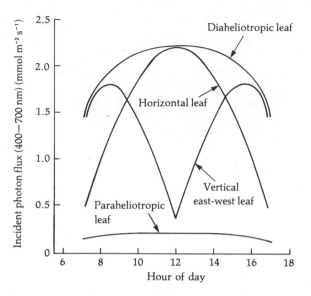

Potential enhancement of solar radiation (*S*) received by diaheliotropic leaves and possible drought avoidance by paraheliotropic leaves can be inferred from Figure 14-15.

Reflectivity (*R*) is enhanced by a surface with high albedo. Possession of a white or light-colored leaf surface increases *R*. Desert holly (*Atriplex hymenyletra*) in the Chenopodiceae provides a good example of how one desert shrub species is adapted to increase reflectivity by different means and thereby reduce heat load and drought stress. Unlike the cacti, which have virtually no leaves that are photosynthetic, or ocotillo, which is drought deciduous, desert holly is an evergreen perennial of the hot desert. Mooney et al. (1977) found that the adaptation of this plant is to change the characteristics of the leaves during the year. Those leaves produced during the cool season are nearly twice the size of the summer leaves. Furthermore, young leaves are covered with expanded, hydrated salt bladders, which collapse as the summer progresses. The salt solution contained in the bladders crystallizes, giving the leaf a white appearance and increasing the reflectance (*R*) of the leaf surface. Leaf reflectance (*R*) (at 550 nm) is inversely correlated with leaf water content. Leaves with 60% leaf moisture exhibit a reflectance above 75% at some wavelengths. At all wavelengths, their reflectance is greater than that of leaves with 89% leaf moisture.

The leaves of desert holly are held at a steep angle. Field measurements of 50 leaves yielded a mean angle of 70° from the horizontal, with random orientation to the azimuth. Although the angle would seem to be more adaptive in summer than during the very active spring period, it does not change seasonally. This tough shrub is a C_4 plant, but it photosaturates at a rather low light level. Thus, photosynthesis is not reduced, even in spring, by the steep angle. The angle does provide protection by reducing heat load and by allowing greater

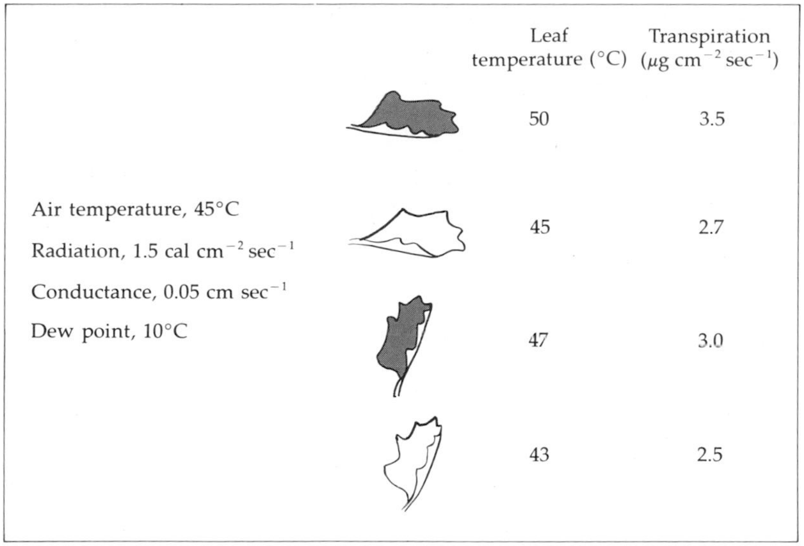

	Leaf temperature (°C)	Transpiration (μg cm^{-2} sec^{-1})
	50	3.5
Air temperature, 45°C	45	2.7
Radiation, 1.5 cal cm^{-2} sec^{-1}		
Conductance, 0.05 cm sec^{-1}		
Dew point, 10°C	47	3.0
	43	2.5

Figure 14-16 Leaf temperature and transpiration of leaves of differing absorptances and angles as determined by the energy equation. The wavelength absorptance (ratio of absorbed to incident radiation) is 0.5 for a typical desert holly *(Atriplex hymenelytra)* leaf (gray) and 0.25 for a summer desert holly leaf (white). Leaf conductance to water vapor transfer was measured on summer leaves. Environmental conditions are those characteristic of Death Valley at midday during the summer. Leaf orientation is horizontal or at 70° as indicated. Leaf size is 3.9 cm^{-2}. (From H. Mooney, J. Ehleringer, and O. Björkman 1977. The energy balance of leaves of the evergreen desert shrub *Atriplex hymenelytra. Oecologia* 29:301–310.)

interception of light during the early morning and late afternoon, which are the times when relative humidity is most favorable (Figure 14-16).

Angle of leaf, reduced leaf size, and high reflectance from the leaf surface enhance the capacity of this desert shrub to remain photosynthetically active even during the hottest summer months, adaptations that allow evergreenness.

Success of any taxon, measured by distribution or abundance or both, may be as greatly influenced by lack of heat as by high heat. Nobel (1980*a* and *b*, 1982) investigated the relationship between northernmost limits and low temperature tolerance in certain cacti, particularly as that tolerance relates to morphology, tissue cold sensitivity, and cold hardening. In columnar cacti, minimum surface temperatures evidently occur at the apex. Increasing stem diameter, increasing shading by apical spines, and increasing apical pubescence increase that surface temperature minimum and thereby provide protection against cold. Morphological differences such as these explain the presence of *Ferocactus acanthodes* in colder regions, as compared to *F. wislizenii*, which is restricted to warmer areas. Similarly, *Carnegiea gigantea* extends farther north than does its southern counterpart, *Stenocereus gummosus,* due to greater protection against cold afforded by morphological differences.

Coryphantha vivipara var. *rosea* has an elevational limit about 600 m higher than does its counterpart, var. *deserti*. Nobel attributes this to differences in tissue cold sensitivity. He also found that certain cacti respond to decreasing day/night air temperatures by cold hardening. The ability to cold harden (and to withstand subzero temperatures) allows cacti to be successful in regions of considerable wintertime freezing (Nobel 1982).

Response to Ultraviolet-B Radiation

Solar radiation in the UV-B range (280–320 nm) constitutes a minor percentage of the total shortwave flux on the surface of the planet. However, it is potentially stressful to young plant tissues because the short wavelengths are capable of causing deleterious changes in sensitive mesophyll cells.

We might expect to find a correlation between the increased percentage of UV-B along a latitudinal gradient and some increase in protective response by the plant. To the contrary, young leaves on six tropical rainforest species exhibit *lower* reflectance in the UV-B region of the spectrum than do mature leaves of the same species (Lee and Lowry 1980), and young leaves are vulnerable. However, the young leaves are protected: They contain high levels of anthocyanins and total phenols, both of which are strongly absorptive of deleterious UV radiation.

Leaf UV optical properties were examined along a latitudinal gradient of UV-B radiation (Robberecht and Caldwell 1980). At latitudes with low UV-B radiation, mean epidermal transmittance of UV-B was >5%; in regions with high UV-B radiation, the species examined transmitted <2%. Various pigments, including flavenoids and other phenols, were apparently responsible for the lower transmittance, as they absorb strongly in the UV-B ranges, transmitting in the visible portion of the spectrum. In both these studies (Lee and Lowry 1980; Robberecht and Caldwell 1980), there is evidence that UV-B absorbing pigments and phenols have adaptive value in plants that grow in regions of high UV-B radiation flux.

Thermoperiodism

There are many plants that require a day-night temperature difference for optimal growth. For instance, red fir (*Abies magnifica*) and Jeffrey pine (*Pinus jeffreyi*) respond positively to a **thermoperiod** of 13°C, that is, a diurnal difference of 13°C. A positive response to such a temperature regime is termed **thermoperiodism.** Hellmers (1966) investigated thermoperiodism in coast redwood (*Sequoia sempervirens*) seedlings and determined that a thermoperiod of 4°C elicits a response. However, that response was not significantly different from the growth of the plant under a constant temperature regime.

The coast redwood does not form dormant buds but can continue to increase in height in response to favorable light and temperature regime. This is in contrast to trees that occupy inland, high-elevation sites, such as Jeffrey pine and

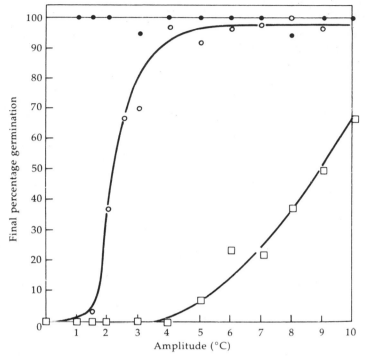

Figure 14-17 The germination response to various amplitudes of diurnal temperature fluctuations in seeds of *Carex otrubae* (●). *Rumex sanguineus* (○), and *Rorippa islandica* (□). The depression in temperature coincided with a dark period of 6 hours. During the photoperiod of 18 hours, all seeds experienced a constant temperature of 22°C. (From Grime and Thompson 1976. *Annals of Botany* 40:795–799.)

red fir. These species do set terminal buds, and this restricts upward growth, even though the plant may continue to acquire new biomass. A thermoperiod requirement is adaptive in the pines and firs, since they live in areas of great diurnal temperature fluctuation. The coast redwood does not require a thermoperiod; it does not live in an area with a large diurnal temperature difference.

Thermoperiodism in woody plants has a counterpart in seed response to a fluctuating diurnal temperature regime. Figure 14-17 shows the effect of varying temperatures on seed germination for three species (Grime and Thompson 1976*). *Carex otrubae*, a sedge, requires no temperature fluctuation for germination, but the other two species show a positive response to temperature depression during the dark period. Sorrel (*Rumex sanguineus*) achieved 50% germination at a thermoperiod of 2.5°C, and yellow-cress (*Rorippa islandica*) did so at 9°C.

---

*This report describes a useful compact germination incubator designed to provide controlled fluctuations in temperature automatically.

Dormancy

Dormancy is a period of inactivity for seeds or plant organs; it can be broken only when certain environmental requirements have been met. A seed that is supplied with moisture, oxygen, and the proper photoperiod but still does not germinate is said to be *dormant*. A key feature in the maintenance of seed dormancy is the need for oxygen. Plant tissues are predominantly aerobic, and the exclusion of a large molecule such as oxygen can therefore help to maintain dormancy in seeds. However, the morphology of a dormant bud does not provide for oxygen exclusion. Thus, the correlation between the maintenance of an anaerobic environment and the maintenance of dormancy may not apply to buds.

Dormancy is a mechanism whereby plants avoid periods of unfavorable weather or periods of high competition. During times of a short photoperiod and a lack of heat or sunlight or liquid water, growth—whether of an established plant or a new one—is very expensive, if not impossible. The protection afforded by the dormant state, especially in extra-tropical regions, is very significant to the plant, and so dormancy is an adaptive feature. The dormancy of apical buds of many woody plants in temperate climates is initiated by short days interacting with cold temperature; this is a photoperiodic response mediated hormonally.

Environmental cues, such as lengthening photoperiod and increasing temperatures, trigger the resumption of activity in more favorable times. The correct interpretation of these cues is vital to the dormant plant or seed. If a dormant beech, for instance, began to leaf out in response to unseasonably warm weather in early November, subsequent frost would kill or injure the new soft tissues, at great cost to the plant. The environmental stimulus that enables the plant to perceive the passage of winter is cold temperature. Degradation of inhibitors and concomitant biosynthesis of growth-stimulating substances take place during the cold period and early spring, and the seed or plant is then prepared to respond to more favorable weather in an appropriate manner.

Stratification

Stratification, the layering of seeds in a moist medium that is then kept at a low temperature, is known to enhance germination. Most plants growing at higher latitudes possess seeds that require cold temperatures for germination. That is, the seeds are dormant prior to exposure to cold. Actual germination following release of dormancy may be in response to a variety of environmental cues, including changing photoperiod, increasing heat and water availability, and a changing hormonal regime within the seed. Germination may also rely upon a physical or chemical disruption of the seed coat, a process called *scarification*.

Vernalization

The flowering of winter rye (*Secale cereale*) and of other cereal plants is influenced by exposure to cold temperature during the time of germination. The time required for flowering of winter rye may be reduced by half, from 14 weeks to 7 weeks, if the seeds are planted in autumn and exposed to winter cold rather than being planted in spring. This cold exposure, copied from nature and used commercially, is called **vernalization,** from the Latin word for spring, *vernus.* Seeds of winter strains of certain cereal grasses, if kept near freezing temperatures during germination, will flower the same summer even though not planted until late spring. We repeat an often-stated but very important qualifier: Even with vernalization, the appropriate photoperiod is necessary for flowering to be initiated.

Sumorization

Heat cracking is necessary to rupture the seed coat of some fire-adapted plants that possess resistant seeds. But what is the influence of temperatures below those of a fire? We mentioned the high temperatures of desert soils. The seeds of most desert annuals are subject to this heat soon after dissemination, and it probably promotes seed maturation in these ephemerals. Capon and Van Asdall (1966) found that germination of eight species of desert annuals native to the Mojave and Sonoran Deserts was enhanced by heat pretreatment. This treatment is called **sumorization,** after the Anglo-Saxon word *sumor,* for summer.

Maximum germination for these species manipulated by Capon and Van Asdall was reached by the fifth week of storage at 50°C (Table 14-3). Untreated seeds and those stored at 4°C showed poor germination percentages, especially the three Sonoran species tested. Of these, none showed more than 12% germination without heat pretreatment. Seeds of all species, including the plantain (*Plantago insularis*), failed to germinate after storage at 75°C.

Summary

Solar radiation is one of the most important aspects of the environment for plants. Heat may be transferred by radiation, convection, or conduction. Energy that fuels this planet comes to us as solar radiation, including ultraviolet light, visible light, and infrared light. Earth is shielded from most of the UV and IR wavelengths by its atmosphere.

All objects absorb and transmit or reflect radiation. Energy gains and losses must balance within a leaf, an organism, and the biosphere itself.

A solar energy budget may be symbolized as

$$S = R + C + G + Ps + LE \qquad \text{(Equation 14-3)}$$

where S = solar radiation; R = reflected or reradiated; C = convection; G = heating of solid objects; Ps = photosynthesis; and LE = latent heat of evaporation. The solar constant (energy reaching earth's outer atmosphere) is 1.94 ly min^{-1}. Locally, radiation balance may be diurnally or seasonally negative, or may be always positive, as in some tropical regions. Energy allocation within a

Table 14-3 Percentage germination of seeds of several species of desert annuals in response to temperature pretreatment. (From "Heat pretreatment as a means of increasing germination of desert annual seeds" by B. Capon and W. Van Asdall, *Ecology* 48:305–306. Copyright © 1966 by the Ecological Society of America. Reprinted by permission.)

Species	Un-treated	Storage at 20°C for		Storage at 50°C for					
		4 weeks	8 weeks	1 week	2 weeks	3 weeks	4 weeks	5 weeks	10 weeks
Mojave Desert spp.									
Coreopsis bigelovii	9	14	30	26	16	24	20	32	2
Eriophyllum wallacei	5	16	24	32	48	30	18	36	2
Euphorbia polycarpa	1	1	2	2	10	12	4	2	0
Geraea canescens	14	14	16	60	74	32	34	48	4
Salvia columbariae	6	4	14	80	26	28	14	16	10
Sonoran Desert spp.									
Lepidium lasiocarpum	0	2	4	53	90	72	61	54	40
Sisymbrium altissimum	10	10	12	12	13	92	73	71	48
Streptanthus arizonicus	0	3	8	21	23	29	50	50	30
Plantago insularis (not leached)	0	0	0	0	0	0	0	0	0
Plantago insularis (leached)	100	—	—	100	—	—	—	—	—

wet mountain meadow differs dramatically from that of, for instance, a desert region, due to plant cover, amount of open soil, and standing water.

The forms of radiant energy that are important to plants include direct and indirect and radiation reflected from other objects. Most radiation received is direct. Leaves absorb in the visible and UV wavelengths. Visible light is absorbed by chlorophyll and carotenes, leaving little visible light under the canopy of a dense forest. There is a latitudinal gradient for UV-B irradiance in the arctic alpine zone. Wavelengths active in photosynthesis are between .4 and .7 μm (PAR).

Radiation is monitored by means of chemical integrators, pyranometers, pyrheliometers, and photographic devices.

Factors that influence variation in temperature include latitude, altitude, topography, proximity to water, cloud cover, vegetation, and slope aspect. Mean annual temperatures decline poleward, but diurnal and seasonal flux increase. Air cools at a certain lapse rate as it moves upslope. Plant cover tends to dampen temperature oscillations locally. South-facing slopes will be warmer, in general, than nearby north-facing slopes in northern temperate latitudes, the situation being reversed in southern temperate latitudes. Phenological events may be retarded on the cooler slope with respect to the warmer.

A plant allocates energy to maintain a balance, losing heat by convection, reradiation, reflectivity, and evaporation of water. Desert ephemerals exhibit heliotropic leaf movements for better irradiance or for drought avoidance. Small leaves aid in loss by convection; white or light-colored leaves enhance reflectivity. Cacti in cold regions exhibit morphological and physiological protection against cold. Thermoperiodism, a requirement for a day-night temperature differential for optimal growth, characterizes certain plants. Some seeds also exhibit a need for temperature fluctuation for germination. Onset of dormancy is triggered, in part, by cold temperature, and cold is used as an environmental cue by plants to perceive the passage of winter. Exposure to cold enhances germination, as well as early flowering, in some seeds and plants. Heat pretreatment enhances germination of some desert annuals.

CHAPTER 15

PHOTOSYNTHESIS

Photosynthetic Pathways

Our understanding of the photosynthetic process and the ecological ramifications of various photosynthetic adaptations has broadened dramatically over the past two decades (Burris and Black 1976). Until recently, it was assumed that higher plants performed photosynthesis only by the Calvin pathway. During the 1960s, a new photosynthetic process, the C_4 photosynthetic pathway, was discovered by Kortshak et al. (1965) in sugar cane and later described in detail by Hatch and Slack (1966). This discovery has stimulated a flurry of research on aspects of the C_4 pathway, from biochemistry to the population and community levels. The photosynthetic properties of a few plants have been characterized, but many more remain to be studied. It is possible that additional pathways will be discovered. Certainly, as research proceeds, we will further modify our concepts of the ecological importance of photosynthetic variations.

Our goal in this section is not to describe the detailed aspects of the various photosynthetic processes, but simply to focus attention on the characteristics of photosynthesis that are most likely to be of ecological significance. If you wish a more complete review of these concepts, most recent textbooks in plant physiology and biochemistry have detailed information.

The photosynthesis process is divided into light reactions and dark reactions. Light reactions convert light energy into chemical energy and are common to all higher land plants. Dark reactions convert carbon dioxide into sugars and starches. In the light reactions (Figure 15-1), 8 to 12 photons of light are absorbed by pigment systems within the chloroplasts. There, the light energy is converted into chemical energy in the form of ATP and NADPH (reduced NADP), which are the sources of energy for the dark reactions. Briefly, the process is as follows. Water is oxidized and is the source of hydrogen to reduce NADP, as well as the source of oxygen released during photosynthesis. ATP is also formed in this process, from ADP and inorganic phosphate.

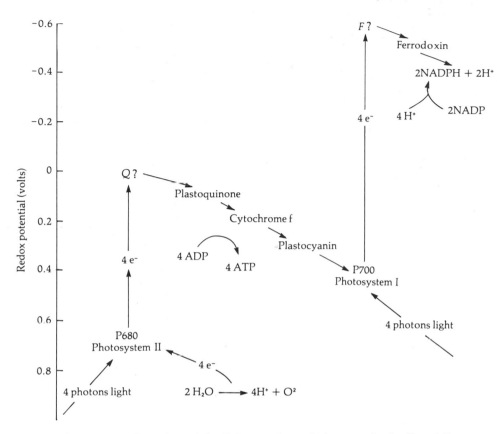

Figure 15-1 Overview of the light reactions of photosynthesis. Q and F are unidentified electron carriers.

$$2H_2O + 2NADP^+ + 3ADP^{-2} + 3H_2PO_4^- \xrightarrow{8\text{--}12\ \text{photons}}$$
$$2NADPH + O^2 + 2H^+ + 3ATP^{-3} + 3H_2O$$

C_3 Photosynthesis and Photorespiration

Prior to the discoveries of Hatch and Slack and Kortshak et al., it was thought that most plants fixed CO_2 only by the pentose phosphate pathway, also called the **Calvin cycle.** This type of photosynthesis is often referred to as C_3 *photosynthesis,* because the first stable product formed is phosphoglyceric acid (PGA), which has a skeleton with three atoms of carbon. A series of reactions follow (Figure 15-2a), in which sugars and starch are formed as the products of photosynthate. The energy necessary for the incorporation of atmospheric CO_2 into photosynthate is obtained from the ATP and reduced NADP formed in the light reactions. Additional ATP from the light reactions provides energy and phosphate to regenerate ribulose bisphosphate (RuBP, a five-carbon sugar), which

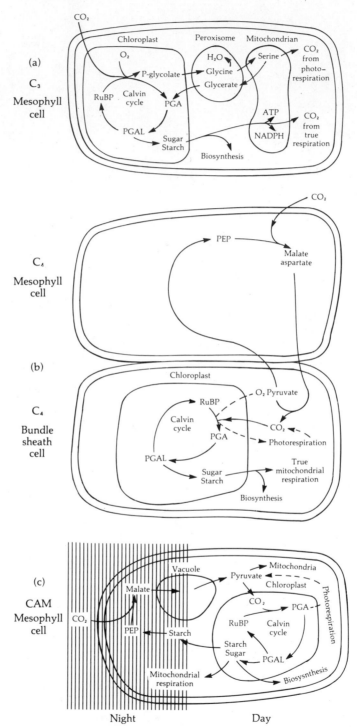

Figure 15-2   Basic dark reactions and product fate of (a) C₃, (b) C₄, and (c) CAM photosynthesis.

Figure 15-3 Cross sections of (a) C$_3$, (b) C$_4$ (Kranz anatomy), and (c) CAM photosynthetic tissues.
P = palisades mesophyll;
S = spongy mesophyll;
M = mesophyll;
B = bundle sheath.

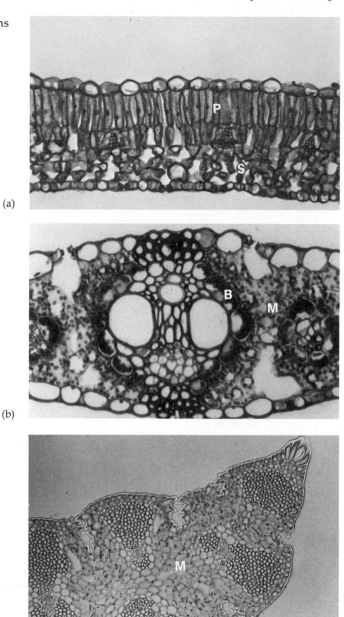

(a)

(b)

(c)

then can combine with CO$_2$ to form PGA. This process takes place during the day within the mesophyll cells and in the cortical cells of stems that contain chloroplasts. Anatomically the leaves of C$_3$ plants are typically divided into palisade and spongy layers, with the chloroplasts widely distributed within the mesophyll (Figure 15-3a).

Coupled with the Calvin cycle is a series of reactions leading to the light-stimulated release of CO_2, known as **photorespiration.** Photorespiration is not the same as true respiration, and both processes may go on independently of each other in the same cell at the same time. True respiration takes place exclusively in mitochondria, oxidizing carbohydrates such as glucose and generating energy in the form of ATP and NADPH. Photorespiration oxidizes organic acids, using up oxygen and releasing carbon dioxide, but it requires more energy than it makes available. From a photosynthetic standpoint, photorespiration is wasteful of the plant's resources; it reduces the rate of net photosynthesis by 30–50% (Goldsworthy 1976; Downes and Hesketh 1968). However, there may be advantages in other aspects of the plant's physiology that offset the reduction in photosynthesis.

Plant physiologists do not currently know what, if any, beneficial effects photorespiration has in plants. Some speculate that photorespiration evolved as a means of dealing with glycolate accumulations or as a regulation for phosphorylated sugar levels. Photorespiration may aid in the transport and interconversion of carbohydrates and nitrogenous compounds (Devlin and Witham 1983; Powles and Osmond 1978).

In the presence of atmospheric levels of oxygen (21%), the enzyme responsible for fixation of carbon dioxide in photosynthesis (RuBP carboxylase) may fix oxygen instead of CO_2 to the receptor RuBP molecule (Figure 15-4). Thus, oxygen and carbon dioxide compete for the same enzyme. For convenience, the enzyme is called *RuBP carboxylase* when it fixes carbon dioxide and *RuBP oxygenase* when it fixes oxygen, but it is the same enzyme. When oxygen is fixed, only half as much phosphoglycerate (PGA) is initially produced; the remaining carbohydrate is converted in steps to glycolate. Through a series of steps in the peroxisome that require energy and oxygen, this is converted to the amino acid glycine. Glycine is then transported to the mitochondria, where two molecules of glycine are combined into one molecule of serine and carbon dioxide is released. Serine can, with additional consumption of energy, be chemically altered back to PGA and transported back to the chloroplast. Thus, some glucose is produced from the products of photorespiration, but the energy gained does not balance that lost through photorespiration. The inhibition of photosynthesis by the fixation of O_2 is related to two factors. First, competition between O_2 and CO_2 reduces the amount of carbon fixed. Second, CO_2 is lost during photorespiration.

C_4 Photosynthesis

Over a dozen families of angiosperms have been shown to fix carbon dioxide by the C_4 dicarboxylic acid pathway. C_4 metabolism may have arisen several times in evolutionary history, as indicated by the conspicuous lack of evolutionary relationship between many C_4 families. Lists of C_4 species have been published in several places (Downton 1975; Krenzer et al. 1975; Smith and Epstein 1971; Teeri and Stowe 1976; Mulroy and Rundel 1977), and the lists keep growing as more C_4 species are discovered.

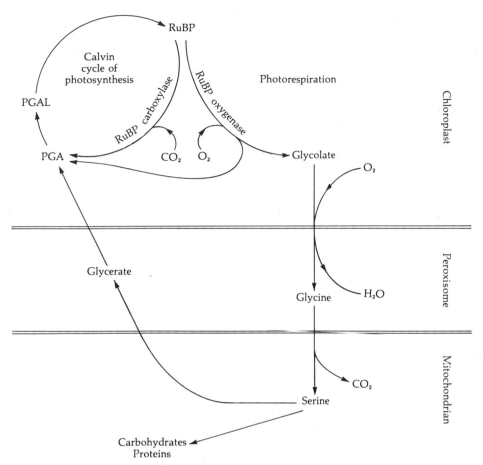

Figure 15-4 Schematic diagram of the relationship between photosynthesis and photorespiration.

In the C_4 pathway, carbon dioxide is fixed by the enzyme PEP carboxylase to form four-carbon acids such as malate or aspartate in the mesophyll cells (Figure 15-2b). These acids then diffuse along concentration gradients from the mesophyll cell into a bundle sheath cell, where CO_2 is released to enter the Calvin cycle (Hatch and Osmond 1976). The three-carbon carrier molecule (pyruvate) returns to the mesophyll cell, where it is converted to PEP to receive another CO_2 molecule. This mesophyll–bundle sheath shuttle thus concentrates CO_2 at the C_3 fixation site. Photosynthesis proceeds within the bundle sheath cells just as it does in the mesophyll cells of C_3 plants. The special anatomical characteristics necessary for these processes to take place are referred to as **Kranz anatomy** (Figure 15-3b). Kranz (wreathlike) anatomy is characterized by the presence of a dense layer of large, thick-walled, chloroplast-containing cells that form a sheath around the vascular bundles and by a lack of differentiation of the mesophyll into a recognizable palisade and spongy layer. Mesophyll cells are often arranged cylindrically around the bundle sheath. Kranz anatomy is easily seen with a hand lens in sections of fresh material. By this means, plants with C_4 photosynthesis can be identified in the field.

The oxygen inhibition noted in C_3 plants is not a problem in plants with C_4 photosynthesis. The lack of oxygen inhibition is apparently due to the high concentration of CO_2 at the site of RuBP carboxylase, which almost eliminates RuBP oxygenase activity and allows C_4 plants to reduce the concentration of CO_2 to nearly zero. CO_2 released by photorespiration is normally refixed by PEP carboxylase in the mesophyll cells. The CO_2 compensation point (the point at which the concentration of CO_2 is sufficient to support net photosynthesis) has been used to distinguish between C_3 and C_4 plants (Downton and Tregunna 1968).

Crassulacean Acid Metabolism (CAM)

Many succulent plants in some 20 families (e.g., Agavaceae, Orchidaceae, Crassulaceae, and Cactaceae) exhibit a carbon fixation scheme referred to as *crassulacean acid metabolism (CAM)*. All CAM plants have succulent photosynthetic tissues but lack the specialized bundle sheath cells that C_4 species have (Figure 15-2c and 15-3c). The stomata of CAM plants are open at night, when carbon dioxide is fixed by PEP carboxylase, forming malic or isocitric acid. In contrast, the stomata of C_3 and C_4 species are open during the day. In CAM photosynthesis, acids are accumulated within the extraordinarily large vacuoles of the cells at night. This results in a pattern of low nighttime pH followed by increasing pH during the day, when photosynthesis proceeds by the Calvin cycle. During the day, when stomates are closed, CO_2 from the organic acids is released, then fixed by RuBP to enter the Calvin cycle. The elevated CO_2 concentrations within the cell probably counteract the inhibitory effect of oxygen by photorespiration (Osmond 1976). Thus, separation of CO_2 uptake and fixation by RuBP carboxylase during the light and dark periods result in the same benefits as the spatial separation of CO_2 uptake and fixation by RuBP in C_4 plants. Not all plants that have these characteristics are obligate CAM plants.

There is considerable variation in the dependence of plants on CAM photosynthesis. Some species with CAM rely on the CAM machinery for essentially all CO_2 uptake. Even these so-called obligate CAM plants have periods of C_3-type CO_2 fixation in early morning and in the late afternoon, especially when environmental conditions are optimal. Most cacti, for example, are considered obligate CAM plants. Under extreme stress, CAM plants can enter a state where there is no external CO_2 exchange but continue to recycle internal CO_2 and show diurnal fluctuations in acidity. Others are facultative CAM plants, which may shift from the C_3 mode of CO_2 uptake to the CAM mode as conditions become more stressful. Age is sometimes a determinant of photosynthetic mode in plants. For example, *Mesembryanthemum crystallinum* is a facultative CAM plant in younger phases but becomes an obligate CAM plant later in the life cycle. These species apparently take advantage of moderate environmental conditions by shifting to the C_3 mode of photosynthesis and maintain nighttime CO_2 fixation in more stressful conditions. Examples of photosynthetic response to environmental conditions are covered later in this chapter.

Welwitschia mirabilis is a unique succulent gymnosperm that grows in the deserts of southern Africa. The literature is not clear concerning the mode of photosynthesis of *Welwitschia*; some evidence suggests CAM, and other data suggest C_3 (Dittrich and Huber 1974; Shulze et al. 1976; von Willert et al. 1982). Recent experimental studies suggest that *Welwitschia* has the capacity for CAM but apparently does not fix atmospheric CO_2 at night. Even under stress, plants grown in the greenhouse or growth chamber do not shift to nighttime CO_2 uptake but do show significant diurnal fluctuation in organic acid content of the leaf tissue. This phenomenon has been referred to as **cycling.** Carbon dioxide released from respiration during the night is fixed by PEP carboxylase into primarily malic acid and stored in vacuoles, where it is released during the day to enter the Calvin cycle. Thus *Welwitschia* has the biochemical capacity for CAM but does not open stomates at night; in the strict sense, it is not a CAM plant (Ting and Burk 1983). It is interesting to speculate that cycling may have been an early step in the evolution of true CAM.

Carbon Isotope Discrimination

The ^{13}C and ^{12}C isotopes of carbon occur in a constant ratio as carbon dioxide in unpolluted air. One would expect that plants would fix carbon isotopes in the same ratio as they occur naturally. However, both carboxylating enzymes (PEP carboxylase and RuBP carboxylase) discriminate against $^{13}CO_2$, resulting in a fractionation of atmospheric carbon in the tissues (Smith and Epstein 1971).

Measurements of tissue or air ^{13}C composition is standardized by expressing the ratio of $^{13}C/^{12}C$ relative to a fossilized carbonate skeleton of the Cephlapod *Belemnitella* as an index:

$$\delta\ ^{13}C\text{‰} = \left(\frac{^{13}C/^{12}C\ \text{sample}}{^{13}C/^{12}C\ \text{standard}} - 1 \right) \cdot 1000 \qquad \text{(Equation 15-1)}$$

The $^{13}C/^{12}C$ ratio can be obtained by burning a tissue sample and subjecting the captured gas to analysis by mass spectrometer.

The carboxylating enzymes differ in the degree of discrimination against ^{13}C, thus providing a means of determining the active enzyme by monitoring $\delta^{13}C\text{‰}$ in plant tissues. The ^{13}C composition of unpolluted air is $\delta^{13}C\text{‰} = -7\text{‰}$ (Keeling et al. 1979). On the average, C_3 plants, which fix CO_2 with RuBP carboxylase, have a $\delta^{13}C\text{‰}$ of -27‰. C_3 plants, therefore, have a ^{13}C composition of -20‰ (2%) less than air. C_4 plants average -11‰. The range for C_3 plants does not overlap with the range for C_4 plants. Therefore, we can distinguish plants with C_3 photosynthesis from those with C_4 by establishing the ^{13}C index ratio. Details concerning the mechanism of discrimination can be found in the recent literature (Farquhar 1983; Farquhar et al. 1982; Berry and Farquhar 1978).

The ^{13}C index for CAM plants varies into the range of both C_3 and C_4, depending on whether a plant fixes more CO_2 in the C_3 mode or in the C_4 mode (Ting and Gibbs 1982). CAM plants fix CO_2 with PEP carboxylase when stomates are open at night and with RuBP carboxylase when stomates are open during the day (O'Leary and Osmond 1980). *Kalanchoe fedtschenkoi* changed ^{13}C index ratio from $-16‰$ to $-33.3‰$, depending on conditions (Bender et al. 1973).

Table 15-1 summarizes and compares 16 traits of C_3, C_4, and CAM photosynthetic pathways.

Table 15-1   Comparison of photosynthetic pathways. (Reproduced with permission from the *Annual Reviews of Plant Physiology,* Vols. 24 and 27. Copyright © 1973 and 1976 by Annual Reviews, Inc.)

Trait	C_3 heliophyte (adapted to a high-light environment)	C_4	CAM
Taxonomic diversity	Very wide: algae to higher plants	No algae, lower vascular plants, or conifers; wide among flowering plants	Some species in about 20 families of flowering plants + Welwitchia
Typical habitat	No pattern	Open, warm, saline (some exceptions)	Open, warm, saline (sometimes cool)
Leaf anatomy	Palisade + spongy parenchyma	No mesophyll differentiation; large bundle sheath; Kranz	No mesophyll differentiation; big cells with large vacuoles
Light saturation point (mmol/ $m^{-2}s^{-1}$)	0.6–1.2	1.6–2.0	like C_3 (?)
Optimum temperature	20–30°C (lower in tundra)	30–45°C (as for light above, can be lower for C_4 species in different habitats)	30–35°C for CAM mode; lower for C_3 mode
Maximum photosynthetic rate			
mg dm^{-2} hr^{-1}	30	60	3 (maximum reported = 13)
mg g^{-1} hr^{-1}	55	100	1 or less

Table 15-1 *Continued*

Trait	C₃ heliophyte (adapted to a high-light environment)	C₄	CAM
Maximum growth rate ($g\ dm^{-2}\ day^{-1}$)	1	4	0.02
Water use efficiency ($g\ CO_2\ kg^{-1}\ H_2O$)	1–3	2–5	10–40
Photorespiration	High	Low	Low
Na required?	No	Yes	No (but salts stimulate CAM mode)
Fixation path and enzyme	$CO_2 + 5\text{-}C \rightarrow 3\text{-}C$ PGA; carboxydismutase or also called ribulose bisphosphate carboxylase	$CO_2 + 3\text{-}C \rightarrow 4\text{-}C$ acids; PEP carboxylase	Still some debate; possibly just as C₄ but enzyme is light-inhibited, thus may be structurally different
Stomate behavior	Open in day, closed at night	Open in day, closed at night	Closed in day, open at night (unless environment shifts plant to C₃-like mode)
Space-time relations	Entire Calvin cycle in any mesophyll cell	Initial fixation in mesophyll, then transfer of acid to bundle sheath for Calvin cycle	Initial fixation at night in any mesophyll cell; storage of acid in vacuole; Calvin cycle during day
Effect of environment on pathway	None	None	Moist, warm night temperature and long day length put plant in C₃ mode
CO_2 compensation point	50 ppm	5 ppm	2 ppm (in dark)
$\delta^{13}C$	−24 to −34‰	−10 to −20‰	Possibly intermediate, although mainly like C₄

Environmental Factors and Photosynthetic Response

Gas Diffusion

The maximum potential rate of CO_2 uptake and the efficiency with which available CO_2 can be fixed are of great ecological importance. Photosynthesis of plants is sometimes limited by the concentration of CO_2 at the site of fixation. Therefore, the rate of diffusion of CO_2 through the stomates and into the plant is a critical factor. Since water leaves the plant when the stomates are open, the balance between water loss and CO_2 uptake is important in determining the relative success of terrestrial plants. The pathway of CO_2 and water diffusion into and out of a leaf may be visualized as a series of steps, any of which can be limiting. The application of this concept to water diffusion is considered in Chapter 19; here we consider the process of CO_2 diffusion.

The steps in the movement of CO_2 can be visualized as a series of resistances and potential gradients defined by a restatement of **Fick's law** for gas diffusion:

$$J_{CO_2} = \frac{\Delta c}{\Sigma r}$$

(Equation 15-2)

where CO_2 flux (or the flux of any diffusing substance J_{CO_2}) depends on the change in concentration from the source to the reaction site (Δc) and the sum of the resistances (Σr) to diffusion. Figure 15-5 depicts the pathway along which CO_2 diffuses from external air to the reaction site within a chloroplast-containing cell. The concentrations and resistances dictate the rate of CO_2 assimilation. Therefore, the consideration of relative values is important in understanding photosynthetic adaptations. **Boundary layer resistance** (r^{bl}) is encountered near the leaf surface, where a zone of air may become depleted of CO_2 and through which CO_2 must move by diffusion rather than by turbulent mass transfer (as it does further from the surface). **Stomatal resistance** (r^{st}) restricts CO_2 diffusion at the leaf surface. **Cuticular resistance** has been shown to be so great in most mature leaves that we may assume that essentially all of the CO_2 which enters a leaf diffuses through the stomates. **Intercellular air space resistance** (r^{ias}) is resistance to diffusion in the gaseous state inside the leaf. The sum of r^{bl}, r^{st}, and r^{ias} is called *leaf resistance*. Resistance encountered from where CO_2 leaves the gas phase in air and is dissolved in the cytoplasm of the photosynthesizing cell and moves to the chloroplast is referred to as **mesophyll resistance** (r^{mes}). Limitations placed on CO_2 fixation by the photosynthetic process itself represent the final resistance to CO_2 transport. These limitations are collectively called **chloroplast resistance** (r^{chl}). Mesophyll and chloroplast resistance can be equal to or greater than stomatal resistance, depending on the plant and environmental conditions.

The concentration of CO_2 in the atmosphere (c^{air}), in the intercellular spaces (c^{ias}), and at the chloroplast surface within the mesophyll cells (c^{chl}), in conjunction with the various resistances, determine the CO_2 flux. In a rapidly photo-

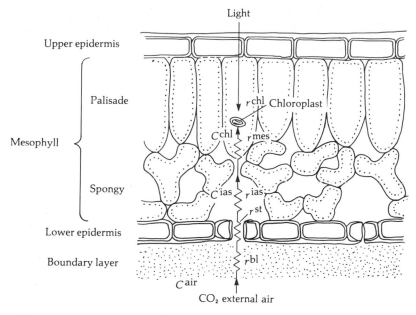

Figure 15-5   Transport pathway, transport resistances, and concentration gradients in a photosynthesizing leaf. See text for definition of terms.

synthesizing cell, CO_2 concentrations at the chloroplast (c^{chl}) are quite low relative to c^{air}, and so a gradient exists. This gradient becomes greater toward the outside of the leaf where average CO_2 concentrations are approximately 340 ppm. Carbon dioxide released by photorespiration and true mitochondrial respiration add to the CO_2 concentration within the leaf. The concentration becomes high enough at night to reverse the gradient (except in CAM plants), and CO_2 passes into the atmosphere.

Stomatal aperture (and therefore stomatal resistance) appears to be responsive to light, intercellular CO_2 concentrations, and leaf water status. Stomatal aperture responds to these factors by active transport of ions into and out of the guard cells, thereby modifying turgidity by changing osmotic concentrations within the cell. For a detailed account of the mechanisms and causes of stomatal response, see Jarvis and Mansfield (1981), Hall et al. (1976), or Raschke (1976).

Different plants have different abilities for utilizing CO_2 at low concentrations and therefore different CO_2 compensation points. The CO_2 compensation point is reached when CO_2 concentration is at the level where photosynthesis equals respiration. In C^4 plants, for example, CO_2 is fixed by PEP carboxylase, which has an extremely high affinity for CO_2, and converted into organic acids that diffuse out of the mesophyll cells. This creates a CO_2 sink in the mesophyll cells, so a steeper concentration gradient can be maintained. C_4 plants can thus reduce the concentration of CO_2 in mesophyll cells to near zero, whereas C_3 plants reach their CO_2 compensation point at much higher levels (Table 15-1).

Figure 15-6 Comparison of the photosynthetic reactions of *Atriplex glabriuscula*, a C_3 plant, and *Atriplex sabulosa*, a C_4 plant, to various intercellular CO_2 levels. (Modified from Björkman et al. 1975. Courtesy of the Carnegie Institution of Washington.)

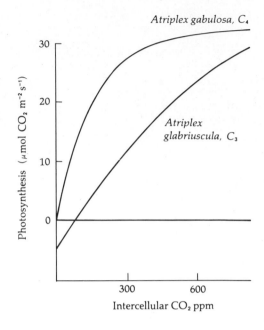

Figure 15-6 shows the results of studies conducted by the Carnegie Institution of Washington, in which C_3 and C_4 plants were subjected to various intercellular CO_2 concentrations (c^{ias}). The ability of the C_4 plant to maintain a positive photosynthetic rate at near zero CO_2 concentration is apparent. It is also interesting to note that at high (~700 ppm) intercellular concentrations of CO_2, C_3 and C_4 plants had nearly identical photosynthetic rates. C_4 plants have a higher fixation rate under conditions where intercellular CO_2 concentrations are low. High stomatal resistance decreases water loss while also reducing intercellular concentrations of CO_2. Since C_4 plants have a lower CO_2 compensation point, they can continue positive net photosynthesis at high stomatal resistances. This ability to conserve water and maintain positive net photosynthesis gives C_4 plants an advantage in arid, warm environments.

Light Utilization

It has been long recognized that some plants are adapted to high-light environments (**heliophytes**) and some to low-light environments (**sciophytes**). Individuals of the same genotype grown in contrasting light conditions exhibit morphological and physiological differences, just as do individual leaves on a single plant that are differentially exposed to light. Plants growing in the shade tend to have larger, thinner leaves, with larger, less dense mesophyll cells, longer internodes, and less overall pubescence. These adaptations increase efficiency of light utilization, increase area for light interception, reduce reflection, and are associated with a variety of biochemical adaptations to the low-light environment

Figure 15-7 Typical light saturation curves for sun-adapted and shade-adapted leaves of C_3 plants. Arrows indicate saturation intensity: $\sim\frac{1}{3}$ full sun for sun leaves, $\sim\frac{1}{6}$ full sun for shade leaves.

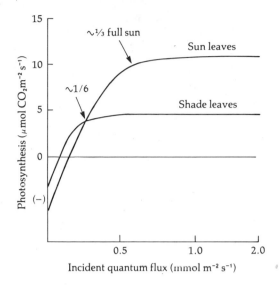

(Taylor and Pearcy 1976). Figure 15-7 shows a typical response of sun-adapted and shade-adapted leaves to increasing incident radiation in the 400–700 nm range (PAR). Note that shade leaves require less light to reach their CO_2 compensation point, and they saturate at lower levels than do sun leaves. The ability of sciophytes to maintain a positive carbon balance in the shade is not necessarily due to a capacity to photosynthesize more rapidly than heliophytes in low light. For example, Loach (1967) found that shade tolerant and shade intolerant species had essentially equivalent rates of photosynthesis when grown in low light. One key to differential success in shade was a difference in dark respiration rate. Shade tolerant species had a lower dark respiration rate and therefore a lower light compensation point, allowing them to maintain a positive carbon balance (positive net photosynthesis) even at very low gross photosynthetic rates. The idea that the dark respiration rate is one key to survival in the shade has been supported by several other authors (Logan 1970; Willmot and Moore 1973).

The photosynthetic apparatus of arctic vascular plants is seldom light-saturated, because of the low angle of the sun in the far north. This is in spite of a lower light compensation point (PAR where photosynthesis and respiration are equal) and a lower light saturation point for arctic as compared with temperate species (Tieszen 1978). The low light compensation point allows continuous positive photosynthesis during the constant light period of midsummer. The photosynthetic qualities of arctic plants are similar to those of shade plants (Chapin and Shaver 1985). Light is the common limiting factor in arctic plants, where the diurnal changes in photosynthesis are more closely related to light than to other environmental factors. By contrast, alpine species have higher light compensation and saturation values, but, because of the high-light environment, show a daily course of photosynthesis that responds more directly to temperature.

Figure 15-8 (a) Response of photosynthesis in warm desert plants to changes in quantum flux (400–700 nm) for *Amaranthus palmeri* (a C_4 summer annual), *Camissonia claviformis* (a C_3 winter annual), *Encelia farinosa* (a C_3 drought deciduous perennial), and *Larrea tridentata* (a C_3 evergreen perennial). Measurements were made under normal atmospheric conditions and at a leaf temperature of 30°C, except for *Amaranthus palmeri*, which was measured at 40°C. (From *Physiological Ecology of North American Plant Communities*, by J. Ehleringer, 1985. Chapman and Hall, Publishers.)

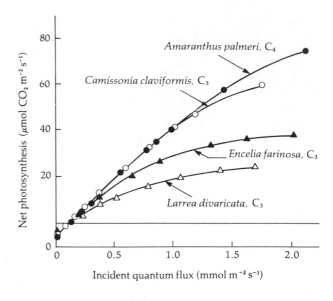

C_4 plants typically have higher light saturation levels than C_3 species (Figure 15-8). However, it is now apparent that at least some C_3 and C_4 plants do not saturate even at full sunlight. *Camissonia claviformis*, a C_3 desert annual, has a very high photosynthetic rate for a C_3 plant (Mooney et al. 1976). The reason for this very rapid CO_2 fixation rate lies in low stomatal resistance and high levels of RuBP carboxylase. As a result, *Camissonia* responds almost linearly to increasing light up to full sunlight. *Encelia farinosa* and *Larrea tridentata* are other species reported to have a similar capability; they are also desert plants (Ehleringer and Bjorkman 1978; Ehleringer et al. 1976; Cunningham and Strain 1969). *Amaranthus palmeri* has the highest photosynthetic rate yet measured (82μmol m^{-2}s^{-1}) in a terrestrial plant (Ehleringer 1983). Another C_4 desert species, *Tidestromia oblongifolia*, with maximum photosynthetic rates of 56 μmol m^{-2}s^{-1}, is comparable to the C_3 annual, *Camissonia claviformis*. It is clear that the C_4 pathway does not, in itself, give plants the ability to have high photosynthetic rate in very high-light environments. It is not clear, however, why other C_3 plants have not evolved this capacity.

PAR is frequently limiting to desert plants with CAM (Nobel 1982*a*). Consequently, cladodes (the flattened stem joints of platyopuntias) of desert cacti tend to grow oriented in a way that maximizes light interception during the season when other environmental factors are optimal for photosynthesis (Nobel 1982*b*). Thus, even in the high-light desert environment, some species are light-limited.

Leaves in the overstory canopy of a plant community not only receive full intensity sunlight, they receive it continuously. Farther down in the canopy, leaves receive a complex mosaic of light: short exposures to nearly full sunlight separated by long periods of very dim light. The pulses of intense sunlight are caused by holes in the canopy through which direct sunlight passes only when the sun is directly overhead. If there is wind, the pulses become shorter and less predictable in time because branches and leaves are first blown into, then out of, the canopy gap. As the seasons progress, the sun's path changes, consequently a sunfleck area in spring is not usually a sunfleck area in late summer. Sunflecks, then, are highly variable. Can plants make use of such—almost random—resources?

Recent research in tropical forests strongly indicates the answer is "yes" and that sunflecks may even contribute a majority of the light energy utilized in growth by understory saplings, herbs, vines, shrubs, and trees. In Hawaii, Pearcy and Calkin (1983) found that 40–60% of the daily carbon gain by seedlings of the trees *Euphorbia forbesii* and *Claoxylon sandwicense* came from sunflecks. Growth rates, as well as photosynthetic rates of those species were highly correlated with daily accumulative duration of sunflecks: Plants receiving 60 minutes of sunflecks a day had a five-fold to three-fold more rapid growth rate than plants receiving 20 minutes a day (Pearcy 1983).

Brief sunflecks of 5 to 10 second duration can be utilized with surprisingly high efficiency; further, the rate of net photosynthesis and its efficiency are both stimulated several-fold as the leaf experiences a sequence of sunflecks (Chazdon and Pearcy 1986*a* and *b*). There is some evidence to show that early successional species in these tropical forests, which require full sunlight for optimal growth, do not exhibit the efficiencies and acclimation potentials described above for shade-tolerant trees.

Water Stress

Several authors (e.g., Boyer 1976; Hsiao 1973) have reviewed the impact of water stress (plant water status when evaporative demands exceed water supply) on the photosynthetic process. There is general agreement that the most immediate effect is an increase in resistance due to stomatal closure (r^{st}). This restricts the movement of CO_2 and water. Stomatal closure may be due to a decrease in leaf water status or a response to soil water depletion not associated with leaf water status (Bates and Hall 1981). It is difficult to measure mesophyll resistance (r^{mes}) and chloroplast resistance (r^{chl}) directly without the interference of stomatal changes. However, there is good evidence that nonstomatal inhibition of photosynthesis does occur in many species (Palta 1983). Johnson and Caldwell (1975) measured CO_2 resistance in arctic and alpine species and found that only a small part of the increased resistance measured in drying plants was due to stomatal closure. Bunce (1977) examined species from habitats with varying levels of available water and reported that increases in mesophyll resistance to CO_2 uptake occurred as the leaves dried. Further evidence of the effect of water stress on photosynthesis comes from studies of lichens conducted by Lange and Kappen (1972) on antarctic and desert lichens. Their results show that photosynthetic rates in these plants vary with changes in water availability even though there are no stomates. The actual biochemical relationship between water stress and photosynthesis remains to be determined.

C_4 and CAM species have an enhanced ability to utilize light while restricting water loss. This higher **water use efficiency** (WUE) means that less water is lost while fixing a molecule of CO_2 in C_4 and CAM plants than in C_3 species. CAM plants conserve water by closing stomates during periods of high temperature and low humidity, when the evaporative demands of the air are greatest. This is also the time when the greatest amount of water is at the evaporative surface of the mesophyll cell wall (Nobel 1974*a*). The mesophyll cell surface is always saturated with water, but more water is present at higher temperatures. For example, with leaf temperatures of 25°C and 5°C in *Agave deserti*, the saturation water vapor concentrations would be 23 g m^{-3} and 6.8 g m^{-3} respectively. With an ambient water vapor concentration of 4 g m^{-3}, the water loss would be seven times higher at 25°C than at 5°C. Stomates are open in CAM plants when both internal and external conditions are conducive to water conservation.

WUE for C_4 plants is nearly double that for C_3 plants. This is due in part to the ability of C_4 plants to reduce internal concentrations of CO_2 (c^{ias}), as discussed in a previous section. This steepens the concentration gradient of CO_2 between the outside air (c^{air}) and the internal air space (c^{ias}). The steep concentration gradient offsets increases in resistance caused by reduced stomatal opening, and CO_2 flux is maintained while H_2O loss is reduced. Water flux is reduced more than CO_2 flux because r^{st} is a greater percentage of the total resistance for H_2O than for CO_2. That is, r^{bl}, r^{st}, and r^{ias} are the only resistances important for water, whereas CO_2 also encounters r^{mes} and r^{chl} (Nobel 1974*a*).

High WUE is one factor that leads to increased numbers of C_4 and CAM species in hot, dry environments. This does not mean that C_4 plants have a significant competitive advantage over C_3 species in arid environments (Syvertsen et al. 1976). Cover and biomass of C_4 and CAM species seldom exceed those of C_3 species, even in arid habitats. However, the relative contribution of C_4 and CAM species to the flora and to cover and biomass increases in arid habitats (Wentworth 1983). The adaptability of C_4 plants to arid regions is evidenced by the overwhelming predominance of C_4 species in the summer annual populations in the hot desert areas of the southwestern United States. Caldwell et al. (1977) considered the growth and gas exchange characteristics of a C_3 and a C_4 shrub in cold desert regions of Utah and found no advantage for the C_4 species. Even though the C_4 shrub was able to photosynthesize during the summer, it had a lower overall rate of CO_2 fixation during the more moderate springtime, so the annual total CO_2 accumulation was no greater than for the C_3 shrub.

Some CAM plants respond readily to water levels and will open stomates during the day if ample water is available. Hartsock and Nobel (1976) heavily watered transplanted *Agave deserti* for 12 weeks under laboratory conditions, causing a shift in stomatal movements so that 97% of the CO_2 uptake occurred in the daytime. Artificial watering in the field did not cause a similar change. However, when water is not limiting, desert *Agave* has a postdawn period of reduced stomatal resistance and a concurrent period of rapid CO_2 fixation similar to that in C_3 plants (Nobel 1976).

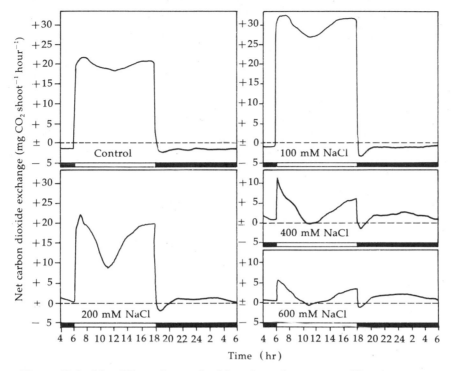

Figure 15-9　Net CO_2 exchange for *Mesembryanthemum crystallinum* at various levels of salinity. (From Winter 1975 as redrawn by Winter and Luttge 1976, "Balance between C_3 and CAM pathway of photosynthesis," *Ecological Studies* 19:323–334. Springer-Verlag, Heidelberg.)

Certain CAM plants will revert to complete and continuous stomatal closure when exposed to water stress (Hanscom and Ting 1978). These plants still have nocturnal increases in acidity because internal CO_2 from respiration is recycled. Thus the loss of carbon by respiration is at least partially counteracted by the phenomenon of recycling.

Water stress caused by salinity also induces photosynthetic response, at least in the facultative CAM plant *Mesembryanthemum crystallinum* (ice plant). Salinity of 200 mM or greater NaCl causes *Mesembryanthemum* to switch from daytime to nighttime CO_2 fixation (Figure 15-9). Also, there is an apparent switch from C_3 to CAM as the leaves get older (Winter and Luttge 1976). *Mesembryanthemum* also has nocturnal CO_2 fixation in naturally saline environments in Israel.

Temperature Responses

Temperature influences photosynthetic responses in various ways. We have already mentioned the ability of C_4 plants to carry on positive photosynthesis at

Figure 15-10 Quantum yield for CO_2 uptake in C_3 species *Encelia california* and C_4 species *Atriplex rosea*, as a function of leaf temperature. Quantum yield was measured in air of 325 ppm CO_2 and 21% O_2. (From J. Ehleringer, *Oecologia* 31:255–267. Published by Springer-Verlag, Heidelberg.)

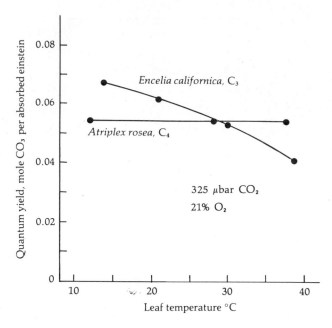

higher temperatures than C_3 plants. The most apparent influence of temperature is the limitation of all enzymatically catalyzed reactions. We would expect that reaction rates would increase gradually to optimum temperatures and then decrease sharply at higher temperatures where enzymes denature. However, membrane phase changes apparently reduce photosynthetic rates at temperatures well below the point of enzyme denaturation (Raison et al. 1980). Photorespiration is much more responsive to changes in temperature than is photosynthesis. A major reason for the different temperature response of C_3 and C_4 species is that photorespiration rates increase as temperature increases. Thus, a C_3 plant may have higher quantum yield (amount of CO_2 absorbed per unit of light) at lower temperatures but will lose that advantage because of greater photorespiration at higher temperatures (Figure 15-10). There is, however, much more involved in understanding temperature responses of photosynthesis, because plants are able to acclimate to previous temperature regimes. In addition, photosynthesis has ranges of temperature tolerance that are related both to the geographic location of the population and to the age of the tissue. Berry and Bjorkman (1980) have recently reviewed photosynthetic temperature responses.

The very sensitive responses of plants to temperature are illustrated in a study by Fryer and Ledig (1972), in which they measured the photosynthetic temperature optima of balsam fir seedlings. The seedlings were grown from seeds collected along a 731–1463 m elevational gradient of the White Mountains of New Hampshire. The photosynthetic temperature optimum changed 4.3°C in

500 m of elevation change. This is very close to the change in mean temperature of 3.9°C in 500 m, as recorded by weather stations along the gradient. This rather precise relationship between photosynthesis and temperature is characteristic of habitats with a short growing season. The restricted time available for growth makes the highest possible photosynthetic rates critical for success.

Where the time for acclimation is greater, we observe temperature acclimation of the photosynthetic apparatus over time. For example, McNaughton (1973) compared Quebec and California ecotypes of *Typha latifolia* and found that the Quebec populations had a narrower range of tolerance for temperature and a much more restricted capability for acclimating to temperature changes than the California plants. The Quebec ecotype also had an increase in photosynthetic temperature optimum with age, which was not present in the California population. The California plants can shift the temperature optimum depending upon the season and maintain this ability throughout the life of the leaf, whereas the Quebec plants have lower optima in the younger stages and a more precisely fixed response in mature leaves. Pearcy (1976) reported that desert ecotypes of *Atriplex lentiformis* have a much greater capacity for temperature acclimation than the ecotype from a more constant coastal environment. The mechanism of temperature acclimation in desert plants, such as *Atriplex lentiformis* (Pearcy and Harrison 1974; Osmond et al. 1980), *Simmondsia chinensis* (Al-Ani et al. 1972), *Atriplex polycarpa* (Chatterton 1970), and *Larrea tridentata* (Strain and Chase 1966), appears to involve, in part, an adjustment in respiration rate. Instead of continuing to increase with temperature, dark respiration rates tend to be constant with increasing temperature. In stress situations that restrict photosynthesis, a positive carbon balance can be maintained by the reduction of respiration rates at temperatures in excess of 45°C in some species. Recall that similar adjustments were the mechanism for survival in sciophytes where gross photosynthesis was limited by light. We can generalize that plants from areas with shorter growing seasons or from habitats which are more predictable will have a reduced ability to acclimate to temperature changes and will tend to have a narrower range of tolerance to temperature.

Temperature is also a critical factor regulating CAM photosynthesis. Nighttime CO_2 uptake is dependent upon low temperature. For example, Neales (1973) found that the normal night–day stomatal rhythm of *Agave americana*, a CAM plant, was inverted when night temperatures reached 36°C (Figure 15-11). Kluge (1974) measured no night fixation of CO_2 when plants were subjected to 25°C day and 30°C night temperatures. Nobel (1976) supported these findings for *Agave deserti*, in which stomatal resistance increased fivefold when leaf temperatures increased from 5°C to 20°C. There is, therefore, strong evidence that the success of CAM plants in water-limiting environments is dependent on low night temperatures. This conclusion is further supported by the actual geographic distribution of CAM plants. They do not usually live in water-limiting areas with high night temperatures. The biochemical reason for low night temperature requirements in CAM plants remains a mystery.

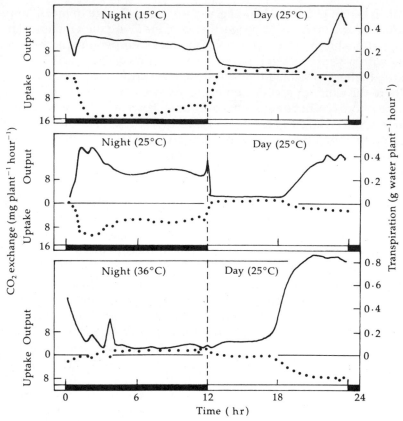

Figure 15-11 Effect of three night temperatures on the patterns of CO_2 exchange (dotted line) and water vapor exchange (solid line) of *Agave americana*, by day and night. (From "The effect of night temperature on CO_2 assimilation, transpiration, and water use efficiency in *Agave americana* L" by T. F. Neales. *Australian Journal of Biological Sciences* 26:705–714, 1973.)

Methods of Photosynthetic Research

The most direct and accurate way of determining photosynthetic rates of land plants is to measure the rate of exchange of CO_2. There are two principal ways that gas exchange rates are determined in terrestrial studies. The most widely used method incorporates an **infrared gas analyzer** (IRGA) to measure the flux of CO_2 to or from a plant or part of a plant that is sealed in an environmentally controlled chamber. Photosynthesis and dark respiration can be continuously monitored through time, and the environmental conditions within the chamber can be regulated. This allows the measurement of plant responses to individual or multiple environmental factors. The other common way of measuring CO_2 uptake is by briefly exposing photosynthetic tissues to an atmosphere containing radioactively labeled CO_2 and measuring the amount of $^{14}CO_2$ fixed per unit time. Later in this section, we will discuss the methods of determining photosynthetic rates by both IRGA and $^{14}CO_2$ techniques. Detailed descriptions of the processes and theoretical considerations are available in a manual of methods edited by Sestak et al. (1971).

The most suitable method of measuring photosynthesis depends on the questions the data will be used to answer. Both methods have limitations. If you wish to have a continuous record of plant response and be able to alter environmental conditions artificially, the gas analyzer is more versatile and more accurate than repeated exposure to $^{14}CO_2$. The most severe limitations of laboratory IRGA systems are the small number of different plants that can be measured and the lack of portability for field use. Ehleringer and Cook (1980) discuss the utility of a field sampling technique where CO_2 depletion is measured by repeated sampling of air surrounding a leaf sealed in a cuvette. The samples are then injected into an IRGA. Since the IRGA is not attached to a plant, several plants can be monitored by taking samples from different cuvettes. This overcomes many of the problems of portability and sample size of the laboratory IRGA. A highly portable and rapid measure of photosynthesis is possible using newly developed miniature IRGA systems, such as the one pictured in Figure 15-12, and $^{14}CO_2$ techniques. If field measurement of large numbers of different momentary samples is important, these field techniques are most satisfactory.

There are, however, errors inherent in the $^{14}CO_2$ measurement process, limiting its application to comparative studies or to surveys where absolute rates

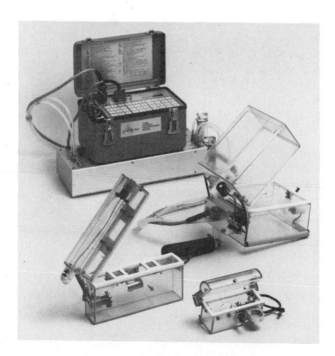

Figure 15-12 Photograph of Licor LI-6000 portable photosynthetic system. (Courtesy of Licor, Inc.)

of photosynthesis are not critical. A major source of error relates to the idea of isotope discrimination characteristics noted earlier in our discussion of C_3 and C_4 photosynthesis. Some plants discriminate against $^{14}CO_2$ (Yemm and Bidwell 1969), possibly because of differences in diffusion rates. Whatever the reason, one cannot be sure that $^{14}CO_2$ is absorbed at the same rate as $^{12}CO_2$ under normal conditions. There are additional errors caused by the dilution of $^{14}CO_2$ by $^{12}CO_2$ released by respiration, which modifies the ratio of labeled and unlabeled carbon at the photosynthetic site. Therefore, factors that influence respiration rates will alter measurements obtained by this method. Respiration measurements can only be inferred by measurements of $^{14}CO_2$ dilution by $^{12}CO_2$; such measurements are not as dependable as those obtained by infrared gas analysis.

Laboratory IRGA Gas Exchange Systems

Gas exchange systems used to monitor CO_2 exchange rates vary widely according to the application and, more often, according to the financial resources available. We will discuss a basic open-flow design that can be modified for specific purposes (Figure 15-13). The gas is pumped through a series of tubes, usually of stainless steel, in which flow rates are carefully monitored. Three flow paths are incorporated so that conditioned gas flows directly to the analyzer, through the plant chamber or through a bypass, and then to the analyzer. The purpose of the bypass is to assure that all modifications being made to the sample gas are due to plant activity. It is also convenient to have bypass valves on environmental conditioning components so that any malfunction can be easily identified and corrected.

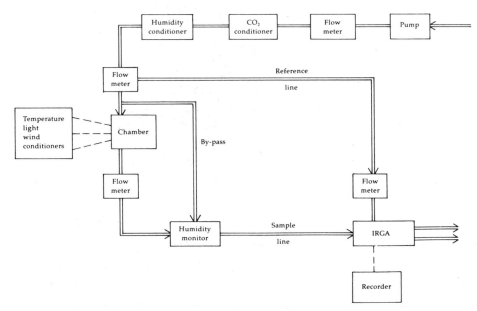

Figure 15-13   Flow diagram showing the major components of an open flow gas exchange system.

The plant or leaf cuvette is simply a sealed chamber into which monitoring probes are inserted for constant measurement of environmental conditions in the chamber. More sophisticated systems have electronic circuitry that feeds information to the conditioning systems, where rates are adjusted to maintain conditions. Some field systems are made so that conditions within the chamber track changes in ambient conditions. The components that condition the air vary widely in sophistication. In a study where CO_2 concentrations of the atmosphere are relatively constant, CO_2 conditioning may not be necessary. It is important in all systems, however, to be able to control and measure gas flow rate, photosynthetically active radiation (PAR), temperature, wind speed, and humidity within the chamber. Humidity changes occurring as the air moves through the chamber, accurately measured, give reliable measurements of transpiration rates.

Additional values such as dark respiration, intercellular CO_2, and stomatal and mesophyll resistances can be measured or calculated from data obtained by manipulating the conditioning components and monitoring plant response.

$^{14}CO_2$ Absorption Technique

Shimshi (1969) described an apparatus (Figure 15-14) that can be used in the field to make rapid measurements of photosynthesis. A plexiglass chamber (usu-

Figure 15-14 $^{14}CO_2$ photosynthesis measuring apparatus: (a) photograph of a typical leaf chamber; (b) diagrammatic representation of the entire apparatus. (Photo courtesy of Charles Lambert.)

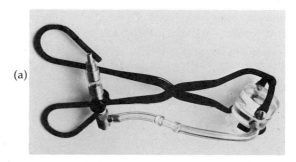

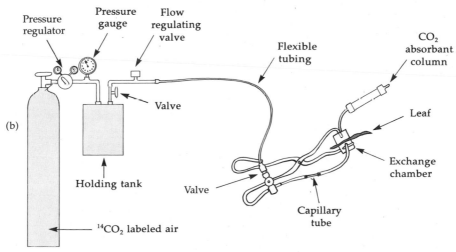

ally fitted with a circulating fan to reduce boundary layer resistance) is sealed on the surface of a leaf for 10–20 seconds, and air containing $^{14}CO_2$ in a known quantity is passed over the leaf. A CO_2 absorbant must be incorporated into the line to prevent releasing $^{14}CO_2$ into the atmosphere. After the leaf has been exposed to the label, a known area of the exposed portion of the leaf is then punched out and immediately killed. Leaf discs are then returned to the laboratory, where the amount of $^{14}CO_2$ fixed is determined by liquid scintillation techniques. Sestak et al. (1971) summarized this method and several other techniques of photosynthetic research, and their book should be consulted before attempting to measure photosynthesis.

Summary

There are three photosynthetic pathways in plants, designated C_3, C_4, and CAM. C_3 plants are the most successful and widespread of all plants, being common in all major climates. The efficiency of C_3 photosynthesis is lowered by photorespiration—the process wherein O_2 substitutes for CO_2, causing the formation of glycolic acid. C_4 plants are most successful in hot, arid environments because low mesophyll resistance and a low CO_2 compensation point permit them to fix CO_2 despite a high stomatal resistance. The high stomatal resistance reduces water loss, which results in a more efficient use of water by C_4 plants. CAM plants are succulents that have adapted to arid environments by absorbing CO_2 at night, when water loss is lower. CAM plants have the highest water use efficiency but are restricted by very low rates of photosynthesis and growth and the requirement for cool nighttime temperatures. C_4 species have intermediate adaptations to arid environments. They are most important in hot, dry habitats. C_3 species, in general, have a lower water use efficiency, but many species are successful in habitats supporting CAM and C_4 species. It is important to keep in mind that these are generalizations; success in any given environment depends on a wide variety of adaptations other than the photosynthetic pathway (Stowe and Teeri 1978; Eickmeier 1978). Alternate photosynthetic pathways may be a mechanism to partition resources between species and reduce competition by dividing the limited resources in time or space.

Primary limitations are imposed on photosynthesis by resistances to CO_2 uptake, water stress, light, and temperature. The rate of CO_2 uptake is determined by the boundary layer resistance, stomatal resistance, resistance as CO_2 dissolves into the liquid phase within the mesophyll cell, and the rate of fixation by photosynthesis. Different plants have different abilities to use CO_2 in low concentrations. For example, C_4 plants can reduce intercellular CO_2 concentrations to near zero, whereas C_3 plants reach CO_2 compensation at much higher levels. Low intensities of light limit photosynthesis, and high intensities may surpass the capacity of the photosynthetic apparatus. However, some desert plants have been found in which the photosynthetic apparatus does not saturate even at full light. The most immediate effect of water stress is stomatal closure, which increases resistance to CO_2 uptake. There are significant changes in the diurnal pattern of stomatal movements in CAM plants, which, under certain conditions, may switch between CAM and C_3 pathways. Plants are able to acclimate to a variety of temperatures and, as a result, show different temperature optima at different seasons of the year.

The two most common ways of measuring gas exchange in plants are systems incorporating an infrared gas analyzer and a technique that measures the rate of uptake of $^{14}CO_2$. The IRGA gas exchange system is expensive, but it is the most accurate technique. It can be used, with proper conditioners, to measure plant response continuously over numerous combinations of environmental factors. Radioactive techniques are most useful for comparative studies, since absolute rates of photosynthesis are difficult to obtain by this method.

CHAPTER 16

FIRE

F ire must be included as an evolutionary force in the development of communities, particularly those subject to lightning, as pointed out by Komarek (1964):

> Lightning is of such frequency and magnitude that there are not many lands, if any on earth, that at some time or other have not been subjected to reoccurring lightning-caused fires. These have occurred at frequent enough intervals to have lasting effect on plant and animal communities.

Lightning is the major cause of natural fires in almost every community. There are an average of 100 lightning strikes to the earth every second, 24 hours a day, 365 days a year (Komarek 1964). This totals over 3 billion strikes per year. The energy of these strikes is impressive; some may range well into hundreds of thousands of volts, with a current of as much as 340,000 amperes.

Fire has influenced the evolution of the various species of the forests, grasslands, and the xeric shrub communities (chaparral, fynbos, maqui, matorral) of the mediterranean climate regions of the world. With the possible exceptions of the wettest or coldest regions of the earth, great tracts of land have been subject to periodic fires for millenia (Komarek 1963; Spurr and Barnes 1973). Fires may be caused by lightning, by sparks from falling rocks, by volcanic activity, by spontaneous combustion, and by human activity. Of these, only the first and the last may develop any periodicity and thus act as a consistent evolutionary force in the community. In the Old World and, to a lesser extent, the New World, periodic burning by humans has played a part in the evolution of the biota. The occurrence of fire results in certain changes in the environment. Our goal should be understanding, as best as we can, why such changes come about, and in what manner fire is responsible for them.

Human History

Humans have used fire as a manipulative tool for at least 0.5 million years (Stewart 1963). Even before people learned to start fire, humans used and carried fire from place to place. We can infer that grassland and forest fires may have been inadvertently caused by those whose practice was to leave a fire banked at their home base, to avoid calamity in case the fire being carried went out. Primitive humans living today are keen observers of minutiae, as undoubtedly our human and prehuman ancestors were. They observed the consequences of fire: Game is driven, though casually and without panic; the capture of insects, rodents and reptiles is facilitated; visibility is enhanced; travel is easier; and forage is renewed. What had been a serendipitous accident surely became deliberate, purposeful arson. In the tropics, grasslands can be maintained by fire at the expense of trees if there is a periodic dry season. Thick forests held little of use for stone-age peoples; grasslands and savanna were of greater value.

In the forests of the northeastern United States, Native Americans probably made deliberate use of fire near camps or villages. There is little evidence that they systematically burned large tracts of forest lands. Their burning was instead confined to the local area. However, their presence and their use of fire for whatever reasons increased fire frequency above the levels of lightning-caused fires and therefore selected for fire-adapted plants (Russell 1983).

In western North America, fire was used by Native Americans to aid in the production of wild seeds, tubers, berries, nuts, and even wild tobacco. Winter ranges for deer were burned in early summer to prevent intense wildfire in late summer, which would have destroyed the primary forage, wedgeleaf ceanothus (*Ceanothus cuneatus*). The Native American practice of burning forested areas was recorded by various early explorers and naturalists, such as Galen Clark, guardian of Yosemite, Dr. L. H. Bunnell, a member of the 1851 Yosemite discovery party, and Joaquin Miller (Biswell 1974).

The role of fire is known in virtually every country, and certainly in those countries in temperate and tropical latitudes. Whether to maintain a wilderness, to increase forest production, to enhance the habitat for game, or to increase grassland productivity, the naturalist, scientist, and manager seek to understand the vital role that fire plays in the ecosystem. In this chapter we will discuss the role of fire in southeastern forests, grasslands, and mediterranean regions. (For a comprehensive treatment of the role and use of fire, see Wright and Bailey 1982.)

Classes of Fire

The various types of fires are divided into three main classes, based on stratum and intensity: (1) surface, (2) crown, and (3) ground fires. Fire behavior and fire intensity are affected by meteorological conditions (e.g., wind and temperature), by fuel loading, and by fuel moisture and soil moisture.

Surface Fires

Surface fires are generally cool, fast-moving fires. They do not build up high temperatures at the plant and ground levels because they are usually fed by lightweight fuels that are quickly converted to ash. Consequently, the basal portions, root stocks, and tubers are not harmed in grasslands and shrublands. In forested regions, cones may be opened, bark and needles scorched, and seedlings and saplings killed, but few mature trees will be severely damaged.

Crown Fires

Crown fires occur in the upper parts of trees and so, by definition, take place only in forest situations. They may result from lightning storms or from intense surface fires fueled by heavy accumulations of litter and debris. This type of fire may occur with surface fire. Crown fires often kill and consume mature trees, dropping boles and branches to ignite and further spread the fire at lower levels (Phillips 1974). In some forests, a clean burn is necessary to open the canopy and thus allow seedling survival in full sunlight. For example, in Canada and the Lake States, nearly pure stands of jack pine *(Pinus banksiana)* often burn during hot dry periods, spectacularly crowning, generating fire storms, and killing all aboveground vegetation. Periodic fires of such intensity maintain the jack pine in vigorous growth. This requirement for intense fires is also thought to be true of the bishop pine *(Pinus muricata)* and of the knobcone pine *(Pinus attenuata)* in California (Aho 1975, personal communication; Kuhlman 1977).

Ground Fires

Ground fires are termed "retrogressive agents" by Vogl (1974). In grasslands, ground fires consume soils down to mineral substrate, creating depressions that then sometimes become ponds. Although of infrequent occurrence, ground fires can be very destructive, not only of roots, tubers and rhizomes, but of the organic matter in the soil itself. Thus, recovery of the plant community may take tens to thousands of years following such a catastrophe. Ground fires are known to be important only in histosols, such as the Okefenokee Swamp.

Fire Effects on Soil

In general, fire has the following effects on soil: Soil temperatures are higher, some soil nutrients and organic matter may be lost, water holding capacity and soil wettability are changed, and populations of soil microbes are altered. The impact of fire on soil will be considered as it affects the chemical and physical aspects of soil and its biotic components.

The effect that fire has on the soil and on soil organisms depends on the intensity of the fire. This intensity, in turn, depends on the heat yield, the availability of fuel, and the rate of spread of the fire. This relationship has been symbolized as

$$I = H \times W \times R$$

where I = intensity (kcal $\sec^{-1}$ m^{-1}), H = heat yield (kcal g^{-1}), W = available fuel (g m^{-2}), and R = rate of spread (m $\sec^{-1}$) (Brown and Davis 1973; Byram 1959; Alexander 1982). The chemical constitution and various physical characteristics of the fuel influence its heat yield. For example, the pine needle litter of a yellow pine *(Pinus ponderosa)* forest has a high surface-to-volume ratio and is not compacted, so an abundance of oxygen is available. This type of fuel carries a fire readily but will produce a relatively cool fire. Alexander (1982) defines frontal fire intensity as the "energy output (kilowatts) being generated from a strip of the active combustion area, one meter wide, extending from the leading edge of the *fire front to the rear of the flaming zone*" (emphasis added).

In part, productivity determines the available fuel. Western coniferous forests may produce as much litter as 30 kg ha^{-1} day^{-1} (Hartesveldt and Harvey 1967). Rates of decomposition of litter and fire frequency are the other determinants of fuel availability. The rate of spread of the fire depends on wind speed, nature of the fuel, humidity, and temperature. The primary determinants of fire behavior are fuel, weather, local landforms, and the ignition pattern (Figure 16-1). Management of a fire-prone ecosystem requires some estimate of available fuel, but quantitative assessment of available fuel in almost any ecosystem is

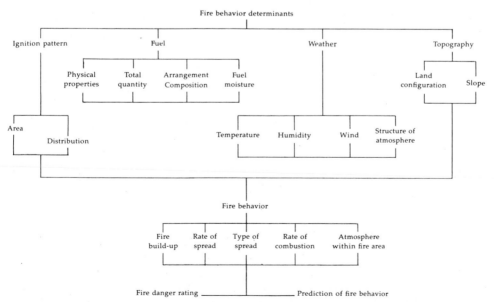

Figure 16-1 Determinants of fire behavior. (From Wakimoto 1977. Courtesy of R. H. Wakimoto.)

costly in time and money. Wakimoto (1977) has developed a method of fuel inventory for chaparral, utilizing linear regression techniques that may, with use, reduce some of the expense. He found that cover (crown area) appears to hold the most promise as a single, easily measured variable that can be used as a basis for predicting total biomass.

Fire Effects on Organic Matter

The heating of soil is of less significance than the action of the fire on the organic matter in the soil. A very intense burn will destroy all the organic matter at the soil surface. Some destructive distillations will be effected at levels within the soil column. At temperatures of 200–300°C, 85% of the organic substances will be destroyed. Nitrogen compounds are subject to volatilization and will be released and lost by distillation at temperatures of 100–200°C. Other nutrients, such as calcium, sodium, and magnesium, will be released and deposited on the soil even at moderate temperatures and rapidly recycled into the biota (DeBano et al. 1977). One consequence of the loss of organic matter by burning is the concomitant loss of the high cation exchange capacity that characterizes organic matter, thus removing a holding substrate for nutrients.

Organic matter decreases bulk density (the weight of the soil solids per unit volume of total soil) of the soil, because it is lighter than a corresponding volume of mineral soil, and because it gives increased aggregate stability to soil. The heat of an intense fire may break down these aggregates, leading to a loss of soil structure and to a lowered rate of water infiltration. It has been shown that, in prescribed burning in the southern Sierra Nevada, the soil carbon is not lost during burning, nor is the bulk density changed significantly. The carbon content is inversely correlated with the bulk density of the soil: The higher the carbon, the lower the bulk density. However, a ground fire, in which almost all of the organic matter is destroyed, would change both carbon content and bulk density (Agee 1973). There is a reduction in the thickness of the humus layer due to compaction after a fire. This is measurable, and especially evident, at the base of trees (Viro 1974).

Chemical Considerations

In general, soil pH will be higher after a fire than before. The degree of change is based on the nature of the cations released by burning and on the acidity of the soil before the fire. Forest soils often have a low pH (3–6), due to the high acidity of the litter. Thus, the magnitude of change when soluble basic cations are released by burning will be greater in forest soils than in chaparral soils (DeBano et al. 1977). The activity of free-living, nitrogen-fixing bacteria is enhanced, both by the addition of nutrients by fire and by the higher pH due to release of mineral bases in the soluble ash.

Physical Effects of Fire on Soil

Soil Acts as an Effective Insulator   Even the ash layer that results from the combustion of the litter can insulate the soil. Heat energy may be transferred downward by conduction, convection, and vapor flux (DeBano et al. 1977). The rate of this transfer is affected most by the levels of soil moisture, although there are many other factors involved. Temperatures will not rise above 100°C in a given layer until all the water has been evaporated; this evaporation requires a great deal of energy, slowing the transfer of heat downward. Some typical temperatures in chaparral, in intense, moderate, and light burns, are shown in Figure 16-2. Note the dramatic attenuation, even in intense burns, as little as 2.5 cm below the soil surface. Even under a hot fire, temperatures do not exceed 200°C 2.5 cm down, and are below 100°C under a light burn 2.5 cm down. In other words, a light surface fire would heat only the top fraction of a cm of mineral soil to near the boiling point.

In Redwood Mountain Grove, Kings Canyon National Park, temperature determinations were made by use of Tempilaq© in burn piles (Hartesveldt and Harvey 1967). Tempilaq© consists of strips of a special paint that fuses at a predetermined temperature. The recorded temperatures are presented in Figure 16-3. Notice that, even near the center of large burn piles, the temperature is less than 66°C at 25 cm depth and less than 177°C at 12.7 cm depth. Grassland fires tend to be cooler than forest fires, with soil surface temperatures in grassland surface fires reaching temperatures of 120°C, and the air temperatures rising to

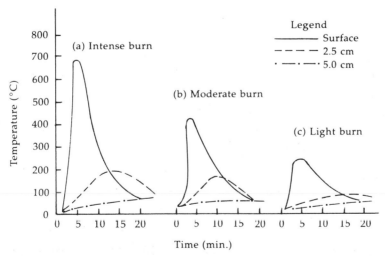

Figure 16-2   Typical temperatures at the surface and downward in the soil during (a) an intense, (b) a moderate, and (c) a light burn in chaparral. (From DeBano et al. 1977. Fire's effect on physical and chemical properties of chaparral soils. In *Proc. Symp. Env. Cons. Fire and Fuel Mangmt. in Medit. Ecosyst.*, pp. 65–74.)

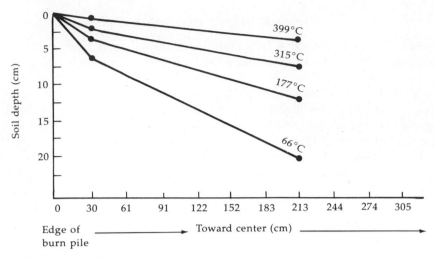

Figure 16-3   Temperatures recorded in a pile burn in a giant sequoia forest. Note the insulating value of the soil, even in the areas that approach the center of the burn. (From Hartesveldt and Harvey 1967. The fire ecology of sequoia regeneration. In *Proc. Tall Timbers Fire Ecol. Conf.*, pp. 65–78. By permission of the Tall Timbers Research Station, Tallahassee, FL.)

600–700°C. Air temperatures in forest fires may exceed 1100°C. In contrast to the Redwood Mountain Grove results, soil temperatures in grassland fires may rise only 10–15°C above ambient conditions in soils only 1–5 cm down, and for only short periods of time (Daubenmire 1968*b*). The nature and availability of fuel probably account for observed differences.

Soil Moisture and Fuel Moisture   Soil moisture levels are affected by fire, but not in a simple fashion. In certain communities, living moss and humus can absorb large amounts of water, effectively preventing rainwater from reaching tree roots. The removal of that layer by fire should allow infiltration of more water. However, the layer also prevents evaporation, so that water holding capacity is lower and evaporation is higher on burned plots than on unburned ones (Viro 1974). Burned sites also have higher temperatures than unburned sites, due to the increased absorption of sunlight by blackened soil surfaces. This is especially true in grasslands. The higher temperature initially increases evaporation, but quick drying breaks capillary connections and thus reduces evaporation.

The influence of fire on water runoff and infiltration is temperature dependent. Following extremely hot fires that completely volatilize hydrophobic substances and thus increase wettability (see next section), there may be greater infiltration of water into the soil. However, after less intense fires, runoff is increased, due to lowered infiltration. Increase in the sediment yield is usually observed too. This may correlate with, for instance, the loss of soil structure just mentioned, increased nonwettability, or both of these. Increased water runoff and sediment yields are significantly correlated with the decrease in the forest floor litter and duff observed following a forest fire. Although heavy surface fuels and aerial fuels (attached branches, epiphytes, etc.) are often unaffected, the weight, depth, and water holding capacity of fine surface fuels are reduced by fire (Agee 1973).

Fuel moisture is estimated by means of **fuel sticks**. These are wooden rods that weigh exactly 100 g when oven-dry. After exposure to ambient conditions, they are reweighed in the field, and the assumption is made that additional weight (over the original weight, expressed as a percentage) is an approximation of the fuel moisture levels in available fuels at the site.

Soil Nonwettability   In certain communities, hydrophobic substances prevent the movement of water into the soil. This condition is called **nonwettability** and is characteristic of ponderosa pine *(Pinus ponderosa)*/incense cedar *(Calocedrus decurrens)* forest, white fir *(Abies concolor)*/giant sequoia *(Sequioadendron giganteum)* forest, and the chaparral community of California. Hydrophobic substances that reside in decomposing plant parts accumulate on the soil surface and in the upper part of the soil column in the years between burns. Temperature gradients, established during fires (Figure 16-3) may alter the translocation of these substances into the lower soil layers. The nonwettable layer is thus moved lower in the soil column. The depth of the layer, and thus the depth of water infiltration, is determined by fire intensity, fuel availability, and physical condition of the soil (DeBano et al. 1977). Water can move into the soil above the nonwettable layer and cause sheet erosion. This phenomenon is sometimes seen in chaparral of southern California following a fire and subsequent rainfall.

Effects of Fire on Soil Biota

Ahlgren (1974) pointed out that the ecology of soil organisms following a fire is not well understood. Even so, numerous studies have been made of soil microflora and fauna in reference to fire (Dunn and DeBano 1977; Ahlgren 1974; Parmeter 1977; Wright and Tarrant 1957; Wright and Bailey 1982). The effect of fire is usually to reduce fungi but to increase populations of soil bacteria and actinomycetes. Any changes in the microbiota are most obvious in the upper soil levels.

One may infer that pathogens will be destroyed by fire, if these pathogens sporulate on litter and if that litter is consumed by fire. Damping-off fungi, seedling root rot fungi, and organisms that decay seeds will be removed from thoroughly heated seedbeds. The brown spot disease pathogen *(Septoria acicola)* of longleaf pine *(Pinus palustris)* is controlled by burning. Thus, the success of seedlings in a burned-over area may be attributable in part to the removal of soil pathogens or their substrates (Parmeter 1977).

There are, however, a number of early post-fire species of ascomycetes, mostly **pyrophilous** (fire-loving) discomycetes such as *Pyronema*. The succession of fungi on burned lands has been studied and is thought to be analogous to succession of higher plants on disturbed lands (Ahlgren 1974).

Nitrogen losses due to volatilization of nitrates, ammonia, and amino acids (as well as through alterations in forms of both organic and inorganic nitrogen), may be recovered, both on grasslands and in the chaparral, by early dominance of the post-burn community by herbaceous legumes. Examples of these legumes are such nodulated plants as *Trifolium, Lotus,* and *Lupinus,* which dominate post-burn regions in southern California. Nitrifying bacteria, such as *Nitrosomonas* and *Nitrobacter,* are very sensitive to fire. Their populations are completely destroyed at 140°C and evidence great mortality even at 100°C. Following a burn, the recovery of such nitrifying bacteria is slow (Dunn and DeBano 1977).

Forests of the Southeastern United States

The region of the United States that stretches from the Appalachian Mountains to the Atlantic is very old geologically. When the northern parts of the continent were glaciated, this southeastern portion became a refuge for many species of plants and animals. Fire has been a major factor in the evolution of many communities in this area.

The complex patterns of weather, influenced by topography and by proximity to the ocean, have created a banded mosaic of communities in the southeastern United States. The frequency, size, and intensity of lightning fires vary accordingly. Komarek (1968) has divided the Southeast into two provisional lightning fire bioclimatic regions. These are defined by the variations in fire weather and other meteorological phenomena, and on the basis of certain prominent forest types. The two regions are the eastern deciduous forest (the region of the Appalachian Mountains characterized by deciduous hardwood forests, located from the Appalachian highlands westward) and the southern pine forest. Space does not permit the close examination of all the communities of these regions, but we will look at one in more detail: the longleaf pine community.

The Southern Pine Forest Lightning Fire Bioclimatic Region

This bioclimatic region includes two biogeographic regions: the coastal plain forest and the coastal plain pine savannas and prairies. In the former are found loblolly, pond, sand, slash, and longleaf pine forests *(Pinus taeda, P. serotina, P. clausa, P. elliotii,* and *P. palustris).* In the latter are found pine-bluestem *(Andropogon* spp.*),* pine-wiregrass *(Aristida stricta, Sporobolus curtissi),* pine-cane savannas, the "Red Hills" savanna, and the coastal marshes and prairies (see Komarek 1974 for more details).

The longleaf pine–bluestem savanna has a grass and forb understory that is actually relict prairie. The open, parklike forestscapes of pre-European settlement were, and are, maintained by fire. Though there are occasional evergreen broad-leaved hardwood trees and shrubs, the grassland understory (little bluestem, *Andropogon scoparius,* and slender bluestem, *A. tener*) is a community rich in herbaceous plants, including some 200 species of grasses.

Figure 16-4 The grass stage of longleaf pine. The long, slender needles protect the perennating bud against fire.

Longleaf pine *(Pinus palustris)* and the role of fire in its ecology has been very well studied. Following germination, most pines grow rather uniformly. Longleaf pines reach a height of several centimeters in just a few weeks, but then upward growth ceases. The young seedling sprouts a pompon of long, slender, and drooping needles, which surround the stem apex (Figure 16-4). This "grass stage" endures for three to seven years, and during that time the plant allocates most of its energy to the production of a sturdy root system, with abundant stored food.

The pines of the Southeast, with few exceptions, require mineral soil as seedbeds, and most of the seedlings are shade intolerant. In addition, longleaf pine is very susceptible to a fungal disease called *brown spot disease.* The causal agent, brown spot fungus *(Septoria acicola),* matures during the summer months. Multitudes of spores are released to the autumn rains, and they are splashed on low-growing plants, including seedlings of various kinds. Control of the fungus is gained by burning the infected needles before new growth resumes the next spring. The dry needles are burned, as are the aboveground portions of competing hardwoods and herbaceous plants. But, the "grass stage" of longleaf pine protects the meristematic regions from heat damage, and the young seedling lives.

If the young tree can survive a full year before fire comes, it is then very tolerant of fire for another one to six years. The longleaf is the most fire tolerant of the pines that Biswell (1958) studied [longleaf and slash in Georgia and ponderosa and sugar pine in California *(Pinus taeda, P. elliotii, P. ponderosa, P. lamber tiana)*]. There is a sensitive period, once the seedling resumes upward growth (terminating the grass stage), until it is about 2 m tall. The mature pines possess the heavy, protective bark necessary to prevent damage to sensitive growing tissue by fire; in fact, even young saplings are rather well insulated with corky bark.

Fire, then, functions in several ways to promote longleaf pine forests. It clears the mineral soil for seed germination and seedling growth; it reduces competition from hardwoods and herbaceous plants; it eliminates brown spot blight; and, we infer, it opens the forest canopy to allow growth of these shade intolerant young trees.

Prescribed Burning

We have referred to prescription burning in previous sections of this chapter. A formal definition is given by Biswell (1977) as follows:

> Prescribed burning is the skillful application of fire to natural fuels under conditions of weather, fuel moisture, soil moisture, etc., that will allow confinement of the fire to a pre-determined area and at the same time will produce the intensity of heat and rate of spread required to accomplish certain planned benefits to one or more objectives of silviculture, wildlife management, grazing, hazard reduction, etc.

In forestry, the main purpose of prescribed burning is to enhance the reproduction and growth of certain trees, by manipulating the major environmental factors that influence reproduction and growth: light, temperature, water, and mineral nutrients. Foresters may desire a change in the relative abundance of species in a forested region, such that desirable species are favored and competing species eliminated or at least put at a disadvantage. Facilitation of the reproduction of valued species and the provision of optimum conditions for growth are the aims of prescribed burning in the forest.

Beginning in the 1920s, burning was used for range improvement and for upland game management. Ultimately, it was applied to forestry. By the 1940s, prescription burning in the Southeast was regularly applied as a management tool in longleaf pine (*Pinus palustris*) and slash pine (*P. elliotii*) forests. By the late 1950s, fire was also used as a silvicultural tool for loblolly pine (*Pinus taeda*) and the shortleaf pines.

For prescribed burning to be successful, there must be enough fuel to carry a fire and, ideally, the fuel should be a "material that burns readily and uniformly soon after a rain" (Biswell 1958). This allows a surface fire; heavier fuels that might contribute to a crown fire would still be too wet to burn. In Georgia, grasses—pineland threeawn (*Aristida stricta*), Curtiss dropseed (*Sporobolus curtissi*), and bluestems (*Andropogon* spp. and *Schizachyrium* spp.)—and some pine needles carry the fire. The loose, well-aerated, high surface-to-volume ratio fuels will dry quickly enough to burn in one to two days after a rain.

Grassland Fire

Both grassland climates and grassland vegetation favor fire. These arid and semiarid regions are usually subject to prolonged drought. The periods of lowest moisture correspond temporally with those of the highest temperatures. The prevailing growth form of grassland species is hemicryptophyte (see Figure

1-3). The prairie of North America, which begins south of 55°N latitude and continues to south of 30°N latitude is a prime example of grassland. Both grasses and forbs are persistent under the stress of recurring fires. Shrub tops, however, are kept in juvenile condition by annual burning. Summer and early fall lightning fires favor grasses over woody species (Higgins 1984).

Causes of Fire in Grasslands

Fire requires ample fuel, abundant oxygen, and an ignition source. Before the time of overgrazing, fences, roads and the like, there were miles of continuous, flammable vegetation on the prairie, which allowed fires to sweep across, virtually unimpeded by structures or firebreaks of any kind. A series of thunderstorms can travel great distances, igniting fires in its path. Under primeval conditions, these fires could simply spread until they burned out, either because the fuel was gone or because the weather changed (Komarek 1965). Fires caused by lightning are primarily summer fires and probably constitute the most frequent type of fires in the grasslands.

Higgins (1984) studied lightning-caused fires in the Northern Great Plains regions. Of 293 fires, 73% occurred during July and August; most of the rest were recorded in April, May, June, and September. That period encompasses the growing period (frost-free days) for the region. Although most fires were suppressed and burned only a few hectares, there were fires in each month from May to August that burned greater than 40 ha.

Early and modern humans also cause fires, and, along with lightning, have the capacity for a rhythmic periodicity in the fires they cause. The intentional burning of grasslands, probably the first land management tool, would have been rapidly rewarded. After a few days of good regrowth, practically any burned grassland will act as a powerful attractant for animal life. Actually, deliberate burning serves a second function, perhaps more meaningful to us today, by reducing the danger of wildfire. Wildfire may spread quickly and destructively over large areas that have been protected from fire for a long period (Lemon 1968; Wright and Bailey 1982).

Interaction of Fire and Vegetation

There are benefits that accrue to those grassland species that survive fire. Silica content of the grasses increases, which enhances resistance to decay and thereby provides more fuel for the next fire. The more immediate advantage is the increased protection afforded the seeds that remain above the hottest part of the fire on fire-resistant seed stalks, thus ensuring survival of the seeds (Vogl 1974). Komarek (1965) pointed out that there are many species of fire-adapted grasses that have very hard stems with a high silica content among the genera *Aristida, Andropogon, Imperata,* and *Setaria.* The high silica content served to select mammals that evolved ever-growing teeth with ridges that were sharpened by use. These mammals eventually dominated the grassland community.

The corollary to the attribute of decay-resistant foliage is the immobilization of nutrients in plant tissue. Frequent fire is the agent of soil nutrient renewal, releasing the nutrients from plant tissue and recycling them into new biomass (Mutch 1970).

Anderson (1965), writing about Kansas prairie forbs, discussed the benefits of fire for the bluestem *(Andropogon* sp.) range. The indirect benefits of a late spring fire are (a) it prevents early annuals from reseeding; (b) it removes the green tops of weedy perennials; and (c) it thus reduces competition. There are also direct benefits: scarification* of certain resistant seeds, so that germination is possible, and the possibility that certain insect species, like certain plant species, may respond differentially to range burning.

Mechanisms for Survival

After a fire, one of the following usually occurs: (a) the plant resumes life from a protected crown comprising perennating buds at the soil surface, or (b) there is increased reproduction from seeds, rhizomes and stolons, all of which may aid in the survival of grassland plants.

Hemicryptophytes   Most of the plants on the prairie are **hemicryptophytes**. This means that, at least once a year, the aboveground portion of the plant dies back. Prairies are windy places, and the winds tend to dry the aerial portions of the plants. In addition, the plants' low life form allows sunlight penetration, which hastens drying. Thus, when the vegetation is dormant, there is often abundant flammable material at the ground level. Ideal conditions for a fire exist: dry, uncompacted, fine fuels, with plenty of available oxygen.

These fine fuels will be ashed quickly, and high temperatures do not build at ground level. Rather, the flames pass quickly through vegetation, the highest temperatures occur at the top of the flames, and the heat is dissipated by the prairie wind. Soil surface temperatures in grassland fires reach only about 120°C, and in soil 1–5 cm deep, there is a temperature increase of only 10–15°C above ambient soil temperatures. There is evidence that 60°C is the lethal temperature for the shoots of most land plants. However, leaf meristems (perennating buds) of many grasses are 40 mm or so below the soil surface. Pineland threeawn *(Aristida stricta)* and Curtis dropseed *(Sporobolus curtissi)* have such buried meristems, which are additionally protected by closely packed, persistent leaf sheaths that do not burn readily. The living portion of the plant is virtually unharmed by the fire and will resume growth in the favorable time of the year in a nutrient regime actually enhanced by fire (Daubenmire 1968*b*). The decay-resistant aboveground plant parts contain minerals that are recycled by fire.

Scarification is used here in the sense of any agent that ruptures the integrity of the seed coat, leaving it permeable to water and oxygen.

Table 16-1 Heat tolerances of seeds of various grasses. (From Daubenmire 1968*b*. Ecology of fire in grasslands. *Advances in Ecological Research* 5:209–266. Copyright by Academic Press Inc. (London) LTD.)

Taxon	Common name	Maximum temperature with survival	Duration of heating
Avena fatua	Wild oat	104–116°C	5 min
Bromus mollis	Soft chess	116–127°C	5 min
Bromus rigidus	Ripgut	93–104°C	5 min
Stipa pulchra	Needle grass	116 127°C	5 min
Bromus mollis	Soft chess	>200°C	2 min

Increased Reproduction from Seeds Enhanced reproduction may occur as a direct response to fire: Certain grassland species have resistant seeds that require scarification before germination will occur. Legumes, such as species of *Astragalus* and *Trifolium*, produce seeds that require scarification before seed dormancy will be broken. Bermuda grass *(Cynodon dactylon)* exhibits enhanced reproduction from seed following a fire because of increased seed set. However, many grasses do not possess fire-resistant seeds. Table 16-1 summarizes heat tolerances of seeds of various grass species. Note that the survival times are brief, but tolerances range well into the temperatures attained by surface fires at the soil surface in grasslands (Daubenmire 1968*b*).

Management

There is evidence of increasing dominance of woody plants in some grasslands. This may be an invasion or may simply reflect changes in the density of shrubs already present. The change may be attributed to, for instance, the absence of hot fires, increased water availability (developed for livestock use), reduced fuel availability, and reduced herbaceous competition for shrub seedlings. The latter two factors are due to grazing and overgrazing (Vogl 1972; Box 1967). In California, some 3,000,000 ha of woodland-grassland is used for grazing. Since 1948, ranchers have prescribed burning an average of 40,000 ha yr^{-1} to destroy invading woody plants and to enhance the growth of palatable grasses (Biswell 1963). For a comprehensive review of the role of fire in the management of rangelands and forest lands, see Wright and Bailey (1982).

Six hundred thousand ha of National Forest land have been judged suitable for conversion to grassland through manipulation, whether by fire, mulching, chipping, or burying, and some 380,000 ha have been judged suitable for conversion to timberland. The conversion to grassland is being attempted in a series of steps, including preparation for prescription burning, seeding, and, finally, control of brush sprouts and seedlings (Doman 1967).

Mediterranean Climatic Regions

The mediterranean climatic region is found between 30° and 45° north and south latitudes, on the west coast of continents, and in the Mediterranean Sea region. The mediterranean climate is defined by four characteristics: (a) moderate precipitation concentrated in the cool winter months, and summer drought; (b) in summer, long periods of sunny days and cloudless skies; (c) marine-moderated atmosphere influence year-round; and (d) warm to hot summers and mild winters (McCutchan 1977; see also Chapter 8, pp. 177 and 179).

Within that regime, certain areas are occupied by **sclerophyllous** (hard-leaved) plants and by thin, stony soils with little organic matter; these areas also have less than 750 mm of precipitation annually. In California, such sites are occupied by a dense scrub called chaparral (originally a Basque word for scrub oak, *chabarro,* spelled by the Spanish *chaparro,* and used by them to designate the mediterranean type scrub of California) (Figure 16-5). In corresponding regions elsewhere are the *matorral* of Chile, the *fynbos* of the Cape of Good Hope, the *phrygana* of Greece, the *garrigue* or *garigue* of dry, calcareous soils of France, known as *macchia* or *macquis* on siliceous soils, and the *tomillares* of Spain (Biswell 1974). In each area, the physiognomy of the vegetation is very similar, as are the plant adaptations to fire. Adaptive traits of plants in mediterranean regions will be considered. We will first discuss the chaparral of California and then consider some higher-elevation coniferous forests.

Fire-Adapted Plants

Exactly what is meant by the phrase *fire-adapted plants*? Gill (1977) listed the adaptive traits for mediterranean regions as follows: (a) fire-induced flowering; (b) bud protection and sprouting subsequent to fire; (c) in-soil seed storage and fire-stimulated germination; and (d) on-the-plant seed storage and fire-stimulated dispersal. Other authors (Lotan 1974; Harvey et al. 1980) have commented on the adaptiveness of such traits as thick bark, evanescent branches, rapid growth, and early maturity.

Fire-Induced Flowering   The phenomenon of increased flowering, and concomitantly enhanced seed-set, would appear to be advantageous in two ways: (a) those plants that do produce increased numbers of seeds would be at a competitive advantage over plants not so adapted; and (b) the seedbed, which receives more light due to canopy removal and has a deposit of soluble mineral ash nutrients, provides an additional opportunity for enhanced growth and reproduction.

Several of the bulb-forming perennials, which may have been present but not numerous in the chaparral between burns, exhibit enhanced flowering following a fire. These include death camas (*Zygadenus fremontii*), soap plant (*Chlorogalum pomeridianum),* purple-head brodiaea (*Dichelostemma pulchella),* and several species of mariposa lily *(Calochortus* spp.) (Muller et al. 1968). In other

Figure 16-5   (a) Typical chamise chaparral community. (b) The shrubs are dense, compact, and finely branched.

(a)

(b)

mediterranean regions, geophytes also exhibit this trait. Bermuda grass *(Cynodon dactylon)* and fireweed *(Epilobium angustifolium)* are known to exhibit enhanced flowering in response to fire.

The seeds of grasses do not require scarification for germination, and the annual grasses will usually not predominate in the first postburn year but appear only around rocks and in other protected regions. However, due to enhanced seed set, by the third year grasses will be very abundant, even dominant (Biswell 1974; Table 16-2). There are examples of fire-enhanced flowering outside the mediterranean region. In Florida, late spring and early summer fires induce vigorous flowering response in several species of grasses (Abrahamson 1984).

Table 16-2   Density of certain herbs that appeared 1–4 years after a burn of chaparral in the Highland Springs area of California. Density is given as number of plants per 28 m². (From J. R. Sweeney, 1956. University of California Publications in Botany 28:143–206. By permission of the University of California Press, Berkeley, CA.)

Category	Species	Year 1	Year 2	Year 3	Year 4
Species that peak the first year	*Mimulus bolanderi*	23	3	7	3
	Oenothera micrantha	19	3	4	2
	Antirrhinum vexillocalyculatum	15	4	6	0
	Emmenanthe penduliflora	280	0	3	3
	Silene antirrhina	25	5	12	10
	Malacothrix floccifera	47	8	14	23
	Mentzelia dispersa	31	11	19	14
	Mimulus rattanii	16	0	0	0
Species that peak the second or third year	*Barbarea americana*	9	21	34	17
	Cryptantha torreyana	19	62	844	127
	Gilia capitata	14	7000	0	3
	Lotus humistratus	132	738	33	17
Species that peak the fourth year or later	*Bromus mollis*	3	43	217	498
	Bromus rubens	19	23	147	1499
	Festuca megalura	20	24	417	737
	Chlorogalum pomeridianum	4	5	3	3
Species that remain constant	*Penstemon heterophyllus*	1	1	1	1

Bud Protection and Subsequent Resprouting   Buds near the base of the shrub may survive a fire and be released from dormancy by the death of shoots above them. Sprouting results as a hormonal response or with the onset of winter rains (Figure 16-6). The sprouting may be from the stem (epicormic) or may be from a basal burl or **lignotuber**. Protection for the stem bud is provided by the bark. Buds of lignotubers are protected by the soil.

In the genus *Arctostaphylos,* there are several examples of species that possess basal burls. The ubiquitous chamise *(Adenostoma fasciculatum)* also possesses a well-developed lignotuber. There are many other species that sprout following a fire, as detailed by Wright and Bailey (1982).

In-Soil Seed Storage and Fire-Stimulated Germination   There are some species whose seeds remain dormant in the soil for many years if there is no fire. For example, it is believed that the seeds of snowbrush *(Ceanothus velutinus)* might remain viable in forest litter for up to 575 years (Zavitkovski and Newton 1968).

Figure 16-6 Stump sprouting. Latent axillary buds are released from inhibition when fire removes the apical meristem.

Various authors discuss the use of heat treatment, or some other form of seed cracking, to induce germination of resistant seeds (Biswell 1974; Sweeney 1956 and 1967; Quick 1959; Sampson 1944). Chamise *(Adenostoma fasciculatum)* responds to fire with phenomenal germination of stored seeds; after a burn, as many as 3000 chamise seedlings m^{-2} may be counted.

    Recently, there has been an increasing interest in the role of charred or ashed wood in the germination of chaparral herbs and suffrutescents. Keeley et al. (1985) examined germination behavior in a variety of chaparral species and concluded that the presence of suffrutescents in burned-over areas is due to seed germination stimulated by powdered wood or heat treatment or both. Annuals in chaparral regions often have polymorphic seeds, with some exhibiting normal germination behavior but a significant portion requiring either heat treatment or exposure to charred wood or both. There is little germination in "fire annuals" without some environmental cues relating to fire.

    At Redwood Mountain, in Sequoia National Park, prescribed burning on three plots yielded the data shown in Table 16-3. The plot designated as Burn #1 was the lightest fire, and Burn #3 was the hottest. The greater numbers of shrub seedlings are in areas of lighter burns, which provide heat for cracking but do not kill the seeds. Hartesveldt and Harvey (1967) suggest several hypotheses for higher rates of sequoia seedling survival in areas of very hot burns, including the destruction of pathogens by the fire and the removal of competition. A control plot did not show a single shrub or sequoia seedling (Kilgore and Biswell 1971).

    Plants with very resistant seeds will often appear in great numbers following a fire (Sweeney 1956; Armstrong 1977). Examples of these fire followers are shown in Table 16-2. Notice the decline of many of the herbs in the post-burn years.

Table 16-3   Sequoia and deerbrush seedling response to 1969 burning at Redwood Mountain in Sequoia National Park. (Recalculated from Kilgore and Biswell 1971. Reprinted by permission of *Califonia Agriculture Magazine*, University of California.)

Plot	Size (hectares)	Mature sequoia*		Sequoia seedlings		Deerbrush seedlings	
		No. per plot	No. per hectare	No. per transect	No. per hectare	No. per transect	No. per hectare
Burn #1	1.52	11	7.2	138	18,567	120	16,145
Burn #2	2.47	28	11.3	337	45,343	53	7,131
Burn #3	2.53	58	22.9	737	99,163	4	539
Totals	6.52	97		1,212		177	
Means			13.8		54,357		7,938
Control	2.14	31	14.5	0	0	0	0

*Trees more than 183 cm dbh.

Many of these same plants will germinate from seed if a light fire burns over a seedbed, mimicking a fire in nature. Otherwise, they lie dormant until some scarifying agent provides the vehicle for resumption of activity.

Thus, chaparral shrubs exhibit two responses to fire: (1) seeders die and come back as seedlings, and (2) sprouters remain alive and come back as sprouts from burls or root crowns. Long fire-free periods favor seeders. The resistant seeds, produced in greater or lesser amounts each year, are then stored in the soil. They remain dormant in the soil until scarification (fire is the most efficient method) or exposure to charred wood induces germination in a nearly optimal environment. During the periods between fires, seedlings grow to maturity within 5–8 years (Biswell 1974) and produce abundant seed, and the stage is set for another cycle.

This division is not a strict dichotomy, for some sprouters are also seeders, though some species are seeders only. There are also plants that sometimes sprout and sometimes do not, including bitterbrush (*Purshia tridentata*) and mesquite (*Prosopis velutina*). There are also degrees of sprouting response, governed in part by the time or season of burning (Humphrey 1974). Sprouters have distinct advantages. Those with lignotubers quickly resume activity following a burn. Even before seasonal rains, chamise (*Adenostoma fasciculatum*) and manzanita (*Arctostaphylos* spp.) will send up sprouts several cm tall by drawing from stored nutrients and water in the basal burl. During the first post-burn season, greenleaf manzanita (*A. patula*) will send up luxuriant growth, exploiting mineral and moisture resources via an already established root system. Those chaparral plants with both resistant seeds and a basal burl are dominant in certain fire-maintained plant communities.

On-the-Plant Seed Storage and Fire-Stimulated Dispersal   There are coniferous plants that retain viable seeds for many years or even decades. In the condition called **serotiny** (literally, late to open), the cones of some conifers remain on the tree and remain closed until mechanical breakage severs the vascular connection with the parent plant or remain closed by resin until the heat of a fire opens them. Serotiny is characteristic of many different plants, such as jack pine *(Pinus banksiana)*, some subspecies of lodgepole pine *(P. contorta)*, bishop pine *(P. muricata)*, knobcone pine *(P. attenuata)*, pitch pine *(P. rigida)*, Monterey cypress *(Cupressus macrocarpa)*, and giant sequoia *(Sequoiadendron giganteum)*. Some subspecies of lodgepole pine require temperatures of 45–50°C (or even higher) to break the resinous bond between the cone scales. Summer soil temperatures may suffice to open low-hanging cones. Knobcone pine, however, requires 200°C for opening the cones. By contrast, giant sequoia cones remain closed, due to cone scale turgidity, and open in response to the severance of vascular connections (Lotan 1974; Vogl 1973; Harvey et al. 1980).

Serotiny is rarely exhibited by every tree in a stand. However, in regions frequently disrupted by fire, a high percentage of serotinous-coned trees would be expected. In an even-aged stand of lodgepole pine of recent fire origin (which was contiguous with an uneven-aged, undisturbed stand), 58% of the trees bore serotinous cones. The uneven-aged stand showed only 38% serotiny. Frequent fire must play a very dramatic role in selection. The genetics of serotiny is not well understood, but we can infer that a stand of trees established from fire-opened, serotinous cones would show a high degree of serotiny (Lotan 1974).

Populations of lodgepole pine in the northern Rocky Mountains contain individuals that bear serotinous cones and others that bear nonserotinous cones. There are no significant differences in these subpopulations in height, basal area, basal area growth rates, or crown ratio. The nonserotinous cones may, however, have more seeds per cone than do the serotinous cones. This difference apparently results from the greater investment by trees bearing serotinous cones in protection of the enclosed seeds (Muir and Lotan 1985).

The Chaparral Community

Mutch (1970) hypothesized that communities maintained by fire possess vegetation that is more flammable than communities not adapted for fire. The woody plants of the chaparral exhibit such flammability and are characterized as follows: intricate branching pattern, resulting in a large surface-to-volume ratio; periodic die-back, resulting in highly flammable twigs; and possession of volatile oils (see Philpot 1977 for more details).

In the mediterranean region, centuries of manipulation—deliberate burning and overgrazing—have modified the natural situation greatly. Certainly, the chaparral of California has also been influenced by human manipulation. We have already noted the California Indians' practice of burning browsing areas in early summer to prevent destructive late summer wildfires. There were drastic changes made by miners, sheepmen, and loggers in some areas of the state. The differences between the mediterranean region and California are probably of degree, not of kind. Both areas also have a long history of volcanism, which would provide fires, but not on a periodic basis.

The Post-Burn Chaparral Community   In the first few years after a fire in a chaparral community, there will be a noteworthy succession of plants, dominated in the first few years by herbaceous plants. Often, many of these are annuals not seen except in the post-burn community—fire followers. However, it should be noted that seedlings and sprouts of woody plants will be present, even the first season following the fire. This situation has given rise to the term **autosuccession** (see Chapter 10), a phenomenon studied by many authors (Sweeney 1967; Biswell 1974; Philpot 1977; Keeley 1977; Keeley et al. 1985). By the sixth year following fire, the shrub canopy is once again closed, and few herbs appear below it.

Why do the annuals disappear? Several causes have been hypothesized. Many of the annuals have seeds that require scarification, which is provided most effectively by fire. In addition, shrubs, present from the early post-burn time, overtop the low-growing herbaceous plants, reducing the light intensity at the soil level. Photosynthesis is reduced in the lowered light regime, and the annuals do not survive. Also, it is thought that many of the shrubs are producers of allelopathic chemicals that inhibit the growth of other plants (see Chapter 6).

Giant Sequoia Forests

The forests of the Sierra Nevada, especially those that are giant sequoia forest, were characterized primarily by surface fire prior to the time of fire suppression. The fire ecology of giant sequoia (*Sequoiadendron giganteum*) has been studied extensively (Biswell 1977; Hartesveldt 1964; Kilgore 1972; Agee 1973; Hartesveldt and Harvey 1967; Harvey et al. 1980). These studies indicate that fire is essential to the continued vigor of the sequoia–mixed conifer forest, and that it plays these primary roles:

1. Reduction of fuel, decreasing wildfire hazard
2. Recycling of nutrients
3. Destruction of pathogens such as damping-off fungi
4. Maintenance and development of a mosaic of vegetation age classes and types
5. Enhancement for wildlife
6. Modification of insect life in the forest
7. Preparation of a suitable seedbed on mineral soil in a favorable light regime
8. Maintenance of a subclimax community
9. Opening of serotinous cones.

Although fire suppression was incorporated into National Park policy as early as 1886, there has been a reintroduction of fire as let-burn and prescription burns in many National Parks, including Sequoia and Kings Canyon National Parks.

The giant sequoia groves are not pure stands of the big trees, but associations of conifers that include white fir *(Abies concolor)*, incense cedar *(Calocedrus decurrens)*, and sugar pine *(Pinus lambertiana)*. These occur at mid-elevation on the western slopes of the Sierra Nevada of California.

The giant sequoia is a fire-adapted species and possesses characteristics that enhance its ability to survive fires. The mature trees are well adapted to resist surface fire damage. The thick, fibrous bark protects the living tree. This protection is enhanced by the possession of **evanescent** lower branches. As they are shaded by the forest canopy, they drop off and do not remain as a "fire ladder," which would otherwise give fires access to the tree's crown. Rapid growth is another fire-adapted feature of the big trees. The young saplings grow quickly up into the light, outgrowing the competition and raising the canopy above the level of surface fires.

Many, if not most, of the mature trees bear fire scars, and many have scars of many fires. It has been shown that between 1760 and 1900 fires in the Mariposa Grove of Yosemite National Park averaged one every seven to eight years. In fact, up to 85–95% of the tree can be burned without resulting in the tree's death (Hartesveldt 1964). A giant sequoia stands near the entrance station of Sequoia and Kings Canyon National Parks showing evidence of heavy burning but no indication of loss of vigor (Figure 16-7).

Figure 16-7   Giant sequoia at the entrance to Sequoia and Kings Canyon National Parks. The fire scar covers more than one-third of the base of the tree and extends nearly 30 m up the trunk, yet the tree is vigorous and sturdy.

Figure 16-8   A closed cone of the giant sequoia. This cone was cut by a chickaree, in Giant Forest, Sequoia National Park. The fleshy scales are eaten by the rodent, and seeds may be dispersed in the process.

Figure 16-9   Comparison of a white fir seed (left) and a giant sequoia seed (right). White fir is shade tolerant, and the greater amount of stored nutrients in the seed allows the root of the seedling greater growth than would be allowed in the giant sequoia seed.

Cone production in giant sequoias is phenomenal. At any one time, a mature tree might have 40,000 cones, 25,000 of which would be green, photosynthetic, and closed (serotinous) (Figure 16-8). However, the remaining 15,000 cones would be open and with few seeds. Cone production begins at a surprisingly early age (around 8 years); and the cones mature within 2 years. The seeds remain viable for as long as 22 years. However, the seeds are tiny; it would take over 200,200 to weigh a kilogram. Food reserves are sufficient to allow for germination, but roots cannot penetrate a thick layer of litter to reach soil for water and minerals. Notice the comparison of white fir seeds and giant sequoia seeds (Figure 16-9). The greater amount of stored food in the former should allow for greater root elongation, through forest litter, to reach mineral soil.

Though the reproduction of giant sequoia is fire-dependent, there is an almost continual, if slight, shedding of seeds due to the activities of various animals. These seeds will germinate, given a suitably moist medium. There are seedbeds available from time to time: root pits of fallen trees, sandbars in streams, snow avalanche paths, and areas cleared by human activity. Thus, at least in theory, there would always be several age classes of giant sequoia. However, in the upper Mariposa Grove, Hartesveldt (1964) counted no more than thirty sequoias that survived past the seedling stage between 1934 and 1964.

Seedling mortality is high in the first year, often more than 86% (Hartesveldt 1964; Harvey et al. 1980). Desiccation is the primary cause of death. Heat canker, damping-off fungi, insect damage, and falling debris can also cause the seedling to die. The combination of environmental disturbance and enrichment that seems to provide optimum conditions for seedling survival is derived from periodic fire in the forest: penetration of sunlight to mineral soil; reduction of competition; reduction of pathogens; and the opening of serotinous cones, allowing as many as 200 seeds per cone to fall to the receptive soil below.

The large size and the longevity of the giant sequoia trees enable them to dominate a forest and to linger on as relict individuals long after the climax forest (shade-tolerant white fir-incense cedar) has begun to dominate the understory. These climax species are shorter and denser than giant sequoia, and they may serve as fire ladders to carry a fire to the crown of a giant sequoia. Fire suppression favors the climax species. Frequent fires remove them, both from competition and as a fire hazard, and favor the maintenance of giant sequoia.

Summary

Fire has undoubtedly occurred at frequent enough intervals in most communities to be a major force in the development of plant and animal communities. Forest, grasslands, chaparral, and other communities of the mediterranean regions show the influence of fire. Of all the causes of fire, only lightning and human activity have the capacity for periodicity and therefore exert a consistent effect upon the evolution of organisms.

Humans have used fire to manipulate the environment to their own advantage since before history began. Native peoples in various lands have used fire to enhance hunting, improve visibility, provide forage, etc.

Fires may be divided into three classes: surface fire, which is rather low in intensity and is not destructive of overstory vegetation; crown fire, which is a spectacular blaze that consumes whole trees, opening the canopy completely; and ground fire, which is termed a *retrogressive agent* because of the catastrophic effect it has on a community.

The effects of fire on the soil include some loss of soil nutrients due to volatilization, changes in water-holding capacity by reducing organic content, and impact on soil microbiota. Fire behavior is determined by character of fuel, weather patterns, humidity, and temperature. Fire destroys some organic matter, vaporizes some, and ashes the rest. In addition, soil structure is changed by fire. Soil pH will usually be higher following a burn. Soil insulates against the heat of a fire. Even a few centimeters down, the temperature does not exceed 100°C under a light burn.

Water runoff is increased by burning, as is sediment yield. Soil wettability is increased by hot fires. Fungal diseases may be controlled by burning. Bacterial populations are often favored over most fungi. For example, nitrogen losses due to volatilization are counterbalanced by the enhancement of the regime for root-nodule bacteria in certain plants.

Lightning fires have been a common occurrence for millenia in the southeastern United States, and some coniferous forests there depend on fire for maintenance. The longleaf pine is most highly fire resistant but is also fire dependent for its maintenance. Fire is used in the Southeast in many pine forests and pine savannas as a silvicultural tool. In general, the effects of fire are to reduce competition, to prepare a mineral seedbed, to reduce fungal disease, and to open the forest canopy.

The grasslands are maintained by fire. The adaptations of grassland plants, most of which are hemicryptophytes, allow them to survive and even thrive in the presence of periodic fire. Fire recycles nutrients and reduces shrubs to the juvenile stage. The fire environment, including lightning and heat, provides a very powerful selective force in the grassland. Mechanisms for survival in grasslands include increased reproduction from seeds as well as resprouting from perennating buds protected at the soil's surface.

The mediterranean regions support vegetation that is inherently flammable and usually well-adapted to fire. The adaptations include: fire-induced flowering; bud protection and sprouting subsequent to fire; in-soil seed storage and fire-stimulated germination; and on-the-plant seed storage and fire-stimulated dispersal.

The giant sequoia of the Sierra Nevada provides an excellent example of adaptation to fire. With its serotinous cones, small seeds, fire-resistant bark, and long life, this impressive tree actually requires periodic fire or it will be succeeded ecologically by a white fir–incense cedar forest.

CHAPTER 17

SOIL

T he soil is the common ground between the living and the nonliving world. We could have logically begun this book with a foundational chapter on soils. The importance of the medium for growth cannot be overstated. However, plant ecology is much like any other science in that we could begin almost anywhere and work through a cyclical progression of information with effectiveness. We can now use the knowledge gained from other chapters to support our discussion of soils as they influence plant distribution. We will consider soil development, profiles, texture, and structure. Following a brief discussion on soil physics and chemistry, we will consider the taxonomy of soils.

There are four separate components of soil: mineral grains, organic matter, water, and air. The mineral substrate provides anchorage, pore space for storage of water and air, and nutrients on an exchange basis. Organic matter refers to the plant and animal residues in various stages of decomposition, as well as to the cells, tissues, and exudates of soil organisms. Organic matter enhances the intrasystem cycling of nutrients and improves soil structure, pore space, and water storage, among other things. Soil water is the solvent medium for nutrients needed by growing plants, maintaining equilibria among the cations and anions adsorbed* on soil particles, the living plants, and the soil water (soil solution) itself. Soil air contains oxygen for cellular work, carbon dioxide that augments further mineral weathering and biotropic processes, and atmospheric nitrogen for nitrogen-fixing soil organisms.

Soil, along with climate, is a primary agent in plant selection through evolutionary change. The term *soil* comes from the Latin word for floor, *solum.* A soil scientist, or **pedologist,** views soil as "a natural product formed from weathered rock by the action of climate and living organisms" (Thompson and Troeh 1973). The foregoing statement implies that soil is mainly composed of mineral materials, and this is true, for the most part. However, pedologists do not over-

---

*Adsorbed refers to a material being attracted and held to, whereas absorbed refers to a material being taken in and probably held more tightly than material adsorbed.

Figure 17-1 A vertical section of soil, showing relative positions of bedrock, regolith (unconsolidated material), and soil (highly weathered portion of regolith).

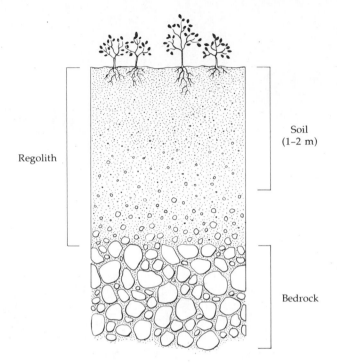

Regolith

Soil
(1–2 m)

Bedrock

look the fact that some soils are formed mainly from the weathering of accumulating organic debris (organic soils). We will use an edaphological definition of soil, considering it primarily as a medium for growing plants: Soil is a mixture of mineral and organic materials, capable of supporting plant life. Implicit in both views is the interaction of soil and life. Each one exists because of, and is formed at least in part by, the other.

The materials in the upper fraction of the earth's crust are exposed to weathering agents more intensely than are the underlying strata. Three levels can be seen in the section of an exposed hillside diagrammed in Figure 17-1, which has loose material resting on underlying rocks. The foundation is termed **bedrock**. Above that is the **regolith**, the unconsolidated mantle of weathered rock and soil material, of which the upper 1–2 meters or less of highly weathered, biochemically altered material is termed **soil**.

The Soil Cycle and Soil Development

The process of soil development may seem to be linear and formational in terms of human time, but in geologic time the process is part of a cycle. The processes that take place in the cycle of soil development include sedimentation or deposition of fragments from clay size up, lithification (stone formation), metamorphism and melting, crystallization, volcanism, and the agents of erosion and

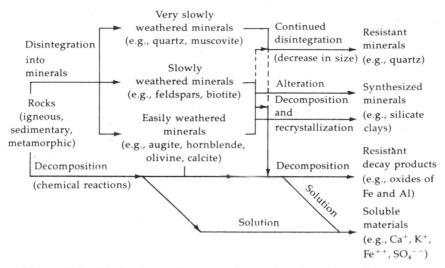

Figure 17-2 Weathering pathways under moderately acid conditions common in humid temperate regions. Major paths of weathering are indicated by the heavier arrows, minor pathways by broken lines. As one would expect, climate modifies the exact relationships. In arid regions, physical breakdown (disintegration) would dominate, and soluble ions would not be lost in large quantities. In humid regions, decomposition becomes more important, especially under tropical conditions. (Reprinted with permission of Macmillan Publishing Co., Inc. from *The Nature and Properties of Soils,* 8th Ed. by Nyle C. Brady. Copyright 1974 by Macmillan Publishing Co., Inc.)

transport. The process of forming soil may be conveniently divided into two phases. The first is rock weathering, and the second is soil formation. Whether the substrate is a granitic dome, a lava deposit, or a limestone cliff, there are certain weathering phenomena that are common to all rock substrates, and we will examine a few of them. Weathering pathways are shown in Figure 17-2. Notice that a variety of materials are produced, but that common pathways yield the same minerals, whether derived from sedimentary, igneous, or metamorphic rocks.

Mechanical Factors

Mechanical erosion, or disintegration, may be attributed to the expansion and contraction of rocks in response to thermal flux. The outer layers are heated differentially compared to the inner rock, creating stress that is relieved by cracking and spreading. The fissures and crevices thus formed, along with those caused by tectonic activity and gravitational tensions, will be filled with water. Expansion caused by the freezing of this water is impressive; normal hydrostatic pressure amounts to more than 10,000 kg cm^{-2}. Ice floats on liquid water, and the water at the air/liquid interface is cooled more rapidly than that within the rock, so that the ice acts as a wedge, forcing still wider cracks into the rock (Figure

Figure 17-3 Cracks in granite caused by the force exerted by the freezing of water, which seeped into minute hairline fractures.

17-3). The stresses generated by the freezing of water are also significant after initial fragmentation of rocks. The regolith, including the soil at the top, continues to be altered physically by this force.

Exfoliation, the peeling off of layers of rock, is attributed to the differential expansion just mentioned and also to the expansion allowed when surface material is removed by erosion. Igneous intrusions that crystallize far beneath the surface are, in a sense, spring-loaded. The removal of the pressure of overlying material allows exfoliation, so layers of rock are released much as the modified leaves peel away from an onion bulb. Half Dome, in Yosemite National Park in California, is an example of exfoliation.

Erosion by wind and water and deposition by these same forces are significant to rock weathering. The force exerted by plant roots must also be counted as a very important physical force, even though plants exert biological and chemical influences as well and also influence subsequent soil formation. Moving ice (or glaciers) transports tremendous amounts of material. In addition, the abrasive materials carried by the ice polish and further reduce the rocks and rock fragments. Alluvial (stream-carried), glacial, aeolian (wind-carried), and lacustrine (lake-deposited) materials form the basis for parent materials of many kinds of soils. Streams carry an enormous amount of material into lakes or into rivers that ultimately empty into the sea. The Mississippi River is estimated to discharge in excess of 660,000 kg of soil into the Gulf of Mexico each year. The size of alluvium may range up to house-sized boulders (Figure 17-4), depending on the volume and velocity of the water. By contrast, aeolian materials are fine, usually dust-like particles that abrade, polish, and further erode rocks in their passing.

Figure 17-4 Alluvium forming a fan originating at the mouth of a desert canyon. The sparseness of the vegetation allows a clear view of the sizes of material moved; some of the "particles" are huge boulders. (Courtesy of Dr. John S. Shelton.)

Chemical Factors

Chemical processes of weathering (decomposition) are active almost as soon as disintegration has begun. Included under chemical processes are a number of rather simple reactions. Oxidation, especially of iron-carrying rocks, is a very obvious form of chemical weathering. Carbonation, produced by the action of a weak concentration of hydrogen ions in percolating waters and attack by other acids including organic acids, causes decomposition. Some of the organic acids may, however, initiate processes leading to recrystallization of clays and are thus synthetic in effect. The most significant chemical weathering is exerted by water, in one of three ways: solution, the solvent action of water upon the mineral substrate; hydrolysis, the cleavage of a mineral by water; and hydration, the rigid attachment of H^+ and OH^- ions to the material being decomposed. The products of some of these processes, such as oxidation and hydration, usually have a larger volume than the initial minerals, and, if not displaced, contribute to the mechanical forces involved in exfoliation. These processes are all influenced by climate and are very effective in tropical regions.

Soil Formation

Soil formation *(S)* is characterized by Jenny (1941) as dependent on a set of independent variables as follows: climate *(cl)*, organisms *(o)*, topography *(r)*, rock type or parent material *(p)*, and time *(t)*. Jenny's classic equation is as follows:

$$S = f^{(cl,\ o,\ r,\ p,\ t\ \ldots\ )} \tag{Equation 17-1}$$

He defines organisms as the biotic potential of a site, not just the existing organisms, which are indeed interdependent or dependent factors. The biotic potential includes consideration of soil organisms, litter, vegetation type, etc. The soil macrobiota and microbiota interact with the growth medium, altering its capacity to support life. The amount of litter deposited on the soil is determined by the productivity of the community, and the rate of decay or removal of that litter is determined by climate, by the nature of the litter, and by fire frequency. Also, the incorporation of litter-derived materials into the soil, the maintenance of soil porosity and permeability, and ultimately the soil fertility, all rely at least in part on the activities of soil organisms. Vegetation and climate determine what kind of soil is formed. Podzol soils develop beneath coniferous forests in cool, moist climates, and they have a prominent, ashy gray upper horizon. Chernozem ("black earth") soils develop under grasslands in temperate regions. As previously noted, a warm, moist environment hastens both disintegration and decomposition. The concepts relating to plant community succession developed in Chapter 11 also apply to soil succession.

Thus, parent materials interacting with soil biota and climate produce a certain soil with characteristic features. Another very important factor in soil formation is time—time for large mineral grains to be disintegrated, decomposed, and changed to secondary minerals; time for finer and finer mineral grains to accumulate; time for organics and humus to become a significant fraction of the total soil; but, also, time for finer grains and soluble anions and cations to be leached away (Chapter 13).

The composition of parent material largely determines the chemical make-up of its derived soils. For instance, acid soils derive from (a) aluminum-rich bauxites, (b) silica-rich substrates such as sands, slates, and diatomaceous earth, (c) mine tailings of lead and zinc deposits, (d) sulfide-rich, hydrothermically altered volcanic rocks and ash, and (e) estuarine deposits. Soils with poor fertility may develop from ultrabasic (ultramafic) rock types rich in silicates of iron and magnesium. For example, serpentine develops from igneous rocks such as peridotite and dunite, which contain silicates of iron and magnesium. Serpentine soils are high in magnesium and low in calcium (Table 17-1). They may also be high in nickel and chromium and deficient in nitrogen and phosphorous; thus, they are not very fertile.

Table 17-1   Comparison of certain soil nutrients in serpentine and nonserpentine soils.

Nutrient	Nonserpentine: Serpentine
Replaceable Ca (meq 100 g^{-1})	7.4
Replaceable Mg (meq 100 g^{-1})	0.5
Percent N (total, all forms)	14.0
Percent P (total, all forms)	65.0
K (meq 100 g^{-1})	3.8
Ni (meq 100 g^{-1})	0.1
Cr (meq 100 g^{-1})	0.01

Soil may form from transported material or *in situ* parent material (residual soil). Topography influences the stability of a soil body in a variety of ways, including its effect on rainfall patterns and the steepness of slopes. Resistant parent materials may result in steep slopes; parent rocks or materials that erode easily produce wide valleys, rounded hills, and often deep soils. Steep slopes may continually expose new surfaces to weathering and soil formation, as alluvium or talus (material moved by gravity) is moved downslope; resident soils, if present at all, are young and shallow. Gentle slopes restrict transport and often enhance development of deep, strongly developed soils.

Soil Profiles

Soil Profile Development

Soils, especially deep soils of temperate latitudes, develop very characteristic layers or horizons, more or less differentiated and discernible from each other if a pit or trench is dug into the soil. A section encompassing all the horizons of the soil and extending into its parent material is called a **soil profile** (Figure 17-5). The rock weathering that we discussed earlier is mainly a physical process.

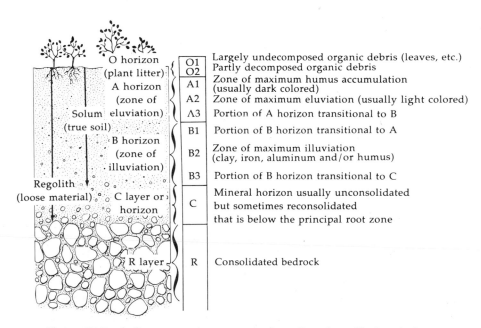

Figure 17-5   A diagrammatic representation of a soil profile (vertical section of the soil), showing soil horizon nomenclature and symbols for mineral soils. Additional subdivisions of the O horizon, not shown, are used in organic soil. A and B horizons taken together are referred to as the **solum** or true soil. (Reprinted by permission of McGraw-Hill Book Co. from *Soils and Soil Fertility* by Thompson and Troeh 1973).

Table 17-2   The more important original and secondary minerals found in soils. The original minerals are also found abundantly in igneous and metamorphic rocks. Secondary minerals are commonly found in sedimentary rocks. (Reprinted with permission of Macmillan Publishing Co., Inc. from *The Nature and Properties of Soils*, by Nyle C. Brady. Copyright 1974 by Macmillan Publishing Co., Inc.)

Name	Formula
Original minerals	
Quartz	SiO_2
Microcline ⎫ Orthoclase ⎭	$KAlSi_3O_8$
Na plagioclase	$NaAlSi_3O_8$
Ca plagioclase	$CaAl_2Si_2O_8$
Muscovite	$KAl_3Si_3O_{10}(OH)_2$
Biotite	$KAl(Mg, Fe)_3Si_3O_{10}(OH)_2$
Hornblende*	$Ca_2Al_2Mg_2Fe_3, Si_6O_{22}(OH)_2$
Augite*	$Ca_2(Al, Fe)_4(Mg, Fe)_4Si_6O_{24}$
Secondary minerals	
Calcite	$CaCO_3$
Dolomite	$CaMg(CO_3)_2$
Gypsum	$CaSO_4 \cdot 2H_2O$
Apatite	$Ca_5(PO_4)_3 \cdot (Cl, F)$
Limonite	$Fe_2O_3 \cdot 3H_2O$
Hematite	Fe_2O_3
Gibbsite	$Al_2O_3 \cdot 3H_2O$
Clay minerals	Al silicates

*These are approximate formulae only because these minerals are so variable in their composition.

Soil formation, by contrast, is largely biochemical weathering and the gradual mixing of organic material with inorganic material.

As the upper layers of soil are exposed to the roots of plants, to animal activities, to the activities of both soil microflora and microfauna, and to accelerated chemical weathering, mineral grains are altered. Primary or original minerals are transformed into **secondary minerals**. Some primary and secondary minerals are shown in Table 17-2. Recall from Figure 17-2 that quartz is a very resistant primary mineral. Primary minerals are found in abundance in igneous and metamorphic rocks, and secondary minerals are often found in sedimentary rocks, as we would expect.

The presence of secondary materials and abundant organics and changes due to interactions between soil and soil biota result in differentiation of soil horizons. Two of the processes are significant enough that they provide names for the principal (master) horizons of mineral soils in which they occur. The A horizon includes the surface soil in which surface or near-surface organic matter accumulates. It is also referred to as the **zone of eluviation,** due to the movement of materials downward from the A horizon, mostly by leaching. The B horizon includes the subsurface part of the soil, or subsoil, which may be a **zone of illuviation,** where the materials lost from A are accumulated. The illuvial materials may include iron and aluminum compounds, clays, and humus. Humus is the dark brown or blackish stable fraction left over after most of the organic residues have been decomposed.* Alternatively, the B horizon may be a **zone of alteration,** where little material accumulates from A but significant changes in structure or secondary mineral formation take place.

Soil horizon nomenclature is shown in Figure 17-5. The O (organic) horizon and its subdivisions may be largely or totally lacking in certain desert soils but may be the only horizons existing in organic soils. The R layer underlies many, but not all, upland or residual soils, where it may displace the C horizon. A young soil may have A and C horizons, but no B horizon. This lack of a B horizon results when there has been no translocation or eluviation of clay from the A horizon downward; such a profile is characteristic of the chernozem or "black earth" profiles in steppe regions.

Soils can be identified in the field based on the differentiation of the horizons, the texture, the color of the principal materials, and even which horizons (or portions) are lacking. A field sheet for recording soil characteristics is shown in Figure 17-6, and Figure 17-7 is a photograph of the soil described in the field sheet. Note the site characteristics that are recorded: elevation, slope, aspect, erosion, and the like. This soil is an Inceptisol, identified in part by its lack of a high clay content in the B horizon. There is no C horizon, and the lower B horizon rests directly on the R horizon, or bedrock.

Zonal, Azonal, and Intrazonal Soils

Although we will discuss the classification of soils in detail in a later section, it is appropriate to mention here the three orders in the 1938 system of soil classification (Baldwin et al. 1938). These include zonal soils, whose characteristics and profiles are mostly influenced by climate and vegetation (e.g., desert soils, lateritic (red tropical) soils); intrazonal soils, which exhibit unique soil profiles due to unusual parent material or topography (e.g., saline soils, bog); and azonal soils, which have little or no profile development (e.g., recent alluvial soils, rapidly eroding soils). The appearance of a soil profile largely determines the fit of a soil into a soil classification scheme.

---

*The formation of humus is a complex topic that we will discuss in a later section.

FIELD SHEET FOR RECORDING SOIL CHARACTERISTICS

No. _8_

Soil Type _Henneke variant_
Location _9 km S of Coloma NE 1/4 SW 1/4 Sec 1 T 10 N, R 9 E   MDBM_
Geographical Landscape _Upland_
Elevation _326 m_   Slope _20%_   Aspect _S-SW_   Erosion _slight-moderate_
Groundwater _deep_   Drainage _mod. well_   Alkali _none_
Mode of Formation _Primary serpentinitic_   Parent Material _Serpentine_
Climate _MAT 14°C : Jan 4.5°C, Jul 23°C.   MAP ≈ 89cm   FFD 223 @ 0°C_
Natural Cover _Chamise, Digger Pine, Ceanothus_   Soil Region _VIII_
Profile Group _VII - II_   Higher Categories _loamy-skeletal, serpentinitic,_
Genetically Related Soil Series _Del Piedra (an alfisol);_   _thermic, Typic Xerochrept_
Dukabella (higher, timbered with Jeffrey pine)

PROFILE SKETCH	COLOR	TEXTURE	STRUC-TURE	CONSISTENCE	REAC-TION	MISC: Roots, Pores, Clay films, Concretions
A 11 0-13 cm a s	7.5YR 4/4	g l	1 f gr	so, vfr; so, po	7.3	3 vf, 3f, 2m, 2 co rts 3 vf, 3f, 2m, 1 co pores
A 12 13-20 cm a w	5YR 4/6	g l	2 m gr	so, vfr; so, po		2 vf, 2f, 1m, 1 co rts 2m, 1 co, 3 vf, 1f pores
B 1 20-28 cm c w	5YR 4/6	g l	1 c abk → 2 c gr	so, vfr; ss, ps		1 vf, 1f, 2m, 2 co rts 2 vf, 2f, 2 m pores
B 21 28-51 cm c w	5YR 3/4		1 m abk → 2 c gr	so, fr; ss, ps		1 f, 1m, 1 co rts 2 fi, t; 2 vf i pores
B 22 51-68 cm c i	5YR 4/4		1 m sbk → 2 c gr	sh, fr; ss, ps		1f, 1m, 2 co rts 2f, 2 vf pores
R					↓	

Natural Land Division _E 8 - 3 m_
Soil Rating (Storie index) _11_   Soil Grade _5 (6)_

Figure 17-6 Field sheet for recording soil characteristics. a s = abrupt smooth; a w = abrupt wavy; c w = clear wavy; c i = clear irregular; 7.5 YR = Yellow-Red hue and 4/4 = value and chroma (in Munsell notation); g l = gravelly loam; f = fine; m = medium; c = coarse; gr = granular; abk = angular blocky; sbk = subangular blocky; so = soft; ss = slightly sticky; sh = slightly hard; vfr = very friable; po = non-plastic; ps = slightly plastic; fr = friable; vf = very fine; co = coarse; rts = roots; t = tubular pores; i = interstitial pores. For structure, 1 = weak and 2 = moderate. For roots and pores, 1 = few, 2 = common, and 3 = many.

Figure 17-7 Soil profile of Henneke variant. The soil is classified as loamy-skeletal, serpentinitic, thermic, Typic Xerochrept, an Inceptisol derived from serpentinitic materials.

Physical Properties of Soil

Soil Texture

Soil texture is defined by the percent by weight of sand, silt, and clay. That is, only mineral components are considered in determining soil texture, even though soil is also composed of organic matter, water, and air, with organic materials contributing up to 5% or more of the total. The United States Department of Agriculture classification of soil particles places the upper limit for silt at 0.05 mm in diameter and the lower limit at 0.002 mm. Particles greater than 0.05 mm in diameter are considered sand, and particles less than 0.002 mm in diameter are considered clay (Brady 1974). Although almost any soil contains pebbles and cobbles that are larger than the coarsest sand (2 mm diameter), such coarse fragments play but a minor role in soil function unless the bulk of the soil is composed of these fragments. In the field, a good estimate of texture can be made by dampening and handling a small amount of the soil. The particles that feel grainy and gritty are sand; those that are silky, like moist talcum powder, are silt. If the sample is sticky and if a self-supporting, flexible ribbon can be extruded between the thumb and finger, there is a high percentage of clay. If it is not sticky or only slightly sticky, and no ribbon can be extruded, there is a low percentage of clay. Different soil textures can be recognized by various combinations of the foregoing.

In general, the structural or skeletal support of the soil is provided by the sand fraction, or at least by the largest soil particles, and the finer particles help

Figure 17-8 An idealized group of clay particles representing the **card house** effect in soil structure, in which positive charges on edge positions are attracted by negative charges on the broad surfaces. The negative charges provide exchange sites for cations and are thus a storage facility. (Reprinted by permission of McGraw-Hill Book Co. from *Soils and Soil Fertility* by Thompson and Troeh 1973.)

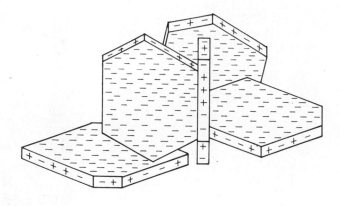

to store nutrients and to bind particles together into aggregates. Size classes are assigned arbitrarily, recognizing that minerals (and the building blocks of minerals) form a continuum of sizes, ranging from single tetrahedra* to large and very complex mineral crystals. Material that is without identifiable structure in the fine clay fraction is called **allophane.**

If mineral grains are not spherical, then their diameter is taken to be an average of their maximum and minimum dimensions. Sand grains tend to range from blocky and irregular to subround in shape, and silt particles are similar but smaller. Clay particles are very unlike sand grains; their plate-like shape is reminiscent of mica (Figure 17-8).

A more precise determination of soil texture can be made in the lab. Particles larger than 2 mm are sieved out of a given sample, which is then treated to oxidize and remove the organic matter. This is usually done with hydrogen peroxide, which also has the effect of rupturing the binding of clays by organic matter. The sample is then dried and weighed. The remaining materials are carefully flushed and shaken through a series or nest of sieves that allow particles 1 mm, 0.5 mm, 0.25 mm, 0.1 mm, and 0.05 mm to pass through. The total weight of sand and its distribution can be determined by this process. The fraction passing through the sieves contains silt and clay together, and these can be separated by using a hydrometer or pipette method of analysis.[†] Such methods, which are based on the rate of particle settling, allow the separation of the silt and clay fractions.

Once the percentage of weight of separate components has been determined, the use of a soil texture triangle (Figure 17-9) facilitates textural class determination. Assume, for instance, that our soil contains 30% clay, 40% silt, and 30% sand. This soil would be called *clay loam*. Make note of two features: (a) The designation is an area, not just a point on the triangle. (b) There are four

---

*A *tetrahedron* is a three-dimensional figure with four triangular sides and an oxygen ion at each of the four points.

†For a more complete discussion, consult Thompson and Troeh 1973.

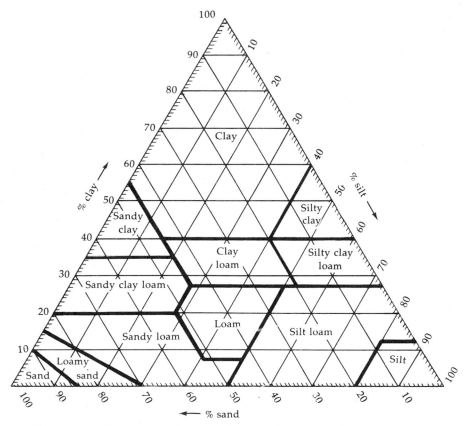

Figure 17-9 Relationship between the particle size distribution of a soil and its class name. The points corresponding to the percentages of silt and clay present in the soil are located on the silt and clay lines, respectively. Lines are then projected inward from % silt parallel to the clay side of the triangle and inward from % clay parallel to the sand side. Note that the third particle size (in this case, sand) is defined by the two projected lines. The name of the compartment in which the lines intersect is the texture class of the soil in question. (Reprinted by permission of McGraw-Hill Book Co. from *Soils and Soil Fertility* by Thompson and Troeh 1973.)

terms commonly applied to soil texture: clay, silt, sand, and loam. *Loam* designates a mixture of sand, silt, and clay that exhibits the properties of each fraction about equally. Thus, our sample exhibits the clay fraction more strongly than does loam, but is still a loamy soil, hence *clay loam*. Table 17-3 lists the general terms used by the United States Department of Agriculture to describe soil texture.

To understand the function of these various soil fractions, we must comprehend two more physical properties of soil: porosity and permeability. *Porosity* refers to the total space that is not occupied by soil particles. *Permeability* deals with the ease of passage through soil of gases, liquids, and plant roots. In gen-

Table 17-3 General terms used by the U.S. Department of Agriculture to describe soil texture in relation to the basic soil textural class names. (Reprinted with permission of Macmillan Publishing Co., Inc. from *The Nature and Properties of Soils*, by Nyle C. Brady. Copyright 1974 by Macmillan Publishing Co., Inc.)

General terms		Basic soil textural class names
Common names	Texture	
Sandy soils	Coarse	Sand Loamy sand
Loamy soils	Moderately coarse	Sandy loam Fine sandy loam
	Medium	Very fine sandy loam Loam Silt loam Silt
	Moderately fine	Clay loam Sandy clay loam Silty clay loam
Clayey soils	Fine	Sandy clay Silty clay Clay

eral, finer-textured soils have greater porosity, with a pore space of perhaps 40–60% of the total, and sandy soils have lesser porosity, with perhaps 35–50% pore space.

Water moves into and drains out of a sandy soil with much greater ease than for a finer-textured soil, which means that sandy soils are more permeable than fine-textured soils. This is because water has strong adhesion and cohesion capacity. Water molecules form a film **(hygroscopic water)** about each clay particle and easily bridge the gap between particles; the cohesion between water molecules is sufficient to maintain the bridge. Such soils are nutritionally rich but can be a problem agriculturally, due to low permeability and poor aeration. Hygroscopic water also forms a tight film about sand grains, but the surface-to-volume ratio is very low, there are fewer grains than in clays, and the large pore size does not allow efficient cohesion of water molecules, which therefore drain away under the influence of gravity. (See Chapter 18 for a full discussion of water potential and soil moisture.)

Clay particles and bits of organic matter within the soil are referred to as **micelles**. Micelles are negatively charged on their surfaces. The capacity of micelles to provide the primary soil nutrient storage results from (a) the negative charges on the micelles and (b) their very large surface-to-volume ratio. The hypothetical

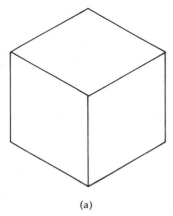
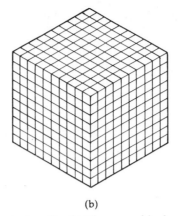

(a)   (b)

Figure 17-10   (a) A block with sides 1 mm in length. (b) The same block
sliced vertically and horizontally into blocks with sides 0.1 mm in length.
(Reprinted by permission of McGraw-Hill Book Co. from *Soils and Soil
Fertility* by Thompson and Troeh 1973.)

dicing of a mineral grain (Figure 17-10) allows us to envision the difference in
the surface-to-volume ratio of a sand grain with sides 1 mm in length and the
same volume of clay particles with sides 0.001 mm in length. The sand grain has
a volume of 1 mm^3 and a surface area of 6 mm^2. If we slice this into bits with
sides 0.1 mm in length, there will be 1000 bits, each one with 6×10^{-2} mm^2
surface area, or a total of 60 mm^2, which is an increase of ten times more than
the surface area of the parent sand grain. Slice again, so that each soil particle
has sides 0.01 mm in length, and the surface area increases to 600 mm^2. Slice
again, so that clay particles have sides 0.001 in length, and surface area now
equals 6000 mm^2, which is 1000 times that of the parent grain.

In summary, the sand fraction provides skeletal (structural) support, aeration
and permeability for the soil. The clay fraction provides water holding and nutrient
storage capacities (see also the section on soil chemistry). Silt also provides water
holding capacity and, through weathering, provides plant nutrients and material
to form clays, to replace losses from leaching.

Soil Structure

Primary soil particles are arranged into secondary units, called **peds.*** The
nature of the arrangement or **aggregation** of the peds is called *soil structure*. Soil
aeration is greatly enhanced by good soil structure, as is the movement of water
and plant roots through the soil.

Aggregate stability, and the shapes and sizes of aggregates, are the most
important characteristics of soil structure. **Stability** means the capacity of the

---

*A ped is a unit of soil structure such as an aggregate, crumb, prism, block, or granule, formed by
natural process.

Table 17-4   Classification, description, and illustration of various types of soil structure. (Reprinted by permission of McGraw-Hill Book Co. from *Soils and Soil Fertility* by Thompson and Troeh 1973 (modified).)

Classification	Description	Illustration
Structureless		
Single grain	Each soil particle independent of all others.	
Massive	Entire soil mass clings together, no lines of weakness.	
Structured		
Granular	Primary soil particles grouped into roughly circular peds, such that there is space between them, as they do not fit tightly together. Enhances permeability.	Granular
Platy	Peds with horizontal dimensions greater than vertical. Common at soil surface and often associated with lateral movement of water.	Platy
Blocky	Peds approximately equal in vertical and horizontal dimensions, but fit well together, unlike granular peds. May be angular or subangular.	Angular blocky Subangular blocky
Prismatic	Peds taller than wide, common in B horizons of well-developed soils. May be columnar, due to eluviation (old age or high sodium content will produce columnar peds of prismatic peds).	Prismatic
Structure destroyed		
Puddled	Soils disturbed when wet, puddled or run together; structure is destroyed, pores collapse.	

peds to retain their structure—that is, to absorb water and not disintegrate. There are various agents that bind these aggregates. Fungal mycelia, microbial exudates, hydrogen and calcium ions, and certain clays promote stability; sodium ions and expansion and contraction of certain clays promote instability.

The size and shape of peds are peculiarly significant to the success of plants and other soil-dwelling organisms. Table 17-4 presents illustrations and descrip-

Figure 17-11 Soil profile of Blacklock sand, a podzol soil in the Pygmy Forest, Mendocino County, California. The horizon boundaries are wavy rather than smooth. Note the highly bleached A2 horizon, beneath which is a B2h horizon. Though not apparent here, it is darkly stained with humus. The B2 SIM is a silica cemented hardpan, underlain by the B2t claypan. The entire profile encompasses only about one meter in depth.

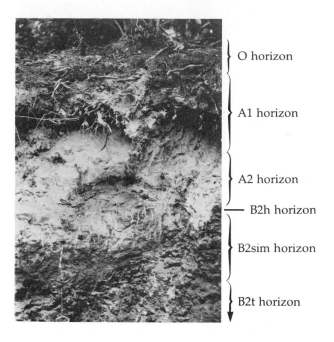

O horizon

A1 horizon

A2 horizon

B2h horizon

B2sim horizon

B2t horizon

tions of several common soil aggregates and the classification of soil structures that possess them. Notice that granular soils are particularly suitable for plant growth, because both water and air can circulate between the peds. The presence of good structure—granular peds, high degree of aggregate stability—is especially desirable in the topsoil. If the texture of the B horizon is sandy, then structure is of little consequence, because drainage is assured. However, clay soils in the B horizon can create a "claypan" or hardpan, which has a deleterious effect on plant life. An excellent example of this is the Blacklock podzol soil (Typic Sideraquod) of the Pygmy Forest in Mendocino County, California. Jenny et al. (1969) note with some amusement the lack of agreement among naturalists as to the cause and effects:

> naturalists indulged in an apparent *circulus vitiosus*. On field trips the professors of botany would tell their students that the podsol soil is the cause of the unusual assortment of plant species, whereas the visiting professors of pedology (soil science) would attribute—in the light of classical podsol theory—the soil horizon features to the acid-producing vegetation. While it is true that a species individual responds to its soil niche, it is also true that it modifies that niche, which, in turn, reacts upon the individual.

This Blacklock soil is one of the most acid soils known, with a pH of 2.8–3.9 and a characteristic bleached white surface horizon underlain by an iron-cemented hardpan (Figure 17-11). Leachate from the acid conifer litter cannot drain through the hardpan but continues to remove nutrients from the already depauperate surface soil.

Soil Chemistry

One of the most important aspects of soil chemistry is the state of and the relationship between chemical elements in soil systems, as soluble ions in the soil solution, as adsorbed ions on charged particles, and as constituents of mineral and organic particles. Soil chemistry deals primarily with reactive materials, which are the chemicals that are of significance to living things. Reactive materials tend toward equilibria in a changeable environment. There is virtually no correlation between the absolute and relative amounts of an element present in the soil and the significance of that element to life. Oxygen (49.5%), silicon (25.8%), and aluminum (7.5%) are the most abundant, the second most abundant, and the third most abundant elements, respectively, in the earth's crust. Yet, they are relatively inaccessible due to the bound nature of their chemical compounds. By contrast, the most important elements in soil chemistry (Table 17-5) make up less than 16% of the earth's crust materials. Soil pH, cation exchange capacity, and the nature of the interface between the solid phase and the liquid phase are all important aspects of soil chemistry.

Soil pH

Typical soils exhibit pH values that range from 4 to 8, although certain soils may have values higher or lower. For instance, saline soils commonly exhibit pH values ranging from 7.3 to 8.5. In southern Death Valley, California, saline soils support a saltbush scrub of desert holly (*Atriplex hymenelytra*), Parry saltbush (*A. parryi*), and honeysweet tidestromia (*Tidestromia oblongifolia*). The distinctly alkaline soils of the broad desert valleys of the Great Basin region of western North

Table 17-5 Elements important in soil chemistry and their chemical symbols and principal ions. (Reprinted by permission of McGraw-Hill Book Co. from *Soils and Soil Fertility* by Thompson and Troeh 1973.)

Element	Principal ions	Element	Principal ions
Aluminum	Al^{+++}	Manganese	Mn^{++}, MnO_4^-
Boron	$B_4O_7^{--}$	Molybdenum	MoO_4^{--}
Calcium	Ca^{++}	Nitrogen	NH_4^+, NO_2^-, NO_3^-
Carbon	CO_3^{--}, HCO_3^-	Oxygen	With other elements
Chlorine	Cl^-	Phosphorus	$H_2PO_4^-$, HPO_4^{--}
Cobalt	Co^{++}	Potassium	K^+
Copper	Cu^{++}	Sodium	Na^+
Hydrogen	H^+, OH^-	Sulfur	SO_4^{--}
Iron	Fe^{++}, Fe^{+++}	Zinc	Zn^{++}
Magnesium	Mg^{++}		

America support a shadscale scrub vegetation dominated by shadscale saltbush (*A. confertifolia*) and bud sagebrush (*Artemisia spinescens*). A pH of less than 6.6 is considered an acid soil, and vernal pools have pH values of perhaps 6.4–6.85 (a vernal pool is an ephemeral body of water present in spring). Very ancient soils, especially those covered by coniferous forest in humid regions, have very low pH values. The pygmy cypress (*Cupressus pygmaea*) grows in the Blacklock podzol soil of the Mendocino Forest, which we noted has an extremely acid soil (pH of 2.8–3.9). The highly weathered Blacklock podzol soils are half a million years old. They consist primarily of very resistant quartz grains and contain very small amounts of nutrients for plant growth. In general, leaching tends to remove basic cations and thus lower the pH of the soil. Strong podzols with a pH less than 4.0 develop in as little as 10,000 years in glacial materials.

The influence of plants on the soil pH is very complex. Nutrients (both cations and anions) are brought up from subsurface soil by plants and then deposited with litter on the surface. Grasses tend to use more bases, and grass cover may act to keep the soil pH from dropping. Coniferous litter, on the other hand, provides acid leachate, which effectively lowers the pH. Fewer cations are held on exchange sites, and more cations are released by acid weathering. Figure 17-12 shows the changing availability of plant nutrients in response to changing pH values.

Agricultural soils may develop a low pH due to fertilizer applications that add nitrogen and sulfur. Lime is often applied to acid soils to make them more basic. Very alkaline soils, on the other hand, may be treated with gypsum to lower pH. In most soils, the pH cannot be changed very much because of very effective soil pH buffering.

Although the direct effects of soil pH on plant growth are very limited, the indirect effects are numerous and significant. Toxicity of certain metals, such as aluminum and manganese (which are more soluble at lower pH), would be a direct effect of pH value. The most important indirect effect of soil pH is its influence on nutrient availability (Figure 17-12). Rates of weathering, availability of nutrients such as nitrogen, phosphorus, and sulfur, and the leachability of nutrients such as potassium are all influenced by soil pH. The availability of nutrients with low solubilities, such as phosphorus, is dramatically influenced by pH changes. Calcium phosphates are less soluble as the pH climbs, and iron and aluminum phosphates are less soluble as the pH drops. A near-neutral pH, between 6.5 and 7.5, is best for phosphorus availability—indeed for the availability of most nutrients needed for plant growth.

Humus Formation

The action of microorganisms on organic materials eventually converts them into humus. The organic materials become very finely divided and nearly black in color. By virtue of their tiny size, these particles can coat soil particles to the point that the soil appears black when it contains, for example, as little as 5% organic matter. Root exudates, earthworm excreta, and the like may provide a significant contribution to this nonliving, finely divided organic matter. However,

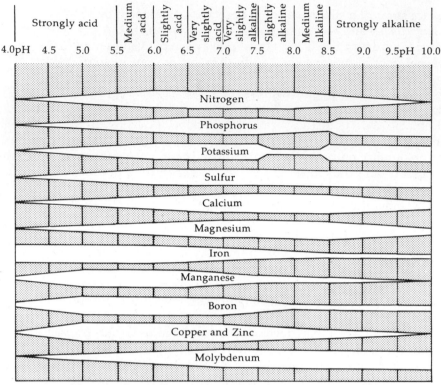

Figure 17-12 is a chart showing the relationship between pH and plant nutrient availability. The pH scale runs from 4.0 to 10.0 with categories: Strongly acid, Medium acid, Slightly acid, Very slightly acid, Very slightly alkaline, Slightly alkaline, Medium alkaline, Strongly alkaline. Bars are shown for: Nitrogen, Phosphorus, Potassium, Sulfur, Calcium, Magnesium, Iron, Manganese, Boron, Copper and Zinc, Molybdenum.

Maximum availability is indicated by the widest part of the bar

Figure 17-12   A schematic illustration of the relationship between pH and plant nutrient availability. (Reprinted by permission of McGraw-Hill Book Co. from *Soils and Soil Fertility* by Thompson and Troeh 1973.)

probably 90% of humus is attributable to microorganisms, microbial tissue, and microbial activities on litter.

Humus, an intermediate product of decomposition, may decompose at an annual turnover rate of less than 3% in temperate regions. It comprises not only the resistant fraction of the organic matter but also the microbes that have accomplished the decomposition of the more decomposable fraction. In a climax community or even a stable seral one, the amount of new humus formed each year may approximate that which is released by decomposition. The significance of slowly decomposing humus is that it provides a secure supply of continually available nutrients for plant growth.

Although the composition of humus varies, 10–15% of its weight is made up of easily identifiable organic compounds: polysaccharides such as cellulose and its decomposition products; polyphenols such as tannins; proteins and their decomposition products; and a host of hydrocarbons, organic acids, alcohols, esters, and aldehydes. The rest of the humus is not so easily identified but is characterized as humic materials, many of which contain reactive groups—carboxyl, amine, phenolic hydroxyl groups—combined with chains and rings, forming rather large and complex molecules.

Cation Exchange Capacity

Cation exchange capacity (CEC) is a measure of the number of negatively charged sites on soil particles that attract **exchangeable cations,** that is, positively charged ions that can be replaced by other such ions in the soil solution. Three factors strongly influence the cation exchange capacity of a soil: its clay content, the kinds of clay minerals or amorphous colloids (allophane) it contains, and the humus content. We have already commented on the significance of the surface-to-volume ratio to mineral storage. The negative charge, which provides for nutrient storage on clay mineral particles, is a result of, for instance, substitution of ions within clay mineral structures or ionization of hydrogen ions from OH groups, which also produces negatively charged surface sites.

Tropical Rainforest Soils

Let us consider the interaction of vegetation, soil texture, rainfall regime, temperature regime, and soil microorganisms, in reference to cation exchange in a particular ecosystem, the tropical rainforest. Soils of the tropical rainforest (TRF) are usually highly weathered and **laterized**—there is little or no horizon development in the reddish-brown loam, which is comprised of aluminum and ferric sesquioxides (Al_2O_3, Fe_2O_3). Lateritic soils (Oxisols) are very acidic and nutrient poor. The litter layer is extremely thin because mobilization of nutrients occurs at a very rapid pace in the warm, moist rainforest. Even wood is quickly recycled by subterranean termites.

The luxuriant growth of vegetation (52.5 t ha^{-1} yr^{-1} gross productivity) belies the extremely poor nutrient content of most TRF soils. Walter (1979) states that the entire nutrient reserve required by the forest is contained in the phytomass.

What mechanism allows for litter recycling, so that mobile nutrients are not leached away by the inevitable rains? Work by Went and Stark (1968) showed that feeding roots of TRF trees are directly connected to litter. These rootlets exploit the hyphae of mycorrhizal fungi to obtain nutrients. As noted in Chapter 7, these are mutualistic fungi, interacting with the roots and increasing their host's ability to extract nutrients from the forest litter. Recall that many fungi are saprobic, obtaining nutrients by absorption. Due to this highly adaptive relationship, there is little chance that free, mobile ions released from litter biomass will be lost through leaching.

As long as the rainforest is not disturbed, the lack of a soil storage compartment is unimportant. The nearly complete recycling of materials allows for stability for hundreds or even thousands of years. Yet deforestation through burning, clearing of the land for cultivation, or logging, removes the only significant storage facility, the phytomass, and quickly allows the mobilized ions to move out of the system. The luxuriant growth of the virgin forest is never (in human terms) renewed.

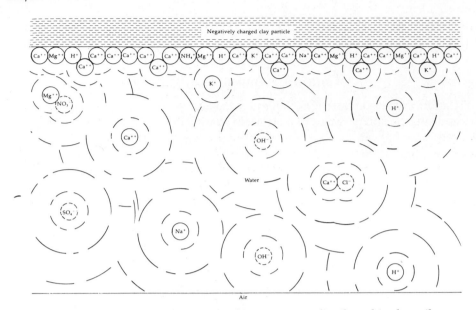

Figure 17-13   An illustration of how ions are distributed in the soil solution near a negatively charged clay particle. (Reprinted by permission of McGraw-Hill Book Co. from *Soils and Soil Fertility* by Thompson and Troeh 1973.)

Soil Solution

There is no way that soil water can be withdrawn and studied as a separate entity; it must be understood *in situ,* in relationship to its function as the liquid medium that renders the soil a habitable environment. The readily available ions are present in the soil solution and maintain an equilibrium with adsorbed ions. These, in turn, maintain an equilibrium with absorbed mineral ions (Figure 17-13). Plant removal of nutrients from the soil solution lowers the concentration of those nutrients in the rhizosphere; ions move toward the rhizosphere down a chemical gradient. This shift, due to solubility constants of the ions involved, enhances the release of adsorbed ions, and so forth.

For example, plant uptake of phosphorous—roots removing P from the soil solution—establishes a concentration gradient in the soil solution such that the movement of P is toward the root. Under optimum conditions, daily uptake of P in grassland may be as much as 50 times the quantity of P in the solution pool, the pool itself being replenished from the labile pool (that is, from adsorbed ions). When conditions of temperature and water are exactly right, the labile pool has the potential of replacing the solution pool 250 times daily. As we experience time, there is neither gain nor loss of phosphorus in grasslands, and the cycle itself is probably governed as much by biotic factors as by abiotic factors.

With continued uptake, the concentration of P in the cytoplasm of root cells may exceed concentrations by up to two orders of magnitude in ambient soil solution. It is well known that mycorrhizae greatly enhance uptake of P. This enhancement is attributed to the tremendously enlarged contact zone of root and soil, mediated through hyphal fungi. A generalized phosphorus cycle is depicted in Figure 13-3. Phosphorus participates in a sedimentary cycle, which means that local deficits may occur.

Soil Taxonomy

The naming of soils allows one to impart a great deal of information simply by invoking the name of a certain soil. A name is applied to an individual kind of soil that has a definable existence on the landscape. A unit of soil, large enough such that its profile and horizons can be studied and its properties defined, is called a **pedon.** This is the smallest practical volume that can be properly referred to as a soil. A pedon may range from 1 m^2 to 10 m^2 in area, depending on the continuity or cyclic discontinuity on the soil's horizons within short linear distances (1 to 7 m). Pedons of soils with continuous horizons have the smallest area (1 m^2). Groups of similar adjacent pedons, termed **polypedons,** form the basis for soil survey and soil classification.

For names to be used, a classification system must be scrupulously followed and names must be assigned on the basis of certain specified properties. Characteristics that are used to define a soil are termed **differentiating characteristics** or **differentiae.**

Soil scientists strive for a classification scheme that reflects a natural relatedness; the scheme of naming is called **soil taxonomy.** *Classification* is the broader term and may include schemes of ordering for specific practical purposes, whereas taxonomy is concerned with relatedness among groups. Nevertheless, any taxonomic system is an arbitrary creation of its author, reflecting certain biases and devised to fill a particular need.

History

The term *soil* is very old, and humans have been concerned with soil as a medium for plant growth for millenia. In the 1870s, a Russian school led by Dokuchaiev developed and introduced a revolutionary new concept of soil and a system of naming soils (Soil Survey Staff 1975).

> Soils were . . . independent natural bodies, each with a unique morphology resulting from a unique combination of climate, living matter, earthly parent materials, relief, and age of land form. The morphology . . . reflected the combined effects of the particular set of genetic factors responsible for its development.

The Russian soil taxonomy was adopted first in Europe, then in this country early in this century. As more soils were described and other information became available, the weaknesses of the Russian system became increasingly apparent. In 1938, a system was developed in the United States and published under the title "Classification of Soils on the Basis of Their Characteristics" (Baldwin et al.

1938). Then, starting in the 1950s, a taxonomic scheme was developed, modified, improved, and presented as a series of approximations, each one more refined and more workable than the last. Six of these approximations were tested by the Soil Survey Staff of the U.S. Department of Agriculture, various U.S. land grant universities, and interested European individuals and agencies. The seventh approximation was published and issued on a limited basis in 1960 at the Seventh International Congress of Soil Science at Madison, Wisconsin, for wider testing and evaluation (Soil Survey Staff 1960). It was generally referred to as the Seventh Approximation or "the brown book." In 1965, a modified form was officially adopted for soil classification in this country by the National Cooperative Soil Survey. With additional modification, it was finally published for unrestricted use as a comprehensive system in 1975 by the Soil Survey Staff of the U.S. Department of Agriculture as Agricultural Handbook 436, Soil Taxonomy—A Basic System of Soil Classification for Making and Interpreting Soil Surveys. It is this system that we will discuss here. (For current information regarding placement of various soils into this taxonomic scheme, see Classification of Soil Series, 1986 USDA Soil Conservation Service publication.)

To be useful, a soil taxonomy should have the following attributes.

1. The definition of each taxon should have the same meaning for each user.
2. The taxonomy should be a multicategoric system, with many taxa in the lower categories.
3. The taxa should refer to real soils, known to occupy geographic areas.
4. Differentiae should be soil properties that can be discerned in the field, inferred from other properties that are observable in the field, or taken from the combined data of soil science and other disciplines.
5. The taxonomy should be modifiable with a minimum of perturbation to the system.
6. The differentiae should maintain pristine soil and manipulated soil that is its equivalent in the same taxon whenever possible.
7. The taxonomy should provide taxa for all the soils of a landscape or at least be capable of such provision (Soil Survey Staff 1975).

Differentiating Properties of Soils

The properties of a polypedon should serve as differentiae. Soil color and soil horizons have long been used for differentiae. Properties of soil that interact with other properties should be carefully considered when differentiae are selected to define soil taxa.

There are six defined diagnostic surface horizons or **epipedons.** An epipedon is a horizon, or horizons, forming in the upper part of the soil profile. We will name these but not describe them, due to limitations of space. A subsurface horizon (often B, but perhaps part of A) is also of diagnostic value, and the names, descriptions, and genesis or mode of formation of these horizons are also important in soil classification. The diagnostic epipedons are mollic, anthropic, umbric, histic, plaggen, and ochric. The diagnostic subsurface horizons include argillic, agric, natric, spodic, placic, cambic, oxic, and duripan, among others.

Other properties used in defining a soil include an abrupt textural change, mineralogical composition, amorphous material dominating the exchange complex, microrelief, lithic contact, organic soil materials, particle-size classes, and so on.

The Structure of Soil Taxonomy

The USDA soil taxonomy system recognizes a hierarchy of categories. Each category includes a set of taxa. The series is the unit of classification. One of the strengths of the system lies in its capacity for internal modification, as new information is gathered, without disruption of the rest of the system. The hierarchy is as follows:

Categories	*Example of taxa within categories*
Order	Entisol
Suborder	Orthent
Great Group	Cryorthent
Subgroup	Typic Cryorthent
Family	Loamy-skeletal, carbonatic Typic Cryorthent
Series	Swift Creek

Most of the formative elements in the names of soil orders (Table 17-6) are derived from Latin or Greek words appropriate to that soil order, and the soil order name is formed with the suffix *-sol,* from the Latin *solum* for soil. The

Table 17-6 Formative elements in the names of soil orders. (From Soil Survey Staff 1975. Courtesy of the U.S. Dept. of Agriculture.)

Name of order	Formative element in name of order	Derivation of formative element	Pronunciation of formative element
Alfisol	Alf	Meaningless syllable	Ped*alf*er
Aridisol	Id	L. *aridus,* dry	Ar*id*
Entisol	Ent	Meaningless syllable	Re*cent*
Histosol	Ist	Gr. *histos,* tissue	*Hist*ology
Inceptisol	Ept	L. *inceptum,* beginning	Inc*ept*ion
Mollisol	Oll	L. *mollis,* soft	M*oll*ify
Oxisol	Ox	F. *oxide,* oxide	*Ox*ide
Spodosol	Od	Gr. *spodos,* wood ash	*Od*d
Ultisol	Ult	L. *ultimus,* last	*Ult*imate
Vertisol	Ert	L. *verto,* turn	Inv*ert*

formative elements *alf* and *ent* are not from Greek or Latin, but are meaningless syllables. Alfisols are variable, often found in continually wet, cold, or seasonally dry climates. Entisols are often young soils with little or no horizon development. The derivations of the other formative elements are listed in Table 17-6.

The names of suborders are a combination of the formative element of the parent order as the suffix with a prefix that suggests the diagnostic properties of the soil. For example, *alb* (from *albus*, white) with *oll* (from mollisol, soft) names a suborder of the Mollisol order with a white horizon—*Alboll*. Similarly, a great group name consists of the name of the suborder coupled with a prefix containing one or two formative elements to refer to definitive properties. For instance, *natralboll* indicates a Mollisol with an *albic* (white eluvial) horizon and a *natric* horizon (an illuvial subsoil with excess exchangeable sodium).

The great group name is combined with one or more modifiers to generate the subgroup name. *Typic* Natralboll would denote the subgroup thought to typify the great group. Families are named with polynomials consisting of the subgroup name modified by adjectives in a specified order that are names of classes describing the soil's particle size mineralogy, reaction, and temperatue classes, among other factors. Series names are place names, usually taken from the region where the soil was first described. In the field, profiles that do not match the descriptions of officially recognized series may be designated as **variants** of the existing, closely related series. With time and additional field information, a variant may be defined as an official soil series with its own unique name.

Summary

The components of soil are mineral grains, organic matter, water, and air. They provide, respectively, plant anchorage and nutrients, intrasystem cycling, solvent medium, and oxygen and nitrogen. The upper layers of soil weather and change in response to abiotic and biotic factors. The first phase of soil formation is rock weathering, and the second is biochemical weathering.

Soil formation relies on climate, organisms, topography, parent material, and time, as does soil profile development, which encompasses development of soil horizons. Soils can be classed as zonal, azonal, and intrazonal.

Soil texture refers to the content of the soil, by weight, of the sand, silt, and clay fractions. Skeletal support and permeability are provided by the largest particles; clay provides water and nutrient storage; silt aids in water storage and weathers to produce additional nutrients and clay.

Micelles (clay particles and bits of organic matter) provide the primary storage for nutrients in the soil. The nature of the arrangements of peds and aggregate stability are the most significant characteristics of soil structure.

Equilibrium among ions in the soil solution, adsorbed ions, and absorbed ions are concerns of soil chemistry. Most soils exhibit a pH range from about 4 to 8. Ancient soils may be very acid. Climate, vegetation, parent material, and relief determine soil pH, which in turn affects vegetation and rates of weathering.

Soil microorganisms convert organic matter to humus, a finely divided, nearly black fraction that decomposes at a stable equilibrium rate in climax communities. Cation exchange capacity is influenced by the presence and the kinds of clay minerals, allophane, and humus. Clays with expanding lattice provide more efficient cation exchange than do nonexpanding clays. Some tropical soils have little or no horizon development, and are ancient, nutrient poor soils with little exchange capacity. The phytomass associated with these soils provides the bulk of their nutrient storage.

Soils are classified by their differentiating characteristics. Soil taxonomy is a classification scheme designed to reflect a natural relatedness among soil groups. To be effective, a taxonomy must have definitions for each taxon that have the same meaning for each user. The system currently used in the United States was developed by the Soil Survey Staff of the U.S. Department of Agriculture, Soil Conservation Service. It relies on the properties of the polypedon for differentiating characteristics and has a very complex but flexible multicategoric system. Names for the soil taxa are formed with syllables selected from terms that reflect the nature of the soil.

CHAPTER 18

WATER

The Soil–Plant–Atmosphere System

Most terrestrial plants are periodically exposed to water stress. Our goal in this chapter is to examine the nature of water movement in the soil–plant–atmosphere continuum and to examine the nature of water stress. We will consider the environmental factors that influence water availability in the soil, water loss to the atmosphere, and the characteristics that control plant response to changes in soil and atmospheric water. Terrestrial plants are exposed to different water regimes both seasonally and diurnally. If they are to survive, they must have mechanisms for adjusting to changes in environmental water. We will also discuss methods of measuring soil and plant water status.

The Water Potential Concept

Over the past two decades, the terminology relating to plant water, soil water, and water vapor in the atmosphere has changed dramatically. Prior to that, nearly all such data were in relative values such as percent soil moisture. Percentage values convey information concerning the amounts of water present but do not give indications of the direction of movement or the response of a plant to that amount of water. These problems led to the formulation of the water potential concept, in which the status of water is expressed in thermodynamic terms. All matter tends to move from areas of high free energy (high capacity to do work) to areas of lower free energy. **Water potential** (ψ) is a measure of the free energy of water in comparison to the free energy of pure water. The ψ of pure water has been assigned a value of 0 MPa. Megapascal is a pressure unit directly relatable to energy per unit mass (0.1 MPa = 1 bar = 0.987 atmospheres = 10^6 ergs g^{-1}). The chemical energy of water in the biosphere is lower than pure water; therefore, water potential values are expressed as negative numbers.

Water potential breaks down into several components that influence the chemical energy of water. Water potential depends on three predominant forces:

$$\psi = \psi_m + \psi_\pi + \psi_p \qquad \text{(Equation 18-1)}$$

Matric potential (ψ_m) is a measure of the reduction in free energy caused by the attraction of the solid surfaces for water molecules. The matrix is made up of macromolecules, soil particles, cell walls, or other surfaces to which water adheres and that do not go into solution. Such attractions always reduce the free energy of water, so matric potential is always negative. **Osmotic potential** (ψ_π) is a measure of the reduction in free energy (always negative) attributable to electrical attraction, reorientation, and increased entropy resulting from solutes in solution. In living systems, a selectively permeable membrane must be present, since no potential difference could exist without a restriction on the movement of solute molecules (Campbell 1977). **Pressure potential** (ψ_p) is determined by the positive or negative hydrostatic pressure present in the system. Within living cells, ψ_p is usually positive, but it is negative in xylem when transpiration is occurring. Total water potential is a negative value except in fully turgid cells, where ψ_p may balance the negative ψ_π and ψ_m potentials (Kramer 1969). The terminology of plant water relations has changed very rapidly, and the literature has been inundated with different symbols for the same quantities. Table 18-1 provides a comparison of the most commonly used terms and symbols.

Humidity and Vapor Pressure

Understanding the processes of evaporation and transpiration (**evapotranspiration**) is basic to determining the interchange of energy and water between the plant and the environment. Large amounts of energy are involved in any change in the state of water. For example, it takes most of the solar energy falling on 1 cm^2 on a clear summer day to evaporate a cubic centimeter of water. This rapid use of energy during the evaporation of water from plants (**transpiration**) causes an important cooling of the leaf surface. Plants grown in very high humidity can develop an inhibitory heat load in the leaves because of a reduction in water loss (details and further discussion of this process may be found in Chapter 14). Court (1974) estimated that two-thirds of the precipitation falling on the conterminous United States evaporates—a massive energy sink. Conversely, when water condenses or freezes, large amounts of energy are released. This energy is in the form of heat; it is called **latent heat.** When temperatures reach **dew point** (the temperature at which atmospheric water condenses), enough latent heat is released to reduce the rate of cooling significantly. Plants growing near bodies of water can escape frost damage because of this phenomenon (Rosenberg 1974).

Rates of evapotranspiration are indicators of water vapor transfer and energy exchange rates at leaf or soil surfaces. The rate of water loss is determined in part by the water vapor content of the atmosphere. When, at a constant temperature, pure water evaporates into a closed space, an equilibrium will occur at saturation vapor pressure. **Saturation vapor pressure** is the maximum possible partial pres-

Table 18-1   Terminology for the osmotic quantities as used by several authors. (Adapted from *Plant Physiology* by F. Salisbury and Cleon Ross. © 1969 by Wadsworth Publishing Co., Inc. Used by permission of the publisher.)

This text and others[a]	Meyer et al. and others[b]	Bonner and Galston[c]	Levitt[d]	Steward[e]	Salisbury and Parke[f]	James[g]	Fogg[h]
Osmotic Potential ψ_π (or π)	Osmotic Pressure OP	Osmotic Concentration OC	Osmotic Potential O	Osmotic Pressure P	Osmotic Potential ϕ	Osmotic Pressure O.P.	Osmotic Pressure P_i (internal) P_0 (external)
Pressure Potential ψ_p (or P)	Turgor Pressure (equal and opposite to Wall Pressure) TP	Turgor Pressure TP	Wall Pressure p	Wall Pressure W	Pressure P	Wall Pressure W.P.	Turgor Pressure T
Water Potential ψ	Diffusion Pressure Deficit DPD	Diffusion Pressure Deficit DPD	Osmotic Equivalent E	Suction Pressure S	Enter Tendency E	Suction Pressure S.P.	Diffusion Pressure Deficit S
Water Potential Difference $\Delta\psi$			Osmotic Potential Difference P				
$\psi = \psi_\pi + \psi_p$	$DPD = OP - TP$	$DPD = OC - TP$	$E = O - p$	$S = P - W$	$E = \phi - P$	$S.P. = O.P. - W.P.$	$S = (P_i - P_0) - T$

[a] Gardner, W. R. 1965. Dynamic aspects of water availability to plants. *Annual Review of Plant Physiology*. 16:323–342.
Kramer, P. J., E. B. Knipling, and L. N. Miller. 1966. Terminology in cell water relations. *Science* 153:889–890.
Salisbury, F. B. and C. Ross. 1969. *Plant physiology*. Belmont, CA: Wadsworth.
Slatyer, R. O. 1967. *Plant-water relationships*. New York: Academic Press.
Taylor, S. A. and R. O. Slatyer. 1962. Proposals for a unified terminology in studies of plant-soil-water relations. *UNESCO Arid Zone Research* 16:339–349.
[b] Devlin, R. M. 1966. *Plant physiology*. New York: Reinhold Publishing Corporation.
Ferry, J. R. and H. S. Ward. 1959. *Fundamentals of plant physiology*. New York: Macmillan.
Kramer, P. J. and T. T. Kozlowski. 1960. *Physiology of trees*. New York: McGraw-Hill.

Kozlowski, T. T. 1964. *Water metabolism in plants*. New York: Harper and Row.
Meyer, B. S., D. B. Anderson, and R. H. Bohning. 1960. *Introduction to plant physiology*. Princeton, NJ: D. Van Nostrand Co., Inc.
[c] Bonner, J. and A. W. Galston. 1952. *Principles of plant physiology*. San Francisco, CA: W. H. Freeman and Co.
[d] Levitt, J. 1954. *Plant physiology*. Englewood Cliffs, NJ: Prentice-Hall, Inc.
[e] Steward, F. C. 1964. *Plants at work*. Reading, MA: Addison-Wesley.
[f] Salisbury, F. B. and R. V. Parke. 1964. *Vascular plants: form and function*. Belmont, CA: Wadsworth.
[g] James, W. O. 1963. *An introduction to plant physiology*. Oxford: Clarendon Press.
[h] Fogg, G. E. 1963. *The growth of plants*. Baltimore, MD: Penguin Books, Inc.

sure of water that can be held in the air at a given temperature. Air is seldom saturated, so we need a means of expressing levels below saturation. One simple measurement of water vapor content of the atmosphere can be obtained with psychrometers, which consist of two ventilated thermometers. One is dry and measures air temperature; the other is wet and measures the reduction in temperature caused by evaporative cooling. Figure 18-1 shows the relationship between

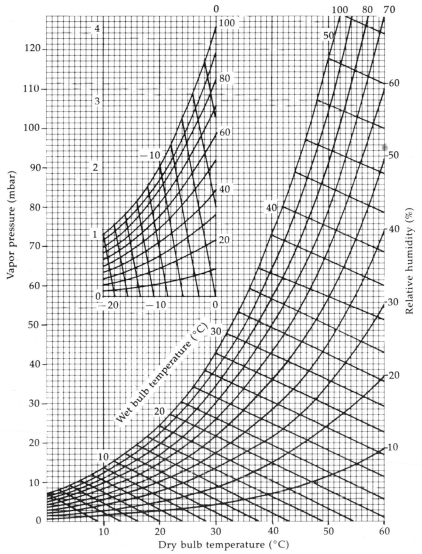

Figure 18-1   The relationship between vapor pressure, relative humidity, and wet and dry bulb temperature at sea level. (From Campbell, G. S. 1977. *An Introduction to Environmental Biophysics.* By permission of Springer-Verlag, New York.) See inset for values below 0°C.

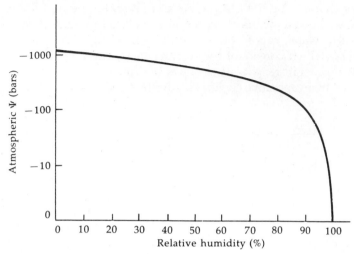

Figure 18-2   The relationship between atmospheric water potential (ψ) and relative humidity at 25°C (1 bar = 0.1 MPa).

vapor pressure, relative humidity, and wet and dry bulb temperatures. **Relative humidity** can be constant over a range of air temperatures and vapor pressures, making it a meaningless term unless air temperature is specified. **Saturation deficit,** a more meaningful measure of the evaporative power of the air, is the difference between the actual vapor pressure and the saturation vapor pressure at the same temperature. For example, in Figure 18-1, the saturation vapor pressure (100% relative humidity) at 20°C is 22.5 mbar; at 50% relative humidity, vapor pressure is 11.2 mbar, leaving a saturation deficit of 11.3 mbar. The saturation vapor pressure at 30°C is 40.5 mbar, and at 50% relative humidity, vapor pressure is 17.4 mbar, resulting in a 23.1 mbar saturation deficit. The potential evaporation is much greater at 30°C than at 20°C, even though the relative humidity is, in both cases, 50%. It is important to understand that factors determining evaporation are not restricted to the physical state of the atmosphere but also reflect the water potential of the evaporating surface. The importance of the evaporative power of the air is apparent in Figure 18-2, where at 25°C the relative humidity must be above 97% to approach the ψ of even the most xerophytic plants. Consequently, there is almost always a ψ drop between leaves and the atmosphere.

The Soil–Plant–Atmosphere System

Transpiration

A continuous path for water exists, from the soil, through the root, stem, and leaf, and into the atmosphere. The factors that affect this pathway are shown in Figure 18-3. For water to move through the plant, it is necessary that $\psi_{soil} > \psi_{root} > \psi_{stem} > \psi_{leaf} > \psi_{air}$. Even with a suitable ψ gradient, resistance to flow is

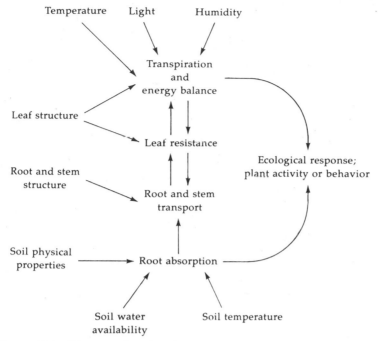

Figure 18-3 Diagram showing the interactions of the physical and biotic factors that are the most important in determining the ecological response of plants to water.

inherent in the system (Slatyer 1967; Meidner and Sheriff 1976). The pathway of water through the plant follows the scheme illustrated in Figure 18-4. Water enters the root primarily through root hairs, not only because they have greater permeability than older roots, but because they have a much greater surface area for absorption. Older roots lack root hairs and become suberized (impregnated with suberin), which renders them less permeable to water. From the root hairs, water passes along the course of least resistance through cell walls and between root cells to the endodermis. Here the water and dissolved material are forced to move through the cell membrane, because the casparian strip makes the cell wall impermeable and attaches the membrane to the cell wall. Passage through the endodermis offers considerable resistance, but once the water (and minerals) enter the vascular cylinder and the xylem, resistance is significantly lower. Vascular strands divide into very fine segments upon entering the leaves and are distributed throughout the mesophyll, so that almost every cell is within one or two cells of vascular tissue (Esau 1965), but no vascular tissue is exposed directly to the intercellular spaces adjacent to stomates. From the vascular tissue, water continues to travel in the liquid phase through the cell walls and between mesophyll cells to substomatal cavities, where it vaporizes and travels out through the stomatal pore into the atmosphere. This final loss of water from the plant **(transpiration)** is the most significant force in the movement of water, because

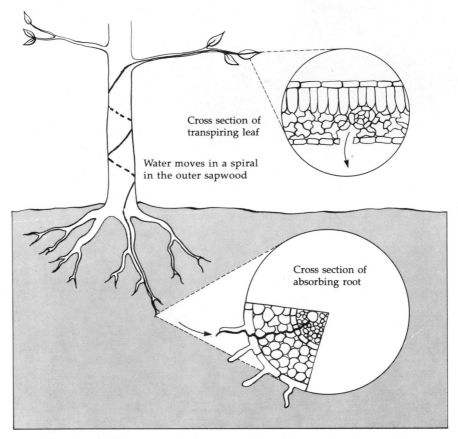

Figure 18-4   Diagrammatic representation of the course of water flow through the soil–plant–atmosphere continuum.

here the water potential gradient is the steepest. Slatyer (1967) indicated that a common set of water potential measurements along the gradient might be soil −0.1 MPa, stem −1.0 MPa, leaf −1.5 MPa, atmosphere −100 MPa. The most effective regulation of plant water status is associated with the stomates because of their location in the steepest portion of the gradient. Stomates, because of their ability to open and close, represent the major controlling factor of water flux from the plant to the air.

Energy Balance

The **energy balance** of a leaf depends on net radiation (R_n) (cal cm^{-2}s^{-1}), sensible heat transfer* (H) (cal cm^{-2}s^{-1}), and heat absorbed by the vaporization of water (*LE*):

$$R_n + H + LE = 0 \qquad\qquad \text{(Equation 18-2)}$$

---

*The combined terms *C* and *G* of Equation 14-1.

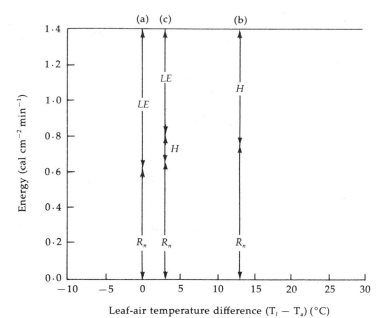

Figure 18-5 Energy exchange diagram for an idealized leaf of downwind width 10 cm, with air temperature 25°C and wind velocity 200 cm sec^{-1}. The total energy absorbed by the leaf is assumed to be 1.4 cal cm^{-2} min^{-1}. (Modified from Slatyer 1967. *Plant-Water Relationships.* By permission of Academic Press.)

where L is latent heat of vaporization (590 cal g^{-1}) and E is transpiration (g cm^{-2}s^{-1}). The transpiration rate is:

$$E = \frac{c^{ias}c^{air}}{\Sigma r} \qquad \text{(Equation 18-3)}$$

which is the water vapor concentration gradient between the leaf intercellular air spaces (c^{ias}) and the air (c^{air}) divided by the sum of the resistances to flow along the transfer path (Σr). This is related to a form of Fick's law that we used to describe CO_2 flux in Chapter 15.

We can envision leaf energy balance by considering the fate of 1.4 cal cm^{-2}min^{-1} of energy absorbed by a hypothetical leaf (Figure 18-5). The diagram shows the relative importance to energy dissipation of net radiation, sensible heat transfer, and transpiration. When leaf and air temperatures are equal, the absorbed energy is effectively dissipated by transpiration and radiation emission (a). When transpiration is stopped due to stomatal closure (b), leaf temperature increases and energy is dissipated by sensible heat transfer and radiation emission. Situation (c) represents the usual case, where absorbed energy is dissipated by the combined factors. Temperature and water loss are, therefore, intrinsically related. Their relative values, together with the photosynthetic capacity of a plant (which they directly influence), are a subject of profound importance in physiological ecology.

Where water is limited, there is often a tradeoff between CO_2 uptake for photosynthesis and water loss by transpiration. Plants with high **water use efficiency** (WUE) will have an advantage in these arid habitats. WUE is the ratio between net photosynthesis (g CO_2) and the amount of water (kg) consumed in transpiration. The nature and control of factors influencing WUE are the subject of the remainder of this chapter.

Leaf Resistance

We noted earlier that the steepest water potential gradient in the soil–plant–atmosphere continuum occurs as water leaves the plant in the vapor phase. The most important resistances influencing water loss are, therefore, associated with the leaf. **Diffusive resistance*** to water vapor transfer in a leaf (r^{leaf}) may be divided into an external component (Figure 18-6), referred to as the **boundary layer resistance** (r^{bl}), and an internal component consisting of

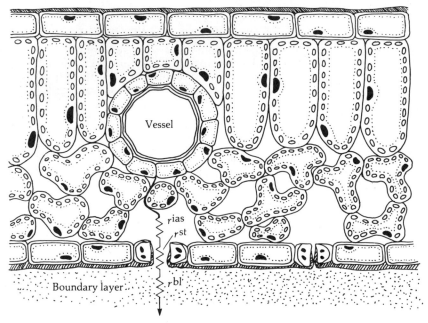

Figure 18-6 Cross section of a leaf showing the resistances associated with water loss. r^{ias} is the resistance to vapor movement in the substomatal cavity, r^{st} is the resistance offered by the stomatal pore and guard cells, and r^{bl} is the resistance offered by the boundary layer on the leaf surface.

---

*Resistance units are s cm^{-1} and express the same resistance to diffusive flow as equivalent path lengths of air of unit cross-sectional area divided by the diffusive coefficient of water in air (2.57×10^{15} m^{2}s^{-1} at 20°C). Conductance (cm s^{-1}), the inverse of resistance ($1/r$), is frequently used to evaluate barriers to flow.

(a) resistance to movement through intercellular air spaces and the substomatal cavity in the vapor phase (r^{ias}), and (b) resistance to movement through the stomatal pore itself (r^{st}). An alternate path of leaf water loss is referred to as **cuticular transpiration.** Here the water travels through the cuticle in the vapor phase, having vaporized at the epidermal cell wall surface. The importance of cuticular transpiration varies with the species, the level of secondary thickening of epidermal cell walls, and characteristics of the waxy cuticle, but resistance is so high that water loss via this path is commonly inconsequential in comparison to stomatal transpiration.

Resistance to vapor movement in intercellular spaces and in the substomatal cavity (r^{ias}) depends on the distance the vapor must travel. The value is more or less constant for a species, depending primarily on stomatal location and density and leaf thickness.

Variation in **stomatal resistance** (r^{st}) represents the most significant regulator of transpiration because r^{st} is located at the steepest part of the ψ gradient, and the guard cells are sensitive to environmental factors. Stomates open and close in response to differences in guard cell turgidity stimulated by variations in light, leaf water potential, CO_2 concentrations, and humidity (Figure 18-7). Turgidity changes are related to osmotic potential variations in the guard cells caused by the movement of K^+ into and out of the cells. The mechanism that stimulates the transport of K^+ and accompanying anions is not clearly understood, though several hypotheses have been proposed. (For a review, see Allaway and Milthorpe 1976 or Hsiao 1976.) In any event, when potassium ions enter a guard cell, they cause increased guard cell turgidity and opening of stomates. Water stress will stimulate production of abscissic acid (ABA), which drives the K^+ pump in the other direction, causing stomatal closure. The level of water stress necessary to cause ABA production varies with the drought tolerance of the species. For any species, there is a critical ψ_{leaf} at which the stomatal aperture will decrease dramatically, increasing resistance to ∞. Resistance in fully open stomates may range as low as 0.4 s cm^{-1}.

The effect of atmospheric vapor pressure on stomatal aperture has only recently been satisfactorily documented (e.g., Lange et al. 1971; Schulze et al. 1972, 1973; Camacho-B et al. 1974; Hall and Kaufmann 1975; Smith and Nobel 1977a). Again, the mechanism of stomatal response has not been established. A possible explanation was proposed by Lange et al. (1971), based on the concept of **peristomatal transpiration.** Guard cells lose water more readily than other epidermal cells and would therefore respond directly to changes in vapor pressure. Confirmation of the peristomatal transpiration hypothesis remains a subject of scientific inquiry. Gradual closing of stomates, as evaporative demands of the atmosphere increase, may serve to maximize WUE in plants by increasing the time necessary to reach the critical ψ_{leaf}.

Boundary layer resistance (r^{bl}) is considered as part of leaf resistance because it is influenced by morphological characteristics of the leaves. Resistance to diffusion is caused by a layer of air adjacent to the surface of the leaf that, as with laminar flow over any flat surface, does not readily mix with passing air. Vapor must pass through this potentially saturated layer by molecular diffusion before

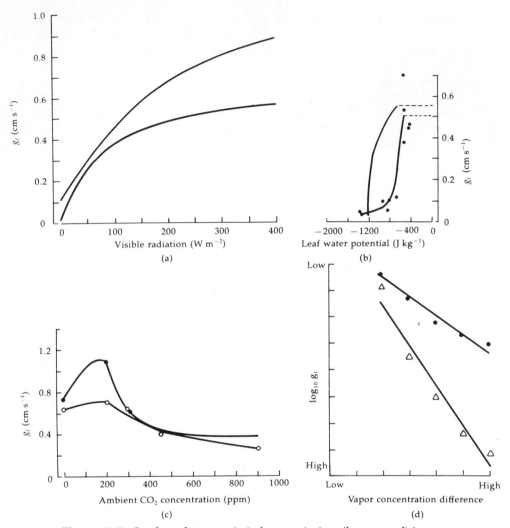

Figure 18-7 Leaf conductance (g_1) changes (primarily stomatal) in response to variations in (a) visible radiation, (b) leaf water potential, (c) ambient CO_2 concentration, and (d) vapor concentration difference between leaf and air for pairs of cultivated plants. (Modified from Burrows and Milthorpe 1976. Stomatal conductance in the control of gas exchange. In *Water Deficits and Plant Growth*, vol. IV, ed. T. T. Kozlowski. By permission of Academic Press.)

moving into the free atmosphere. The thickness of the boundary layer depends on leaf size and wind speed. Greater wind speeds cause turbulence closer to the leaf surface and reduce the depth of the boundary layer. Smaller leaves have thinner boundary layers because of increased convective exchange between the surface and the free air. Surface texture, leaf shape, and leaf orientation affect air turbulence near the surface and, therefore, the thickness of the boundary layer.

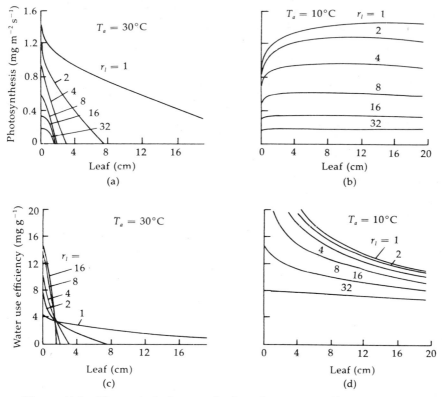

Figure 18-8 Theoretical photosynthetic and water use efficiency responses for leaves of different sizes and for different levels of resistance (r_l) when exposed to ambient temperatures (T_a) of 30°C and 10°C. (From Campbell, G. S. 1977. *An Introduction to Environmental Biophysics.* By permission of Springer-Verlag, New York.) WUE = $mgCO_2$ fixed per g H_2O transpired.

Optimal Leaf Form

Taylor (1975) and Campbell (1977) considered the interactions of physical factors, such as temperature and light, and biological regulation of water loss (leaf resistance, r^{leaf}) on photosynthetic output and water use efficiency (WUE) of plants. Figure 18-8 shows the idealized photosynthetic and WUE responses of leaves of various sizes with different diffusive resistances when exposed to full sunlight. At high air temperatures (30°C in Figures 18-8a and 18-8c), larger leaves have lower net photosynthetic rates, especially when diffusive resistance is high (e.g., $r^{leaf} = 32$ s cm^{-1}). This is a result of overheating; at low resistances (e.g., $r^{leaf} = 1$ s cm^{-1}), evaporative cooling maintains temperatures that support positive photosynthesis even in leaves 16 cm wide. However, this evaporative cooling requires large amounts of water and the water use efficiency is therefore very low (<4 mg g^{-1}). There is a clear advantage for leaves less than 1 cm wide and with a diffusive resistance above 6 s cm^{-1} in a warm, dry environment.

Larger leaves have higher photosynthetic rates than smaller leaves in cool air (10°C in Figure 18-8b), because they are at temperatures above ambient and are therefore closer to the optimum for photosynthesis. Figure 18-8d shows that water use efficiency drops as leaf size increases, but, at comparable diffusive resistances, efficiency remains higher than at air temperatures of 30°C (Figure 18-8c). This theoretical model shows that the interactions of temperature, water, and photosynthesis should lead to plants with small leaves in warm, arid environments. This is, in fact, the case, and may help explain the great success of leguminous trees and small leaved plants in arid regions. Pinnately compound leaves with small leaflets are common in desert leguminous trees. They are apparently a significant advantage in desert habitats because of reduced heat load and greater WUE. In cooler environments, large leaves are warmer and, with low diffusive resistance, have high photosynthetic rates. There is a gradient of increasing leaf size from warm arid to cool moist habitats.

Theoretical photosynthetic and WUE responses to different light intensities and diffusive resistances are presented in Figure 18-9 for leaves of intermediate size (3 cm). When ambient temperatures are cool (10°C in Figure 18-9b), maximum photosynthesis is attained at full sunlight (500 W m^{-2}), because leaf temperatures remain moderate, but when temperatures are higher (30°C in Figure 18-9a), leaf temperature increases beyond optimal, and intermediate light intensities are the most effective. Full sunlight in conjunction with a 30°C air temperature results in low WUE. In fact, unless diffusive resistance is less than 4 s cm^{-1}, leaf temperature surpasses the tolerance of our hypothetical plant (Figure 18-9c). High WUE is maintained across a range of light intensity when ambient temperatures are low (10°C in Figure 18-9d), again because leaf temperatures remain moderate. Many leaves (especially larger ones) of plants in arid or semiarid habitats are oriented vertically. Vertical orientation allows interception of maximum solar radiation in early morning and late afternoon, when temperatures are lower, and reduces interception at midday, when temperatures are high. Vertical orientation would thus maximize both photosynthesis and WUE in our hypothetical plant.

While these relationships are for an idealized situation, they provide results that we can confirm, in general, by observing real plants. Understanding adaptations of specific plants with regard to these variables will allow us to interpret better the habitat responses observed in nature.

Diffusive Resistance and Transpiration Measurements

The most popular method of measuring leaf resistance is by the use of a diffusive resistance porometer (Figure 18-10). A porometer measures water loss into a small chamber enclosing part or all of a leaf. Resistance values may be standardized by attaching the porometer to an apparatus with a known diffusive distance (δ) in cm. We can then calculate resistance to water vapor transfer as:

$$r^{leaf} = \frac{\delta}{D} \hspace{4cm} \text{(Equation 18-4)}$$

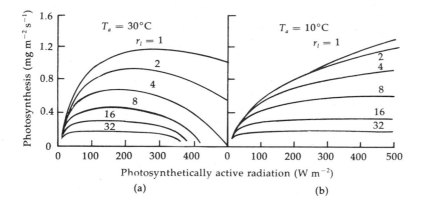

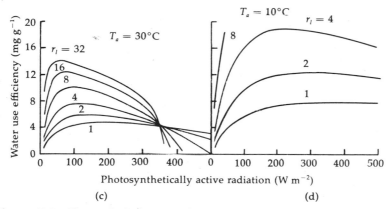

Figure 18-9   Theoretical photosynthetic and water use efficiency
responses at different light intensities and for different levels of
resistance (r_1) when exposed to ambient temperatures (T_a) of 30°C and
10°C. All leaves are intermediate sized (3 cm). Leaf temperature is
allowed to vary in these simulations. (From Campbell, G. S. 1977. *An
Introduction to Environmental Biophysics.* By permission of Springer-Verlag,
New York.)

Figure 18-10   A diffusive resistance
porometer, the Li-Cor LI-1600 Steady
State Porometer. (Courtesy of Li-Cor,
Inc.)

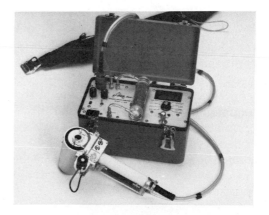

where D is the diffusion coefficient of water vapor at a given temperature (cm^2 s^{-1}). The units of resistance are s cm^{-1}. r^{leaf} can also be calculated when transpiration rates (J_{wv}) and ambient humidity are known:

$$r^{leaf} = \frac{\Delta c_{wv}}{J_{wv}}$$

(Equation 18-5)

Here, Δc is the concentration gradient between mesophyll cell surface (assumed to be saturated) and the air. Humidity is measured by hygrometers with lithium chloride or aluminum oxide sensors. Thermocouples are incorporated because of the interaction between vapor pressure and temperature (Figure 18-1). A fan to circulate air reduces the effect of the boundary layer, which may be inordinately large in the small, closed porometer chamber (Nobel 1983). Older diffusive resistance porometers measured the time necessary to increase water content of a dry atmosphere to a predetermined level. Newer steady-state porometers circulate conditioned air through the chamber and maintain humidity at ambient or other preset levels. This gives estimates of resistance under ambient conditions rather than maximum water loss into fully dried air. Such measurements are more accurate because stomatal aperture may change in response to the low humidity of older porometer chambers (Appleby and Davies 1983). Kanemasu et al. (1969), Beardsell et al. (1972), and Parkinson and Legg (1972) have proposed a variety of porometer designs.

Less expensive estimates of leaf resistance can be made using cobalt chloride paper (Meidner and Mansfield 1968; Milthorpe 1955), or calculation from stomatal dimensions using infiltration techniques or direct measurements. (For a review, see Burrows and Milthorpe 1976.)

Transpiration rates from detached leaves and potted plants are easily made by gravimetric analysis; repeated weighing of the parts over time gives an estimate of water lost. For detached leaves and whole branches, a potometer is used. Figure 18-11 illustrates one form of potometer, which manometrically measures water flux through an excised leaf, branch, or shoot system of a plant. More accurate measurements are possible when transpiration measurements are incorporated into a gas exchange system (see Chapter 15).

Stem and Root Transport

Resistance to mass transport of water in vessels and tracheids is low in contrast to the resistances encountered in other phases of the continuum (i.e., root endodermis and mesophyll). Heine (1971) reviewed the development of concepts relating to conductivity in woody plants and stressed the importance of anatomy in determining water flux. Conductivity depends not only on the ψ gradient but also on lumen (central open portion of the cell) area and length of xylem elements. Carlquist (1975) has shown a relationship between element dimensions and the amount of available water. This close relationship between vascular element size and the environment is apparently related to the negative pressures that develop in xylem.

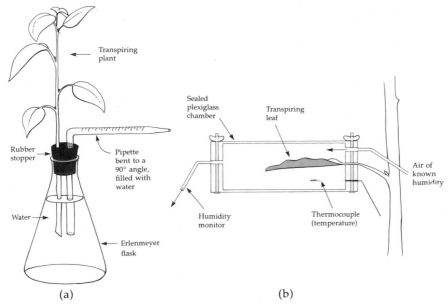

Figure 18-11 (a) Potometer for the measurement of transpiration of plant shoots or leaves. (b) A gas exchange chamber for the measurement of transpiration in continuously flowing air.

The ψ gradient in xylem elements is an expression of a gradient in pressure potential, since solute concentration and matric influences are more or less constant throughout the xylem water column. As transpiration proceeds, resistance prevents the instantaneous replacement of transpired water, thereby establishing a negative hydrostatic pressure (tension) in the xylem. This negative pressure is a natural component of vascular water. It is intensified when transpiration demand is high and replacement is slow because of low ψ_{soil} or when high resistance to water flow increases the lag between inflow and outflow. Adhesion of water molecules to cell walls and cohesion between water molecules maintain the integrity of the water column. The more negative the pressure (lower ψ), the greater the stress placed on the cells and the water column. Smaller vascular elements are, therefore, associated with tissues exposed to the greatest stress (Carlquist 1975). Rundel and Stecker (1977) report that this relationship is clearly evident in tracheid diameters of a 90 m nontranspiring Sierra redwood (*Sequoiadendron giganteum*) in Kings Canyon National Park, California (Figure 18-12a). This is not surprising, since the weight of water in the conducting tissues of tall trees would lead one to expect that upper xylem tissues would be subjected to highest levels of stress even in the absence of transpiration (Figure 18-12b).

Water uptake and transport in nonvascular tissues of roots are sensitive to the water potential gradient (as is vascular transport) but are also sensitive to temperature changes. This effect is related to changes in endodermal cell mem-

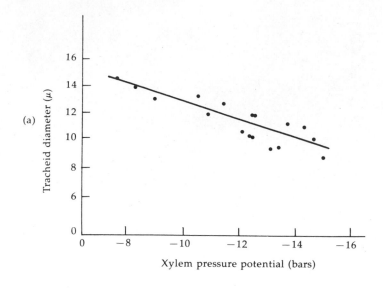

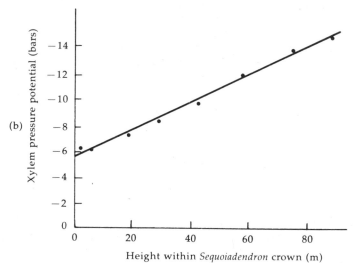

Figure 18-12   (a) The relationship between predawn xylem pressure potential and tracheid diameter in *Sequoiadendron giganteum;* (b) the relationship between height and predawn xylem pressure in *Sequoiadendron giganteum.* (From Rundel, P. W. and R. E. Stecker. 1977. Morphological adaptations of tracheid structure to water stress gradients in the crown of *Sequoiadendron giganteum. Oecologia* 27:135–139.)

brane permeability with temperature, rather than the direct physical effect of temperature on water movement. Water uptake in response to temperature is subject to modification by acclimation and varies with species and habitat (Anderson and McNaughton 1973). For example, soil temperature was identified by Fernandez and Caldwell (1975) as a potential stimulus for root activity in Great

Basin Desert shrubs. Initiation of rapid growth and water absorption in *Atriplex confertifolia* occurred as the soil profile warmed to progessively deeper levels. Cessation of root activity was associated with soil water depletion, so that activity in any profile zone was limited to one or two weeks of adequate water supply following the attainment of sufficiently warm temperatures. This mechanism prolongs the time that soil water is available by effectively partitioning the water for use during different periods.

Soil Moisture

Soil water potential ultimately determines the availability of water for plant growth. It is important, then, to understand the relationship between the physical properties of soils discussed in Chapter 17 and the availability of water for plants. We will discuss soil properties that influence retention, infiltration, and movement of water, but first we need to define several terms basic to understanding soil water relations.

Moisture Status in Soils: Basic Definitions   Water entering the soil will fill most pore spaces and drain downward through pores in response to gravity. We can add the term ψ_g to our initial water potential formula to represent gravitational potential:

$$\psi_{soil} = \psi_m + \psi_p + \psi_\pi + \psi_g \qquad \text{(Equation 18-6)}$$

Gravitational potential is significant only in saturated soils. Water moving through soil in response to gravitational forces is gravitational water. Once gravitational water has moved from the soil or is dispersed in the soil, the soil is left at field capacity (FC). **Field capacity** is the amount of water that can be retained by a soil after gravitational water has been removed. Field capacity for a particular soil is a constant, depending on soil texture, soil structure, organic content and $\approx \frac{1}{3}$ bar.

The water available for plant use from soils at field capacity depends on the species, the physiological state of the individual, and environmental conditions. The lower limit of water availability is designated as the **permanent wilting point** (PWP). PWP is reached when ψ_{soil} is equal to or below the minimum osmotic potential of the plant, so that water cannot be removed from the soil. The plant wilts and remains wilted even if placed in a cool, dark, humid chamber. Addition of soil water may allow recovery. The amount of water between FC and PWP is called **available water** or **capillary water,** since it is retained against the pull of gravity by capillary action and matric forces. Plants are able to utilize capillary water to varying extents but cannot remove all water from the soil. Below the PWP, water is held so firmly to the soil matrix that it is removable only under conditions surpassing biological tolerance. The most important components of water potential in nonsaturated soils are matric potential and, in some cases such as saline soils, osmotic potential. PWP for mesophytes is often near -1.5 MPa.

The relationship between FC, soil water content, and ψ_{soil} depends to a large degree on soil texture (Figure 18-13). Clay soils hold large volumes of water at

Figure 18-13   Typical water characteristic curves for sand, sandy loam, and clay soils (1 bar = 0.1 MPa). (Modified from Slatyer 1967. *Plant-Water Relationships.* By permission of Academic Press.)

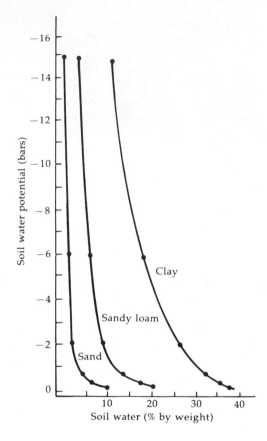

relatively lower water potentials than do coarse-textured soils. This is because of the greater total pore space and greater particle–water contact space in fine-textured soils.

Movement of Soil Water   Availability of water within a soil depends, in part, on how quickly and how far water will move upward from the water table toward an absorbing root or into the soil during a rainstorm. **Infiltration** and movement of soil water depend on factors that determine soil water potential. Water will infiltrate into a dry soil faster than a wet soil, if they have comparable texture and structure, because the gradient in water potential is greater in the dry soil. The flux density of water through a saturated soil (J_w) is:

$$J_w = -K\frac{\Delta\psi_{soil}}{\Delta z}$$
(Equation 18-7)

where K is the **hydraulic conductivity** (the capacity of a soil to transport water in $g\ cm^{-2}s^{-1}$); $\Delta\psi_{soil}$ (MPa) is the difference in water potential of the soil between two points along which water moves; and z is the distance between the two points in the soil used to determine $\Delta\psi_{soil}$. As ψ_{soil} drops, there is a significant

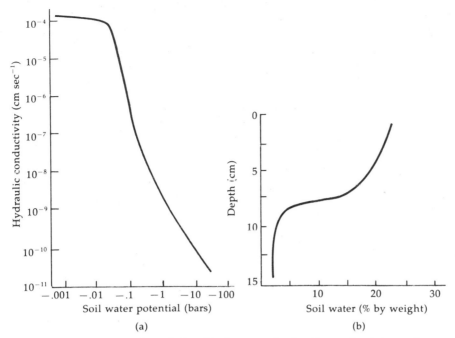

Figure 18-14 (a) The relationship between hydraulic conductivity and soil water potential in a loam soil (1 bar = 0.1 MPa). (b) The water content profile of an initially dry loam soil at one day following irrigation. (After Slatyer 1967. *Plant-Water Relationships.* By permission of Academic Press.)

drop in hydraulic conductivity (Figure 18-14a), especially in sandy soils where the continuity of the water is broken. In finer-textured soils, hydraulic conductivity also decreases, but to a lesser extent because the continuity of water surfaces are maintained at lower ψ_{soil}. After gravitational water has been dispersed, the upper zone of the soil profile is at field capacity, and then there is a narrow zone of soil with a very steep drop in water content; the soil below that is essentially dry (Figure 18-14b). Water may rise from an underground water table in limited quantities. The rate of movement depends on the water potential gradient.

Lateral movement of water through the soil to zones of absorption by roots depends on the water potential gradient and the hydraulic conductivity of the soil. Root water potential must be lower than ψ_{soil} for absorption to occur. This water potential difference can be quite low when ψ_{soil} is −0.5 MPa or higher, whereas at −1.5 MPa, a significantly greater drop in potential is necessary for absorption. This is a result of the decrease in hydraulic conductivity as ψ_{soil} drops. The distance through which water will move toward the dry zone surrounding a root depends on the plant's tolerance for stress and the soil properties that influence hydraulic conductivity. Water infiltration into the soil is a critical factor in areas of steep slopes or intense rainfall. Sandy soils generally have rapid infiltration, with finer soils restricting the entrance of water in proportion to clay

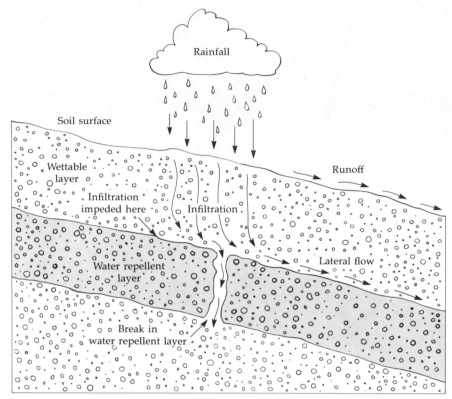

Figure 18-15   A surface profile of a typical chaparral soil following a fire where the surface is wettable but a subsurface water repellent layer causes lateral flow of the surface soil. (After "Water repellent soils and wetting agents as factors influencing erosion" by J. S. Krammes and J. Osborn. In *Proceedings of Symposium on Water-Repellent Soils.* L. F. DeBano and J. Letey, eds. 1969. University of California, Riverside.)

content. An exception exists for sandy soils in certain situations, where organic compounds from plants form "skins" over soil particles so that, rather than penetrating, the water droplets bead up on the soil surface. Such soils are called **hydrophobic** or **water repellent soils.** They have been observed under a wide variety of species and environmental conditions (Bond 1964; Krammes and DeBano 1965; Adams et al. 1970; DeBano and Letey 1969). Water repellency increases surface runoff and post-fire erosion on chaparral soils in southern California. When chaparral burns, the water repellent zone moves into the soil (Figure 18-15), leaving a wettable zone at the surface. Post-fire rains penetrate and saturate the wettable zone, causing the surface soil to slip downslope. Not all species in a habitat have the same influence on wettability, and few communities have been characterized for wettability patterns. Soil wettability is another environmental factor that affects water availability and therefore regulates the niche relations of plant species.

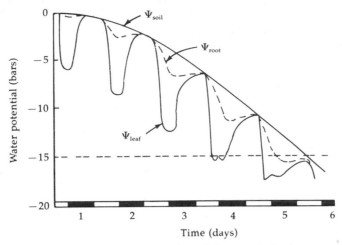

Figure 18-16 Schematic representation of changes in leaf water potential (ψ_{leaf}), root surface potential (ψ_{root}), and soil mass water potential (ψ_{soil}) as transpiration proceeds from a plant rooted in initially wet ($\psi_{root} \approx 0$) soil. The same evaporative conditions are considered to prevail each day. The horizontal dashed line indicates the value of ψ_{leaf} at which wilting occurs. (From Slatyer 1967. *Plant-Water Relationships.* By permission of Academic Press.)

Significance of Plant Water Stress   Slatyer (1967) devised a hypothetical scheme (Figure 18-16) to show the relationship between ψ_{soil}, ψ_{root}, and ψ_{leaf} as soil water is reduced from field capacity to below the permanent wilting point. Significant diurnal patterns of ψ_{plant} are present even when ψ_{soil} is near field capacity. The diurnal range must get larger as the soil becomes drier to maintain sufficient uptake from soils in which both water potential and hydraulic conductivity are decreasing. Following stomatal closure in response to darkness, ψ_{plant} gradually equilibrates with the soil so that just before dawn, $\psi_{soil} \approx \psi_{leaf}$. The time necessary for equilibration increases as ψ_{soil} decreases, until $\psi_{soil} \approx \psi_{leaf}$ for only a brief period. Temporary wilting and midday stomatal closure may reduce transpiration at the point along the drying curve where ψ_{leaf} = leaf osmotic potential, ψ_{π}. Mesophyll cells reach the point of incipient plasmolysis ($\psi_p = 0$) when $\psi_{soil} = \psi_{\pi}$ (the PWP), because a ψ gradient sufficient to extract water from the soil is no longer possible. Osmotic potential is not fixed within an individual plant; a wide range of values are found. We will discuss some examples of osmotic adjustment in the next section.

Plant responses to water stress have been reviewed in detail by Hsiao (1976). We will briefly mention the physiological responses to water stress and later consider some specific adaptations to avoid or deal with stress. Water is involved in and essential to all physiological processes. The photosynthetic decline as water stress increases is primarily due to stomatal closure or other direct effects of water stress on photosynthetic reactions. When stress is great enough, photosynthesis, respiration, protein synthesis, and most other processes involving chemical reactions may be severely reduced because of protein (enzyme) denaturation. Most physiological processes decrease gradually with increasing stress to the point where all turgor is lost, then cease beyond that point. Under moderate stress, biochemical processes may proceed, but growth may be reduced because of lack of sufficient turgor pressure to cause cell enlargement. Lower overall cell size may increase the drought tolerance of the new tissues (Cutler et al. 1977). Newly formed cells may enlarge only when water becomes available, in which case growth much exceeds that expected by the rate of formation of new cells (Oechel et al. 1972). Water stress also increases the viscosity of phloem sap, as well as reducing mass transfer in the xylem, thus limiting transport of nutrients, photosynthate, and hormones.

Solute Concentration and Plant Water Status

Osmotic concentration of cell sap is a key quantity needed to understand variations in the level of stress necessary to induce wilting. **Osmotic concentration** depends on cell solute content and the amount of intracellular water. Only recently (Tyree and Hammel 1972) has the importance of such measurements been recognized and interpreted with respect to ecologically meaningful adaptations of plants (Cheung et al. 1975). Consequently, data on osmotic parameters of native plants are few (e.g., Monson and Smith 1982; Bennert and Mooney 1979; Osonubi and Davies 1978; Roberts and Knoerr 1977;

Cheung et al. (1975) contrasted *Ginkgo* and *Salix* (Figure 18-17) as an example of plants showing very different osmotic qualities. *Ginkgo* has a lower (more negative) osmotic potential (point *A* for *Ginkgo*, Figure 18-17) and would consequently remain turgid at much higher stress than would *Salix*. Plasmolysis (tur-

Figure 18-17   A plot of water potential versus osmotic water loss as predicted by the pressure–volume curves of single *Ginkgo biloba* and *Salix lasiandra* leaves. *A* is the estimate of original osmotic potential; *B* is the osmotic potential at incipient plasmolysis. (From Cheung, Tyree, and Dainty 1975. Reproduced by permission of the National Research Council of Canada from the *Canadian Journal of Botany,* Volume 53, pp. 1342–1346, 1975.)

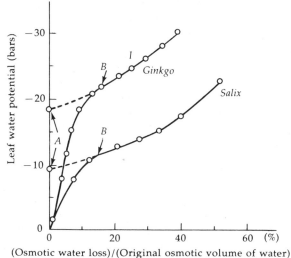

(Osmotic water loss)/(Original osmotic volume of water)

gor pressure = 0 MPa) occurs at point *B*, which is about -2.2 MPa for *Ginkgo* and -1.2 MPa for *Salix*. The steep slope for *Ginkgo* shows that very little water needs to be lost to reduce ψ_{leaf} greatly. *Ginkgo*, the more xerophytic of the two species, can remove soil water to a much lower ψ_{soil} than *Salix*, yet maintain a higher level of tissue hydration. This capacity is partly due to the more rigid (nonelastic) cell walls of *Ginkgo*. Removal of small amounts of water from cells with rigid walls reduces turgor pressure rapidly. In *Salix*, the more flexible cell walls shrink, thus maintaining turgor pressure but creating only a minimal gradient for soil water extraction. *Salix*, therefore, must grow in more mesic environments if it is to maintain a high level of tissue hydration. The ability to compare species water requirements and responses at this fine scale using information derived from pressure–volume curves offers exciting possibilities for expanding our knowledge of physiological ecology.

The capacity to reduce turgor pressure to negative values would give xerophytes further ability to extract soil water to very low water potentials. Negative turgor pressures are theoretically possible but have not been shown, beyond reasonable doubt, to exist (Kyriakopoulos and Richter 1976; Tyree 1976). Such a phenomenon would help explain how some desert plants are able to reduce soil water potentials to less than -10 MPa.

Osmotic adjustment is an effective acclimation device in response to water deficit (Monson and Smith 1982; Roberts et al. 1980; Bennert and Mooney 1979; Osonubi and Davies 1978; Cutler and Rains 1978; Jones and Turner 1978; Cutler et al. 1977). Both stomatal response and wilting point depend on solute concentration, which increases in plants subjected to gradual drying or repeated periods of drought. The chaparral shrub *Ceanothus greggii* shows both seasonal and diurnal adjustments in osmotic potentials that allow it to maintain positive turgor during the extreme drought of late summer (Figure 18-18). In August and Sep-

Figure 18-18   Seasonal courses of midday water potential (dashed line) and midday turgor loss point (solid line) for the chaparral shrub *Ceanothus greggii*. (Redrawn from Bowman and Roberts 1985. *Ecology* 66:738–742.)

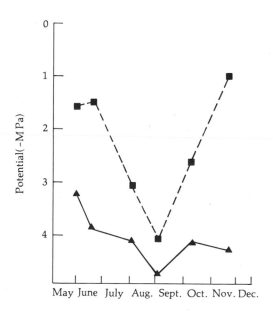

tember, midday water potentials fall well below the point of zero turgor of early May. Without osmotic adjustment, these plants would experience severe dehydration of tissues during midsummer days (Bowman and Roberts 1985). The increase in solute concentration may be due to increased solute content or to smaller solvent volumes in smaller cells formed under conditions of stress. Osmotic adjustment in *Ceanothus greggii* (Figure 18-18) is due to changes in the volume of osmotic water present in the tissues, whereas other chaparral plants such as *Quercus dumosa* and *Arctostaphylos glandulosa* accumulate solutes as the means of acclimating to the dry season. Regardless of the mechanism operating in a particular species, many plants reflect previous conditions of water availability.

Plant and Soil Water Status Measurements

Recent technological advancements have dramatically improved our capabilities of measuring plant and soil water status. Most new techniques are developed around energy differences expressed by rates of evaporation as measured by heat flux from a wet thermocouple, or the negative xylem pressure created by transpiration. Nonelectronic methods based on the direction of water movement and total water content may also be used when more sophisticated equipment is not available.

Pressure Chamber

Scholander et al. (1964) developed the pressure chamber now widely used in ecophysiological research, and Waring and Cleary (1967) showed the great utility of **xylem pressure potentials** in solving ecological problems. This method is based on the fact that negative pressures (tensions) exist in the xylem. When stems are severed, the pressure necessary to force water back to the cut surface is equivalent to the negative pressure in the xylem prior to cutting. A typical pressure chamber apparatus is diagrammed in Figure 18-19. An advantage of the pressure chamber lies in the large number of samples that can be measured in a short time. The apparatus has also been made portable so that data may be taken on plants in their native surroundings. Ritchie and Hinckley (1975) discussed in detail the applications, calibration, and data interpretation of pressure chambers. This review should be consulted before using the pressure chamber technique.

Pressure–volume relationships can be determined with a pressure chamber by placing a fully hydrated, excised shoot into the pressure chamber, incrementally increasing the chamber pressure, and recording the volume of sap expressed at each pressure. When the linear portion of the resulting curve is extended, it crosses the ordinate at a value (*A* in Figure 18-20) equal to the inverse of the original osmotic potential. By extending the linear portion of the curve to the abscissa, the original osmotic volume of water can be estimated. The inverse of the tissue water potential at the point of zero turgor is read from the ordinate as the point where the curve becomes linear. Osmotic potential indicates the amount

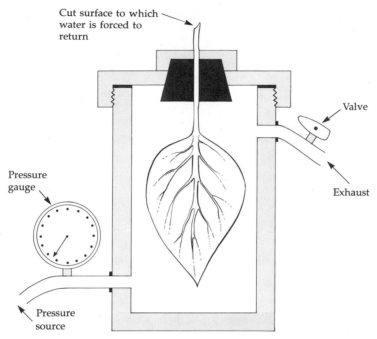

Cut surface to which
water is forced to
return

Valve

Pressure
gauge

Exhaust

Pressure
source

Figure 18-19   Schematic diagram of a pressure chamber used to measure
plant water potential.

Figure 18-20   A sample
pressure–volume curve.
The inverse of the balance
pressure is plotted on the
ordinate, and the volume
of water expressed from
the leaf or branch is
plotted on the abscissa. *A*
is the inverse of the
original osmotic potential
(0.94 MPa^{-1}). *B* is the
inverse of the water
potential at incipient
plasmolysis (0.66 MPa^{-1}),
and *C* is the volume of
osmotic water originally
in the tissues (0.75 ml). *A*
and *C* are determined by
extending the linear
portion of the curve to
the ordinate (*A*) and to
the abscissa (*C*). *B* is the
point where the curve
becomes linear.

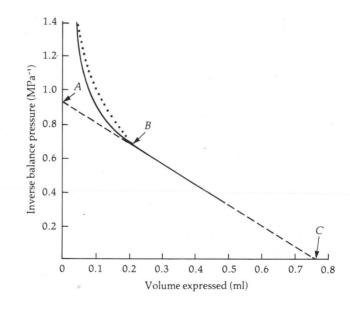

of solutes present per unit volume of cellular water and sets the limit on the magnitude of turgor pressure that can develop. Pressure–volume curves can also be used to determine cell wall elasticity by observing the steepness of the nonlinear portion of the curve. Where cells are very rigid, the slope is very steep (above point *B* in Figure 18-20). More elastic cells will show a more gradual decrease in water potential for each unit of water expressed from the tissue (the dotted line in Figure 18-20). Tyree and Hammel (1972) and Cheung et al. (1975) have interpreted pressure–volume curves in ecological terms. Stress tolerant plants would be expected to have a higher osmotic potential and less cell wall elasticity than more mesophytic species.

Pressure chamber values represent the total water status of the plant and can be used to determine ψ_{plant} responses to environmental factors, both diurnally and seasonally. These data, in combination with pressure–volume curves, provide an effective means of measuring plant water status and stress tolerance.

Psychrometry and Hygrometry

Psychrometers were developed to measure relative humidity of the air. With the development of the water potential concept, it was found that ψ could be determined from measurements of relative humidity. The relationship is:

$$\psi = \frac{RT}{V} \ln wv \qquad \text{(Equation 18-8)}$$

where R is the ideal gas constant, T is the absolute temperature, V is the volume of a mole of liquid water, and wv is the relative humidity. **Thermocouple psychrometers** measure temperature depression caused by evaporation from a wet surface. (See Figure 18-1 to review the relationship between wet bulb temperature and humidity.) They can detect small changes in relative humidity in a closed chamber (Spanner 1951; Richards and Ogata 1958). Sample material is placed in a small, sealed chamber containing a thermocouple junction. When the air around the thermocouple is in equilibrium with the sample, the thermocouple is electrically cooled to below dew point. The film of water collected on the thermocouple will evaporate and cool the thermocouple at a rate inversely proportional to the water potential of the air in the sample chamber. Figure 18-21 is a diagrammatic sketch of a thermocouple psychrometer apparatus that can be used for soil or leaf disc measurements. Various chambers have been developed for specific purposes. Psychrometers with screen or ceramic covers may be inserted into the soil for ψ_{soil} determinations. Thermocouple psychrometers can be inserted into a clamp and attached to leaves or stems as a nondestructive measure of water potential. Nondestructive methods are new; the influence of the rather extended time necessary for equilibration needs further testing (Campbell and Campbell 1974; Brown and McDonough 1977; Zanstra and Hagenzieker 1977).

It is possible to determine vapor pressure in sample chambers like those used in psychrometry by measuring the dew point. **Dew point hygrometers** contain a thermocouple or mirror with a film of water, just as in a psychrometer,

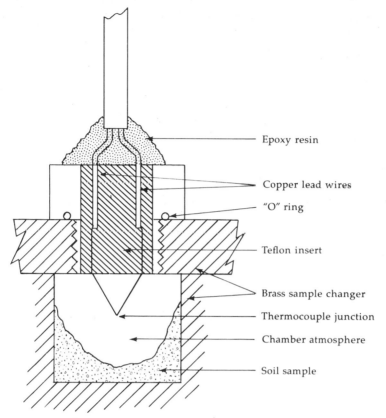

Figure 18-21 The Spanner (Peltier) thermocouple psychrometer and a psychrometer chamber, shown with a soil sample. The entire assembly is immersed in a water bath to maintain constant temperature. (From Brown 1970. Measurement of water potential with thermocouple psychrometers: construction and applications. U.S. Dept. of Agriculture Forest Service Research Paper INT–80.)

but the hygrometer maintains the film by staying at dew point temperature. Dew point hygrometry is a continuous measurement of humidity. It is frequently utilized to measure water vapor in circulating air streams such as those in a gas exchange apparatus (see Chapter 15).

Dye Method

A simple and inexpensive method of measuring plant water potential is based on the change in density of a solution, sometimes referred to as the **Shardokov dye method.** Two series of replicate sugar solutions are prepared to span a range of osmotic potentials. Leaves or leaf discs are placed in one series of solutions. The second series is colored by methylene blue or another appropriate dye. After the tissue has been exposed to the solution for an hour or so, a drop of the colored solution is introduced into the center of the replicate solution that contains the tissue. If the drop moves up, the ψ_{plant} was lower than the osmotic potential of the solution; if it moves down, the ψ_{plant} was higher. The ψ_{plant} is equal to the osmotic potential of the solution when the drop diffuses in all directions. Tables of osmotic potential versus molality of solutions are available. For details and precautions, see Knipling (1967), Knipling and Kramer (1967), and Barrs (1968).

Relative Water Content

Relative water content (RWC) provides a simple means by which information concerning the water status of plants can be obtained. RWC is an expression of the current plant water status relative to what it would be if the plant were fully hydrated.

$$RWC = \frac{\text{fresh weight} - \text{dry weight}}{\text{fully turgid weight} - \text{dry weight}} \times 100 \qquad \text{(Equation 18-9)}$$

Fully turgid weight is obtained by exposing the tissue to water until it no longer increases in weight. RWC is an accurate measure of water status but gives no information concerning plant water status with respect to soil or atmosphere, nor any indication of the plant's response to water stress.

Other Measures of Plant and Soil Water Status

Barrs (1968) has reviewed the extensive literature on water status methodology in plants, and Rawlins (1976) has done the same for soils. Several more useful references and more specific information can be found in Brown (1970) and Brown and van Haveren (1972).

Summary

Water potential is a measure of the free energy of water. Knowing the water potential of a system gives information concerning the potential for movement of water to a neighboring system of known water potential. Water potential is determined by four properties of the system. (a) The matric potential varies with the amount of surface area affecting water molecules. (b) The osmotic potential depends on the solutes in the system. (c) The pressure potential is a measure of energy in the system due to changes in pressure. (d) Gravitational potential, an important component of water potential in saturated soils, is a measure of gravitational pull on water in the system.

Atmospheric water vapor concentrations are important influences in the plant environment. Evapotranspiration, the loss of water by evaporation and transpiration, is highest in air with a high saturation deficit. Transpiration in plants is controlled by stomates located at the point in the soil–plant–atmosphere system where the most dramatic drop in water potential occurs—the leaf surface. Evaporative cooling associated with transpiration is an important mechanism of temperature regulation in plants. The resistance to water loss of leaves depends on (a) boundary layer resistance as water vapor diffuses through still air at the leaf surface, (b) stomatal resistance as the stomatal pore is opened and closed in response to internal and external conditions, and (c) resistance in the internal air space of the leaf as water diffuses from the saturated mesophyll cell walls toward the stomate. The size, shape, and orientation of leaves determine, in part, their effectiveness in a particular environmental setting. Transpiration and diffusive resistance measurements can be made in gas exchange systems, using steady-state diffusive resistance porometers, or (less expensively) by chemical means.

Transport of water through roots and stems depends on atmospheric water potential, soil water potential, and characteristics of the vascular system. Water potential is highest in the soil and decreases continuously to the atmosphere for an actively transpiring plant. Smaller vessels have greater resistance. They are common in plants from arid habitats and at the tops of tall trees, where new vascular tissues are formed under water stress.

Water held in soil is available for plant growth to varying degrees, depending on soil texture and structure, characteristics of the plant, and soil salinity. Movement of water in the soil depends on the water potential differences along the path of movement and on the hydraulic conductivity of the soil. Plants growing in soils at field capacity show small diurnal fluctuations in tissue water potential as transpiration exceeds absorption during midday. As the soil dries, the magnitude of the diurnal fluctuations becomes greater until plants reach wilting during midday. When soil water potential is above plant osmotic potential, tissue water potential will equilibrate with soil water potential during the night. Depletion of soil water to the point of permanent wilting has occurred when soil water potential is equal to tissue osmotic potential.

The osmotic potential of tissue determines the degree of water stress the plant can tolerate without serious decreases in tissue water. Plants exposed to dry environments tend to have higher solute concentrations in the tissues, allowing them to extract water from the soil under situations of low soil water potential. Some plants acclimate to changes in water availability by actively modifying tissue osmotic potential.

Methods of measuring soil and plant water include the pressure chamber, psychrometry, dew point hygrometry, the dye method, and relative water content. Many of these techniques depend on recently developed technology, so few studies have been conducted that utilize the full potential of these instruments. Much remains to be done if we are to understand the wide range of water-related adaptations of terrestrial plants.

CHAPTER 19

WATER
Environment and Adaptations

P lant ecologists have expended a great deal of effort on considerations of water, since no other single environmental factor can be directly related to so many plant responses. We noted the significance of water in determining productivity in Chapter 12, in influencing biogeochemical cycles in Chapter 13, as an influence on leaf temperature in Chapter 14, and, if we were to look carefully, in almost every other chapter in this text. In this chapter, we hope to place the diverse effects of water on plants into perspective. This chapter covers the general aspects of water availability to plants, factors determining the distribution of water over the landscape, and the vegetational patterns that are correlated with the distribution of water.

Water in the Environment

Consider water availability in various environments, and the contrasts between a cypress swamp in Georgia and the Sonoran Desert of Arizona are clear. Less apparent yet significant variations in water availability occur in most plant communities. It is the spatial and temporal variations in the hydrologic cycle that determine the amount and effectiveness of precipitation in plant communities. The cyclic nature of water is shown in Figure 19-1. Condensation of atmospheric water vapor in the form of precipitation, fog, or dew places water into the biosphere, where it is temporarily used or stored in bodies of water. Regardless of its immediate fate, processes of evaporation and transpiration eventually cycle the water molecules back to the vapor state.

Fog and Dew

Dew, the condensation of water on objects in the environment, is a source of water for plants, the importance of which has been much debated. In a lucid

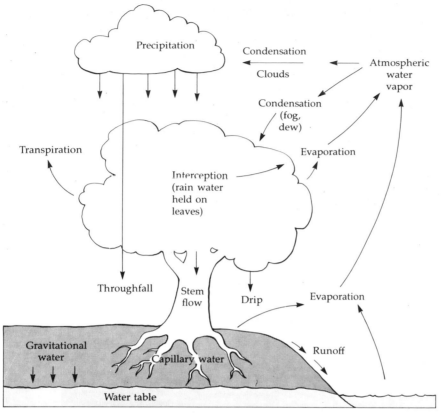

Figure 19-1   Schematic diagram of the hydrologic cycle, showing potential pathways of water flux in an ecosystem.

review of the literature on dew, Stone (1963) reported that there was no evidence suggesting that the presence or absence of dew acted as a limiting factor, but that it was a contributing factor to competitive success. More recent studies support Stone's conclusions and document conditions in which the presence of dew may promote survival.

Dew forms when the temperature of a surface in the environment falls below the temperature at which water will condense from air (the dew point). The formation of dew is promoted under conditions conducive to rapid cooling. Condensation of dew reaches the greatest levels under clear skies (clouds reduce reradiation), low wind speeds, high vapor pressures (which reduce the amount of cooling necessary), and beneath vegetation with open branching and leaf area spread over considerable heights. Lloyd (1961) found that no dew formed under a closed forest canopy, presumably because reradiation was reduced. Dew formation is also enhanced with cool surfaces, poleward-facing slopes, in flats and valleys with cold air drainage (Chapter 14), and in areas subjected to cold ocean air when conditions are not such that fog will form. Even under optimal conditions, the theoretical maximum dewfall during a single night is less than 1 mm (Slatyer 1967; Stone 1963).

Soil water may be redistributed by vaporization and condensation, similar to dew formation, within the soil profile. This is referred to as **distillation** by Stone (1963). Periods of high ambient temperatures heat the upper layers of soil. The heat then moves downward through the soil. When ambient conditions cool, a cool zone of air follows the heated zone into the soil. Air in the interstitial spaces of the soil becomes warmed, and water in the heated zone vaporizes and moves upward in the soil. When the warmed, high-moisture air mass encounters the cool soil particles above, the warmed air can be cooled to dew point. Water from lower in the soil has moved upward in the vapor state and condensed back to the liquid state in the upper zones of the soil. Syvertsen et al. (1975) used this phenomenon to explain daily fluctuations in soil water potential in the 20–40 cm depth rooting zone of desert shrubs. Vaporization below, and condensation in, the rooting zone redistributes the soil water to areas where it may be absorbed.

Dew may also collect on plant shoots. The amount of dew harvested depends on species morphology, but little quantitative data are available concerning dew harvest by various species. Shure and Lewis (1973) noted this difference in ability to harvest dew accidentally while introducing radioactive tracers into stems of old-field weeds. Wells placed on stems to hold tracers were found full of dew water on common ragweed (*Ambrosia artemisiafolia*) but empty on wild radish (*Raphanus raphanastrum*) and lamb's quarter (*Chenopodium album*). Ragweed collected 2.5–3.2 ml plant^{-1} day^{-1} for an estimated 1100 ml m^{-2}mo^{-1}. This consistent addition of soil water during the growing season may be the factor that gives ragweed the competitive edge during the first year of succession in the northeastern United States.

Stone also found that dew present on leaves had a positive influence on *Pinus ponderosa* seedlings subjected to drought. Stressed seedlings lived an average of 3 months if sprayed with water, whereas those not sprayed lived only 2 months. Unsprayed seedlings used twice the water of sprayed seedlings. Similar advantages were found for corn stressed to near wilting (Duvdevani 1964). Therefore, dew may be important in reducing water requirements by reducing the leaf–air ψ gradient even when it never enters the soil.

Fog, especially along coastal areas (Figure 19-2), has long been recognized as an important ecological factor. Fog forms when warm, moist air (a) passes over cold water or land (as in the coastal communities of the Pacific Northwest), (b) cools as it ascends a mountain mass (as in tropical cloud forests), or (c) is cooled by rapid reradiation from soil (as in inland valleys). Fog formation is greatly enhanced by the presence of condensation nuclei. Condensation nuclei are frequently salt particles released by the bursting of bubbles from surf, white caps, etc. (Boyce 1951), or dust, pollen, and other particulate matter in the atmosphere.

The importance of **fog precipitation** is most pronounced near the edge of a forest, because the water droplets are rapidly removed by obstacles in the environment. This is why fog gauges are rain gauges with a cylindrical, vertical screen extending above the opening, forming an obstacle that causes water droplets to enter the rain gauge.

Figure 19-2 Early summer coastal fog in southern California.

The giant coast redwood forests of northern California are frequently said to exist because of fog. Oberlander (1956) measured summer fog precipitation on the San Francisco peninsula and found that 4.8–150 cm of water was added to the soil during a single month without rainfall. Azevedo and Morgan (1974) measured 123–388 cm (depending on the species) of fog precipitation in coastal California forests during 28 summer days. Davis (1966) suggested a relationship between ocean fog and the distribution of spruce-fir forests on the coast of Maine. Vogelmann et al. (1968) measured a 66.8% increase in precipitation due to fog at 1100 m elevation, but little or no increase below 850 m, in the Green Mountains of Vermont. Fog is a potentially significant source of moisture, increasing precipitation 1.5–3 times in many habitats.

There may be effects of dew and fog other than providing additional moisture. For example, the removal of foliar metabolites from plants may result in rapid nutrient recycling, especially of K and Ca (Henderson et al. 1977; Tukey and Mecklenberg 1964; Azevedo and Morgan 1974). Tukey (1966) measured carbohydrates, amino acids, and organic acids in plant leachates; del Moral and Muller (1969) measured allelopathic substances in plant leachates.

The direct absorption of water from the surface of leaves into the plant does occur but is greatly restricted by a number of factors. The water potential at the surface of mesophyll cells is relatively high, so only a small energy gradient is established toward the leaf interior even when pure water is on the leaf surface. Stomates are usually closed during periods of dew deposition, thus severely restricting water vapor entry, especially in xeric-adapted plants (Vaadia and Waisel

1963). Surface tension of water droplets prevents liquid water from moving through the small stomatal openings. Slatyer (1967) concluded that absorption of surface water on leaves could never fulfill more than a small portion of the water requirements of plants. At best, it may speed nocturnal recovery of turgor pressure in wilted plants and may delay the onset of stress after sunrise.

Precipitation

Water droplets formed around condensation nuclei are called *fog* when in contact with the earth; they are called *clouds* when a layer with no condensation appears between them and the earth. As these water droplets become large enough to respond to gravity, precipitation will fall as rain. At temperatures below freezing and when conditions are such that the water arrives at the earth's surface as ice, the precipitation is called *snow*. Under certain circumstances, sleet or hail may result from summer storms, but these forms of precipitation are of little consequence as sources of water for plants (although they may cause considerable physical damage on a local basis).

Condensation of atmospheric moisture sufficient to result in precipitation is stimulated by any of three sets of atmospheric conditions. Temperature generally decreases with altitude, so any condition where warm, moist air rises may cause precipitation. This type of precipitation is classified as either orographic precipitation (oreos = mountain) or convectional precipitation. **Orographic precipitation** occurs when an air mass is forced upward as it moves across a mountain mass. **Convectional precipitation** occurs during the summer, as warm humid air rises abruptly from the earth's surface because of its low density. As this air reaches high altitudes, intense but usually brief summer showers or thunderstorms occur. Cool-season precipitation in temperate areas is usually associated with a low pressure center moving easterly and along cold polar air masses. **Cyclonic precipitation** occurs as air masses move counterclockwise (in the Northern Hemisphere) around the low pressure center, become warmed on the advancing front, ascend over cool local air masses, cool, and cause large areas of sustained precipitation. Because of the seasonally predictable location of cool polar air, high pressure areas, and pathways of cyclonic storms, seasonal patterns and amounts of precipitation are quite predictable in those areas where the dominant cause of precipitation is cyclonic.

Rainfall   The spatial and temporal distribution of rainfall is very complex and has great effects on the productivity, distribution, and life forms of the major terrestrial biomes. The factors determining the distribution and availability of rainfall and the response of plants to these variations have occupied plant ecologists for decades.

Vegetation may reflect patterns of orographic precipitation. In California, for example, the deciduous coastal sage scrub community gets less precipitation than the higher-elevation chaparral. In very high mountains, such as the Sierra Nevada, air masses continue to move upward after the majority of water has precipitated, so subalpine and alpine communities get much less precipitation than middle-

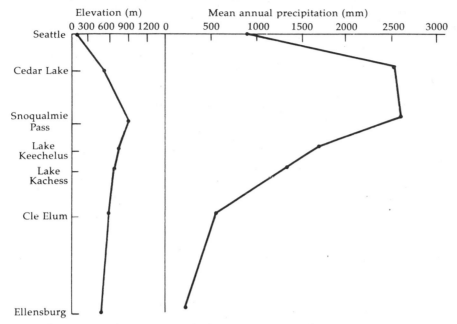

Figure 19-3   Precipitation along a cross section of the Cascade Range in the vicinity of Snoqualmie Pass, Washington (47° 25′ N. lat.); the distance from Seattle to Ellensburg is approximately 152 km. (From Franklin and Dyrness 1969, *Natural vegetation of Oregon and Washington,* U.S. Dept. of Agriculture Forest Service Research Paper PNW-80.)

altitude associations. Moisture-laden air masses frequently pass around high peaks through depressions and breaks in mountain ranges, leaving these peaks essentially devoid of orographic precipitation.

Once these air masses cross the summit of a mountain mass, the air expands and warms, greatly increasing its water holding capacity. The effect is to increase potential evapotranspiration and to decrease precipitation on the lee side. This low precipitation belt is referred to as a **rainshadow**. The pattern of precipitation in the Cascade Mountains of Washington is influenced primarily by topographic factors; it is illustrated in Figure 19-3.

Topography also influences the hydrological cycle by affecting potential evapotranspiration. Chapter 14 describes the variations in solar radiation incident upon slopes of various aspects. Wentworth (1981) incorporated solar radiation with elevation and slope parameters to derive an index to correlate with vegetation mosaics. Significantly less total solar radiation is incident on poleward slopes, leading one to suspect that more water would be present on these slopes because of lowered evapotranspiration. However, several studies (e.g., Griffin 1973, Ng and Miller 1980) show that plants on poleward slopes could be under greater stress during the dry season than those on warmer slopes. Presumably, these poleward slope plants have a reduced capacity to restrict water loss compared with those on other slopes and therefore deplete the available water supply more quickly.

Humid air masses may accumulate significant amounts of water vapor from several sources. The most important source is the oceans, where salt particles form condensation nuclei, resulting in cloud formation. Ocean proximity is thus important in determining amounts of precipitation. Proximity to the ocean does not, however, always assure a mesic environment. In certain geographic situations, such as in Baja California, Peru, and southwestern Africa, coastal deserts are formed when cold air masses (cooled by passing over cold ocean currents) contact a warm land mass; the air masses warm and cause high rates of evaporation. Conversely, warm ocean currents create warm air masses and are the source of very high levels of coastal precipitation when these heavily water-laden air masses contact cooler land masses, as in the Pacific Northwest.

Snow   Snowfall has significant ecological impact in many areas of the world—as the prime source of water to replenish underground supplies, as a source of soil moisture, as an insulator in areas of extreme cold, and as a physical force important in controlling the distribution of vegetation. Snow varies considerably in water content per unit volume, but typically 8–12 cm of snow is equivalent to 1 cm of rainfall. Increases in soil water from snow depend on the rate of melt. Sudden increases in temperature may cause rapid snowmelt, so that much will be lost as runoff, as in a very intense rainstorm. Slow melting will increase infiltration and water availability to plants in the immediate vicinity. Soil temperatures under snow seldom fall much below 0°C, because the insulating capacity of the snow traps heat radiating from the earth, thus protecting plants from extreme cold. Protection from wind is also important where temperatures are low and winds are desiccating. The physiognomic modification called **krummholz** (Figure 19-4) is due to desiccation and physical damage caused by wind and blowing ice crystals. In Wyoming, *Picea engelmannii* and *Abies lasiocarpa* needles directly exposed to winter winds experience greater water stress than leeward needles or those protected by snow. Hadley and Smith (1983) found a high correlation between leaf water status and leaf mortality in these timberline conifers. Patterns of vegetation in subalpine forests are also dependent upon snow. Meadows within subalpine forests are often the result of deep snow accumulation and the resulting shorter growing season. Snow avalanches frequently clear large areas of subalpine forest, thereby modifying the vegetational mosaic of steep mountain slopes.

Snow has been more thoroughly studied in alpine tundra habitats than in other communities, because of the very great importance of snow in determining the distribution and abundance of plants in that habitat. Summer rainfall is typically very low in alpine communities, so the snow cover provides the major source of soil moisture for these plants. In fact, most scientists agree that the related factors of snow cover and soil moisture are the primary limiting factors for vascular and nonvascular plants in alpine communities (e.g., Billings and Bliss 1959; Johnson and Billings 1962; Buttrick 1977; Hrapko and LaRoi 1978; Komarkova and Webber 1978; Flock 1978). As shown in Figure 19-5, a constant

Figure 19-4   Krummholz vegetation at timberline in the Rocky Mountains. Flagged appearance is caused by the abrasive and desiccating effects of wind and blowing ice crystals. (Photo courtesy of Harold Bradford.)

supply of water below snowbeds results in meadow communities in drained areas and sedge-sphagnum communities in concave areas. The higher, steeper slopes above snow accumulations and melt are usually boulder fields with little vegetation, because of the combined influence of maximum winter exposure and summer desiccation. The importance of snow as an ecological factor results from both its influence on water supply and its role in physical modification of the habitat. In alpine habitats in Colorado, plant productivity is highest in areas protected by moderate amounts of snow during the winter, where snow melt occurs early in the season, and where soil moisture is high. Snow banks that persist well into summer retard development of plants, even though soil moisture may be high. Flowering time in alpine communities is closely related to the persistence of the snow cover (Greenland et al. 1984). The distribution of alpine populations of *Kobresia bellardii* (*myosuroides*) depends on the longer growing season provided by light snow (and early snow melt) in the Rocky Mountains. Some snow is necessary for protection from wind (Bell and Bliss 1979).

Precipitation Effectiveness

Knowledge of the total amount of precipitation falling on a community does not always give a clear picture of the availability of water to plants. The season, atmospheric condition, precipitation type, intensity, annual variation, soil con-

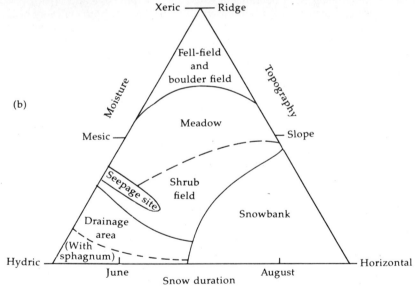

Figure 19-5 (a) Uneven patterns of snow accumulation in a Rocky Mountain alpine community. (b) Relationship between topography, moisture, and snow duration gradients in northern British Columbia alpine communities. [(a) Photo courtesy of Harold Bradford. (b) From Buttrick 1977. Reproduced by permission of the National Research Council of Canada from the *Canadian Journal of Botany*, Volume 55, pp. 1399–1409, 1977.]

dition, and vegetation physiognomy influence the availability of precipitation and its distribution within the habitat. Many attempts have been made to determine **site water balance** (or **water budgets**), based on measurements of rainfall, runoff, percolation, and evapotranspiration (Rosenberg 1974). However, the accurate measurement of all the parameters necessary to estimate these properties is technologically difficult. Luxmoore (1983) was able to obtain detailed simulations of water fluxes in an oak-hickory stand in Tennessee, using a computer model. The data base included measurements of soil water, meteorological phenomena, and plant water status.

The conditions discussed earlier that influence atmospheric vapor pressure are important determinants of water loss through evaporation. In fact, measurements of temperature, saturation deficit, and wind have been combined to give a meaningful picture of biologically significant variations in climate. Significant evaporation occurs from free water surfaces, such as droplets intercepted by the vegetation canopy, and in slow-draining depressions. Evaporation from exposed soils is limited to the upper 10–20 cm, but there is a certain amount of water vapor transfer from lower levels, especially in desert environments. A similar transfer to the vapor state may occur from the surface of snow or ice. This process, called **sublimation**, takes place as ice is transformed to the vapor state without going through a liquid phase. Sublimation, like evaporation, reduces the effectiveness of precipitation.

Seasonal distribution of precipitation determines, in part, how much of the water is available to the vegetation. An annual precipitation rate of 70–80 cm can support either certain deciduous forest communities or the strongly drought-adapted broad sclerophyll vegetation of mediterranean climates. The distinguishing factor is that almost all of the precipitation in mediterranean climates falls during the cool season, so that chaparral plants must endure annual extended drought during the hottest months. In the semi-arid mediterranean climate, summer drought occurs regardless of the amount of winter precipitation or the vegetative cover (Miller et al. 1983). If the precipitation were more evenly distributed or concentrated in the warm season, a more mesic vegetation could be expected. Where a large portion of the precipitation falls as snow, its effectiveness depends on the melt rate. When temperatures warm suddenly, a rapid melt will render most of the snow ineffective—much of the moisture will be lost as runoff, particularly if the soil is frozen or already saturated.

Intensity of rainfall is an important determinant of runoff and is related to seasonality. Typically, summer showers in temperate areas are of very short duration and may be so intense that most water will run off the surface before the slower soil infiltration process can take place.

Predictability of precipitation is also reflected in the flora and vegetation. Many deserts average enough annual precipitation to support more mesic plants, but rainfall is very unpredictable; periods between effective precipitation may extend up to several years. Plants are therefore adapted to respond quickly to precipitation and also to tolerate periods of drought. Where droughts are predictable, seasonality patterns in plant response are also predictable, and little

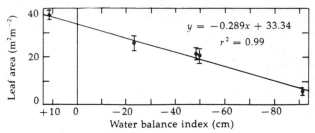

Figure 19-6   Relationship between water balance index and leaf area in five forest zones of western Oregon. Zones plotted are, from left to right: *Picia sitchensis, Tsuga heterophylla,* interior (Willamette) valley, east slope mixed conifer, and *Juniperus occidentalis.* Bars show range of observed leaf areas in each vegetation zone. (From "Leaf area of mature northwestern coniferous forests: Relation to site water balance," by C. C. Grier and S. W. Running 1977. *Ecology* 58:893-899. Copyright © 1977 by the Ecological Society of America. Reprinted by permission.)

impact on the vegetation is apparent. In mesic environments, however, periodic unpredictable droughts may have severe impacts on the vegetation.

Many attempts to measure site water balance have been made (e.g., Penman 1950; van Bavel 1966; Major 1977). Grier and Running (1977) devised a site **water balance index** that correlates closely with leaf area in coniferous forest communities of western Oregon (Figure 19-6). The index was calculated by adding soil water storage to measured growing season precipitation and then subtracting open pan evaporation. This approach may be very instructive in those areas where low nutrient levels, severe temperature, physical damage, and high levels of runoff are not important factors. Major (1977) has compiled water balance information for 12 sites on a transect across the Sierra Nevada (Figure 19-7). Water surpluses are greatest at middle elevations; summer water deficits there are less than at lower-elevation sites in the Central Valley or in the rainshadow of the Sierra Nevada, where higher temperatures increase potential evapotranspiration.

Interception, Throughfall, and Stem Flow

The distribution of rainfall under a forest canopy depends on canopy structure. It influences the amount of water entering the soil at the base of the trunk (**stem flow**) and the amount of rainfall coming through the canopy (**throughfall**). Lawson (1967), Leonard (1961), and Henderson et al. (1977) agreed that 10–20% of precipitation is intercepted by forest canopies (somewhat less during the dormant season). Henderson et al. (1977) compared amounts of throughfall in various forest types and found no significant difference between types. Most of the intercepted water is lost to evaporation, since little is absorbed by the leaves. Stem flow causes a very uneven distribution of soil water by concentrating precipitation near the trunk of the tree. Geometric shapes that encourage stem flow may be a competitive advantage in water-limiting environments.

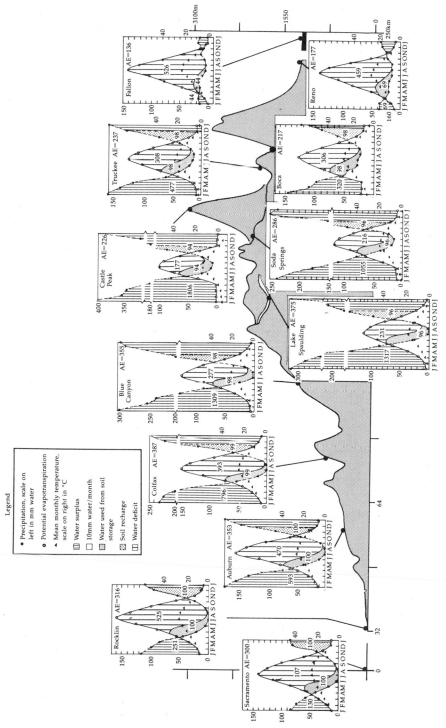

Figure 19-7 A transect over the Sierra Nevada from Sacramento (38.5°N), over Donner Summit (39.3°N), to Reno and Fallon, Nevada (39.5°N), showing water balances. Vertical exaggeration 14.5×. (From Major 1977. In *Terrestrial Vegetation of California*. Edited by Barbour and Major. Copyright 1977 John Wiley and Sons, Inc. Reprinted by permission.)

Special Adaptations

Osmotic Balance and Toxicity in Saline Habitats

Halophytes are specifically adapted to widely varying external and internal concentrations of salt. Most saline habitats (Figure 19-8) contain sodium chloride as the primary salt, but sodium, magnesium, and calcium sulfates, magnesium and potassium chlorides, and sodium carbonate are also significant in many situations. Glycophytes (plants that are not salt tolerant) generally exclude salt from entering the roots. High concentrations of salt at the root surface may soon reverse the chemical energy gradient, causing plant tissues to dehydrate (Greenway and Munns 1980). Ownbey and Mahall (1983) found that the glycophyte *Raphanus sativus* adjusted osmotically to high salinity but suffered long-term salinity damage to processes associated with root conductivity. Thus intolerance in some glycophytes may be related to toxicity or nutrient imbalance. Maintaining the chemical energy gradient necessary for water uptake requires that osmotic concentrations in halophytes be higher than in glycophytes. Even though protoplasmic salt tolerance is higher in halophytes than in glycophytes, there is a limit to this tolerance. Mechanisms that restrict cytoplasmic salt concentrations are necessary to halophyte survival.

Regulation of salt concentrations in the tissues of halophytes may be classified into three basic categories (Albert 1975). First, some halophyte species restrict the rate of salt uptake by increasing rates of ion accumulation, only for brief periods or in certain habitats. Ungar (1977) has shown that saltbush (*Atriplex trinagularis*) accumulates salt in direct proportion to the salt content of the soil. Salt uptake is restricted or excluded by root membranes of some species, such as mangrove (Scholander 1968). When salts are excluded, cellular osmotic potential is adjusted by increasing the amounts of organic metabolites in order to maintain the free energy gradient necessary for uptake. Various nitrogen compounds (amino acids and amines), carbohydrates, and organic acids accumulate

Figure 19-8 Halophytes growing in a desert depression. Note the salt accumulation at the soil surface. (Photograph courtesy of Alan Romspert.)

in cells of halophytes. Such osmotic adjustments occur in ice plant (*Mesembryanthemum crystallinum*), a salt tolerant succulent that switches from C_3 photosynthesis to CAM and accumulates malate when growing in a saline medium (Luttge et al., as cited in Flowers et al. 1977). Brownell and Crossland (1972, 1974) have shown that some C_4 and CAM plants require or are stimulated by sodium ions in the soil. Such evidence has led to speculation that the C_4 and CAM biochemical pathways were an adaptation to accumulate organic acids in order to counterbalance the osmotic influences of saline environments (Laetsch 1974; Flowers et al. 1977).

Salt dilution and salt extrusion are the other ways in which halophytes regulate salt concentrations in their tissues. **Succulence** caused by the accumulation of water in the tissues serves to dilute salts in cells, thereby reducing solute concentrations (Jennings 1968). Rapid growth of new tissues also has a diluting effect, since salts are distributed to greater volumes of tissue (Greenway and Thomas 1965). **Salt extrusion** is another common mechanism by which excess salts are removed from the plant body. Some species (e.g., *Atriplex* spp.) have salt glands that expel salt onto the surface of leaves, where it can be washed from the plant by rain, dew, or fog condensation (Figure 19-9). Vesicular bladder hairs are another mechanism whereby salt is deposited on the leaf surface. The hair fills with salt water until it ruptures, releasing the salt onto the leaf. Salt may

Figure 19-9   A scanning electronmicrograph of the leaf surface of saltbush (*Atriplex*) showing salt glands. (From *Probing Plant Structure* by T. Troughton and L. A. Donaldson, 1972. Reed Methuen Publishers, Auckland, New Zealand.)

also be extruded by the shedding of organs that have accumulated salt. Salt evidently induces senility; as the organ ages, nutrients and stored organics are mobilized to younger parts of the plant, followed by abscission of the organ containing the accumulation of salt.

Any one species usually employs a combination of these mechanisms to reduce the salt load. Further details may be obtained from the literature, which was reviewed by Waisel (1972), Ranwell (1972), Reimold and Queen (1974), and Flowers et al. (1977).

Anatomical Adaptations

There is abundant information concerning anatomical responses to water deficits (e.g., Esau 1965; Hanson 1917; Oppenheimer 1960; Daubenmire 1974). Much of it is now being critically reevaluated in light of new technology and plant response information. **Xeromorphic leaves** are generally small, with reduced cell size, thicker blades, smaller, sunken, and more dense abaxial stomates, more pubescence, a thicker cuticle, more heavily lignified epidermal cells, better developed palisade mesophyll, and less intercellular space. It is true that these modifications are found in arid adapted plants, but we can no longer assume that all plants growing in hot, dry environments have similar anatomical adaptations. For example, creosote bush (*Larrea tridentata*), the most common shrub in the warm North American deserts, has no pubescence, stomates evenly distributed on both surfaces, stomates not in crypts, etc. (Barbour et al. 1974).

Pubescence, leaf size, and the ratio of mesophyll cell (internal leaf) surface to outer leaf surface (A^{mes}/A) interact to determine water use efficiency in desert plants (Cunningham and Strain 1969; and Figure 19-10.) Leaf pubescence not only increases the depth of the boundary layer but is even more important as a regulator of light absorptance and, as a result, leaf temperature (Ehleringer et al. 1976; Smith and Nobel 1977b). Small leaves are more efficient energy dissipators and lose less water per unit leaf area than large leaves. The ratio of mesophyll cell surface to outer leaf surface may be less important in regulating water loss than formerly assumed; its prime importance is in the resistance to CO_2 uptake in the liquid phase (see Chapter 15). Smith and Nobel (1977b) studied the desert

Figure 19-10 Schematic of the influences of pubescence, leaf length, and A^{mes}/A on transpiration and photosynthesis. (From "Influences of seasonal changes in leaf morphology on water use efficiency for three desert broadleaf shrubs" by W. K. Smith and P. S. Nobel, 1977. Ecology 58: 1033-1043. Copyright © 1977 by the Ecological Society of America. Reprinted by permission.)

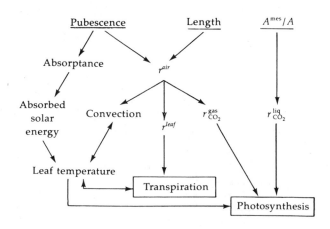

shrub *Encelia farinosa*, which develops leaves that vary seasonally in size, pubescence, and A^{mes}/A. Leaves produced in hot, dry conditions are small and highly pubescent and have a high A^{mes}/A ratio. With cooler, more mesic conditions, new leaves develop that show decreased A^{mes}/A and pubescence in conjunction with increased size and absorptance. Such morphological changes increase potential photosynthesis over a variety of environmental conditions. Water use efficiency was thus maximized in summer months and decreased during periods of cooler temperatures and more rainfall. Pubescence functioned to reduce absorptance, thus lowering leaf temperature in conjunction with reduced leaf size. Minimum r^{leaf} occurs during the wetter periods, maximizing photosynthesis when transpirational loss is not critical. *Mirabilis teniuloba* coexists with *E. farinosa* but lacks the capacity to modify r^{leaf} seasonally. Its active photosynthesis is restricted to a shorter period, and the photosynthetic rate is lower during favorable periods (Figure 19-11).

Figure 19-11 (a) Seasonal transpiration, (b) net photosynthesis, and (c) water use efficiency for *Encelia farinosa* (triangles) and *Mirabilis tenuiloba* (circles), based on field measurements. (From "Influences of seasonal changes in leaf morphology on water use efficiency for three desert broadleaf shrubs" by W. K. Smith and P. S. Nobel, 1977. *Ecology* 58: 1033-1043. Copyright © 1977 by the Ecological Society of America. Reprinted by permission.)

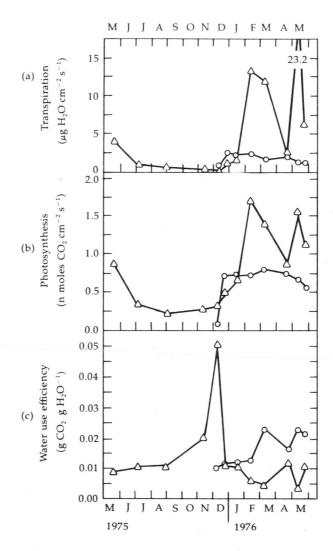

Succulence was discussed in relation to halophytes and with respect to CAM, but we have not yet addressed the water relations of desert succulents. Park Nobel (1977) studied the barrel cactus (*Ferocactus acanthodes*) in the Sonoran Desert. Perhaps the most important aspect of the stored water in succulents lies in the ability of plants to continue a diurnal stomatal opening after soil water potential falls below plant water potential. In the barrel cactus, this occurred at a high (for a desert plant) value of -0.45 to -0.47 MPa. Stomates continued to open for another 50–60 days, remaining closed only after soil water potential reached -9.0 MPa (more negative values are associated with less available water). Following a 7-month period of no major rainfall, plant water potential fell to only -0.62 MPa (close to the permanent wilting point, as osmotic potential was nearly the same). Such high levels of tissue hydration allowed stomatal response to 19 mm rainfall within 24 hours, with tissue rehydration and full stomatal opening in 48 hours. Cacti have a high water use efficiency because they are CAM and store a volume of water large enough to continue CO_2 uptake for long periods after the soil water potential drops below the permanent wilting point. A question which deserves further study is: What keeps the water from diffusing out into the soil when the plant water potential is over 8 MPa higher than soil water potential?

Life Form Responses and Habitat Selection

Evergreen Plants

Plants that maintain leaves during periods of water stress must have a high degree of tolerance for desiccation. The obvious advantage of being evergreen is that when water again becomes available, there is no lag while new tissues are formed; existing tissues quickly rehydrate and become active. Many evergreen species do, however, shed some leaves during periods of water stress. Presumably, this serves to reduce surface area and water loss (Oppenheimer 1960; Mooney and Dunn 1970*b*; Evanari et al. 1971; Orshan 1972).

Drought-Deciduous Plants

Some species avoid water stress by becoming dormant during the dry season (Mooney and Dunn 1970*b*; Mooney 1977*b*). The coastal scrub community of California (Figure 19-12) is an example of an entire community that avoids the hot, dry mediterranean summer by shedding leaves. These plants show maximum activity in the cool, wet winter months. Ocotillo (*Fouquieria splendens*) is a desert shrub that ephemerally becomes deciduous, perhaps four or five times a year. The leaves are simply not drought tolerant; they drop as stress increases, returning within a week after new rains. (Cannon 1905*a*; Scott 1932). For more details on ocotillo, see the desert scrub section of Chapter 20.

Paloverde (*Cercidium floridum*) is an example of a drought-deciduous plant that lives in desert areas in a leafless condition but has the advantage of green

Figure 19-12   Drought-deciduous coastal scrub community of southern California. (Photograph courtesy of Ted L. Hanes.)

stem tissue. This gives paloverde and other similar desert plants the advantage of evergreenness and the advantage of increased photosynthetic area during cool, moist periods (Adams and Strain 1968, 1969; Adams et al. 1967).

Phreatophytes

Phreatophytes (literally, "well plants") are usually found in riparian (stream-side) habitats. Even in arid zones, they may develop into very large trees (Figure 19-13). They are restricted to habitats of permanent underground water supplies. Cottonwood (*Populus fremontii*), willow (*Salix* spp.), sycamore (*Plantanus racemosa*), fan palm (*Washingtonia filifera*), and salt cedar (*Tamarix* spp.) are examples of phreatophyte trees that avoid many of the rigors of arid environments by having roots in constant contact with the fringe of capillary water above a water table. These and many associated shrubs are examples of *obligate* phreatophytes. Other species are apparently able to take advantage of ground water when present but can tolerate periods of low water availability. These *facultative* phreatophytes (e.g., mesquite (*Prosopis glandulosa*) and saltbrush (*Atriplex polycarpa*)) frequently occur in desert depressions where water and salts accumulate. Because of their high water requirement, many facultative and obligate phreatophytic species must be tolerant of rather high levels of salinity. Horton (1977), Horton and Campbell (1974), McDonald and Hughes (1964), Campbell and Dick-Peddie (1964), and Vogl and McHargue (1966) have considered various aspects of phreatophyte ecology.

Figure 19-13   Riparian community with large phreatophyte trees in the Chihuahuan Desert of the southwestern United States. (Photograph courtesy of Patrick H. Boles.)

Not all plants associated with watercourses are phreatophytic, but in a given location they are able to survive only in areas with more soil moisture. Riparian plants may also be those able to survive low oxygen conditions in flooded or highly saturated soils, which eliminate upland species from flood plains. In mountain areas, a species may be restricted to riparian habitats at lower elevations and extend on to slopes only at higher elevations, where more water is available. In desert areas, some species (e.g., *Chilopsis linearis*) are winter deciduous and relatively intolerant of midsummer conditions. They survive only near watercourses, where the additional moisture allows survival between spring and fall growth periods (Odening et al. 1974).

Ephemerals

Ephemerals are annual plants that germinate in response to periodic phenomena and are able to complete their life cycle during short periods of mesic conditions. Ephemerals are the most common life form in severe desert situations (Whittaker and Niering 1975), where they are able to survive because they grow only during periods of moderate temperatures and soil water availability. Germination and mortality of desert ephemerals are largely independent of photoperiod, since they are controlled by soil water and temperature (Went and Westergaard 1949; Went 1948; Tevis 1958*a,b*). Beatley (1974, 1969, 1967, 1966) has studied the relationship between the ephemerals of the Mojave Desert and the environmental factors triggering germination and death. Details of these studies can be found in Chapters 5 and 20. C_3 and C_4 plants occupy different temporal habitats—summer ephemerals are typically C_4, and winter ephemerals are typically C_3 in the southwestern desert areas (Mulroy and Rundel 1977; Syvertsen et al. 1976).

Summary

The success of plants depends on a multitude of factors. In terrestrial environments, limitations related to water are always one such factor. Environments differ in the relative amounts of rain, snow, dew, and fog precipitation they receive and in the seasonal distribution of available water. These factors interact with habitat and phenological development of plants to determine the effectiveness of precipitation. The distribution of 1000 mm of precipitation in an environment with a precipitation/evaporation ratio of 2 is presented in Figure 19-14.

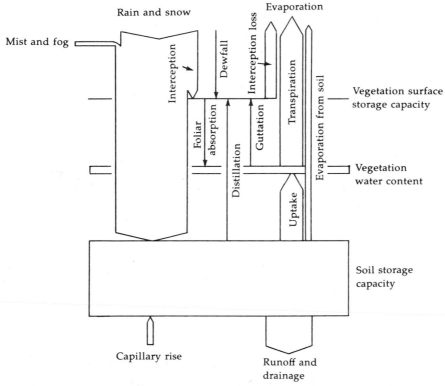

Figure 19-14 Diagram summarizing the hydrological cycle in a vegetation–soil system; the components are approximately to scale for an annual rainfall of 1000 mm and evaporation of 500 mm. (Guttation is the exudation of liquid water from leaves.) (From Rutter 1975. "The hydrological cycle in vegetation" in *Vegetation and the Atmosphere*, vol. 1, ed. J. L. Monteith. By permission of Academic Press, London.)

Large amounts of precipitation are stored in the soil or lost as runoff and drainage; minor amounts are intercepted by the vegetation canopy (most of which evaporates); and most of the soil water is vaporized through the vegetation as transpirational water loss. Evapotranspirational water flux depends on the amount of water present and on the evaporative power of the air.

Dew is water that condenses from the air onto objects in the environment with a temperature at or below dew point. Factors promoting dew formation include clear skies, still air, and high humidity. Some plants collect considerable amounts of water in the soil around the plant by having morphological features that funnel the water to the ground. Dew on the leaves does little to increase leaf water content but does much to reduce the evaporative demands of the atmosphere. A process known as *distillation* redistributes soil water by vaporization deep in the soil, upward movement of the vapor, then condensation in the upper areas of the soil during cool periods. This process may be a significant source of water for desert plants.

Fog forms when warm, moist air cools by moving over cool land areas along the coast, ascending a mountain mass in the tropics, or rapid reradiation of heat from soil in inland valleys. Fog precipitation is an important source of water in coastal forests of the Pacific Northwest and in tropical cloud forests. Fog drip carries various organic compounds released as foliar metabolites, contributing to nutrient recycling and the transfer of allelochemics.

Precipitation forms when condensation drops in clouds become large enough to fall earthward. Precipitation arrives at the earth's surface as rain, hail, or snow, depending on turbulence in the clouds and ambient temperature. There are three types of precipitation, called *orographic* (associated with air rising over mountains), *convectional* (associated with convectional air currents rising from the heated ground), and *cyclonic* (associated with large air masses circulating around a low pressure area). Orographic rainfall is highest on the windward side of mountains. The leeward side of mountains is an area of very low rainfall because of the rainshadow caused by the descending of air. Convectional storms are usually associated with the summer season in areas of humid air. Cyclonic storms are generally associated with weather fronts moving west to east in the Northern Hemisphere. Snow is an important source of soil water and an important insulating factor for plants in very cold environments. Krummholz of timberline are formed by the exposure of plants to the desiccating, abrasive action of wind. Low-growing plants are protected from exposure by the blanket of snow. The distribution and phenology of alpine plants is very closely related to the distribution and duration of snow cover.

There are many special adaptations related to the spatial and temporal distribution of water in the environment. Halophytes are specifically adapted to habitats with high salt content. Glycophytes are nonhalophytic plants. Glycophytes usually exclude salt at the root surface but, when exposed to high salt habitats, may dehydrate because of osmotic imbalance or die due to nutrient imbalance or toxicity. Halophytes adjust to salt by modifying the internal osmotic concentration to compensate for external salt, diluting tissue salt in the high water content of succulent tissues, or extruding salt from the vegetative surfaces. Leaves of drought-adapted plants tend to be small, have reduced cell size, thick blades, and small and sunken stomates, be highly pubescent, and have a thick cuticle and a high ratio of internal to external surface. Succulence also provides stored water in some xeric plants.

The life form of many plants is determined by water-related adaptations. Evergreen leaves are generally tolerant of desiccation and can survive dry periods, being ready to photosynthesize immediately upon rehydration. Drought-deciduous species shed their leaves during dry periods. Many of the desert drought-deciduous species have photosynthetic bark to subsidize energy balance during dry periods. Phreatophytes grow where their roots are in constant contact with the fringe of capillary water above a water table. These plants can have deciduous leaves in very low-rainfall habitats because of a constant underground water supply. Ephemerals are annual plants that adapt to drought by very sensitive germination requirements, which restrict germination to periods of ample water. Summer desert ephermerals usually have the C_4 photosynthetic pathway, which increases water use efficiency.

CHAPTER 20

MAJOR VEGETATION TYPES OF NORTH AMERICA

T he North American continent has a moderately rich flora and a very rich diversity of habitats. The result is a complex mosaic of vegetation types and plant communities. One particular vegetation map for the conterminous United States shows over one hundred community types (Küchler 1964), but surveys of individual states or regions may show 50–100 community types because the level of detail is more refined than that used for the U.S. map. Since we choose to take a very broad survey view, we will discuss only five vegetation types: **tundra, conifer forest, deciduous forest, grassland,** and **desert scrub,** taking account of regional differences in each case. Our objective is to present enough information on community structure, community dynamics, and autecology of dominants to underscore the uniqueness of each of the five types.

These are the five major vegetation types covering the North American land mass. Every reader is likely to find one of these types nearby. Other types, not discussed here, are also likely to be found nearby, possibly even dominating the local region. Such types might be chaparral, fresh water marsh, coastal scrub, woodland, savanna, or tropical forest. Since our preparation of the first edition, several syntheses of North American vegetation have been published: texts by Daubenmire (1978) and Vankat (1979) and edited volumes by Chabot and Mooney (1985) and Barbour and Billings (1987). Much of the new material we have added here comes from those books, and they are cited in the References section at the end of this chapter. As shown in Figure 20-1, most of North America can be mapped in one of the five categories we do discuss.

Tundra

Tundra vegetation occurs in regions beyond timberline that have cold, usually moist, climates with short growing seasons. The vegetation is low, often only

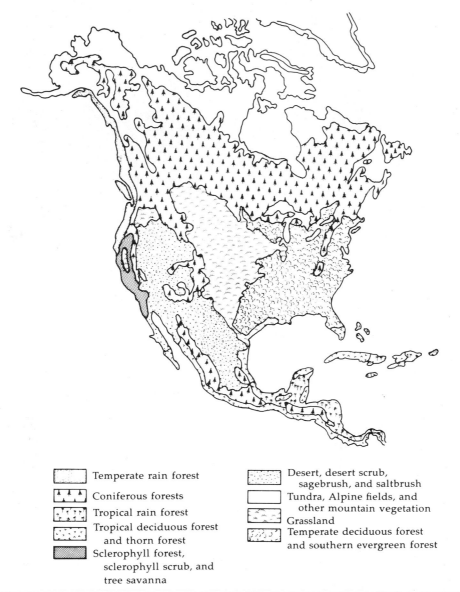

Temperate rain forest		Desert, desert scrub, sagebrush, and saltbrush
Coniferous forests		Tundra, Alpine fields, and other mountain vegetation
Tropical rain forest		Grassland
Tropical deciduous forest and thorn forest		Temperate deciduous forest and southern evergreen forest
Sclerophyll forest, sclerophyll scrub, and tree savanna		

Figure 20-1   Major vegetation types of North America. (From *Environmental Conservation*, 3rd ed. Dasmann. Copyright 1972 John Wiley and Sons, Inc. Reprinted by permission.)

10 cm tall, and is dominated by perennial forbs, grasses, sedges, dwarf shrubs, mosses, and lichens (Figure 20-2). In the arctic, on the average, only the top 0.5 m of soil thaws in summer. The permanently frozen subsoil impedes drainage, root growth, and decomposer activity. In the Lapp and Russian languages, tun-

(a)

Figure 20-2   Aspect views of two types of arctic tundra. (a) dwarf heath-cottongrass tussock tundra, a form of low arctic vegetation. Heath plants are mainly evergreen shrubs in the families Ericaceae, Empetraceae, and Diapensiaceae; cottongrass, shown flowering here, is *Eriophorum vaginatum*. (b) cushion-plant-herb-cryptogam polar semidesert vegetation in the high arctic. (Photos courtesy of James B. McGraw.)

(b)

dra means a marshy, unforested area, and these are its most prominent characteristics.

There are two major types of tundra in North America. **Arctic tundra** occurs at high latitudes in Canada and Alaska, but at low elevations. **Alpine tundra** occurs at higher elevations in mountain ranges farther south. The vegetation is similar in both types of tundra. Many species occur in both, but environmental differences (such as maximum summer temperatures, extremes of day length, and intensity of solar radiation) lead to important ecological differences. North of the tundra, or above it in mountains, is a barren region of permanent ice and snow called the **nival** zone. South of the arctic tundra, or below the alpine tundra, is the **boreal conifer forest** (also called the **taiga**) or subalpine forest, which will be discussed later in this chapter. Arctic tundra is far more extensive than alpine tundra, and it covers 20% of the North American continent.

Physical Features

The tundra's climate is rigorous and cold, and relatively few species of plants have evolved tolerances to withstand it. The short growing season is 50–90 days long on the average, and the mean temperature of the warmest month is less than 10°C. Maximum air temperature, especially in the alpine tundra, can be much higher than 10°C, reaching above 25°C. In addition, leaves near the ground surface can be 10–20°C warmer than air temperature. Summer days in the alpine tundra exhibit greater fluctuations in temperature than in arctic tundra. The thin mountain air and lack of cloud cover allow rapid reradiation of heat at night, permitting frost to occur throughout the growing season. As much as 70% of summer solar radiation is expended on evaporating water, leaving little to warm soil or air.

Day length may be shortened to zero hours in the arctic tundra, and during winter net radiation is negative. Winter conditions are very similar for both tundra types. Minimum temperatures below −50°C, a modest amount of snow cover, and strong winds of 15–30 m sec^{-1} (30–60 mph) are common. Annual precipitation varies greatly from region to region, but alpine tundra averages 100–200 cm, which is considerably more than the 10–50 cm arctic areas receive. The latter annual precipitation may seem low, but evapotranspiration is also quite low, so the P/E ratio is far above 1. The arctic tundra may be latitudinally divided into two climatic and vegetational zones: low arctic tundra and high arctic polar semidesert. They differ dramatically in climate, flora, and vegetation (Table 20-1).

Table 20-1 Some differences between low arctic tundra and high arctic polar semidesert. (Modified from L. C. Bliss, "Arctic tundra and polar desert biomes." In *North American Terrestrial Vegetation*, M. G. Barbour and W. D. Billings (eds.). Copyright © 1987 by Cambridge University Press. By permission.)

Trait	Tundra	Semidesert
Growing season (months)	3–4	1.5–2.5
Mean July temperature (°C)	8–12	3–6
Mean July soil temp. (°C)	5–8	2–5
Vascular plant flora	700	350
Common growth forms	Dwarf shrubs, grass-like	Cushion plants, grass-like
Plant height (cm)	5–500	2–100
Root: shoot ratio	3:1	1.5:1
Bryophytes	Common	Abundant
Lichen growth forms	Fruticose, foliose	Crustose, foliose
Total plant cover (%)	80–100	20–80
Net primary production (g per square m per yr)	80–200	25–85

Figure 20-3   Profile of a typical tundra soil with permafrost. The plant is drawn to scale. AO = litter.

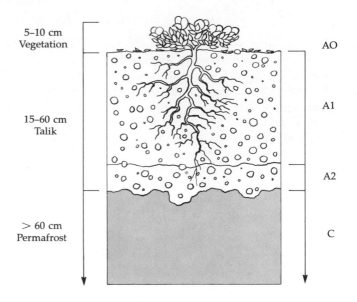

5–10 cm
Vegetation

15–60 cm
Talik

> 60 cm
Permafrost

AO

A1

A2

C

Two soil orders dominate the tundra: **histosols** (bog soils, mucks, and organic soils with more than 20% organic matter) and **entisols**. The entisols have little soil profile development, and their traits reflect the character of the parent material more than that of the macroenvironment. A typical entisol profile is shown in Figure 20-3. The top 15–60 cm of soil thaws in summer and freezes in winter. It is coarse-textured and brown from undecomposed organic matter and contains all the root mass of the vegetation above. Beneath this topsoil (**talik**) is permanently frozen parent material (**permafrost**) 400–600 m thick. If the vegetation is killed or removed, if the talik is eroded or compacted, or if heat is supplied to the ground, as from a buried pipeline or a house resting on the surface, then the upper part of the permafrost will melt, creating an unstable soil, further erosion or subsidence, and long-term vegetational change. Even one pass over tundra in summer with a normal wheeled vehicle will leave long-term scars. Bechtel Corporation developed a truck that ran on enormous, broad, air-filled tires to permit nondestructive travel on tundra during construction of the famous oil pipeline (Figure 20-4) from Valdez to the Arctic Slope in Alaska. Whenever possible, the pipeline was built aboveground, and care was taken to prevent heat from "leaking" into the ground along the supports. Once damaged, tundra is slow to heal. Roadcuts made in alpine tundra in the Rocky Mountains 50 years ago still show only half the plant cover of nearby, undisturbed tundra.

Freeze/thaw cycles, over the course of thousands of years, produce a pattern to the landscape that is most visible from the air (Figure 20-5). The ground surface is broken into polygons, each 10 cm to 100 meters across. The polygon edges may be marked by accumulations of rock, ridges of soil, or dense vegetation. The centers may be depressed and filled with standing water in summer, or they may be raised into hillocks several decimeters high. **Patterned ground** develops most extensively on gentle slopes.

Figure 20-4 The Alaska pipeline, here above the ground and heavily insulated. Care has been taken in planning the supports so that heat is not transferred from the pipe to the ground.

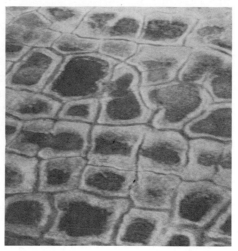

Figure 20-5 Aerial photograph of frost polygons in the tundra. These polygons are up to 100 m across.

Microrelief affects tundra vegetation markedly, because (a) it will determine how late into summer snow will lie on the ground, first covering vegetation, then supplying it with water; (b) it will determine the degree of protection that the plants will receive from ice-blasting winter winds; and (c) it will have a bearing on the depth of the talik, and hence on the amount of root space and the adequacy of drainage. Ridge tops support very little plant cover because winter winds remove protecting snow cover. Without the snow, abrasive ice crystals carried by the wind can prune plants back to the ground or, at least, winter winds can desiccate twig tissue. In summer, ridge soils dry rapidly. In contrast, as shown in Figure 20-6, gentle, north-facing slopes may accumulate

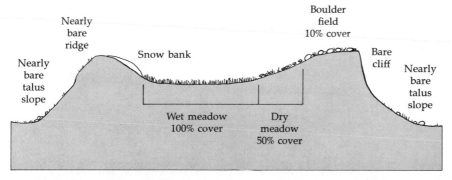

Figure 20-6 Major alpine tundra microenvironments, each with its own characteristic plant community.

snow drifts that remain well into summer, providing melt water to **wet meadows** just below them. Wet meadows exhibit 100% ground cover and dominance by grasses and sedges. Rocky talus slopes and relatively level boulder-strewn areas **(fell fields)** support relatively little cover, scattered in patches where the microenvironment is favorable. Dwarf shrubs, for example, grow tallest in the lee of rocks or other projections. The productivity and biomass of fell fields can be about one-third that of wet meadows, and they resemble the high arctic polar-semidesert (see Table 20-1).

Vegetation

The tundra flora is not rich. The arctic tundra of North America supports only about 700 species, which is only 3% of all angiosperms in the world. Tropical regions of smaller area support tens of thousands of species. The most common families in the tundra are Caryophyllaceae, Asteraceae, Brassicaceae, Cyperaceae, Betulaceae, Diapensiaceae, Empetraceae, Ericaceae, Poaceae, Polygonaceae, Rosaceae, and Salicaceae. Only 1–5% of the species are annual; most are relatively long-lived (20–100 years) herbaceous perennials. Hemicryptophytes—herbaceous perennials with perennating buds just at the soil surface—make up more than 60% of the flora. Trees and tall shrubs are absent. The tundra flora is geologically young, in a regional sense, probably younger than 3 million years. Much of it has been derived from taxa in the highlands of central Asia and the Rocky Mountains.

At its most luxuriant state, as in wet meadows, the vegetation covers 100% of the ground, but it is quite short, 10–40 cm tall, with a leaf area index (LAI) of only 1–2. Most plant biomass is belowground, giving a root:shoot ratio of 3:1. Many perennial forbs are "cushion plants"— their aboveground system is condensed into a compact canopy of small, xerophytic foliage, often with large, showy flowers incongruously scattered over the surface (Figure 20-7). The xerophytic foliage (small size, sometimes pubescent, sometimes succulent, often light colored, often vertically oriented) may be caused by low soil nitrogen levels more than by high light intensity or late summer drought. Although tundra soils are brown from organic matter, the nutrients are largely unavailable to the plants because decomposer activity is inhibited by the cold environment.

In spring, some plants resume growth while still covered by snow and root temperatures are barely above freezing. The hollow-stemmed forb *Mertensia ciliata* is an example of a subalpine snow bank plant able to conduct photosynthesis in the relatively warm, CO_2-rich air inside its stems even before the leaves have expanded. Some lichens may exhibit positive net photosynthesis while covered by 5 cm of snow. Seeds of tundra plants, on the other hand, require relatively high temperatures, close to 20°C, before they can germinate. Flower buds, formed one or more seasons ago, are induced to open by long days. About two-thirds of the flora is insect-pollinated, so the flowers are large and showy. Some flowers are dish-shaped and reflect solar radiation into their centers, creating a favorable thermal microenvironment for pollinators such as bumble bees. Pollinator den-

Figure 20-7 Cushion plant. Note the relatively large size of the flowers. The vegetative part of this plant is only 5 cm high.

sity is low, however, and many species compensate by being facultatively or obligately self-fertilizing. For reasons that are still not clear, most tundra species are polyploids. Seed set is high, the seeds are light, and they have a long life.

During the short growing season, net primary productivity is low (see Table 20-1). Much of this is channeled to roots and rhizomes and stored as lipid or protein. Part of the reason that NPP is low is that the ratio of net photosynthesis to respiration is low, only about 2:1 in contrast to a ratio of 10:1 for tropical plants. In addition, the maximum photosynthetic rate of vascular plants is relatively low, roughly equivalent to that of mesic, temperate zone shade plants. Associated mosses and lichens have maximum photosynthetic rates an order of magnitude lower than those of vascular plants. Temperature optima for photosynthesis is 15–20°C, lower than the 25–30°C optimum of temperate zone plants. Furthermore, there is a broad plateau between 0 and 25°C where photosynthesis is only modestly affected by temperature. UV radiation (wavelength 280–320 nm) may depress photosynthesis and cause epidermal damage, and alpine species or ecotypes appear to be more tolerant of UV exposure than low elevation arctic taxa. A few tropical tundra species have the C_4 photosynthetic pathway.

Dormancy, in late summer, is triggered by lowering temperatures, increasing soil moisture stress, shortening day length, or a changing quality in the solar spectrum that results from a more oblique solar angle (the latter is important only in the arctic tundra). Wet meadow species apparently have little control over stomatal aperture; as the soil dries, plants lose water rapidly, finally becoming dormant at a tissue water potential of only −1 MPa. Dry meadow and fell field plants can tolerate lower water potentials, to below −4 MPa. Winter snow cover helps to maintain a modest tissue water potential. For example, needles of whitebark pine (*Pinus albicaulis*) in the timberline zone of the Sierra Nevada show a water potential of −1.8 MPa in winter when covered with snow, but a water potential of −4.0 MPa when exposed. In fact, the height of trees at timberline may correspond closely with the average depth of winter snowpack.

Timberline

There is a broad ecotone in Canada and Alaska between the arctic tundra and the boreal forest to the south. In places, it may be 300 km wide. Within the ecotone, the forest first thins, no longer retaining 100% cover, and fruticose lichens cover the ground (Figure 20-8), then trees become restricted to patches, and finally trees in the patches become dwarfed, twisted, and shrublike (Figure 20-9). The German word **krummholz,** meaning twisted wood, has been applied

Figure 20-8 Lichen woodland near timberline in central Labrador. The trees are mainly white spruce *(Picea glauca)*; the understory lichen is *Cladonia*. (From Elliott-Fisk 1986.)

Figure 20-9 Krummholz at timberline in the Sierra Nevada (Sonora Pass, about 3400 m). The shrub-like thicket is a pure stand of white bark pine *(Pinus albicaulis)*, about 1 m tall.

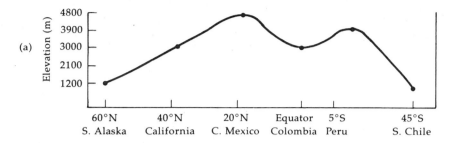

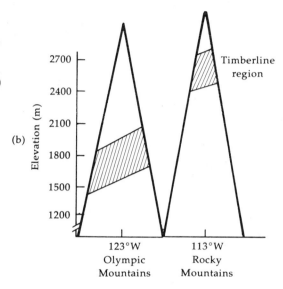

Figure 20-10   (a) The effect of latitude on the elevation of timberline, from the Coast Mountains of southern Alaska, through the Cascade Range, Sierra Nevada, Sierra Madre, and the Andes Mountains. (b) The effect of distance inland from oceanic influence, all at the same latitude, 47°N. The shaded region is the distance between the end of the closed forest and the end of the krummholz. [(a) Redrawn from Daubenmire 1954. "Alpine timberlines in the Americas and their interpretation" in *Butler University Botanical Studies* 11:119–136.]

to both this life form and to the ecotone "forest"; a synonym is **elfinwood**. Surrounding the patches of trees and krummholz is an increasingly continuous matrix of shrubby tundra. Finally, even the patches of krummholz disappear and the tundra proper is entered.

A similar ecotone exists in mountain ranges between alpine tundra and subalpine forest, but the ecotone is much narrower, extending upward through 150–300 m of elevational change. The exact elevation of timberline depends on latitude and distance from a maritime climate, as shown in Figure 20-10. In the Rocky Mountains, between 35 and 50°N latitude, timberline falls 100 m for every 1°. Within one mountain range at one point in latitude, timberline will be lower on the moister, ocean-facing slopes than on the drier, continental-facing slopes (Figure 20-10b).

The location of timberline is also a function of time, changing in response to changing climate. In some parts of Canada, there is evidence that timberline was 100 km north of its present location 8000 years BP (before the present), 300 km north 4000 years BP, 100 km south 2000 years BP, and 100 km north once again, 1000 years BP. The ecotone is a dynamic tension zone within which the basic limitation to tree growth seems to be the amount of warmth received during the growing season. Below some critical level of seasonal heat, it becomes impossible to maintain a large amount of woody tissue. Several climatic markers correspond with this heat boundary: (a) North of treeline, the July mean temperature falls below 10–13°C. (b) The yearly heat sum (degree-days above 0°C) falls below 1000. (c) Summers are dominated by the arctic high pressure system more than 50% of the time. Besides the depression of tree growth, sexual reproduction and successful seed germination become rare. Foliage longevity becomes extreme—up to 30 years for bristlecone pine, with secondary phloem growth throughout that period. In addition, fire is thought to play a large role in the location of timberline in Canada's lichen woodland region, as fire permits the establishment of new tree individuals.

Conifer Forests

Several distinctive types of conifer forest exist in North America, and those covering the most area will be included in this section.

The most extensive conifer forest type is the **taiga,** which stretches across most of Canada and Alaska and dips down into the Great Lakes states and New England to about 45°N latitude. *Taiga* is a Russian word applied to Eurasian conifer forests described as damp, wild, and scarcely penetrable. A synonym is **boreal forest.** The taiga is a monotonous forest, with crowded trees of modest stature and low species diversity in the overstory and understory. This low-elevation forest extends over impressive distances, interrupted often with lakes, ponds, and moss-covered bogs. It can be a quiet forest, not rich in vertebrate animal activity.

Conifer forests extend south of the taiga, at moderate to high elevations in mountain chains. Just below the alpine tundra is a **subalpine forest.** Different species dominate than in the taiga, but its physiognomy is the same. The **montane zones** below this are usually covered by a richer diversity of conifers, and the forests are often more open and the trees larger, more productive, and longer-lived than those of the taiga and subalpine forests. Below the montane zones lie vegetation types dominated by broadleaved trees, scrub, or grassland. Exceptions to this pattern are the mountain ranges that penetrate north into the taiga; in these areas, the surrounding low-elevation vegetation is still conifer-dominated.

Another magnificent conifer forest dominates a wet, temperate, lowland coastal strip that reaches from northern California to Alaska: the **north coast temperate conifer forest,** sometimes called a **temperate rain forest.** Some of the largest trees in the world, and stands with the greatest biomass, exist in this forest.

Other low-elevation conifer forests, all dominated by pines, are found in smaller areas of coastal California, in the Great Lakes area, in New Jersey, and in the southeastern coastal plain of the United States. Some of these grow on droughty, nutritionally poor soils, and many are subclimax types, maintained by fire. These will be discussed briefly at the end of this section.

Physical Features

The taiga is a cold, wet region, with a short growing season only 3–4 months long. During more than 6 months of the year, the mean temperature is below 0°C, snow covers the ground, and net radiation is negative. Maximum summer temperatures reach into the low 20s and winter minimums are in the −50s. The temperature difference between monthly means within a year can be enormous in the interior, continental part of the taiga—a span of 90°C. Annual extremes are dampened near the Atlantic and Pacific coasts (Figure 20-11). Annual precipitation varies between 30 and 85 cm, and snow cover is not deep. Despite the modest amount of precipitation, the P/E ratio is still >2. Most precipitation falls in summer.

About 60–75% of annual solar radiation is absorbed by soil and vegetation, in sharp contrast to the tundra. The northern and southern limits of taiga, which

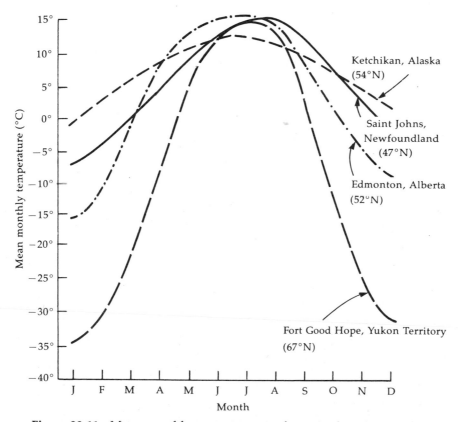

Figure 20-11   Mean monthly temperatures at four taiga locations. Two sites are coastal, and the oceanic influence moderates winter extremes: Ketchikan on the Pacific coast and Saint Johns on the Atlantic coast. Two are interior, with bitterly cold winters: Edmonton and Fort Good Hope.

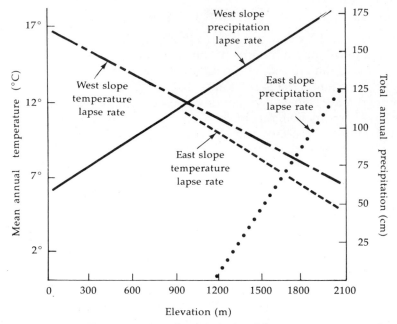

Figure 20-12   Temperature and precipitation lapse rates on west and east slopes of the Sierra Nevada at 38°N latitude. Temperature lapse rates are similar for both slopes, but precipitation lapse rates differ because of a rainshadow effect: The east side is drier. (From Major and Taylor 1977. In *Terrestrial Vegetation of California.* Edited by Barbour and Major. Copyright 1977 by Wiley-Interscience, Inc. Reprinted by permission.)

span 20° of latitude, bracket a region that experiences 30–120 days with a mean temperature above 10°C.

The subalpine zone is generally 5°C warmer throughout the year and experiences 20–50% more precipitation than the taiga. There is a complex environmental gradient that corresponds to elevation in mountain ranges, and this gradient is responsible for the change in conifer species from one zone to the next. Two important environmental factors are temperature and precipitation. Figure 20-12 shows how average annual temperature and average annual precipitation change with elevation in the Sierra Nevada Mountains for west-facing (ocean-facing) and east-facing (desert-facing) slopes. Precipitation increases about 5 cm for every 100 m increase in elevation. The highest elevation shown, 2100 m, corresponds to the red fir forest in the upper montane zone. If higher elevations were shown, you would see that precipitation drops in the subalpine and alpine zones. The east slope is much drier and has a different **lapse rate** (rate of change), because a rainshadow has been created: Winds generally come from the west, drop their moisture on the west slope as air rises, and descend on the east slope as dry winds. The west slope temperature lapse rate shows a drop of 4°C for every 1000 m rise in elevation. Lapse rates differ with different seasons and

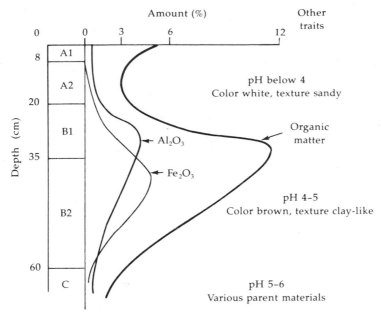

Figure 20-13 Some chemical characteristics of a spodosol (= podzol) soil profile. Note that depths of maximum accumulation of organic matter, aluminum, and iron are slightly different. (From the translation of *Plants, the Soil and Man* by M. G. Stålfelt. Copyright 1972 by John Wiley and Sons, Inc. Reprinted by permission.)

mountain ranges. Considering many factors and locations, montane temperature lapse rates average about 3° per 1000 m.

The north coast forest lies in a wet, equable strip. Annual precipitation is generally between 200 and 300 cm, very little falling as snow. Mean January minimums are above freezing, at about 2°C, and mean July maximums are only about 20°C. Diurnal thermoperiods average only 10°C. A pronounced summer dry period exists from Washington to California, but it is moderated by frequent fog that lowers transpiration stress and also drips off the foliage onto the ground, adding perhaps 20 cm of "rain" in the Californian part of the range.

Taiga soils mostly belong to the **spodosol** soil order, which has been described in Chapter 17. These are acidic soils (pH 3–5) whose upper layers have been leached of clay, organic matter, and many nutrients (Figure 20-13). Conifers are characterized by shallow root systems (root:shoot ratio = 0.2), which lie above zones of nutrient and clay accumulation in the B2 horizon. However, nutrient uptake is improved by mycorrhizae. All taiga conifers possess mycorrhizae (see Chapter 7), and there is experimental evidence showing that tree growth and survival would be minimal without them. Soils are acidic in part because of leaching and because conifer foliage is acidic. The hydrogen ions replace nutrient cations on soil colloids (giving a low percent base saturation), drive some cations into insoluble states, and inhibit bacterial decomposers. Litter tends to accumulate, and as much as 60% of all ecosystem carbon is locked up in humus. Nutrient cycling is further slowed by a layer of feather mosses (typically *Hylocomium splendens* and *Pleurozium schreberi*) up to 30 cm thick, which acts as a nutrient sponge, preventing nutrients from reaching the root zone. Mineral cycling is slow. The half-time for litter decay is about 3–5 years. Permafrost may underlie taiga soils, especially in the northernmost latitudes.

Soils beneath montane and coast forests are generally either podzolic or young and weakly developed (**inceptisol** soil order). If podzolic, they are not as acidic as taiga soils and lack the ashy A2 horizon of taiga soils.

Taiga Vegetation

Taiga vegetation consists of closed forest to the south and more open lichen woodland to the north. Closed forests are dominated by white spruce *(Picea glauca)* and/or black spruce *(P. mariana)*. The most common associates are balsam fir *(Abies balsamea)*, larch *(Larix laricina)*, jack pine *(Pinus banksiana)*, lodgepole pine *(P. contorta)*, and three broadleaf deciduous trees *(Populus tremuloides, P. balsamifera, Betula papyrifera)*. The conifers are closely spaced, with dense canopies (LAI = 9–11), but they are of relatively small stature (20 m tall, 30 cm dbh) and short longevity (100–300 years).

Many species of evergreen and deciduous shrubs and perennial herbs are present, but cover is modest. A ground layer, dominated by feather mosses, lichens, and ferns, is often continuous (Figure 20-14). The forest, then, is essentially a two-layered community.

Near the edge of bogs, the forest canopy thins and eventually only scattered trees of black spruce and larch (also called tamarack) remain, underlain by a

Figure 20-14   The interior of a taiga forest. Note the high ground cover by ferns and other cryptogams.

dense shrub canopy of *Vaccinium, Ledum,* and other ericads. Wetter parts of the bog are covered by *Sphagnum* moss. As discussed in Chapter 11, this moss can advance into more upland sites, creating a wetter, more sterile microenvironment, which leads to retrogressive succession. Droughty, sandy soils often support stands of jack pine *(Pinus banksiana)*.

The boreal forest is a disturbance (disclimax) forest. Small areas are often disturbed by wind-throw from winter storms or by fire from summer dry lightning strikes. Black spruce has semiserotinous cones and regenerates well after fire. Succession often proceeds through meadow, then broadleaf tree *(Populus tremuloides, P. balsamifera, Betula papyrifera)* seral stages before reaching an even-aged spruce-fir climax in a span of about 300 years. These broadleaf trees also form an extensive parkland, up to 150 km wide, in an ecotone where the boreal forest meets the central grasslands in Manitoba, Saskatchewan, and Alberta.

Net annual primary productivity is relatively low (600 g m^{-2}), but this is not surprising, given the short, cool growing season and the low nutrient status of the soil. The eastern deciduous forest shows twice that biomass gain, and the tropical rainforests of Central America show more than three times that gain.

The physiological ecology of taiga species has become better known over the past decade. Black spruce, as a representative of overstory conifers, is conservative in nutrient use. Its needles are retained for up to 15 years, and their photosynthetic activity remains high throughout that life span. Optimum temperature for photosynthesis is 15°C, but there is a broad plateau from 9 to 23°C, so temperature during the growing season is not an important limiting factor. Light saturation for photosynthesis is at ⅓ full sun, and the compensation point is very low, at 1% full sun (12–35 μE m^{-2} sec^{-1}). Saplings are therefore very shade tolerant. Maximum photosynthetic rates are low compared to temperate zone plants and are lower than those of most associated growth forms in the forest (Table 20-2). Black spruce is sensitive to water stress, but moisture is not a limiting factor in most years or habitats.

Table 20-2 Maximum net photosynthetic rates (μmol CO_2 g^{-1} sec^{-1}) of taiga species grouped by life form. (Modified from Oechel and Lawrence. In *Physiological Ecology of North American Plant Communities.* Chapman and Hall, Publishers, 1985.)

Life form	Genera	Rate
Deciduous conifer	*Larix*	0.135
Deciduous shrub	*Alnus, Betula, Rubus, Vaccinium*	0.120
Deciduous hardwood	*Alnus, Betula, Populus*	0.090
Sedges	*Carex, Eriophorum*	0.060
Evergreen shrub	*Ledum, Empetrum, Vaccinium*	0.030
Evergreen conifer	*Picea*	0.025
Lichen	*Cladonia (= Cladinia)*	0.010
Moss	*Dicranium, Hylocomium, Pleurozium, Polytrichum, Sphagnum*	0.009

Farther north, black spruce becomes the typical dominant of open lichen woodlands (see Figure 20-8). The understory lichen is either *Stereocaulon paschale* in the west or *Cladonia (= Cladinia) stellaris* in the east. The climax nature of this woodland is debated; some ecologists believe it is a fire climax rather than a climatic climax. The lichens appear to inhibit tree regeneration mechanically.

Much of the taiga is still botanically and ecologically unexplored. It is largely undisturbed by any human activities, in fact, and this may be due to small factors as well as large ones. The large factors have to do with climatic hardship and the absence of roads. The small factors are called black flies and mosquitoes.

Subalpine and Montane Vegetation

The Appalachian Mountains   The Appalachian Mountains extend from Maine south to Georgia. These mountains are unique in that the montane zones are largely dominated by hardwoods; conifers are restricted to the subalpine zone. In the north, for example in the Adirondack Mountains of upstate New York, the upper limit of the hardwood montane zone is about 900 m; in the southern Smoky Mountains of North Carolina it is about 1600 m.

The taiga dominants, white spruce and balsam fir, do occur in the New England portion of the Appalachians, but they are not dominants. Red spruce (*Picea rubens*) dominates throughout the Appalachians, and it is joined by Fraser fir (*Abies fraseri*) south of Pennsylvania. Broadleaved subdominant trees include mountain ash (*Sorbus americana*), yellow birch (*Betula lutea*), and maple (*Acer spicatum*).

The boreal forest of New England has been exhibiting dramatic die-back of red spruce for the past 25 years. Spruce decline has been reported for the mountains of Vermont, New Hampshire, and New York (and also farther south in North Carolina). Needles become brown and then abscise prematurely, beginning with the youngest and progressing to the oldest. Reproduction stops, basal area growth slows, and trees ultimately die, creating canopy openings that permit wind-throw damage to increase. Mortality is apparent in all age classes. Studies of associated fir and of hardwoods at lower elevations do not consistently reveal similar patterns.

Acid deposition has been implicated as a cause of massive conifer die-back in northern and eastern Europe, but the evidence for North America is equivocal. **Acid deposition**, or acid rain, is caused by gaseous SO_x or NO_x emissions becoming strong acids in solution in precipitation. These gases are produced as by-products of internal combustion (automobiles), coal burning, or the refining of certain metals. Normal rain has a pH of 5.6, due to CO_2 becoming weak carbonic acid in solution, but rain in the northeast has a pH averaging 4.0 (Figure 20-15). Rain of sufficiently low pH can cause leaf lesions and can lower starch and sugar reserves in leaves, increase water stress, reduce nitrogen fixation and nitrification in the soil, inhibit germination, and depress bacterial decomposition. It is not yet clear, however, that the pH of New England rain can induce such changes in species of the boreal forest.

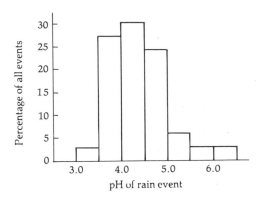

Figure 20-15 Frequency distribution of the pH of rain during the 1981 growing season at Brookhaven National Laboratory, Long Island, New York. No rainfall event had a pH below 3.0 or above 7.0. The overall weighted mean was 4.0. (From Evans 1984.)

The Rocky Mountains   Running obliquely from southeast to northwest, the Rocky Mountains cross 40° of latitude and 30° of longitude. There are regional differences in the tree species that associate at particular elevations, but several species range throughout the Rocky Mountains: Engelmann spruce *(Picea engelmanni)* and subalpine fir *(Abies lasiocarpa)* in the subalpine zone, Douglas fir *(Pseudotsuga menziesii),* lodgepole pine *(Pinus contorta),* and white fir *(Abies concolor)* in the upper montane zone, and ponderosa pine *(Pinus ponderosa)* in the lower montane zone.

The elevation of each zone increases to the south. At any given latitude, the zones are highest on southwest-facing (dry) slopes and lowest on northeast-facing (wet) slopes. In northern Alberta, the upper limit of the subalpine zone is about 2000 m, and the montane zone below begins at 1400 m. The lower montane farther downslope is essentially the same as surrounding low-elevation taiga. Far to the south, in Arizona, the upper limit of the subalpine zone is about 2700 m and the lower montane begins at 1800 m (Figure 20-16).

Regional associates of spruce-fir in the subalpine zone of the northern Rockies include white bark pine *(Pinus albicaulis)* and alpine larch *(Larix lyallii),* the latter retaining a tree form higher into timberline than any other species in North America. Subalpine forest in the central Rockies may include white bark pine, lodgepole pine *(Pinus contorta),* and mountain hemlock *(Tsuga mertensiana),* in addition to spruce-fir. The southern Rockies have limber pine *(Pinus flexilis),* foxtail pine *(P. balfouriana),* and bristlecone pine *(P. longaeva,* which is famous for being the longest-lived tree in the world), in addition to spruce-fir.

The upper montane zone in the Rockies of Montana and Idaho is enriched in tree species because storm tracks coming east from Washington are able to penetrate that far. In this area, only 200 km separate the Rockies from the Cascade Mountains. Dense stands of grand fir *(Abies grandis),* western red cedar *(Thuja plicata),* and western hemlock *(Tsuga heterophylla)* are typical. Only in the drier, eastern part of this region—for example, in Yellowstone National Park—does this zone support the Douglas fir and lodgepole pine forest more typical of the Rocky Mountains as a whole.

(a)

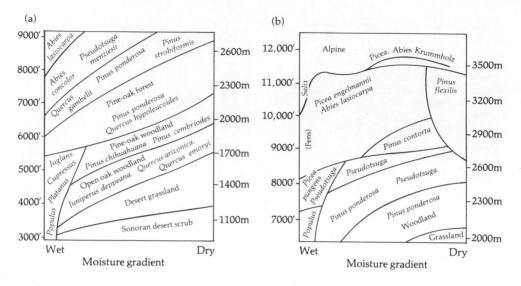

(b)

(c)

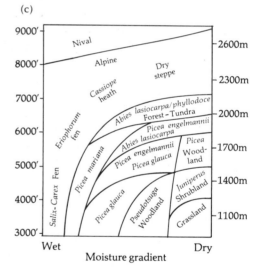

Figure 20-16   Gradient diagrams for major vegetation zones of the Rocky Mountains. (a) Santa Catalina Mountains of Arizona, 32°N latitude; (b) Colorado Front Range, 40°N; (c) Jasper National Park, Alberta, Canada, 53°N. The horizontal axis represents a complex moisture/exposure/aspect gradient, and the vertical axis is an elevational gradient (feet on left, meters on right). (From Peet 1987.)

The lower montane zone is dominated by ponderosa pine and often has a rather open, parklike aspect (Figure 20-17). This open physiognomy and pure dominance have apparently been maintained by natural surface fires with a frequency of one fire every 3 to 10 years. Fire suppression policies over the past 60 years have resulted in denser stands, more brush, loss of grass, and the invasion of other conifers (see Chapter 16).

The Cascade and Sierra Nevada Mountains   The Cascade Range extends along 10° of latitude, from southern British Columbia to Mt. Lassen in northern California. The Sierra Nevada Mountains continue south from Mt. Lassen an

Figure 20-17 The lower montane zone ponderosa pine forest of the Rocky Mountains. Note its relative openness. In the absence of ground fires every 8 years or so, other conifers will invade, and the canopy will close.

additional 6° of latitude, to the edge of the Mojave Desert in southern California. There is considerable floristic shift along that distance—so much so that the northern and southern limits show a Sorensen-type community coefficient of only 30 (100 = identity, 0 = no species in common) (see Chapter 10).

The subalpine zone of the Cascade Range is dominated by mountain hemlock in the wettest, westernmost portions, and by subalpine fir. Associates are numerous and include Engelmann spruce, lodgepole pine, alpine larch, Douglas fir, western larch *(Larix occidentalis)*, western white pine *(Pinus monticola)*, and white bark pine.

Lodgepole pine, white bark pine, and mountain hemlock continue south into subalpine forests of the Sierra Nevada, joined by red fir *(Abies magnifica)*, limber pine, foxtail pine, and—in the extreme south, in the White Mountains—by bristlecone pine. The Sierran subalpine forest is often relatively open, contrasting strongly with the dense, closed forests in the upper montane zone below (Figure 20-18).

The upper montane zone in British Columbia and northern Washington is dominated by western hemlock, western red cedar, and (in the wettest, western portion) Pacific silver fir *(Abies amabilis)*. In the southern Cascades, this zone is dominated by grand fir and is associated with white fir, ponderosa pine, lodgepole pine, Douglas fir, and western larch.

Figure 20-18   The subalpine forest of the central Sierra Nevada, about 3200 m. The open forest is dominated by white bark pine *(Pinus albicaulis)* and mountain hemlock *(Tsuga mertensiana).*

The upper montane zone of the Sierra Nevada is quite distinctive, with an upper portion dominated almost exclusively by red fir and a lower portion called the **mixed conifer forest** (Figure 20-19). The red fir forest has been called a subalpine forest by some ecologists, and it does have some physiognomic similarities to subalpine spruce-fir. It is a dark, quiet, two-layered community, but the trees are relatively large and long-lived compared to typical spruce-fir. Below the red fir zone, between 1800 and 900 m, is a more open forest, richer in trees, shrub, and herbaceous species. Dominance is shared by five trees: ponderosa pine, Douglas fir, sugar pine *(Pinus lambertiana)*, white fir, and incense cedar *(Calocedrus decurrens;* synonym is *Libocedrus decurrens).*

A certain frequency of surface fires seems to maintain the importance of each species, and a fire suppression policy over the past 70 years has resulted in a major shift from the previous balance. Table 20-3 shows that the sapling community is quite different from the overstory community: White fir and incense cedar are 2–4 times as important in the sapling layer as in the overstory; sugar pine is holding its own; and Douglas fir and ponderosa pine are 2–9 times less important in the sapling layer than in the overstory. The mature forest 100 years from now may be a dense white fir forest, quite unlike the forest of the present and past. One of the reasons for the shift has to do with the seedling requirements of each species. Seedlings of ponderosa pine, for example, are known to do best when the litter layer is thin or absent, and in relatively open sun. Seedlings of white fir can tolerate denser shade and a deeper litter layer.

Figure 20-19 Gradient diagram of major Sierra Nevada vegetation, 37–38°N latitude. The horizontal axis represents a complex moisture/ exposure/aspect gradient, and the vertical axis is an elevational gradient (meters). (Redrawn from Vankat 1982.)

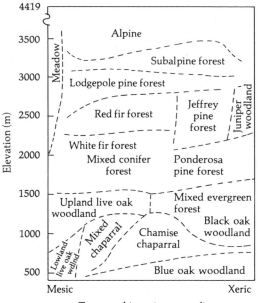

Topographic moisture gradient

Table 20-3 Importance values (IV), average diameter at breast height (cm), and density per hectare of all conifer species in a virgin mixed conifer forest at Placer County Big Trees, California, 1600 m elevation. Importance values (IV = relative basal area + relative density + relative frequency) are for the size classes shown, but average diameter at breast height and density are only for trees >3 cm dbh. Total basal area for trees >3 cm dbh was 49 m² ha⁻¹.

Taxon	IV for three size clases			Average dbh (cm)	Density ha⁻¹
	<3 cm dbh (saplings)	3–40 cm dbh	>40 cm dbh (overstory adults)		
White fir *(Abies concolor)*	132	123	36	8.3	258
Incense cedar *(Calocedrus decurrens)*	62	35	27	15.0	67
Sugar pine *(Pinus lambertiana)*	59	38	69	23.1	87
Ponderosa pine *(P. ponderosa)*	10	47	89	16.6	114
Douglas fir *(Pseudotsuga menziesii)*	37	57	79	19.1	122

Figure 20-20 Bigtree *(Sequoiadendron giganteum)* in the mixed conifer zone of the Sierra Nevada Mountains, about 1400 m.

Scattered within this mixed conifer forest are 75 groves of bigtree *(Sequoiadendron giganteum)*, the most massive organism on earth (Figure 20-20). The largest groves cover more than 800 ha and contain 20,000 trees; the smallest and most northerly grove extends over only 0.4 ha and contains only 6 mature trees. The smaller stands appear to be losing mature trees to death by old age faster than they are gaining saplings. The reasons for the population decline are not known, but the decline has been going on for 500 or more years, so recent human activities are not necessarily the cause (see Chapter 4).

The lower montane zone is dominated by ponderosa pine throughout the Cascade and Sierra Nevada Mountains. It is a relatively open forest and closely resembles the Rocky Mountain lower montane. A well-developed understory of shrubs *(Symphoricarpos, Holodiscus, Chamaebatia, Rosa, Arctostaphylos, Physocarpus, Ceanothus* spp.) or of grasses *(Agropyron, Festuca, Stipa* spp.) can exist. Subdominant oak trees, such as black oak *(Quercus kelloggii)*, Oregon white oak *(Q. garryana)*, and canyon live oak *(Q. chrysolepis)* are common.

As is true for the Rocky Mountains, many hectares are not covered by forest. In the Sierra Nevada, for example, more than half the land is covered by meadow or brushfield. **Brushfields (montane chaparral)** are often found on hot, rocky, dry exposures where tree growth is limited; they may also invade burned or logged areas on better sites that once supported conifer forest (Figure 20-21). In the latter case, montane chaparral is successional and will ultimately be replaced

Figure 20-21   Montane chaparral on a disturbed area on the east side of the Sierra Nevada, about 1900 m. The dominant shrub is tobacco brush *(Ceanothus velutinus)*, and it is suppressing the growth of white fir *(Abies concolor)* coming up beneath it.

by a closed forest. The rate of succession is very slow, and management techniques are being developed to suppress brush by selective herbicides that leave the conifers unaffected.

Temperate Low Elevation Conifer Forests

Pacific North Coast Conifer Forest   This coastal forest extends from the Gulf of Alaska, at 62°N latitude, to the Mendocino coast of California, at 40°N latitude. It is unusual for two reasons. First, hardwoods typically dominate temperate, mesic climates such as this, but here softwood volume is 1000 times hardwood volume. Second, the size and longevity of the dominants is unprecedented. The largest trees of 10 genera are found here: *Abies, Chamaecyparis, Larix, Calocedrus, Picea, Pinus, Pseudotsuga, Sequoia, Thuja,* and *Tsuga.* Overstory trees are commonly 50–75 m tall and more than 2 m dbh (Table 20-4). Aboveground standing crop averages 2000 t ha^{-1}. Longevity is often well beyond 500 years, and sometimes it is beyond 1000 years.

This is the most productive forest in North America, averaging 1.74 kg m^{-2} yr^{-1}, which is nearly twice that of any other conifer or hardwood forest on the continent. Softwoods dominate here, instead of hardwoods, for three reasons:

Table 20-4   Average maximum age, diameter breast height (cm), and tree height (m) for eight common overstory species of the Pacific north coast forest. (Modified from J. F. Franklin, "Pacific Northwest Forests." In *North American Terrestrial Vegetation*, M. G. Barbour and W. D. Billings (eds.). Copyright © 1987 by Cambridge University Press. By permission.)

Species	Age	DBH	HTE
Sequoia gigantea	1250	150–380	75–100
Chamaecyparis nutkatensis	1000	100–150	30–40
Thuja plicata	1000	150–300	60
Pseudotsuga menziesii	750	150–220	70–80
Chamaecyparis lawsoniana	500	120–180	60
Picea sitchensis	500	180–230	70–75
Tsuga heterophylla	400	90–120	50–65
Abies grandis	300	75–125	40–60

1. The potential for positive net photosynthesis outside of the usual growing period is high, favoring evergreens.
2. Water stress during the growing season can be appreciable, favoring the more sclerophyllous evergreens.
3. Evergreens have a higher nutrient-use efficiency.

Except for the Californian part of the range, the most common overstory trees are Sitka spruce *(Picea sitchensis)*, western hemlock, and western red cedar. Common associates include Douglas fir, grand fir, silver fir, and two species of cedar *(Chamaecyparis)*. Douglas fir appears to be a successional species, despite its great size and long life; it does not reproduce well in its own shade and is succeeded by western hemlock. Broadleaf trees, such as red alder *(Alnus rubra)* may be scattered beneath the conifer canopy, and a lush ground layer of shrubs, perennial herbs, ferns, and cryptogams is typical. Like conifers in other vegetation types, Sitka spruce is noted for striking differences in sun and shade needles, adapting it to full sun as an emergent and to deep shade as a sapling. The needles differ in orientation, size, specific leaf weight, stomatal density, net photosynthetic rate, stomatal conductance, and mesophyll conductance.

In California, coast redwood *(Sequoia sempervirens)* becomes the dominant species. Up to 100 m tall and as old as 2000$^+$ years, redwoods are nevertheless seral species, dependent on periodic disturbance by fire or flood in order to maintain their vigor, reproduction, and dominance.

Great Lakes Pine Forest   A mosaic of softwood and hardwood stands spreads over part of the Great Lakes region of the United States, in parts of Michigan,

Wisconsin, and Minnesota. Three pines form nearly pure, even-aged stands: red pine *(Pinus resinosa)*, eastern white pine *(P. strobus)*, and jack pine *(Pinus bank-siana)*. The pines are most often restricted to sandy soils, with beech, maple, yellow birch, and hemlock dominating the surrounding deciduous forest on more mesic soils. This mosaic lies in a tension zone, or ecotone, between the boreal forest and the deciduous forest, and there has been considerable debate about the climax nature of pine stands here (see also Chapter 6). The usual view is that pine is seral to hardwood, and at best the pine forms an edaphic climax on sites too dry to support the hemlock-hardwoods. These stands were extensively cut for lumber in the nineteenth century, and their area today does not reflect past dominance.

New Jersey Pine Barrens   A half million hectares of land in the coastal plain of New Jersey, and limited areas on Long Island and on Cape Cod, have very sandy, acidic soil of low fertility. They support a scrubby forest, about 6 m tall, of pitch pine *(Pinus rigida)* and several oak species. The tree canopy is somewhat open, and there is a well-developed understory of low shrubs, mainly oaks and ericads, and a ground layer of mosses, lichens, and sedges. In part, the poor, droughty soils limit invasion of surrounding oak-hickory deciduous forest, but most ecologists currently believe that a high fire frequency is the major factor favoring the pines. Where fire frequency is 10 years or less, the pines do not reach tree stature, but instead form a shrubby pygmy forest little more than 1 m high. Other parts of the barrens experience a fire frequency of about 20 years.

Despite a large surrounding human population, the pine barrens remain largely a wilderness, a precious resource in the most densely populated state in the United States.

Southeastern Pine Savanna   The coastal plain of North Carolina, South Carolina, Georgia, and Florida often supports an open stand of pines underlain with grass. The principal species of pine are: long leaf *(Pinus palustris)*, short leaf, *(P. echinata)*, slash *(P. eliottii)*, and loblolly *(P. taeda)*. The dependence of this forest type on frequent surface fires, once every several years, has been discussed in Chapter 16. In the absence of fire, pine is seral to hardwood forest.

Deciduous Forests

The deciduous forest offers many contrasts to the boreal forest. The deciduous forest has a lower LAI, permitting more light to reach the forest floor; it has more animal life; the overstory can be very rich in species diversity; and the structure of subdominant canopy layers can be more complex. Because of a longer, warmer growing season, net primary productivity and standing biomass are about twice what they are in the taiga. Litter decay is more rapid, litter half-life being on the order of one year instead of several years. As a result, mineral cycling is more rapid.

Deciduous forests cover about one-third of all land area in the conterminous United States. Some of this area is in the form of narrow fingers of riparian forest that extend west along river courses into nonforested regions, but the great bulk of the forest lies east of 95°W longitude and south of 45°N latitude. Only this eastern portion will be discussed here.

The eastern deciduous forest is not homogeneous throughout its extent. Certain groups of dominant tree species characterize particular regions. Much of what we know about the regional forests comes from the extensive field work done by Dr. Emma Lucy Braun (Chapter 2) prior to World War II. Her gender and diminutive size permitted her to gain the confidence of back-country people, who helped her search for prime, virgin examples of each regional association. Many of the sites she described were logged during World War II, making her 1950 book an important record of what we once had.

Less technical descriptions, recorded by eighteenth-century settlers, also give us a perspective. In Ohio and Pennsylvania, the forest overstory was dominated by 400-year-old oaks, sugar maples, and chestnuts, 1–2 m dbh, with straight boles rising 25 m before the first side branch. Black walnuts, shagbark hickories, and cottonwoods grew in flood plains near rivers, the latter being large enough for travelers such as Daniel Boone to make into dugout canoes 20 m long and more than 1 m across. The dense overstory canopy, together with grapevines that ran up many trunks, created a deep shade in the understory. There was little underbrush, so that it was easy for travelers to wind in and out among the great trunks, even though "one could not shoot an arrow in any direction for more than twenty feet without hitting a tree."

By day, the pristine forest was somber, dark, silent, and gloomy to some. "In the eternal woods it is impossible to keep off a particularly unpleasant, anxious feeling, which is excited irresistibly by the continuing shadow and the confined outlook," wrote one pioneer. The songbirds of our modern forest did not live in that dark forest. At night the forest became unnervingly vocal with calls from wolves, panthers, horned owls, and whip-poor-wills. "It is clear," wrote John Bakeless in his book on the accounts of early explorers, "that no one lamented the disappearance of the picturesque forest, since there were altogether too many trees for comfort."

Braun described nine regional forest associations (Figure 20-22). In this chapter, we will simplify the picture (and take into account some reinterpretations by other ecologists) by condensing those nine into five: **mixed mesophytic, oak-hickory, southern mixed hardwood, beech-maple,** and **hemlock, northern hardwood.**

Physical Features

Annual precipitation ranges from 80 to 150 cm, generally increasing from the northwest towards the east and south. There is no pronounced seasonal wet or dry period. Except in the southern mixed hardwood forest, snow and hard frost in winter are common. Minimum winter temperatures may drop to −30°C. The growing season is 4–6 months long (longer in the south) and maximum temperatures can reach 38°C. Even moderate temperatures of 30°C are uncomfortable because of the high humidity (as much as 10% of summer rain is inter-

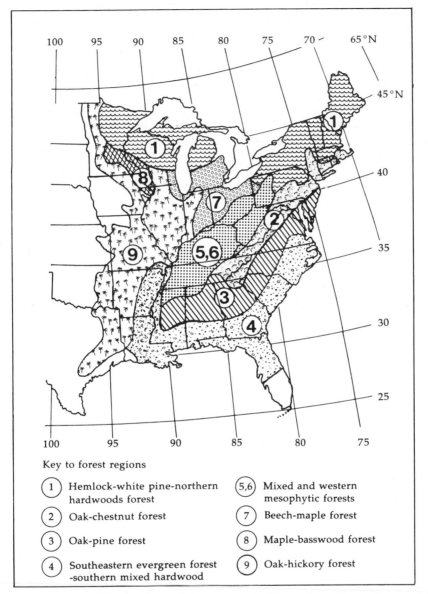

Figure 20-22 Nine regions of the eastern deciduous forest formation recognized by Braun (1950). (From Robichaud and Buell 1973. *Vegetation of New Jersey*. By permission of Rutgers University Press, New Brunswick.)

cepted by foliage and evaporates back to the air). The P/E ratio is greater than 1. Despite the wet summer, however, soil moisture deficits develop that are acute enough to affect herb growth and tree seedling survival.

Two soil orders characterize the eastern deciduous forest. The **alfisol** order extends from north of Virginia and Kentucky to well into an ecotone area with the taiga. Alfisols are also known as **gray-brown podzolics**; they are acidic (pH 5–6), but not as acidic as taiga podzols, partly because the pH of deciduous leaves is 5–7 rather than the pH 4 of spruce and pine. There is no bleached A2 horizon, and the fertility (cation exchange capacity and percent base saturation) is relatively high. Diversity and activity of soil biota are high.

The **ultisol** order dominates the southern part of the forest. Ultisols have not been glaciated, hence they are older and more weathered than alfisols. Plowed fields show a B horizon that is bright yellow-orange as a result of oxidized iron. Ultisols are slightly more acidic than alfisols, and considerably lower in fertility.

Physiognomy and Phenology

The modern deciduous forest has been modified by 200 years of coexistence with farmers, harvesters, and industrialists. Community architecture is no longer so strongly one-layered and impressive as it once was, except in locally protected areas. Today, much of the forest is four-layered (Figure 20-23). The overstory may reach up to 65 m in the best sites, but is more typically 35 m. It is a closed, interlocking canopy, with more than 100% cover. Beneath the overstory is a

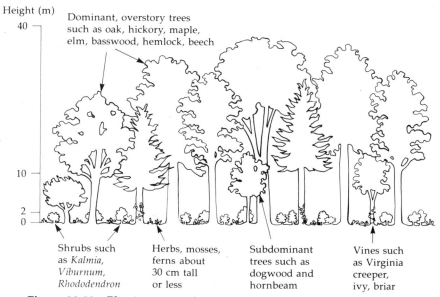

Figure 20-23 Physiognomy of a typical eastern deciduous forest. The dominant overstory trees may reach 60 m or more in height.

second tree canopy that is much more open and consists of saplings and genetically smaller trees about 5–10 m tall, such as dogwood *(Cornus)*. A third layer is made up of scattered shrubs, often members of the Ericaceae, about 1–2 m tall. A fourth layer consists of seasonally active patches of herbs, mosses, and ferns. Vines such as the Virginia creeper *(Parthenocissus quinquefolia)* pass through several layers. Total LAI ranges between 4 and 8, and total biomass averages 30 kg m^{-2}. Annual net primary productivity is about 1200 g m^{-2}.

The aspect of the forest changes dramatically with the seasons, because most species in each stratum are deciduous or die back. If we take New Jersey or Pennsylvania as a reference point, then snowmelt in mid-March coincides with release of spring-flowering herbs from dormancy. At this time, the overstory tree canopy is leafless and more than half full sunlight reaches the ground. By late March to early April, the ground is green from renewed vegetative activity by herbs that have overwintered as bulbs or rhizomes. These herbaceous geophytes account for about 18% of all species in the forest, a proportion three times Raunkiaer's world average. By mid-April, the herbs are flowering and the overstory trees have resumed meristem activity and begun to leaf out. The major control of bud break for eastern deciduous trees is a critical heat sum (degree-days) following a winter cold period. In a few exceptional cases (such as beech, *Fagus*), photoperiod is the environmental control. Over a period of 3–4 weeks, the various tree species begin and complete leaf expansion. Species with ring-porous wood and large diameter vessels (ash, elm, hickory, oak, sumac, walnut) resume cambial activity first, but leaf expansion is delayed. Species with diffuse-porous wood and narrower vessels (alder, birch, cherry, maple, poplar, willow) tend to leaf out first.

As the canopy closes through May, light intensity on the forest floor is reduced to a minimum of 2–5% full sun. Less than 10 ly day^{-1} reach the forest floor at this time, in contrast to 450 ly day^{-1} above the canopy. Moving light flecks temporarily allow up to 70% full sun to reach ground herbs, and there is some evidence that this light may be responsible for half of herb photosynthesis. By July, many of the spring-flowering herbs have died back and become dormant. This dormancy may be induced by competition for moisture, rather than by competition for light. Other herbs, as described below, grow throughout the summer and flower in August and September. Leaves of shrubs and trees turn color in early October and are shed into November. Length of the growing season, from the time tree leaves are half expanded in early May to the time half their chlorophyll is lost in fall, averages 160 days. Time of radial tree growth averages 105 days in extent. Many phenological events may be triggered by environmental changes in the root zone, rather than in the air.

The guild of understory herbs is floristically rich, with 100 common species. About 95% are perennial, and average life span of a ramet is typically 10–20 years. Large, vegetatively spreading genets may reach ages of 100 years. Sexual reproduction is generally delayed past a 3^{+}-year juvenile stage. Most species allocate less than 10% of annual photosynthate to sexual reproduction, but a few expend 20–50% *(Claytonia virginica, Erythronium americanum, Hieracium venosum)*. Few of these forest floor herbs are wind-pollinated. Seed dispersal is often accom-

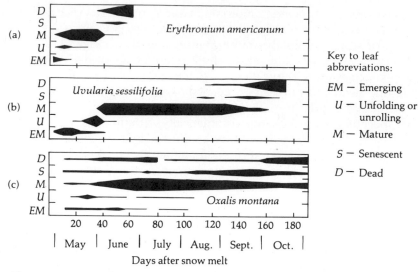

Figure 20-24   Leaf phenology of three understory herbs beneath a northern hardwood forest in New Hampshire. (a) the spring ephemeral *Erythronium americanum*; (b) the summer green *Uvularia sessifolia*; (c) the evergreen *Oxalis montana*. Width of lines indicate number of leaves in each state. (Modified from Mahall and Bormann 1978.)

plished by ants, but in other cases there are no special modifications for dispersal, and seeds do not travel far from the parent. Some species rely mainly on vegetative reproduction.

There are at least three phenological herb patterns: **spring ephemeral, summer green,** and **evergreen**. Spring ephemerals complete most of their growth cycle in a short period between snowmelt and tree canopy closure. Leaf emergence of *Erythronium americanum* (Figure 20-24a), for example, begins 5 days after snowmelt, and all leaves have died 50 days later, by late June. Such herbs are heliophytes with a light saturation point of about 25% full sun. It is possible that their flush of growth in spring is ecologically important in that it conserves nutrients that otherwise would be lost in runoff before tree roots become active.

Summer green herbs begin growth soon after spring ephemerals, but they maintain leaves and active photosynthesis until fall, when their leaves senesce and drop off (see *Uvularia sessifolia*, Figure 20-24b). Summer greens are shade tolerant, with a light saturation point at 11% of full sun. Their maximum net photosynthetic rate is only about 30 μmol CO_2 g^{-1} sec^{-1}, or half that of spring ephemerals.

Evergreens retain their more sclerophyllous leaves for 1–3 years, but they are dormant during the winter. All new leaves may be produced in spring, or there may be a continuous leaf turnover. Flowering usually occurs in late spring or early summer. *Oxalis montana* (Figure 20-24c) is a good example. Its photosynthetic behavior is intermediate between that of ephemerals and summer greens.

Regional studies reveal only weak associations between particular understory and overstory species. The abundance of any one herb is related more to general physical factors such as depth and texture of litter, soil drainage, and sunflecks, rather than to the effects of a specific overstory tree. As overstory trees change from region to region within the deciduous forest, the understory herb community either shows little change or changes at rates that do not correlate with change in the overstory.

Major Ecotones

Taiga-Deciduous Forest Ecotone: The Hemlock-Hardwoods Forest   In the central part of North America, in Manitoba, Saskatchewan, and Alberta, the taiga passes into a spruce-aspen parkland, then an aspen parkland in a wide ecotone with the grasslands. East of this, in an equally wide ecotone, the taiga passes into the eastern deciduous forest. This ecotone centers on 45°N latitude (southern Maine, northern New York, and northern Michigan), then bends north through northern Wisconsin and northeastern Minnesota. As balsam fir, white spruce, and red spruce become less abundant, a mixture of hardwoods and other conifers assume dominance.

From studies done in the Green Mountains of Vermont, the ecotone's location correlates closely with cloudiness and fog in spring and fall. Other factors are shortness of the frost-free period and soil depth. Above 800 m elevation, fog drip abruptly increases by 60% and the number of frost-free days per year decreases by 40%. Once spruce and fir are established, they modify the site to make it unfavorable for hardwoods: Soil pH drops, litter depth increases, and soil moisture increases.

Ecotone hardwoods include sugar maple *(Acer saccharum)*, yellow birch *(Betula alleghaniensis)*, beech *(Fagus grandifolia)*, paper birch *(Betula papyrifera)*, and basswood *(Tilia americana)*. Conifers include three pines *(Pinus banksiana, P. resinosa, P. strobus)* and hemlock *(Tsuga canadensis)*. This **hemlock-hardwoods forest** is the type of forest Thoreau lived in at Walden Pond (Figure 20-25). To the south, it merges into beech-maple or oak-hickory forest. The pines are especially important in the Midwest, and this phase of the ecotone has been called the **Great Lakes Pine Forest.**

Prairie-Deciduous Forest Ecotone: Oak Savanna and Woodland   In an eastward arc, extending from Minnesota to Texas, the deciduous forest gives way rather abruptly to grassland. The ecotone projects east into Illinois, Indiana, Ohio, and Kentucky, permitting a **prairie peninsula** to cover almost a quarter of a million square kilometers east of the Mississippi River (Figure 20-26). The ecotone is very complex. Sometimes it is abrupt, with a transition from a closed oak forest to an open grassland in a matter of less than 10 m. In other places, the forest gradually thins to a woodland of more open trees with a grassy understory, then to a savanna of scattered trees (less than 30% canopy cover), and finally to a grassland in the distance of 50 km. The boundary is sinuous, with interdigitations of grassland and forest. It is also a mosaic, with numerous islands of prairie within the forest and outliers of forest in the grassland. Major tree species

Figure 20-25 The hemlock-hardwoods forest, an ecotone between the taiga and the deciduous forest.

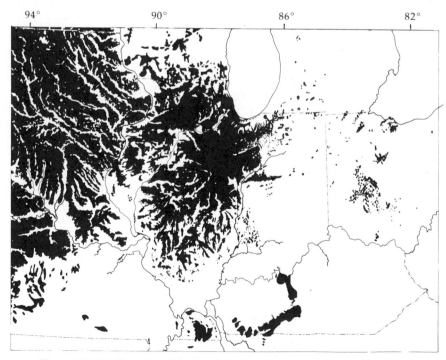

Figure 20-26 Location (black) of the prairie peninsula, as mapped by Transeau (1935). (By permission of the Ecological Society of America.)

of the ecotone are bur oak *(Quercus macrocarpa)*, post oak *(Q. stellata)*, and black-jack oak *(Q. marilandica)*.

There are many physical and biotic factors that change across the ecotone, and it has not been possible for ecologists to determine which is most important. The major factors that correlate with change from forest to grassland are as follows:

1. P/E ratio drops below 1 (except in the prairie peninsula).

2. Annual precipitation drops below 60 cm. (Aridity increases to the west, as does prairie.)

3. Yearly variability in precipitation increases and so does the frequency of drought. Low rainfall cannot be the total reason for forest to give way to grasslands, however, because grassland borders further west support trees under even more arid conditions.

4. Fire frequency increases, favoring grasses (see next paragraph).

5. Finer, deeper prairie soils favor fibrous roots of grasses.

6. Tree mycorrhizal fungi are absent in prairie soils, and tree saplings are at a competitive disadvantage without mycorrhizae.

7. Heavy grazing by bison until the late nineteenth century may have eliminated tree saplings.

8. The tree line may have been pushed east by the Xerothermic period, about 6000 years ago, and not yet regained its climatic limits.

In the pristine grassland, wildfires were common virtually every autumn. The perennial grasses could recover from rhizomes and root crowns below the surface, but the woody species were killed. Where forest cover was dense enough, ground vegetation was too thin to carry a fire. In this way, fire could maintain an abrupt forest-grassland ecotone. In the absence of fire, some Wisconsin studies show that woody species invade prairie at the average annual rate of 0.3 m. Burning also improves the vigor of the grasses, stimulating productivity and flowering two- to threefold due to removal of litter. In the absence of litter, soil temperatures in the root zone are 5°C higher, which stimulates plant growth and decomposer activity.

Some Regional Forest Associations

The **mixed mesophytic forest association** of the Cumberland Mountains and Alleghany Mountains contains the richest collection of overstory dominants of any forest in North America. As many as 20–25 species of overstory and understory trees can be found in one hectare, and dominance fluctuates so greatly from stand to stand that the community cannot be named by even a handful of species. The most common, characteristic trees include sugar maple *(Acer saccharum)*, buckeye *(Aesculus octandra)*, beech *(Fagus grandifolia)*, tulip tree *(Liriodendron tulipifera)*, white oak *(Quercus alba)*, northern red oak *(Q. borealis var. maxima)*, and basswood *(Tilia heterophylla)*. Hemlock may sometimes be present, especially in mesic, protected cove forests of Appalachian foothills (Figure

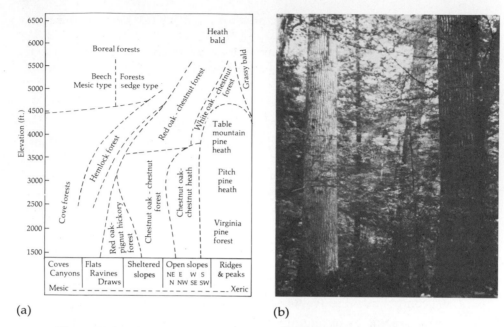

(a)   (b)

Figure 20-27   (a) Gradient diagram of vegetation of the Great Smoky Mountains in the mixed mesophytic forest region. The horizontal gradient is a complex moisture/exposure/aspect gradient; the vertical gradient is elevation (in feet). The hemlock cove forest occurs in a broad elevational range within the most mesic sites. (b) The mixed mesophytic forest, located in a cove of the Blue Ridge Mountains of North Carolina. The large trunks are 1.5 m dbh. [(a) From Whittaker 1956.]

20-27). Understory trees are also diverse and may produce a striking floral display in the spring. They include redbud (*Cercis canadensis*), dogwood (*Cornus florida*), witch hazel (*Hamamelis virginiana*), ironwood (*Carpinus caroliniana*), hornbeam (*Ostrya virginiana*), hackberry (*Celtis occidentalis*), and species of *Rhododendron* and *Magnolia*. Viburnum, sumac, huckleberry, blueberry, and blackberry shrubs are common; midsummer is a berry-picker's delight.

Nearly surrounding the mixed mesophytic association is the **oak-hickory forest,** which has several species each of *Quercus* and *Carya*. In places, tulip tree may be a codominant. A portion of this forest, just to the east of the mixed mesophytic association, was an area once dominated by chestnut (*Castanea dentata*). As briefly described in Chapter 8, a fungal parasite of chestnut was accidentally introduced to the New York area from Europe early in the twentieth century. By the 1930s, the fungus had spread throughout much of the range of chestnut and had killed all mature trees. Apart from an occasional stump sprout and a few trees, the only remains of chestnut today are "ghosts"—slowly decaying but sometimes still standing skeletons of dead trees. Oaks and hickories now occupy the openings in the canopy vacated by chestnuts.

Along the Atlantic coast, just behind the beach and foredune, the oak-hickory forest becomes dominated by short, twisted oaks with relatively small, some-

Figure 20-28 A live oak forest (mainly *Quercus virginiana*) along the North Carolina coast.

times evergreen, sclerophyllous leaves: *Quercus virginiana, Q. laurifolia,* and *Q. myrtifolia.* Below the dense canopy, only about 5 m tall, are many shrubs *(Ilex, Myrica, Vaccinium, Serenoa, Sabal)* and vines *(Vitis, Smilax, Rhus, Parthenocissus).* The pruning effect of wind, and of salt spray carried in the wind, is quite evident (Figure 20-28). The maritime climate is more moderate than for other parts of the deciduous forest. Less than 1% of the hours each year have frost. The unique climate and the evergreen foliage make this community very different from the rest of the eastern deciduous forest.

A **beech-maple forest** extends through much of Ohio and Michigan. Major dominants are sugar maple *(Acer saccharum)* and beech *(Fagus grandifolia).* Common associates include yellow birch *(Betula alleghaniensis),* red maple *(Acer rubrum),* white ash *(Fraxinus americana),* and basswood *(Tilia americana).* The southern limit of this association corresponds to the limit of Wisconsin glaciation. The **hemlock northern hardwood forest,** already discussed in the ecotone section, lies to the north, through much of Pennsylvania, New York, and New England.

The southeast is covered by the **southern mixed hardwood forest** (southeastern evergreen forest in map, Figure 20-22). Dominants are many (up to 40 tree species), including evergreens such as southern magnolia *(Magnolia grandiflora),* the semi-evergreen laurel oak *(Quercus laurifolia),* live oak *(Q. virginiana),* and several pines, and deciduous species such as beech *(Fagus grandifolia),* sweet gum *(Liquidambar styraciflua),* and some hickories. Epiphytes, such as Spanish moss *(Tillandsia usneoides),* are common. A subdominant tree canopy may also be closed, but the shrub and herb strata are patchy. Based on their average importance values, evergreens account for about half of the canopy. From site to site, however, percentage evergreenness can vary from 0 to 100. In general, evergreen conifers are most abundant on sandy, sterile sites, where retention of leaves may have survival significance.

Historical Changes

All of North American vegetation has undergone enormous latitudinal and areal shifts since the last maximum glacial advance 18,000 years ago (BP). The history of the eastern deciduous forest may be typical (see Figure 20-29). At the glacial peak, climatic gradients from tundra through deciduous forest were steeper than at present, only 800 km separating tundra from oak-hickory-pine. The Appalachian Mountains were covered in tundra. Riverlands and associated bluffs extending north from the Gulf of Mexico were probably refugia for the mixed mesophytic association.

As glaciers retreated 10,000 years BP and formed the Great Lakes, grassland and oak savanna encroached from the west. The mixed mesophytic forest dominated its maximal area, through the center of today's eastern deciduous forest. Florida vegetation changed from a sand scrub to an oak savanna. At the end of the Xerothermic, 5000 years BP, the southern mixed forest was still more widely established than at present, the prairie peninsula was apparent, the mixed mesophytic forest had contracted, and the northern hardwoods zone was close to its modern limits.

One can now appreciate that modern forests in the northern part of the deciduous forest region are quite young. Sugar maple in the mountains of New Hampshire, for example, has probably been present for fewer than 8000 years. Considering that such a species has a life-span averaging 250 years, this is a span of only 30 generations. Eastern hemlock, with a life-span three times as long, has been present there for only 10 generations. It is impressive to recognize, then, that such species have evolved ecotypes. Sugar maples in New Hampshire, separated by less than 1 km in elevation and distance and free of any apparent breeding barriers, exhibit differences in photosynthesis, respiration, and specific leaf weight. High-elevation genotypes have a net photosynthesis rate 40–80% higher than that of low-elevation genotypes.

Grasslands

Grasslands cover more area of North America than any other formation. The central grasslands are locked into the interior of the continent (Figure 20-30). They occupy a swath 1000 km wide between about 95°W and 105°W, which tapers to a northern limit in Saskatchewan, at 52°N, and to a southern limit in central Texas, at 30°N.

This enormous strip of grassland is not homogeneous. As precipitation decreases from east to west, the height, growth form, and identity of dominant grasses change. The easternmost portion, including the prairie peninsula, is the **tall grass prairie.** Flowering stalks of some sod-forming grasses here reach 2 m in height. Farther west is the **mixed** (sometimes called the **mid-**) **grass prairie,** where maximum plant height is about 1 m. Still farther west, in the area of lowest rainfall (40 cm), is the **short grass prairie,** dominated by bunchgrasses less than

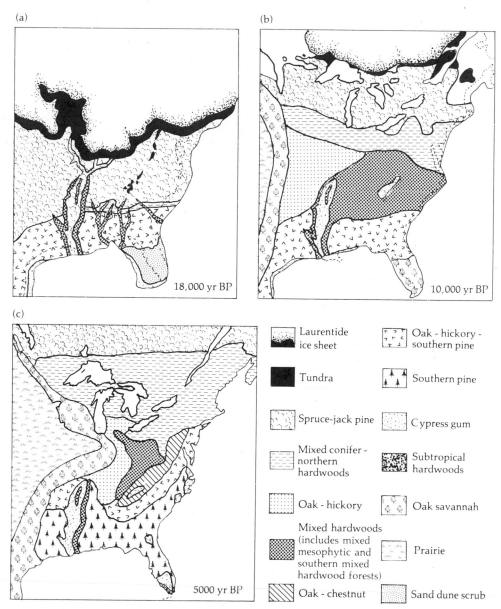

Figure 20-29   Reconstruction of major vegetation types in eastern North America since the last glacial maximum advance: (a) 18,000 years BP (before the present); (b) 10,000 years BP; (c) 5000 years BP. (Redrawn from Delcourt et al. 1983.)

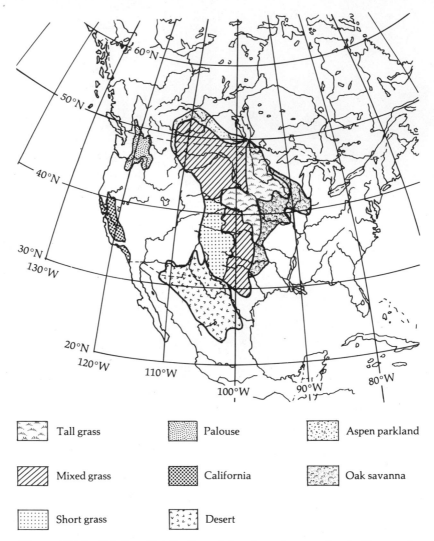

Figure 20-30   Pristine distribution of the six major grasslands discussed in the text. The intermountain grassland is shown as an ectone in Figure 20-38. Modern limits have been modified by cultivation, fire control, and grazing. Much of the desert grassland in this figure overlaps with the Chihuahuan desert. The aspen parkland and oak savanna ecotones are also shown. (Redrawn from Sims 1987, Looman 1983, and Eyre 1963.)

0.5 m tall. The spatial relationship of these grasslands to each other and to forests already discussed is summarized for latitude 37°N in Figure 20-31.

Outlying grasslands include a) the **intermountain grassland,** extending through Wyoming, Idaho, Utah, Nevada, and eastern Oregon, and including the Palouse region of eastern Washington (see sagebrush steppe, Figure 20-38, p. 537); b) the **desert grassland** in Texas, New Mexico, Arizona, and northern Mexico; and c) the **central valley grassland** of California.

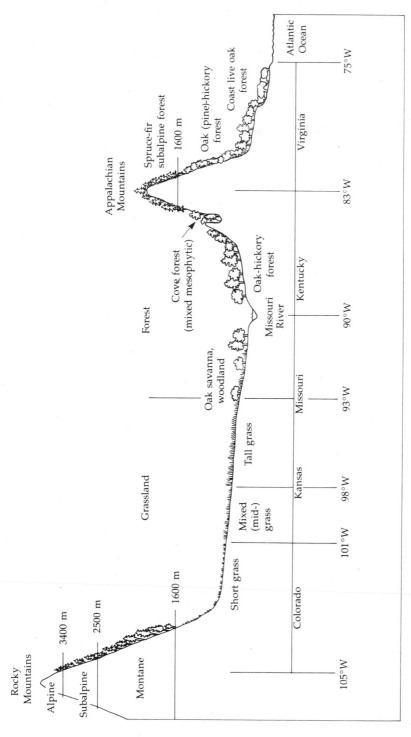

Figure 20-31 Diagrammatic representation of vegetation at 37°N latitude from the Atlantic coast to the Rocky Mountains.

All of the North American grasslands have been extensively modified, primarily through four activities: cultivation, fire suppression policies, grazing of domestic stock, and the accidental introduction of aggressive, weedy, alien plant species.

Some Definitions

Grasslands are herbaceous communities dominated by **graminoids** (grasses, sedges, rushes), but with **forbs** (non-graminoid herbs) present and sometimes seasonally dominant. Trees are absent except for local sites, such as along watercourses.

Grasslands may be dominated by annual grasses, perennial **bunchgrasses,** or perennial **sod-forming** grasses. Annuals are most abundant on dry, overgrazed, or disturbed sites. Some annuals are cool-season types, germinating in late fall and setting seed the following early summer. Cheatgrass *(Bromus tectorum)* in the intermountain area is such an annual. Other annuals are warm-season types. They germinate in spring or summer and complete a much shorter life cycle in a matter of weeks, rather than months. Six-weeks grama *(Bouteloua barbata)*, in the desert grassland, is such an annual.

Annuals do not reproduce vegetatively with runners or rhizomes. Their lateral spread is limited to the production of **tillers** (stems) from buds near the root crown, giving the plant a bushy, clumped appearance (Figure 20-32). Per-

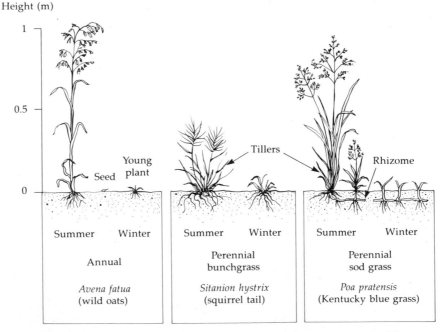

Figure 20-32 Summer and winter aspects of three major grass life forms: annual, bunchgrass, and sod-forming grass.

ennial bunchgrasses, such as purple needlegrass (*Stipa pulchra*) in California, also produce tillers, but the continuation of that process for many years results in a large clump a decimeter or more in diameter. Bunchgrasses alone do not generally produce a community with 100% cover, but the spaces between clumps can be seasonally filled by forbs and other grasses.

Perennial sod-forming grasses, such as big bluestem (*Andropogon gerardii*) in the tall grass prairie, spread laterally by rhizomes. New shoots and roots arise from nodes on the rhizomes in such numbers that a **turf** results. The topsoil is thoroughly penetrated and held together by fibrous root systems, and a continuous sward of shoots covers the surface. Sod-formers are more resistant to grazing than bunchgrasses because they have so many more growing points. Some sod-formers are less aggressive than others, producing only short rhizomes, and some produce rhizomes only under certain environmental conditions. The growth forms of common grasses in North American grasslands are summarized in Table 20-5.

Table 20-5 Common dominant grasses in each of the six major grasslands of North America. A = annual, B = bunchgrass, S = sod-forming grass,* = introduced, not native, ‡ = known to be C_4 (some of those not marked may be C_4).

		Tall	Mixed	Short	Intermountain/ Palouse	Desert	Central California
Switch grass (*Panicum virgatum*)	S	x					
Kentucky bluegrass (*Poa pratensis*)	S*	x					
Indian grass (*Sorghastrum nutans*)	B‡	x					
Quack grass (*Agropyron repens*)	S*	x	x				
Western wheatgrass (*A. smithii*)	S	x	x	x	x		
Big bluestem (*Andropogon gerardii*)	S‡	x	x	x			
Little bluestem (*A. scoparius*)	B‡	x	x	x			
Hairy grama (*Bouteloua hirsuta*)	B or S‡	x	x	x			
Sideoats grama (*B. curtipendula*)	B‡	x	x	x	x		

(Table continues)

Table 20-5   *continued*

		Tall	Mixed	Short	Intermountain/ Palouse	Desert	Central California
Blue grama (*B. gracilis*)	B‡	x	x	x	x		
Needlegrass (*Stipa spartea*)	B	x	x	x			
Needle and thread (*S. comata*)	B		x	x	x		
June grass (*Koeleria cristata*)	B		x	x	x		x
Dropseed (*Sporobolus cryptandrus*)	B		x	x			
Sand bluestem (*Andropogon hallii*)	S			x		x	
Buffalo grass (*Buchloe dactyloides*)	S‡			x			
Vine mesquite (*Panicum obtusum*)	S			x			
Tobosa (*Hilaria mutica*)	B‡					x	
Galleta (*H. jamesii*)	S					x	
Black grama (*Bouteloua eriopoda*)	S					x	
Six-weeks grama (*B. barbata*)	A					x	
Three awn (*Aristida* spp.)	B				x		
Indian rice grass (*Oryzopsis hymenoides*)	B		x	x	x		
Muhly (*Muhlenbergia* spp.)	S or B			x	x		
Sacaton (*Sporobolus wrightii*)	B				x		
Alkali sacaton (*S. airoides*)	B		x	x	x		
Burro grass (*Scleropogon brevifolius*)	S				x		

Table 20-5 *continued*

	Tall	Mixed	Short	Intermountain/ Palouse	Desert	Central California
Arizona fescue (*Festuca arizonica*)	B			x		
Bluebunch fescue (*F. idahoensis*)	B			x		
Red fescue (*F. rubra*)	B			x		
Greenleaf fescue (*F. viridula*)	B			x		
California oat grass (*Danthonia californica*)	B			x		
Bluebunch wheatgrass (*Agropyron spicatum*)	B			x		
Slender wheatgrass (*A. trachycaulum*)	B			x		
Cheatgrass (*Bromus tectorum*)	A*			x		
California brome (*B. carinatus*)	A or B			x		x
Brome (*B. marginatus*)	B			x		
Pinegrass (*Calamagrostis rubescens*)	B or S			x		
Wild rye (*Elymus* spp.)	B or S			x		x
Melic (*Melica* spp.)	B					x
Needlegrass (*Stipa* spp.)	B			x	x	x
Soft chess (*Bromus mollis*)	A*					x
Rip gut (*B. diandrus*)	A*					x
Fescue (*Festuca* spp.)	A					x
Wild oats (*Avena* spp.)	A*					x

Table 20-6   Changes in mean annual temperature (MAT, °C) and mean annual precipitation (MAP, mm) as one crosses north through the parkland ecotone. (From data in Looman 1983.)

	Northern shortgrass prairie	Aspen parkland	Southern boreal forest
MAT	4	1	0
MAP	300–400	400–530	420–540

The term **prairie** is most often applied to a grassland dominated by tall, sod-forming grasses, with 100% cover. **Steppe** is often applied to a grassland dominated by short bunchgrasses, often interspersed with shrubs, with a ground cover that may be less than 100%. In this chapter, we will use **grassland** as a general term, applicable to prairie, steppe, and all intermediates. As precipitation increases, grassland gives way to forest. An ecotone vegetation type is the **savanna.** Savannas still retain a continuous grass cover, but trees are regularly present and contribute up to 30% cover. Two major ecotone savannas are shown in Figure 20-29: the northern aspen parkland and the eastern oak savanna. The oak savanna, described earlier in this chapter, is a north–south strip that increases in width from north (50 km wide) to south (hundreds of kilometers wide). Its position corresponds to an aridity gradient on which the P/E ratio falls below 1 as one moves west.

The parkland ecotone in Canada is characterized first by increasing aspen (*Populus* spp.) cover, then farther north by increasing aspen and boreal conifer cover. Aspen reproduces vigorously from sucker sprouts that arise from below-ground parts not damaged by fire. The width of the aspen belt ranges from 175 km in the east and north to 75 km in the west. The aspen-conifer belt extends an additional 75–300 km away from the grassland. This ecotone is related to a moisture gradient, but more specifically to the relative amounts of snow or rain that fall in the spring months, April–June. These amounts decrease as one moves north, even though annual precipitation increases (Table 20-6). Insufficient precipitation in those months has a negative effect on the reproduction of several dominant grasses.

Physical Features

Climate of the central grassland is continental, with long, cold winters and long, hot summers. Extreme lows in the winter may dip to −40°C, and extreme highs in the summer may reach +45°C. An Illinois tall grass prairie may show a mean monthly temperature for January at −3°C, and 24°C for July. In the mixed grassland of North Dakota, or in the Palouse region of Washington, mean January temperature can be considerably lower, −14°C, while mean July temperature is still high, 21°C. There are, then, large seasonal temperature changes. Winter snow is common, but it is dry and can be blown into drifts, leaving most of the

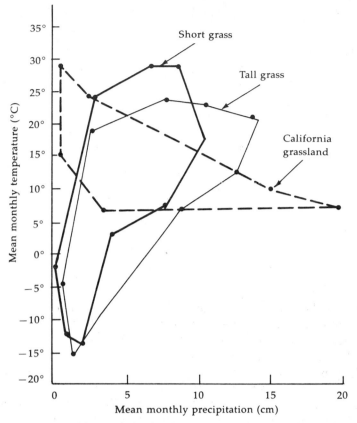

Figure 20-33 Climographs for three grasslands. Note the similarity between tall and short grassland regions and the uniqueness of the California central valley grassland.

ground bare and uninsulated. The growing season for middle latitudes in the central grasslands area is about 240 days.

Annual precipitation for the central grasslands area is generally 40–60 cm, with a strong summer peak. Evapotranspiration is high, and the P/E ratio is less than 1. A pronounced moisture deficit, compared to a minor one in the deciduous forest, characterizes grassland soils in summer. Summer thunderstorms may be violent, accompanied by hail and high winds. Total rainfall can fluctuate two- to threefold in successive years, and periodic droughts (as in the Dust Bowl years of the 1930s) are to be expected. It has been suggested that the distribution limits of some grassland species correlate with occasional extreme lows in rainfall, rather than with long-term means. Precipitation in the desert, intermountain, and Californian grasslands is lower, about 25–45 cm, and the seasonality also differs from that in central grasslands. The desert area has both summer and winter peaks, while the intermountain and Californian areas have winter peaks. Climographs for three grasslands are shown in Figure 20-33. Notice how closely the tall and short grass areas overlap.

Figure 20-34 Relative biomass (percent) of C_3 and C_4 grasses as a function of latitude, for the mid-prairie region of North America. (From Sims 1987.)

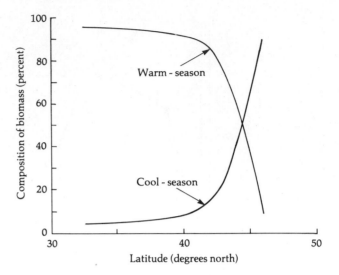

Gradients of both temperature and precipitation correlate with shifts in photosynthetic pathway of the grasses. As one moves north through the mid-prairie region, the fraction of C_4 and C_3 biomass changes (Figure 20-34). North of approximately 44° latitude, cool-season (C_3) grasses predominate; south of this, warm-season (C_4) grasses predominate. This ecotone roughly corresponds with the southern borders of Wyoming and Iowa, and along part of that ecotone elevation falls 400 m within a relatively short distance, modifying both temperature and precipitation. Summer temperatures appear to be especially critical to C_4 species: Locations with mean minimum July temperatures below 8°C have few or no C_4 species.

Grassland soils are quite diverse. In the prairie peninsula and in part of the tall grass area, the prevailing soils are in the ultisol order. A P/E ratio of 1 permits weak podzolization to occur. Soil pH is close to neutral, however, and decay of fibrous root systems gives the soil a dark brown color. There is no bleached A horizon. Where precipitation is less than evapotranspiration, **mollisol** soils become prominent. A layer of calcium carbonate accumulates in the subsoil, deposited at shallower and shallower levels as the P/E ratio declines. The most mesic mollisol is a **chernozem,** with up to 16% organic matter in the nearly black A horizon, a clay loam texture, high cation exchange capacity, very high percent base saturation, pH slightly basic, and carbonate accumulation only at 100 cm depth. *Chernozem* is a Russian word, meaning black earth. The color is a result of richness in organic matter. Chernozems give way to lighter brown soils as one moves through mixed and short grass areas. Intermountain grasslands are also on mollisols, but desert grasslands can occur on **aridisols,** which have little profile development and contain less than 1% organic matter in the root zone. Towards the eastern deciduous forest border, prairie can lie on acidic alfisols.

Central Grassland Vegetation

The **tall grass prairie** has a nearly equal mix of bunchgrasses and sod-forming grasses, which together provide a dense canopy of more than 100% cover and a standing aboveground biomass of 200 g m^{-2}. The LAI is 5–8, about the same as that in the deciduous forest. Most biomass is belowground, and the root:shoot ratio averages 6–7. Annual aboveground net productivity can be more than 1000 g m^{-2}, the great bulk of which is contributed by warm season grasses, some of which are C$_4$ species (e.g., little bluestem and blue grama).

Many species are listed in Table 20-5, but the most typical dominants are big bluestem and Indian grass, both with flowering shoots up to 2 m tall. Associated with them are shorter grasses, such as little bluestem and gramas, and a variety of perennial forbs. As in all the grasslands, forb species outnumber graminoids by 3 or 4 to 1, but they account for only 10–20% of total community productivity. There is considerable latitudinal variation in the associated species, and even at one particular location there are major shifts that reflect the microenvironment. Prairie cordgrass (*Spartina pectinata*), for example, can be prominent in wet depressions, and Kentucky bluegrass can be abundant where the range is heavily grazed.

Most of the tall grass prairie is gone, replaced by today's corn belt, but the records of early travelers can help us reconstruct it. This is from a journal dated 1837:

> The view from this mound . . . begs all description. An ocean of prairie surrounds the spectator whose vision is not limited to less than 30 or 40 miles. This great sea of verdure is interspersed with delightfully varying undulations, like the vast swells of ocean, and every here and there, sinking in the hollows or cresting the swells, appear spots of trees. . . .

The **mixed grass prairie** is perhaps the most floristically complex of the central grasslands. The tall sod-formers of the east generally become restricted to locally heavy soils, and the usual dominants are grasses of medium height overtopping shorter bunchgrasses such as little bluestem, hairy grama, blue grama, and needlegrass. To those who have grown up around forests, a grassland such as this may appear homogeneous and monotonous (Figure 20-35), but careful observation reveals a complex mosaic of communities. The mixed grass prairie is a tension zone or ecotone between the tall grass and short grass regions, and it has had a dynamic recent history. Effects of grazing, fire, and drought have been pronounced. It has unique northern and southern provinces. To the north, such C$_3$ grasses of medium height as western wheatgrass dominate, associated with needle and thread and the C$_4$ blue grama (see Table 20-5 for scientific names). Productivity is typically 100–300 g m^{-2} yr^{-1}. To the south, C$_4$ grasses predominate, and productivity is higher. Cover in the mixed grassland is still 100%, but standing, live aboveground biomass is only half that of tall grass prairie, and only half the annual productivity derives from warm-season grasses.

Figure 20-35   Mixed grassland on rolling hills in North Dakota.

Short grass communities are dominated by bunchgrasses such as buffalo grass, sand bluestem, Indian ricegrass, blue grama, and species of *Muhlenbergia*. A few species are sod-formers. Buffalo grass and blue grama, among others, are C_4 species, but about half the annual productivity (depending on latitude) derives from cool-season C_3 species. Aboveground standing crop is only 50 g m^{-2}. When mixed grass prairie is stressed by heavy grazing or drought, short grass species tend to increase in abundance. Overgrazing has also contributed to recent invasion by a variety of perennial *(Opuntia, Yucca)* and annual *(Bromus, Festuca, Hordeum, Salsola)* weedy species of low palatability. This is the region (especially southwestern Colorado, southwestern Kansas, and the panhandles of Texas and Oklahoma) that spawned the Dust Bowl.

Desert Grassland

Plateaus above 1000 m elevation at the edge of the Sonoran and Chihuahuan deserts were covered until recently by short bunchgrasses, with desert scrub restricted to ravines, knolls, or other locally poor sites. Important grasses were Indian ricegrass, *Muhlenbergia*, black grama, tobosa, and galleta. Its original area amounted to 8% of North American grassland.

A significant fraction of this grassland's acreage has disappeared over the past 100 years. **Desert scrub** has invaded from the locally poor sites and has come to dominate many hectares (Figure 20-36). An 1858 vegetation survey of the Jornada Experimental Range in southcentral New Mexico showed 33,800 ha to have 100% relative cover by grasses. Desert shrubs were absent from those

Figure 20-36 This stand of desert scrub, in New Mexico, was desert grassland a century ago, according to historical land descriptions.

hectares at that time. A resurvey in 1915 showed only 14,400 ha could still be so classified. In 1963, no hectares had 100% relative grass cover (Buffington and Herbel 1965). At the same time, cover by such desert shrubs as creosote bush, mesquite, and tarbush increased twentyfold. A 50$^+$-year study of desert grassland on the Santa Rita Experiment Station in southeastern Arizona showed a similar trend, creosote bush alone increasing thirteenfold between 1904 and 1958 (Humphrey and Mehrhoff 1958). Mesquite *(Prosopis juliflora)* has also spread widely.

Some ecologists have hypothesized that the shrub explosion was due to overgrazing of grasses by cattle. However, even after cattle exclosures were erected in 1931 at Jornada and Santa Rita, shrubs continued to spread in both. Possibly shrub invasion is due to erosion of the thin top soil formerly held in place by the grasses. Other causes suggested for shrub encroachment include fire suppression policies and a slight warming-drying trend in the climate of the southwest region.

Intermountain-Palouse-Willamette Grassland

The northern part of the Great Basin, between the Rocky Mountains and the Cascade Range, is a **shrub steppe** (Figure 20-37). Short bunchgrasses share dominance with cold desert shrubs, especially species of sagebrush *(Artemisia)*. Many dominants of this region are also found in the loess-covered Columbia River Basin of eastern Washington (the **Palouse** area), and in another grassland that used to clothe the Willamette Valley of Oregon. As in the central valley of California (next section), cultivation has eliminated the Willamette grassland, and we must rely on incomplete records to reconstruct the pristine vegetation.

Dominants of these three regions include the following cool-season bunchgrasses: bluebunch wheatgrass, bluebunch fescue, several other fescues, wild rye, sacaton, and California oat grass (see Table 20-5 for scientific names). This intermountain grassland has been invaded and changed by annual cheatgrass,

Figure 20-37   *Artemisia* shrub steppe, Wind River area of Wyoming.

as discussed in Chapter 6. A review of the biogeography of C_4 grasses shows that the intermountain and Californian grasslands are much lower in C_4 species than the central grasslands. Tall and mixed grassland floras are 40% C_4 (grass species only), while the intermountain-Palouse-Willamette-California grassland floras are only 15% C_4.

California Central Valley Grassland

This grassland is located mainly in the Sacramento and San Joaquin Valleys, although historically it was also found in the Los Angeles basin and in several other coastal valleys of central and southern California. (An interrupted grassland along the north coast of California is more closely related to the Palouse and Willamette grassland than it is to the central valley grassland.) Its area totals only 2% of all North American grasslands.

Two hundred years ago, when first viewed by Europeans, the vegetation may have been dominated by cool-season bunchgrasses such as purple needlegrass *(Stipa pulchra)*, nodding needlegrass *(S. cernua)*, wild rye *(Elymus* spp.), pine bluegrass *(Poa scabrella)*, three awn *(Aristida* spp.), June grass *(Koeleria cristata)* and deer grass *(Muhlenbergia rigens)*. These grasses contributed about 50% cover, and the spaces between were filled with great masses of annual and perennial grasses and forbs that flowered in late spring. During the June–October dry period, forbs and grasses alike died back, turning the region golden brown. This grassland once occupied over 5 million hectares, about 13% of the modern state's land area.

Overgrazing, drought, and the introduction of hundreds of weedy annuals have changed this grassland into an annual type, with forage of lower nutritional value. Modern dominants are wild oats, bromes, ryegrasses (*Lolium* spp.), fox-tails (*Hordeum* spp.), and forbs such as filaree (*Erodium* spp.). Cattle exclosure studies indicate that the native bunchgrasses are very slow to reinvade or to increase in cover, given the presence of these aggressive annuals. It appears that the California grassland has been permanently changed, and the weeds are "new natives," to use the expression of one ecologist.

Desert Scrub

Desert scrub is a vegetation type dominated by shrubs with less than 100% cover and generally restricted to semiarid regions receiving 5–25 cm precipitation a year. In North America, desert scrub occupies about 1.2 million square kilometers. Based on climatic, vegetational, and floristic differences, four desert regions and one ectone can be recognized (Figure 20-38).

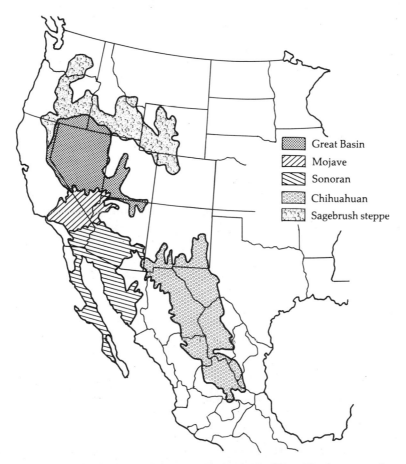

Figure 20-38 Four desert regions of North America. The large northern grassland–desert ecotone, the sagebrush steppe, is also shown. (Modified from MacMahon and Wagner 1985 and West 1987.)

Figure 20-39   Great Basin desert, eastern Oregon, dominated by big sagebrush *(Artemisia tridentata).*

The **Great Basin desert** (also called the **cold desert**) has a considerable amount of snow and hard frost in winter. The intermountain grassland, discussed in the last section, extends through the northern part of this desert, but most of the region is thoroughly dominated by big sagebrush *(Artemisia tridentata,* Figure 20-39). The other three deserts are collectively called the **warm** or **hot deserts** because winters are mild. One shrub dominates all three: *Larrea tridentata* (Figure 20-40), variously called creosote bush, greasewood, gobernadora, or hediondilla. Creosote bush is slow-growing but persistent. It is able to spread vegetatively by a splitting of the root crown followed by adventitious rooting at the base of lateral branches beneath the surface. Clonal circles of shrubs develop that may be hundreds to thousands of years old.

When viewed from a low, oblique angle, plant cover seems high, but when the canopies are projected vertically down, only 10–25% of the ground typically lies beneath perennial cover (extremes are 5–50%). Aboveground biomass is low, averaging 700 g m^{-2}, which is nearly the same as tundra standing crop. LAI is less than 1, and during the dry part of the year, plants are either absent, leafless, dormant, or functioning at a low level. Over the course of a year, then, the fraction of solar radiation trapped in photosynthesis is quite low—as low as 0.003%, which is an order or two in magnitude lower than the efficiency of other vegetation types. Because of all these reasons, net primary productivity is the lowest of all vegetation types examined in this chapter, about 100 g m^{-2} yr^{-1}— 50% lower than tundra productivity.

It is a mistake to think that the low community productivity means low rates of photosynthesis for each species. If productivity is expressed in terms of grams fixed per gram of leaf area instead of per unit ground area, then desert shrubs are fully as productive as shrubs in the mesic eastern deciduous forest (e.g.,

Figure 20-40 Creosote bush scrub, showing a clonal ring of creosote bushes. The diameter of the ring is several meters.(Courtesy of Frank Vasek.)

Larrea $= 0.28$ g CO_2 g leaf^{-1} yr^{-1} versus $0.16-0.37$ for mesic shrubs). Some desert subshrubs and annuals have net photosynthesis rates per unit leaf area comparable to or above those of crop plants under irrigated cultivation, with rates in the range of $60-90$ mg CO_2 dm^{-2} hr^{-1}.

Physical Features

Some temperature and precipitation data for each desert are summarized in Table 20-7. Cold winter temperatures in the Great Basin and Chihuahuan deserts reflect their relatively high elevation on plateaus. The Mojave desert covers a wide amplitude of elevation, from $+1300$ to -86 m (the lowest spot in Death Valley). The Sonoran desert generally is at low elevations. Annual precipitation does not vary much from desert to desert, but its seasonality and form do. The Great Basin and Mojave have winter peaks and experience moderate to considerable snowfall. Portions of the Sonoran have winter and summer peaks of equal intensity but no snow, and the Chihuahuan has a pronounced summer peak only. As in the grassland, fluctuation in rainfall from year to year can be large. The P/E ratio averages 0.3 but may drop below 0.1 at the head of the Gulf of California. Summer temperatures are almost equally high in all four deserts.

Some important topographic features common to all deserts are diagrammed in Figure 20-41. The lower slopes of desert mountains are long and gentle. They are given the Spanish name **bajada.** Their average rise is only 1–5%, which is so gradual that one does not realize the elevation gain until one looks back

Table 20-7   Some climatic features of the four desert regions of North America.

	Great Basin	Mojave	Sonoran	Chihuahuan
Area (km^2)	409,000	140,000	275,000	453,000
Annual precipitation (mm)	100–300	100–200	50–300	150–300
Precipitation falling in summer (% of total)	30	35	45	65
Snowfall (cm; 10 cm snow = 1 cm rain)	150–300	25–75	trace	trace
Winter mean max/min temperatures (°C)	+8/−8	+15/0	+18/+4	+16/0
Hours of frost (% of total)	5–20	2–5	0–1	2–5
Summer mean max/min temperatures (°C)	34/10	39/20	40/26	34/19
Elevation (m)	>1000	variable	<600	600–1400

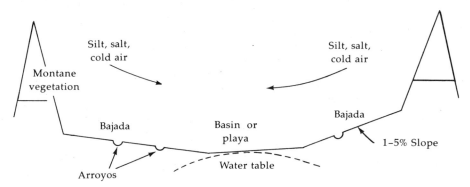

Figure 20-41   Major topographic features of the desert landscape. Typical desert vegetation is restricted to the bajadas.

downslope. Bajadas are composed of coarse alluvium eroded from steeper slopes in the montane zone above. They are crisscrossed with channels, called **arroyos, washes,** or **wadis,** cut by intermittent streams that flow only after heavy rains.

Bajadas support typical desert vegetation, with the highest species diversity of any desert habitat. Evidently, the coarseness of the soil provides a wide range of microhabitats. On a typical Sonoran bajada near Tucson, Arizona, the upper bajada community consisted of 18 perennial species with 35% cover and a species diversity index (H') of 1.2. Near the base of the bajada, on finer soil, the community consisted of only 6 perennial species with 20% cover and a diversity index of 0.7. Along the same gradient, soil texture changed dramatically (Figure 20-42). Throughout the desert, species diversity correlates better with soil texture than with rainfall.

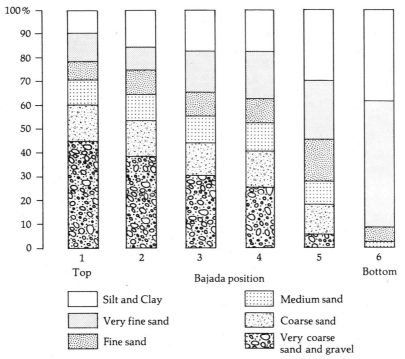

Figure 20-42 Soil particle size distribution for six sites along an Arizona bajada. (Redrawn from MacMahon and Wagner 1985.)

Arroyos have low plant cover because of frequent disturbance, but the larger ones may have scattered trees. These riparian trees are able to survive because their root systems are in contact with greater supplies of water than are available on the surrounding bajadas. Only in areas of significant summer rainfall, as in the Arizona Sonoran desert, do these trees occur on the bajadas. Common examples include smoke tree *(Dalea spinosa)*, desert ironwood *(Olneya tesota)*, mesquite *(Prosopis* spp.), and palo verde *(Cercidium* spp.). All of these are legumes with large seeds that require scarification for germination.

Bajada soils are typical **aridisols,** with little profile development, less than 1% organic matter, sandy loams to loamy sands, pH 7–8.5, and often with an indurated layer of calcium carbonate (called a **caliche** layer) 25–50 cm below the surface. Such a layer impedes root growth, and the shallowness of the caliche affects the species composition of the community above. In some areas, erosional processes produce a **desert pavement**—a surface layer of close-fitting rocks, polished and burnished on their exposed faces. Plant establishment is very difficult here, and plant cover is unusually low. Another surface phenomenon can be a thin but persistent water-repellent layer beneath the canopy of certain shrubs, such as creosote bush. Evidently, material released from living or decaying leaves combines with the soil particles to produce the layer, which may remain for years after the plant is gone. Decay products also accumulate in the topsoil in quantities which differ from species to species. Soils beneath different species may significantly differ in such traits as water holding capacity, texture, the concentration of organic carbon and nitrogen, and the concentration of various anions and cations such as Na, K, Ca, Mg, Cl, SO$_4$, Fe, and Mn. Soils in the open, between canopies, naturally differ from soils beneath canopies. In addition, mosses, lichens, and algae may colonize the soil surface beneath canopies, forming a soil crust that affects permeability and nitrogen content (if the lichens or algae are nitrogen-fixers).

Playas are undrained basins at the base of bajadas. Runoff from the bajadas carries fine textured soil and dissolved salts. This material accumulates in the playa in such a way that there are circular zones of increasing salinity from the edge to the lowest part of the basin. Each zone is characterized by a different community of plants. The innermost zone may be devoid of plants, showing only a crust of salts on the surface. Soil aeration is low because of the fine texture and because a water table may be close to or at the surface of the playa bottom. Some research indicates that bajada species are prevented from colonizing playas mainly because of low soil oxygen, rather than because of the high soil salinity (sometimes more than 35,000 ppm salt, pH 9–11).

An additional microenvironmental trait of playas is that they are colder than surrounding slopes. Cold air at night sinks and collects in the basin, and frost can be more common there. One 10-year study in the Mojave desert of southern Nevada showed that a 60 m drop in elevation from a *Larrea*-dominated bajada slope to an *Atriplex*- and *Lycium*-dominated playa bottom correlated with a drop in mean minimum temperatures from several degrees above freezing to several degrees below freezing. Thus, there are several factors that account for major community differences between playas and bajadas.

Not all community changes can be accounted for by elevational or soil changes. In some level parts of the Great Basin, a mosaic of communities exists with quite narrow ecotones. One community may be strongly dominated by shadscale *(Atriplex confertifolia)*, another by winterfat *(Eurotia lanata)*, a third by salt sage *(Atriplex nuttallii)*, and a fourth by sagebrush. Detailed studies of soil texture, chemistry, and water retention have failed to reveal any changes significant enough to account for the mosaic. On a broader regional scale, these species do tend to separate on gradients of salinity and aridity (Figure 20-43). They may also differ in the amount of snow retained in winter.

Autecology of Major Life Forms

Each vegetation type discussed in this chapter includes species in an array of life forms. Desert vegetation may have a wider array than any other type, depending on how life forms are defined. Forrest Shreve (1942) defined 25 desert life forms (Table 20-8). In the following pages, we will briefly discuss a few of these.

Annuals (Ephemerals)   These plants have a life-span that is shorter than 8 months. Over 40% of the flora is annual, in contrast to Raunkiaer's world normal

Figure 20-43 The general regional distribution of several Great Basin shrubs along gradients of salinity and aridity in Utah. Artr = *Artemisia tridentata*, Cela = *Ceratoides lanata*, Atco = *Atriplex confertifolia*, Atnu = *Atriplex nuttallii*, Save = *Sarcobatus vermiculatus*, Aloc = *Allenrolfea occidentalis*. (Redrawn from Caldwell 1985.)

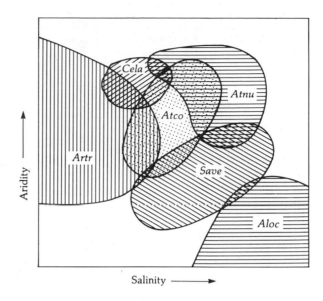

Aridity

Salinity ⟶

of 13%. There are two categories of desert annuals. **Winter annuals** germinate in fall or winter and flower in late spring; **summer annuals** germinate in mid-summer and flower in late summer or fall. Rarely does the same species belong to both groups, partly because germination requirements of winter and summer annuals usually differ. If soil temperature is below 18°C, for example, Mojave desert winter annuals will germinate, but if soil temperature is above 26°C, Mojave summer annuals will germinate. Germination of summer annuals is also enhanced, in some cases, by dry storage for several weeks at 50°C, a requirement which prevents seed germination between the time seeds are shed in fall and a rainy period the following summer.

The diversity of annual species at one season and in one region may be high, and the species may group into distinctive communities that reflect unique microenvironments. Some species grow on bajadas, some at the edge of playas, some in playas; some grow beneath certain canopies and avoid others; some grow only in the open. These patterns may be due to nurse plant requirements (for shade, organic matter), allelopathy, water-repellent layers beneath some shrubs, or salinity tolerances.

Winter annuals in the Chihuahuan and the Mojave desert germinate after fall or winter rains in excess of 10–15 mm. Below this critical limit, germination is absent. The rain may be necessary to leach inhibitors from the seed coat or to provide moisture beyond some critical inhibitional period of time. The density of annuals correlates with increasing rainfall between 15 and 45 mm. Winter annuals grow slowly through winter, are resistant to frost down to −18°C, then grow rapidly in spring as temperatures rise. Flowering, fruiting, and death occur in late May, for a total life cycle of 5–8 months. Herbivory accounts for some mortality, but most seedlings succumb to drought and high temperature. Most do not survive to maturity. Under the best conditions, density may be 1000 m^{-2},

Table 20-8   Plant life forms in North American deserts, as formulated by Shreve (1942). The Sonoran desert is the only desert to contain representatives of all 25 types. (Reprinted by permission from *Botanical Review* 8:195–246. Copyright 1942, F. Shreve and The New York Botanical Garden.)

Life form categories	Examples
Ephemerals:	
Strictly seasonal:	
Winter ephemerals	*Daucus pusillus, Plantago fastigiata*
Summer ephemerals	*Tidestromia lanuginosa, Pectis papposa*
Facultative perennials	*Verbesina encelioides, Baileya multiradiata*
Perennials:	
Underground parts perennial:	
Perennial roots	*Pentstemon parryi, Anemone tuberosa*
Perennial bulbs	*Hesperocallis undulata, Brodiaea capitata*
Shoot base and root crown perennial	*Hilaria mutica, Aristida ternipes*
Shoots perennial:	
Shoot reduced (a caudex):	
Caudex short, entirely leafy:	
Leaves succulent	*Agave palmeri, Dudleya arizonica*
Leaves nonsucculent	*Nolina microcarpa, Dasylirion wheeleri*
Caudex long, leafy at top:	
Leaves entire, linear, semisucculent	*Yucca baccata, Y. brevifolia*
Leaves dissected, palmate, nonsucculent	*Washingtonia filifera, Sabal uresana*
Shoot elongated:	
Plant succulent (soft):	
Leafless, stem succulent:	
Shoot unbranched	*Ferocactus wislizenii, Echinomastus erectocentrus*
Shoot branched:	
Shoot poorly branched:	
Plant erect and tall	*Carnegiea gigantea, Pachycereus pringlei*
Plant erect and low or semi-procumbent and low	*Pedilanthus macrocarpus, Mammillaria microsperma*
Shoot richly branched:	
Stem segments cylindrical	*Opuntia spinosior, O. arbuscula*
Stem segments flattened	*Opuntia engelmannii, O. santarita*
Leafy, stem not succulent	*Talinum paniculatum, Sedum wootoni*
Plant nonsucculent (woody):	
Shoots without leaves, stems green	*Holacantha emoryi, Canotia holacantha*
Shoots with leaves:	
Low bushes, wood soft	*Encelia farinosa, Franseria dumosa*

Table 20-8 *continued*

Life form categories	Examples
Perennials: Plant nonsucculent (woody): *continued*	
Shrubs and trees, wood hard:	
Leaves perennial	*Simmondsia chinensis, Larrea tridentata*
Leaves deciduous:	
Leaves drought-deciduous:	
Stems specialized:	
Stems indurated on surface	*Fouquieria splendens*
Stems enlarged at base	*Idria columnaris, Bursera microphylla*
Stems normal:	
Stems not green	*Jatropha cardiophylla, Plumeria acutifolia*
Stems green	*Cercidium microphyllum, Parkinsonia aculeata*
Leaves winter-deciduous:	
Leaves large	*Populus fremontii, Ipomoea arborescens*
Leaves small	*Olneya tesota, Acacia greggii*

cover 30%, and biomass 60 g m^{-2}, but typically, density is 100 m^{-2} and biomass 10 g m^{-2}.

Mojave summer annuals germinate in August or September after heavy rains, generally are C$_4$ in metabolism, remain small, and mature by the time of autumn frosts. Their life-span is measured in terms of weeks rather than months.

Many winter and summer annuals are able to modify solar radiation striking their leaves by heliotropism. Leaves track the sun's path during the day, such that they remain perpendicular to the sun's rays. Over the course of a day, such leaves receive 38% more radiation than leaves with a fixed horizontal or east–west vertical orientation (see Chapter 15). Some heliotropic leaves fold or cup during periods of drought, and solar radiation is then reduced by an order of magnitude (Figure 20-44), which in turn reduces leaf temperature and transpiration.

Drought-Deciduous Species Some **drought-deciduous** species, such as ocotillo or desert coachwhip *(Fouquieria splendens)* (Figure 20-45), are able to produce several crops of leaves a year, losing each one as a dry period follows a wet period. Most drought-deciduous species, however, produce only one crop of leaves a year and enter a long dormancy following leaf drop. Their leaves are mesophytic in terms of size, anatomy, and shape, making them energetically inexpensive for the plant to manufacture, compared to evergreen leaves. Their large intercellular spaces and thin blades permit rapid gas exchange (and water

Figure 20-44   Ocotillo *(Fouquieria splendens)*, a drought-deciduous species, (left) in full leaf and (right) leafless.

Figure 20-45   Foothill palo verde *(Cercidium)*, a leguminous tree of the Sonoran desert.

loss), so their photosynthetic rate is relatively high, about 2–3 times the rate for evergreens.

Winter-Deciduous Riparian Species   Most desert riparian species are winter-deciduous, a trait common to riparian species throughout all vegetation types in North America. Onset of stem growth, leaf expansion, and blooming is photoperiodically controlled, in contrast to the moisture controls for other desert life forms. Leaves appear in early summer and exhibit some xeromorphic features. During the hot summer, the water potentials developed within these plants are more negative than those of drought-deciduous species. A considerable amount of photosynthesis is accomplished by young, chlorenchymatous twigs in some species.

As already mentioned, many riparian desert species are leguminous trees (Figure 20-45) that produce large seeds with tough seed coats. The scarification they require is easily provided when the seeds are carried for some distance by raging water down an arroyo. Some riparian species are phreatophytes; seedlings must be able to produce a root that can grow fast enough to keep ahead of soil drying down from the surface. One of the fastest reported root growth rates is for mesquite: 102 mm day^{-1}, which is an order of magnitude greater than for typical xerophytes or mesophytes.

Succulents   There are many species of **stem** and **leaf succulents,** in several families, but the cacti are the best known. Recall from Chapter 14 that they have the CAM photosynthetic pathway, with stomates open at night. Consequently, their water use efficiency is very high, 14–40 g CO_2 hg^{-1} H_2O, which is an order of magnitude higher than other desert plants. Internal water stress rarely exceeds -0.5 MPa. They possess a shallow root system that is able to absorb water even from light rains, and in dry periods much of this root system dies. In wet periods, water is stored in large parenchyma cells, swelling the stem; in dry periods this water is used and the stem shrinks. Average root depth is only 8 cm, and succulents can show metabolic response to only 18 mm of rain within 24 hours.

Most succulents are not frost resistant, and some ecologists have correlated their distribution limits to isolines of average consecutive hours of frost. Nevertheless, two species of beavertail cactus (*Opuntia* spp.) do extend throughout the Great Basin desert. Several morphological features significantly improve tissue temperature of cacti at marginal northern sites. These include cladode (flattened stem) orientation to face east–west, increase in stem diameter, apical spines or pubescence, apex orientation, and growth beneath nurse plants. *Opuntia polyacantha* in Wyoming can tolerate temperatures down to $-17°$C, whereas *Opuntia* species in warm deserts to the south can tolerate temperatures only down to $-4°$C.

Although succulents have a slow growth rate, the largest plants in the desert are cacti: the cardón cactus of Baja California and Sonora, Mexico (*Pachycereus pringlei*) can reach 18 m and (presumably) hundreds of years in age. The more widely known saguaro cactus of Arizona and Sonora (*Carnegiea gigantea = Cereus giganteus)* can reach 10 m and an age of at least 200 years. Some populations of saguaro are clearly senescent, dominated by middle-aged and old plants, possibly because of increasing cattle and rodent populations, both of which graze on young plants.

Evergreens   These are true xerophytes because they grow and transpire throughout even the driest part of the year, at most shedding only a fraction of their leaves. Creosote bush *(Larrea tridentata)* and jojoba *(Simmondsia chinensis)* are good examples. For reasons not yet understood, their cytoplasm is able to resist unusual desiccation. Net photosynthesis is possible even when leaf water potential drops to -5 MPa, and root growth continues to -7 MPa. When moisture is not limiting, their transpiration rates are equivalent to those of mesophytes; when moisture is limiting, their transpiration rates are very low.

Their leaves often possess stereotyped morphological features, which ecologists presume are responsible for the low rate of transpiration: for example, thick cuticle, lack of intercellular spaces, sunken stomates, palisade parenchyma beneath both surfaces, small leaves, and vertical leaf orientation. Hairs, salt glands, and waxes on the surface may also serve to reflect light, hence lowering the heat load on the leaf. Many evergreens invest considerable metabolic energy into the production and accumulation of resins and other complex molecules that may have anti-herbivore significance. Some evergreens, such as *Artemisia tridentata* and *Atriplex confertifolia,* exhibit two types of leaves. Small, more xerophytic overwintering leaves are formed in late summer and retained to the following spring, when they are replaced by larger leaves with up to five times higher maximum net photosynthesis rates.

These evergreens give each desert its characteristic appearance, as they dominate the bajadas in terms of cover and biomass. In the warm deserts, CAM succulents and drought-deciduous species may also be important on bajadas, but C_4 plants are significant only in playas. A study of bajada and playa vegetation in the Chihuahuan desert of New Mexico very dramatically documents this pattern (Table 20-9): 90% of playa biomass was contributed by C_4 species, primarily summer annuals, but only 1% of bajada biomass was contributed by C_4 taxa. Evidently, some aspects of C_4 metabolism are ecologically disadvantageous out on the bajadas, regardless of life form.

Great Basin Vegetation

The cold desert is covered by a one-layered shrub community, about 1 m tall and very low in species diversity. The overwhelming dominant is sagebrush, *Artemisia tridentata* (see Figure 20-39). Although species diversity is low in this desert, there is significant ecotypic variation in several dominant taxa. *Artemisia* is represented by twelve species, and *A. tridentata* alone has four ecologically important subspecies (ecotypes). Those ecotypes occupy sites that differ in soil depth, aridity, and temperature (Figure 20-46). Common associates are shadscale *(Atriplex confertifolia),* winterfat *(Ceratoides lanata),* spiny hopsage *(Grayia spinosa),* and Mormon tea *(Ephedra* spp.). Total cover is 15–40%.

Table 20-9 Percent C₃, C₄, and CAM of the total species and of the total biomass in three adjacent communities in the Chihuahuan desert of southern New Mexico. Data were accumulated over the course of several years. (From Syvertsen et al. 1976. By permission of *Southwestern Naturalist*.)

Community	Attribute	Percent			Absolute totals	Dominants
		C_3	C_4	CAM		
Playa bottom	Species	52	48	0	21 species	*Panicum*
	biomass	10	90	0	1243 kg ha^{-1}	*obtusum* (C_4)
Playa fringe	Species	54	44	1	77 species	*Hilaria mutica*
	biomass	42	50	8	11,200 kg ha^{-1}	(C_4), *Prosopis*
						glandulosa (C_3)
Bajada	Species	66	24	10	144 species	*Larrea tridentata*
	biomass	51	1	48	2088 kg ha^{-1}	(C_3), *Yucca elata*
						(CAM),
						Prosopis
						glandulosa (C_3),
						Flourensia
						cernua (C_3)

To the north, or at higher elevations, is the **sagebrush steppe.** In pristine times, it consisted of scattered sagebrush in a matrix of perennial bunchgrasses and seasonal forbs; total cover could exceed 80%. The grasses included western wheatgrass, needlegrass, Idaho fescue, and bluebunch wheatgrass (see Table 20-5 for scientific names). Some of this area is now farmland, and most of the rest is changed because of overgrazing, burning, and the invasion of weedy annuals and native short-lived perennials: rabbitbrush (*Chrysothamnus* spp.), cheatgrass, medusahead (*Elymus caput-medusae*), Russian thistle (*Salsola ibirica*), tumble mustard (*Sisymbrium altissium*), filaree (*Erodium* spp.), and *Halogeton glomeratus*. Sagebrush, which is not palatable, has been purposely reduced in cover by burning (it is not a sprouter). Some of this range has been so degraded that it may not recover even with active management (Figure 20-47). At higher elevations, sagebrush steppe is joined by a 5–15 m tall open canopy of pine and juniper (the **pinion-juniper woodland**).

In more saline areas is a saltbush-greasewood scrub, dominated by several small shrub species of *Atriplex* and *Sarcobatus vermiculatus*, mainly C₄ plants. This vegetation has also been degraded by overgrazing and the invasion of several annual weedy species.

In parts of southern Nevada and eastern Utah, shadscale becomes the dominant, associated with budsage (*Artemisia spinescens*), greasewood (*Sarcobatus baileyi*), and *Ephedra nevadensis*. This has been called the **shadscale zone** of the Great Basin, and its climate is significantly warmer and drier than sagebrush-dominated areas. Ground cover is only 10%.

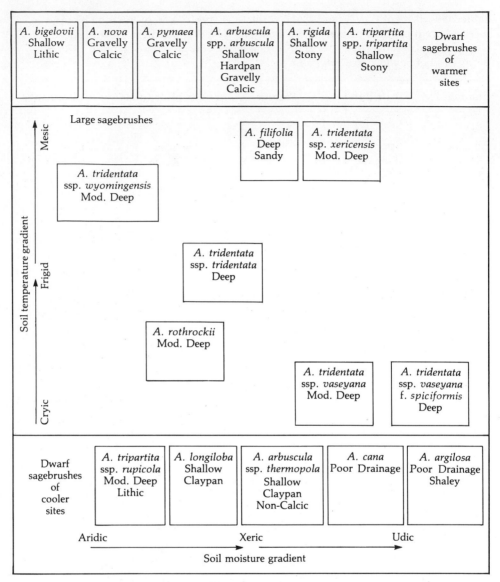

Figure 20-46   Environmental distribution of *Artemisia* species and subspecies in the Great Basin, arranged along gradients of soil moisture, depth, texture, and temperature. (From West 1987.)

Warm Desert Vegetation

The **Mojave** is dominated by a two-layered community: an overstory shrub layer, mostly of creosote bush, and an understory subshrub layer, mostly of burro bush (also called bur sage, *Ambrosia dumosa*) (Figure 20-48). It is a monotonous, open community, with only 5–10% cover. At higher elevations (1200–1800 m),

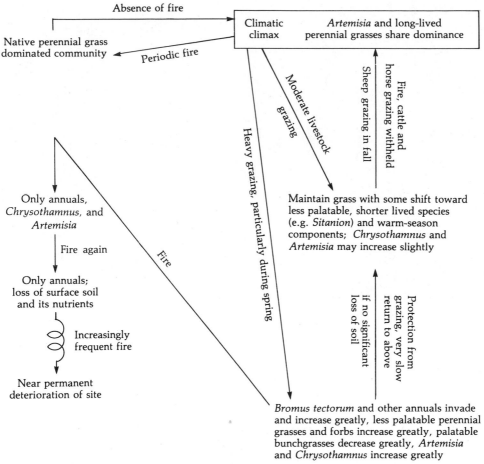

Figure 20-47 Modifications of sagebrush steppe by fire, grazing, and the introduction of weedy annuals in the past 150 years. (From West 1987.)

creosote bush is displaced by blackbrush *(Coleogyne ramosissima)*, Joshua tree *(Yucca brevifolia)*, spiny hopsage, matchweed *(Gutierrezia* spp.), and winterfat. The **blackbrush community** is an ecotone between **creosote bush scrub** below and pinon-juniper woodland above, sharing half of its flora with each.

The **Sonoran desert** is so complex that it has been divided into as many as seven vegetational regions by some ecologists. We will only consider two extremes here. The **Colorado desert** of southern California, southwestern Arizona, and around the northern part of the Gulf of California is very arid, and the vegetation is reminiscent of the Mojave. Creosote bush and burro bush dominate. However, there are many differences in annuals and other life forms between this region and the Mojave. Of 545 Mojave plant species, only half are also found in the Sonoran. Many hectares of the Coachella and Imperial Valleys in California are

Figure 20-48   Typical *Larrea–Ambrosia* community on a Mojave bajada.

now dominated by citrus, date palms, and sugar beets, irrigated by water diverted from the Colorado River.

The other extreme, typical of central Baja California, the **Arizona uplands,** and parts of Sonora, is a complex four-layered community with about 25% ground cover (Figure 20-49). The overstory, 3–5 m tall, is of arborescent cacti and trees. A tall shrub-cactus layer below this is 2–3 m tall and has creosote bush, *Acacia* species, coachwhip, and chollas. A third layer is comprised of subshrubs 1 m or less in height, such as two species of burro bush *(Ambrosia dumosa* and *A. deltoidea)*, brittle bush *(Encelia farinosa)*, and a great variety of cacti (barrel, fish hook, cholla, beavertail). There are approximately five times as many species of cacti in the Sonoran desert as in the Mojave. A fourth layer is dominated by some perennial grasses and many species of summer and winter annuals. Because there are two nearly equal peaks of rainfall in winter and summer in these more mesic parts of the Sonoran, the diversity of species for each annual category is about the same. In contrast, the winter-wet Mojave desert has about six times as many winter as summer annuals.

The **Chihuahuan desert** lacks an arboreal element, but it has great species richness in two shrub layers. Beneath an overstory of creosote bush, coachwhip, mesquite, and *Acacia* are tarbush *(Flourensia cernua)*, guayule *(Parthenium incanum)*, leather plant *(Jatropha* spp.), crucifixion thorn *(Koeberlinia spinosa)*, many cacti, and several distinctive members of the Agavaceae with spinescent, basal leaves (Figure 20-50). The latter include century plant *(Agave lechuguilla* and others), sotol *(Dasylerion wheeleri)*, Spanish bayonet *(Yucca elata, Y. baccata)*, and *Nolina*. Ground cover is relatively high, about 20–25%. Summer annuals can be abundant.

Figure 20-49 Arizona uplands community of the Sonoran desert, with saguaro *(Carnegiea gigantea)* and mesquite *(Prosopis juliflora)* prominent in the overstory.

Figure 20-50 Typical Chihuahuan desert vegetation at Big Bend National Park, Texas, with ocotillo, creosote bush *(Larrea tridentata)*, *Agave* species, and cacti prominent.

This desert is large and complex enough to warrant subdivision into three regions. From north to south, these have been called the **Trans-Pecos, Mapimian,** and **Saladin.** Plant cover, species richness, and taller growth forms increase to the south. Most of our detailed information for Chihuahuan vegetation, however, comes from the Trans-Pecos region.

Chapter 20 References

This is a very personal list of references. Works cited cover the factual material in the text and give a sampling of both current and classical research. Some relatively narrow papers are included, and some broad references are omitted, purely because of our own biases. In short, this list is neither consistent nor exhaustive.

Tundra

Billings, W. D. 1974a. Adaptations and origins of alpine plants. *Arctic and Alpine Research* 6:129–142.

———. 1974b. Arctic and alpine vegetation: plant adaptations to cold summer climates. In *Arctic and alpine environments*, ed. J. D. Ives and R. G. Barry, pp. 403–443. London: Methuen.

Billings, W. D. and P. J. Godfrey. 1967. Photosynthetic utilization of internal carbon dioxide by hollow-stemmed plants. *Science* 158:121–123.

Billings, W. D. and H. A. Mooney. 1968. The ecology of arctic and alpine plants. *Biological Review* 43:481–529.

Black, R. A. and L. C. Bliss. 1980. Reproductive ecology of *Picea mariana* (Mill.) BSP at tree line near Inuvik, Northwest Territories, Canada. *Ecological Monographs* 50:331–354.

Bliss, L. C. 1956. Comparison of plant development in microenvironments of arctic and alpine tundras. *Ecological Monographs* 26:303–337.

———. 1962. Adaptations of arctic and alpine plants to environmental conditions. *Arctic* 15:117–144.

———. 1963. Alpine plant communities of the Presidential Range, New Hampshire. *Ecology* 44:687–697.

———. 1966. Plant productivity in alpine microenvironments on Mt. Washington, New Hampshire. *Ecological Monographs* 36:125–155.

———. 1985. Alpine. In *Physiological ecology of North American plant communities*, eds. B. F. Chabot and H. A. Mooney, pp. 41–65. New York: Chapman and Hall.

———. 1987. Arctic tundra and polar desert biome. In *North American terrestrial vegetation*, ed. M. G. Barbour and W. D. Billings, Chap. 1. New York: Cambridge University Press.

Britton, M. E. 1966. *Vegetation of the arctic tundra.* Corvallis, OR: Oregon State University Press.

Caldwell, M. M. 1968. Solar ultraviolet radiation as an ecological factor for alpine plants. *Ecological Monographs* 38:243–268.

Caldwell, M. M., R. Robberecht, R. S. Nowak, and W. D. Billings. 1982. Differential photosynthetic inhibition by ultraviolet radiation in species from the arctic-alpine life zone. *Arctic and Alpine Research* 14:195–202.

Canaday, B. B. and R. W. Fonda. 1974. The influence of subalpine snowbanks on vegetation pattern, production, and phenology. *Bulletin of the Torrey Botanical Club* 101:340–350.

Chabot, B. F. and W. D. Billings. 1972. Origins and ecology of the Sierran alpine flora and vegetation. *Ecological Monographs* 42:143–161.

Chapin, F. S., III and G. R. Shaver. 1985. Arctic. *Physiological ecology of North American plant communities*, ed. B. F. Chabot and H. A. Mooney, pp. 16–40. New York: Chapman and Hall.

Clausen, J. 1965. Population studies of alpine and subalpine races of conifers and willows in the California high Sierra Nevada. *Evolution* 19:56–68.

Daubenmire, R. F. 1954. Alpine timberlines in the Americas and their interpretation. *Butler University Botanical Studies* 11:119–136.

Dennis, J. G. and P. L. Johnson. 1970. Shoot and rhizome-root standing crops of tundra vegetation at Barrow, Alaska. *Arctic and Alpine Research* 2:253–266.

Douglas, G. W. and L. C. Bliss. 1977. Alpine and high subalpine plant communities of the North Cascades Range, Washington and British Columbia. *Ecological Monographs* 47:113–150.

Elliott-Fisk, D. L. 1983. The stability of the northern Canadian tree limit. *Annals of Association of American Geographers* 73:560–576.

———. 1987. The boreal forest. In *North American terrestrial vegetation*, ed. M. G. Barbour and W. D. Billings, Chap. 2. New York: Cambridge University Press.

Ewers, F. W. 1982. Secondary growth in needle leaves of *Pinus longaeva* (bristlecone pine) and other conifers: quantitative data. *American J. Botany* 69:1552–1559.

Greller, A. M. 1974. Vegetation of roadcut slopes in the tundra of Rocky Mountain National Park, Colorado. *Biological Conservation* 6(2):84–93.

Griggs, R. F. 1938. Timberlines in the northern Rocky Mountains, *Ecology* 19:548–564.

———. 1946. The timberlines of North America and their interpretation. *Ecology* 27:257–289.

Hoffman, R. S. 1958. The meaning of the word "taiga." *Ecology* 39:540–541.

Hustich, I. 1953. The boreal limits of conifers. *Arctic* 6:149–162.

Johnson, D. A. and M. M. Caldwell. 1976. Water potential components, stomatal function, and liquid phase water transport resistances of four arctic and alpine species in relation to moisture stress. *Physiologia Plantarum* 36:271–278.

Kevan, P. G. 1975. Sun-tracking solar furnaces in high arctic flowers: significance for pollination and insects. *Science* 189:723–726.

Klikoff, L. G. 1965. Microenvironmental influence on vegetational pattern near timberline in the Sierra Nevada. *Ecological Monographs* 35:187–211.

LaMarche, V. C., Jr. and H. A. Mooney. 1972. Recent climatic change and development of the bristlecone pine (*P. longaeva* Bailey) krummholz zone, Mt. Washington, Nevada. *Arctic and Alpine Research* 4:61–72.

Larsen, J. A. 1965. The vegetation of the Ennadai Lake Area, N.W.T. *Ecological Monographs* 35:37–59.

Major, J. and S. A. Bamberg. 1967. Comparison of some North American and Eurasian alpine ecosystems. In *Arctic and alpine environments,* ed. H. E. Wright, Jr. and W. H. Osburn, pp. 89–118. Bloomington, IN: Indiana Univ. Press.

Major, J. and D. W. Taylor. 1977. Alpine. In *Terrestrial vegetation of California,* ed. M. G. Barbour and J. Major, pp. 601–675. New York: Wiley-Interscience.

Marr, J. W. 1948. Ecology of the forest-tundra ecotone on the east coast of Hudson Bay. *Ecological Monographs* 18:117–144.

Moldenke, A. R. 1976. California pollination ecology and vegetation types. *Phytologia* 34:305–361.

Oechel, W. C. and W. T. Lawrence. 1985. Taiga. In *Physiological ecology of North American plant communities,* ed. B. F. Chabot and H. A. Mooney, pp. 66–94. New York: Chapman and Hall.

Pewe, T. L. 1966. *Permafrost and its effect on life in the north.* Corvallis, OR: Oregon State University Press.

Polunin, N. 1948. *Botany of the Canadian eastern arctic, III. Vegetation and ecology.* National Museum of Canada Bulletin 104. Ottawa, Canada: National Museum of Canada.

Rickard, W. E., Jr. and J. Brown. 1974. Effects of vehicles on arctic tundra. *Environmental Conservation* 1:55–62.

Shaver, G. R. and W. D. Billings. 1977. Effects of daylength and temperature on root elongation in tundra graminoids. *Oecologia* 28:57–65.

Stålfelt, M. G. 1960. *Plants, the soil and man.* 1972 translation by M. S. Jarvis and P. G. Jarvis. New York: Wiley.

Teeri, J. A. 1976. Phytotron analysis of a photoperiodic response in a high arctic plant species. *Ecology* 57:374–379.

Tieszen, L. L. (ed.). 1978. *Vegetation and production ecology of an Alaskan arctic tundra.* New York: Springer-Verlag.

Conifer Forests

Arno, S. F. and J. R. Habeck. 1972. Ecology of alpine larch (*Larix lyallii* Parl.) in the Pacific northwest. *Ecological Monographs* 42:417–450.

Azevedo, J. and D. L. Morgan. 1974. Fog precipitation in coastal California forests. *Ecology* 55:1135–1141.

Barbour, M. G. 1987. Californian Upland Forests and Woodlands. In *North American terrestrial vegetation,* ed. M. G. Barbour and W. D. Billings, Chap. 5. New York: Cambridge University Press.

Buell, M. F. and W. A. Niering. 1957. Fir-spruce-birch forest in northern Minnesota. *Ecology* 38:602–610.

Chapman, H. H. 1932. Is the longleaf type a climax? *Ecology* 13:328–334.

Cogbill, C. V. 1977. The effect of acid precipitation on tree growth in eastern North America. *Water, Air, Soil Pollution* 8:89–93.

Cooper, C. F. 1960. Changes in vegetation, structure, and growth of southwestern pine forests since white settlement. *Ecological Monographs* 30:129–164.

Cooper, W. S. 1957. Vegetation of the northwest American province. *Proc., 8th Pacific Science Congress* 4:133–138.

Curtis, J. T. 1959. *The vegetation of Wisconsin.* Madison, WI: University of Wisconsin Press.

Dansereau, P. and F. Segadas-Vianna. 1952. Ecological study of the peat bogs of eastern North America. *Canadian J. of Botany* 30:490–520.

Daubenmire, R. F. 1943. Soil temperature versus drought as a factor in determining lower altitudinal limits of trees in the Rocky Mountains. *Botanical Gazette* 105:1–13.

———. 1943. Vegetational zonation in the Rocky Mountains. *Botanical Review* 9:325–393.

Davis. R. B. 1966. Spruce-fir forests of the coast of Maine. *Ecological Monographs* 36:79–94.

D'Itri, F. M. 1982. *Acid precipitation: Effects on ecological systems.* Ann Arbor, MI: Science Publishers.

Elliott-Fisk, D. L. 1987. The boreal forest. In *North American terrestrial vegetation,* ed. M. G. Barbour and W. D. Billings, Chap. 2. New York: Cambridge University Press.

Evans, L. S. 1984. Botanical aspects of acidic precipitation. *Botanical Review* 50:449–490.

Fonda, R. W. and J. A. Bernardi. 1976. Vegetation of Sucia Island in Puget Sound, Washington. *Bulletin of the Torrey Botanical Club* 103:99–109.

Franklin, J. F. 1987. Pacific northwest forests. In *North American terrestrial vegetation,* ed. M. G. Barbour and W. D. Billings, Chap. 4. New York: Cambridge University Press.

Franklin, J. F. and C. T. Dyrness. 1973. *Natural vegetation of Oregon and Washington.* U.S. Dept. of Agriculture Forest Service General Technical Report PNW-8, Portland, OR. U.S. Dept. of Agriculture Forest Service.

Garren, K. H. 1943. Effects of fire on vegetation of the southeastern United States. *Botanical Review* 9:617–654.

Goff, F. G. and P. H. Fedler. 1968. Structural gradient analysis of upland forests in the western Great Lakes area. *Ecological Monographs* 38:65–86.

Habeck, J. R. and R. W. Mutch. 1973. Fire-dependent forests in the northern Rocky Mountains. *J. of Quarternary Research* 3:408–424.

Haines, B. L. 1983. Forest ecosystem SO_4-S input-output discrepancies and acid rain: Are they related? *Oikos* 41:139–143.

Hartesveldt, R. J., H. T. Harvey, H. S. Shellhammer, and R. E. Stecker. 1975. *The giant sequoia of the Sierra Nevada.* U.S. Dept. of the Interior National Park Service Publ. no. 120. Washington, D.C.: U.S. Dept. of the Interior, National Park Service.

Heinselman, M. L. 1963. Forest sites, bog processes, and peatland types in the glacial Lake Aggasiz region, Minnesota. *Ecological Monographs* 33:327–374.

Hutchinson, T. C. and M. Havas (eds.). 1980. *Effects of acid precipitation on terrestrial ecosystems.* New York: Plenum.

Jones, E. W. 1945. The structure and reproduction of the virgin forests of the north temperate zone. *The New Phytologist* 44:130–148.

Kallend, A. S., A. R. W. Marsh, J. H. Pickles, and M. V. Proctor. 1983. Acidity of rain in Europe. *Atmospheric Environment* 17:127–137.

LaRoi, G. H. 1967. Ecological studies in the boreal spruce-fir forests of the North American taiga. *Ecological Monographs* 37:229–253.

Larsen, J. A. 1930. Forest types of the northern Rocky Mountains and their climatic controls. *Ecology* 11:631–672.

———. 1980. *The boreal ecosystem.* New York: Academic Press.

Lassoie, J. P., T. M. Hinckley, and C. C. Grier. 1985. Coniferous forests of the Pacific northwest. In *Physiological ecology of North American plant communities,* ed. B. F. Chabot and H. A. Mooney, pp. 127–161. New York: Chapman and Hall.

Lee, J. J. and D. E. Weber. 1982. Effects of sulfuric acid rain on major cation and sulfate concentrations of water percolating through two model hardwood forests. *J. Environmental Quality* 11:57–64.

Likens, G. E. and T. J. Butler. 1981. Recent acidification of precipitation in North America. *Atmospheric Environment* 15:1103–1109.

Linthurst, R. A. (ed.). 1984. *Direct and indirect effects of acidic deposition on vegetation.* Acid Precipitation Series, Vol. 5. Boston: Butterworth.

Lowe, C. H. and D. E. Brown. 1973. *The natural vegetation of Arizona.* Arizona Resources Information System Cooperative Publ. no. 2. Phoenix, AZ: Arizona Resources Information System.

Lutz, H. J. 1956. *Ecological effects of forest fires in the interior of Alaska.* U.S. Dept. of Agriculture Technical Bulletin 1133. Washington, D.C.: U.S. Dept. of Agriculture.

Maycock, P. F. 1961. The spruce-fir forests of the Keneenaw Peninsula, northern Michigan. *Ecology* 42:357–365.

McCormick, J. 1970. *The pine barrens.* New Jersey State Museum Report no. 2. Trenton, NJ: New Jersey State Museum.

McIntosh, R. P. and R. T. Hurley. 1964. The spruce-fir forests of the Catskill Mountains. *Ecology* 45:314–326.

Moss, E. H. 1955. The vegetation of Alberta. *Botanical Review* 21:493–567.

Oechel, W. C. and W. T. Lawrence. 1985. Taiga. In *Physiological ecology of North American plant communities,* ed. B. F. Chabot and H. A. Mooney, pp. 66–94. New York: Chapman and Hall.

Oosting, H. J. and W. D. Billings. 1951. A comparison of virgin spruce-fir forests in the northern and southern Appalachian system. *Ecology* 32:84–103.

Oosting, H. J. and J. F. Reed. 1944. Ecological composition of pulpwood forests in northwestern Maine. *American Midland Naturalist* 31:181–210.

———. 1952. Virgin spruce-fir of the Medicine Bow Mountains, Wyoming. *Ecological Monographs* 22:69–91.

Peet, R. K. 1987. Conifer forests of the Rocky Mountains. In *North American terrestrial vegetation,* ed. M. G. Barbour and W. D. Billings, Chap. 3. New York: Cambridge University Press.

Potzger, J. E. 1946. Phytosociology of the primeval forest in central and northern Wisconsin and upper Michigan and a brief postglacial history of the Lake Forest formation. *Ecological Monographs* 16:211–250.

Rennie, P. J. 1977. Forests, muskeg, and organic terrain in Canada. In *Muskeg and the northern environment in Canada,* ed. N. W. Radforth and C. O. Brawner, pp. 167–207. University of Toronto Press.

Ritchie, J. C. 1956. The vegetation of northern Manitoba. *Canadian J. of Botany* 34:523–561.

Robichaud, B. and M. F. Buell. 1973 *Vegetation of New Jersey.* New Brunswick: Rutgers University Press.

Rundel, P. W. 1971. Community structure and stability in the giant sequoia groves of the Sierra Nevada. *American Midland Naturalist* 85:478–492.

Rundel, P. W., D. J. Parsons, and D. T. Gordon. 1977. Montane and subalpine vegetation of the Sierra Nevada and Cascade Ranges. In *Terrestrial vegetation of California,* ed. M. G. Barbour and J. Major, pp. 559–599. New York: Wiley-Interscience.

Scott, J. T., T. G. Siccama, A. H. Johnson, and A. R. Briesch. 1984. Decline of red spruce in the Adirondacks, New York. *Bulletin Torrey Botanical Club* 111:438–444.

Shirley, H. L. 1945. Reproduction of upland conifers in the Lake States as affected by root competition and light. *American Midland Naturalist* 33:537–612.

Siccama, T. G. 1982. Decline of red spruce in the Green Mountains of Vermont. *Bulletin Torrey Botanical Club* 109:162–168.

Smith, W. K. 1985. Western montane forests. In *Physiological ecology of North American plant communities*, ed. B. F. Chabot and H. A. Mooney, pp. 95–126. New York: Chapman and Hall.

Stålfelt, M. G. 1972. *Stålfelt's plant ecology*. (Translated from *Plants, the soil and man*, 1960.) London: Longman.

Stallard, H. 1929. Secondary succession in the climax forest formation of Northern Minnesota. *Ecology* 10:476–548.

Stewart, R. 1977. *Labrador*. Amsterdam: Time-Life.

Taylor, D. W. 1977. Floristic relationships along the Cascade-Sierran axis. *American Midland Naturalist* 97:333–349.

Ulrich, B. and J. Pankrath (eds.). 1983. *Effects of accumulate of air pollution in forest ecosystems*. Dordrecht, Holland: Reidel Publishing.

Van Cleve, K. and L. A. Viereck. 1981. Forest succession in relation to nutrient cycling in the boreal forest of Alaska. In *Forest succession, concepts and application*, ed. O. C. West, H. H. Shugart, and D. B. Botkin, pp. 185–211. New York: Springer-Verlag.

Vankat, J. L. 1982. A gradient perspective on the vegetation of Sequoia National Park, California. *Madrono* 29:220–214.

Vogelmann, H. W., G. J. Badger, M. Bliss, and R. M. Klein. 1985. Forest decline on Camels Hump, Vermont. *Bulletin Torrey Botanical Club* 112:274–287.

Zinke, P. J. 1977. The redwood forest and associated north coast forests. In *Terrestrial vegetation of California*, ed. M. G. Barbour and J. Major, pp. 679–698. New York: Wiley-Interscience.

Deciduous Forests

Bakeless, J. 1961. *The eyes of discovery*. New York: Dover.

Bierzychudek, P. 1982. Life histories and demography of shade-tolerant temperate forest herbs: A review. *New Phytologist* 90:757–776.

Bormann, F. H., T. G. Siccama, G. E. Likens, and R. H. Whittaker. 1970. The Hubbard Brook ecosystem study: composition and dynamics of the tree stratum. *Ecological Monographs* 40:373–388.

Boyce, S. G. 1954. The salt spray community. *Ecological Monographs* 24:29–67.

Braun, E. L. 1950. *Deciduous forests of eastern North America*. Philadelphia, PA: Blakiston.

———. 1955. The phytogeography of the eastern United States and its interpretation. *Botanical Review* 21:297–375.

———. 1957. Development of the deciduous forests of eastern North America. *Ecological Monographs*. 17:211–219.

Buell, M. F. and W. E. Martin. 1961. Competition between maple-basswood and spruce-fir communities in Itasca Park, Minnesota. *Ecology* 42:428–429.

Cain, S. A. 1943. The tertiary character of the cove hardwood forests of the Great Smoky Mountains. *Bulletin of the Torrey Botanical Club* 70:213–245.

Christensen, N. L. 1987. The vegetation of the coastal plain of the southeastern United States. In *North American terrestrial vegetation,* ed. M. G. Barbour and W. D. Billings, Chap. 11. New York: Cambridge University Press.

Core, E. L. 1966. *Vegetation of West Virginia.* Charleston, WV: McClain, Parsons.

Curtis, J. T. 1959. *The vegetation of Wisconsin.* Madison, WI: University of Wisconsin Press.

Daubenmire, R. F. 1936. The "Big Woods" of Minnesota. *Ecological Monographs* 6:233–268.

Delcourt, H. R. and P. A. Delcourt. 1984. Ice age haven for hardwoods. *Natural History* 93(9):22–28.

Delcourt, H. R., P. A. Delcourt, and T. Webb III. 1983. Dynamic plant ecology: the spectrum of vegetational change in space and time. *Quarterly Science Review* 1:153–175.

Dyksterhuis, E. J. 1948. The vegetation of western Cross Timbers. *Ecological Monographs* 18:325–376.

Eyre, S. R. 1963. *Vegetation and soils.* Chicago: Aldine.

Forcier, L. K. 1975. Reproductive strategies and the co-occurrence of climax tree species. *Science* 189:808–810.

Graham, S. A. 1941. Climax forests of the upper peninsula of Michigan. *Ecology* 22:355–362.

Greller, A. M. 1980. Correlation of some climate statistics with distribution of broadleaved forest zones in Florida, USA. *Bulletin Torrey Botanical Club* 107:189–219.

———. 1986. Deciduous forest. In *North American terrestrial vegetation,* ed. M. G. Barbour and W. D. Billings, Chap. 10. New York: Cambridge University Press.

Hicks, D. J. and B. F. Chabot. 1985. Deciduous forest. In *Physiological ecology of North American plant communities,* ed. B. F. Chabot and H. A. Mooney, pp. 257–277. New York: Chapman and Hall.

Holt, P. C., ed. 1970. *The distributional history of the biota of the southern Appalachians. Part II: Flora.* Blacksburg, VA: Virginia Polytechnic Institute and State University.

Hutchinson, B. A. and D. R. Matt. 1977. The distribution of solar radiation within a deciduous forest. *Ecological Monographs* 47:185–207.

Keever, C. 1953. Present composition of some stands of the former oak-chestnut forest in the southern Blue Ridge Mountains. *Ecology* 34:44–54.

———. 1973. Distribution of major forest species in southeastern Pennsylvania. *Ecological Monographs* 43:303–327.

Koyama, H. and S. Kawano. 1973. Biosystematic studies on *Maianthemum* (Liliaceae-Polygonatae). VII. Photosynthetic behavior of *M. dialatatum. Botanical Magazine of Tokyo* 86:89–101.

Küchler, A. W. 1964. *Potential natural vegetation of the conterminous United States.* American Geographical Society Special Publ. no. 36. New York: American Geographical Society.

Lechowicz, M. J. 1984. Why do temperate deciduous trees leaf out at different times? Adaptation and ecology of forest communities. *American Naturalist* 124:821–842.

Ledig, F. T. and D. R. Korbobo. 1983. Adaptation of sugar maple populations along altitudinal gradients: photosynthesis, respiration, and specific leaf weight. *American J. Botany* 70:256–265.

Lutz, H. J. 1930. The vegetation of Heart's Content: a virgin forest in northwestern Pennsylvania. *Ecology* 11:1–29.

Maguire, D. A. and R. T. T. Forman. 1983. Herb cover effects on free seedling patterns in a mature hemlock-hardwood forest. *Ecology* 64:1367–1380.

Mahall, B. E. and F. H. Bormann. 1978. A quantitative description of the vegetative phenology of herbs in a northern hardwood forest. *Botanical Gazette* 139:467–481.

Monk, C. D. 1965. Southern mixed hardwood forest of north-central Florida. *Ecological Monographs* 35:335–354.

———. 1966. An ecological significance of evergreenness. *Ecology* 47:504–505.

Mowbray, T. B. and H. J. Oosting. 1968. Vegetation gradients in relation to environment and phenology in a southern Blue Ridge gorge. *Ecological Monographs* 38:309–344.

Muller, R. N. and F. H. Bormann. 1976. Role of *Erythronium americanum* Ker. in energy flow and nutrient dynamics of a northern hardwood forest ecosystem. *Science* 193:1126–1128.

Oosting, H. J. 1942. An ecological analysis of the plant communities of the Piedmont, North Carolina. *American Midland Naturalist* 28:1–126.

Oosting, H. J. and P. F. Bourdeau. 1955. Virgin hemlock forest segregates in the Joyce Kilmer Memorial Forest of western North Carolina. *Botanical Gazette* 116:340–359.

———. 1959. The maritime live oak forest in North Carolina. *Ecology* 40:148–152.

Paillet, F. L. 1982. The ecological significance of American chestnut (*Castanea dentata* (Marsh.) Borkh.) in the Holocene forests of Connecticut. *Bulletin Torrey Botanical Club* 109:457–473.

Potzger, J. E. 1946. Phytosociology of the primeval forest of central-northern Wisconsin and upper Michigan. *Ecological Monographs* 16:211–250.

Potzger, J. E., M. E. Potzger, and J. McCormick. 1956. The forest primeval of Indiana as recorded in the original land surveys and an evaluation of previous interpretations of Indiana vegetation. *Butler University Botanical Studies* 13:95–111.

Quarterman, E. and C. Keever. 1962. Southern mixed hardwood forest: climax in the southeastern coastal plain, USA. *Ecological Monographs* 32:167–185.

Rice, E. L. and W. T. Penfound. 1959. The upland forests of Oklahoma. *Ecology* 40:593–607.

Robichaud, B. and M. F. Buell. 1973. *Vegetation of New Jersey.* New Brunswick: Rutgers University Press.

Rogers, R. S. 1980. Hemlock stands from Wisconsin to Nova Scotia: transitions in understory composition along a floristic gradient. *Ecology* 61:178–193.

———. 1981. Mature mesophytic hardwood forest: community transitions, by layer, from east-central Minnesota to southeastern Michigan. *Ecology* 62:1634–1647.

Siccama, T. G. 1974. Vegetation, soil, and climate on the Green Mountains of Vermont. *Ecological Monographs* 44:325–349.

Siccama, T. G., F. H. Bormann, and G. E. Likens. 1970. The Hubbard Brook ecosystem study: productivity, nutrients, and phytosociology of the herbaceous layer. *Ecological Monographs* 40:389–402.

Skeen, J. N., M. E. B. Carter, and H. L. Ragsdale. 1980. Yellow-poplar: the Piedmont case. *Bulletin Torrey Botanical Club* 107:1–6.

Transeau, E. N. 1935. The prairie peninsula. *Ecology* 16:423–437.

Vogelmann, H. W., T. G. Siccama, D. Leedy, and D. C. Ovitt. 1968. Precipitation from fog moisture in the Green Mountains of Vermont. *Ecology* 49:1205–1207.

Whittaker, R. H. 1956. Vegetation of the Great Smoky Mountains. *Ecological Monographs* 26:1–80.

Williams, A. B. 1936. The composition and dynamics of a beech-maple climax forest (Ohio). *Ecological Monographs* 6:319–408.

Woods, F. W. and R. E. Shanks. 1959. Natural replacements of chestnut by other species in the Great Smoky Mountains National Park. *Ecology* 40:349–361.

Grasslands

Adams, D. E. and L. L. Wallace. 1985. Nutrient and biomass allocation in five grass species in an Oklahoma tallgrass prairie. *American Midland Naturalist* 113:170–181.

Albertson, F. W. 1937. Ecology of mixed prairie in west central Kansas. *Ecological Monographs* 7:481–547.

Albertson, F. W. and J. E. Weaver. 1947. Reduction of ungrazed mixed prairie to short grass as a result of drought and dust. *Ecological Monographs* 16:449–463.

Ayyad, M. A. G. and R. L. Dix. 1964. An analysis of a vegetation-microenvironmental complex on prairie slopes in Saskatchewan. *Ecological Monographs* 34:421–442.

Barry, W. J. 1972. *The central valley prairie.* Sacramento, CA: Dept. of Parks and Recreation.

Beetle, A. A. 1974. Distribution of the native grasses of California. *Hilgardia* 9:309–357.

Bogusch, E. R. 1952. Brush invasion in the Rio Grande Plain of Texas. *Texas J. of Science* 4:85–91.

Borchert, J. R. 1950. The climate of the centeral North American grassland. *Annals of the Association of American Geographers* 40:1–39.

Brown, A. L. 1950. Shrub invasion of southern Arizona desert grassland. *J. of Range Management* 3:172–177.

Buffington, L. C. and C. H. Herbel. 1965. Vegetational changes on a semidesert grassland range from 1858 to 1963. *Ecological Monographs* 35:139–164.

Carpenter, J. R. 1940. The grassland biome. *Ecological Monographs* 10:617–684.

Coupland, R. T. 1961. A reconsideration of grassland classification in the northern Great Plains of North America. *J. of Ecology* 49:135–167.

Daubenmire, R. F. 1968. Ecology of fire in grasslands. *Advances in Ecological Research* 5:209–266.

Diamond, D. D. and F. E. Smeins. 1985. Composition, classification, and species response patterns of remnant tallgrass prairie in Texas. *American Midland Naturalist* 113:294–308.

Dyksterhuis, E. J. 1946. The vegetation of the Ft. Worth prairie. *Ecological Monographs* 16:1–29.

Ellison, L. 1960. Influence of grazing on plant succession of rangelands. *Botanical Review* 26:1–78.

Franklin, J. F. and C. T. Dyrness. 1973. *Natural vegetation of Oregon and Washington.* U.S. Dept. of Agriculture Forest Service General Technical Report PNW-8. Portland, OR: U.S. Dept. of Agriculture Forest Service.

Gardner, J. L. 1951. Vegetation of the creosote bush area of the Rio Grande Valley in New Mexico. *Ecological Monographs* 21:379–403.

Gay, C. W. Jr. and D. D. Dwyer. 1965. *New Mexico range plants.* New Mexico State University Cooperative Extension Service Circular 374. Las Cruces, NM: Cooperative Extension, New Mexico State University.

Gibbens, R. P., J. M. Tromble, J. T. Hennessy, and M. Cardenas. 1983. Soil movement in mesquite dunelands and former grasslands of southern New Mexico from 1933 to 1980. *J. Range Management* 36:145–148 and 36:370–374.

Gleason, H. A. 1923. The vegetational history of the middle west. *Annals of the Association of American Geographers* 12:39–85.

Hadley, E. B. and R. P. Buccos. 1967. Plant community composition and net primary production within a native eastern North Dakota prairie. *American Midland Naturalist* 77:116–127.

Harris, G. A. 1967. Some competitive relationships between *Agropyron spicatum* and *Bromus tectorum*. *Ecological Monographs* 37:89–111.

Hastings, J. R. and R. M. Turner. 1965. *The changing mile.* Tucson, AZ: University of Arizona Press.

Heady, H. F. 1977. Valley grassland. In *Terrestrial vegetation of California,* ed. M. G. Barbour and J. Major, pp. 491–514. New York: Wiley-Interscience.

Hitchcock, A. S. 1950. *Manual of the grasses of the United States.* 2nd ed. U.S. Dept. of Agriculture Misc. Publ. 200. Washington, D.C.: U.S. Dept. of Agriculture.

Humphrey, R. R. 1953. The desert grassland: a history of vegetational change and an analysis of causes. *Botanical Review* 24:193–252.

Humphrey, R. R. and L. A. Mehrhoff. 1958. Vegetation changes on a southern Arizona grassland range. *Ecology* 39:720–726.

Johnston, M. C. 1963. Past and present grasslands of southern Texas and N.E. Mexico. *Ecology* 44:456–466.

Jones, C. H. 1944. Vegetation of Ohio prairies. *Bulletin of the Torrey Botanical Club* 71:536–548.

Kucera, C. L., R. C. Dahlman, and M. R. Koelling. 1967. Total net productivity and turnover on an energy basis for tallgrass prairie. *Ecology* 48:536–541.

Larson, F. 1940. The role of the bison in maintaining the short grass plains. *Ecology* 21:113–121.

Looman, J. 1983. Distribution of plant species and vegetation types in relation to climate. *Vegetatio* 54:17–25.

Mack, R. N. 1981. The invasion of *Bromus tectorum* L. into western North America: an ecological chronicle. *Agro-Ecosystems* 7:145–165.

McMillan, D. 1959. The role of ecotypic variation in the distribution of the central grassland of North America. *Ecological Monographs* 29:285–308.

Old, S. M. 1969. Microclimates, fire, and plant production in an Illinois prairie. *Ecological Monographs* 39:355–384.

Redmann, R. E. 1975. Production ecology of grassland plant communities in western North Dakota. *Ecological Monographs* 45:83–106.

Rice, E. L. and R. L. Parenti. 1978. Causes of decreases in productivity in undisturbed tall grass prairie. *American J. of Botany* 65:1091–1097.

Risser, P. G. 1985. Grasslands. In *Physiological ecology of North American plant communities,* ed. B. F. Chabot and H. A. Mooney, pp. 232–256. New York: Chapman and Hall.

Sims, P. L. 1987. Grasslands. In *North American terrestrial vegetation,* ed. M. G. Barbour and W. D. Billings, Chap. 9. New York: Cambridge University Press.

Sims, P. L., J. S. Singh, and W. K. Lauenroth. 1978. The structure and function of ten western North American grasslands. *J. of Ecology* 66:251–285.

Singh, J. S., W. K. Lauenroth, R. K. Heitschmidt, and J. L. Dodd. 1983. Structural and functional attributes of the vegetation of northern mixed prairie of North America. *Botanical Review* 49:117–149.

Sprague, H. B., ed. 1959. *Grasslands.* Washington, D.C.: American Association for the Advancement of Science.

Teeri, J. A. and L. G. Stowe. 1976. Climatic patterns and the distribution of C_4 grasses in North America. *Oecologia* 23:1–12.

Weaver, J. E. 1954. *The North American prairie.* Lincoln, NB: Johnsen.

Weaver, J. E. and F. W. Albertson. 1956. *Grasslands of the Great Plains.* Lincoln, NB: Johnsen.

Weaver, J. E. and T. J. Fitzpatrick. 1932. Ecology and relative importance of the dominants of tall-grass prairie. *Botanical Gazette* 93:113–150.

———. 1934. The prairie. *Ecological Monographs* 4:109–295.

White, D. 1941. Prairie soil as a medium for tree growth. *Ecology* 22:399–407.

Williams, W. A. 1966. Range improvements as related to net productivity, energy flow, and foliage configuration. *J. of Range Management* 19:29–34.

Young, J. A., R. A. Evans, and J. Major. 1977. Sagebrush steppe. In *Terrestrial vegetation of California*, ed. M. G. Barbour and J. Major, pp. 762–796. New York: Wiley-Interscience.

Young, J. A., R. A. Evans, and P. T. Tueller. 1975. Great Basin plant communities—pristine and grazed. In *Holocene climates in the Great Basin*, Occasional Paper, Nevada Archeological Survey, ed. R. Elston, pp. 186–215. Reno, NV: Nevada Archeological Survey.

Desert Scrub

Adams, S., B. R. Strain, and M. S. Adams. 1970. Water-repellent soil, fire, and annual plant cover in a desert scrub community of southeastern California. *Ecology* 51:696–700.

Anderson, D. J. 1971. Pattern in desert perennials. *J. of Ecology* 59:555–560.

Axelrod, D. I. 1950. *Evolution of desert vegetation.* In Carnegie Institution of Washington Publ. no. 590, pp. 215–306. Washington, D.C.: Carnegie Institution of Washington.

———. 1959. Evolution of the Madro-Tertiary geoflora. *Botanical Review* 24:433–509.

———. 1967. Drought, diastrophism, and quantum evolution. *Evolution* 21:201–209.

———. 1972. Edaphic aridity as a factor in angiosperm evolution. *American Naturalist* 106:311–320.

Barbour, M. G. 1969. Age and space distribution of the desert shrub *Larrea divaricata*. *Ecology* 50:679–685.

———. 1973. Desert dogma re-examined: root/shoot productivity and plant spacing. *American Midland Naturalist* 89:41–57.

Beatley, J. C. 1969. Biomass of desert winter annual plant populations in southern Nevada. *Oikos* 20:261–273.

———. 1974. Phenological events and their environmental triggers in Mojave Desert ecosystems. *Ecology* 55:856–863.

———. 1975. Climates and vegetation pattern across the Mojave/Great Basin Desert transition of southern Nevada. *American Midland Naturalist* 93:53–70.

Benson, L. and R. A. Darrow. 1954. *The trees and shrubs of the southwestern deserts.* Tucson, AZ: University of Arizona Press.

Billings, W. D. 1949. The shadscale vegetation zone of Nevada and eastern California in relation to climate and soils. *American Midland Naturalist* 42:87–109.

Björkman, O., R. W. Pearcy, A. T. Harrison, and H. A. Mooney. 1972. Photosynthetic adaptation to high temperatures: a field study in Death Valley, California. *Science* 175:786–789.

Burk, J. H. 1977. Sonoran desert. In *Terrestrial vegetation of California,* ed. M. G. Barbour and J. Major, pp. 869–889. New York: Wiley-Interscience.

Caldwell, M. 1985. Cold desert. In *Physiological ecology of North American plant communities,* ed. B. F. Chabot and H. A. Mooney, pp. 198–212. New York: Chapman and Hall.

Cannon, W. A. 1919. Relation of the rate of root growth in seedlings of *Prosopis velutina* to the temperature of the soil. *Carnegie Institution of Washington Yearbook* no. 18.

Capon, B. and W. Van Asdall. 1967. Heat pretreatment as a means of increasing germination of desert annual seeds. *Ecology* 48:305–306.

Chew, R. M. and A. E. Chew. 1965. The primary productivity of a desert shrub (*Larrea tridentata*) community. *Ecological Monographs* 35:353–375.

Cloudsley-Thompson, J. L. and M. J. Chadwick. 1964. *Life in deserts.* Philadelphia, PA: Dufour.

Cooke, R. U. and A. Warren. 1973. *Geomorphology in deserts.* Berkeley, CA: University of California Press.

Ehleringer, J. 1985. Annuals and perennials of warm deserts. In *Physiological ecology of North American plant communities,* ed. B. F. Chabot and H. A. Mooney, pp. 162–180. New York: Chapman and Hall.

Flowers, S. 1934. Vegetation of the Great Salt Lake region. *Botanical Gazette* 95:353–418.

Freas, K. E. and P. R. Kemp. 1983. Some relationships between environmental reliability and seed dormancy in desert annual plants. *J. Ecology* 71:211–217.

Gardner, J. L. 1951. Vegetation of the creosote bush area of the Rio Grande Valley in New Mexico. *Ecological Monographs* 21:379–403.

Gates, D. H., L. A. Stoddart, and C. W. Cook. 1956. Soil as a factor influencing plant distribution on salt deserts of Utah. *Ecological Monographs* 26:155–175.

Hastings, J. R., R. M. Turner, and D. K. Warren. 1972. *An atlas of some plant distributions in the Sonoran desert.* University of Arizona Institute of Atmospheric Physics Technical Report on the Meteorology and Climatology of Arid Regions no. 21. Tucson, AZ: University of Arizona Institute of Atmospheric Physics.

Hunt, C. B. 1966. *Plant ecology of Death Valley, California.* U.S. Geological Survey Professional Paper no. 509. Washington, D.C.: U.S. Geological Survey.

Jaeger, E. C. 1957. *The North American deserts.* Stanford, CA: Stanford University Press.

Kemp, P. R. 1983. Phenological patterns of Chihuahuan desert plants in relation to the timing of water availability. *J. Ecology* 71:427–436.

Lowe, C. H. and D. E. Brown. 1973. *The natural vegetation of Arizona.* Arizona Resources Information System Cooperative Publ. no. 2. Phoenix, AZ: Arizona Resources Information System.

Lunt, O. R., J. Letey, and S. B. Clark. 1973. Oxygen requirements for root growth in three species of desert shrubs. *Ecology* 54:1356–1362.

Mabry, T. J., J. H. Hunziker, and D. R. Difeo, Jr. 1977. *Creosote bush.* Stroudsburg, PA: Dowden, Hutchinson, and Ross.

MacMahon, J. A. 1979. North American deserts: their floral and faunal components. In *Arid-land ecosystems,* Vol. 1, ed. D. W. Goodall and R. A. Perry, pp. 21–82. New York: Cambridge University Press.

MacMahon, J. A. and F. H. Wagner. 1985. The Mojave, Sonoran, and Chihuahuan deserts of North America. In *Hot deserts and arid shrublands,* ed. M. Evenari et al., pp. 105–202. Amsterdam: Elsevier Science.

Marks, J. B. 1950. Vegetation and soil relations in the lower Colorado desert. *Ecology* 31:176–193.

McGinnies, W. G., B. J. Goldman, and P. Paylore, eds. 1968. *Deserts of the world.* Tucson, AZ: University of Arizona Press.

Mitchell, J. E., N. E. West, and R. W. Miller. 1966. Soil physical properties in relation to plant community patterns in the shadscale zone of northwestern Utah. *Ecology* 47:627–630.

Muller, W. H. and C. H. Muller. 1956. Association patterns involving desert plants that contain toxic products. *American J. of Botany* 43:354–361.

Mulroy, T. W. and P. W. Rundel. 1977. Annual plants: adaptations to desert environments. *BioScience* 27:109–114.

Niering, W. A., R. H. Whittaker, and C. H. Lowe. 1963. The saguaro: a population in relation to its environment. *Science* 142:15–23.

Nobel, P. S. 1985. Desert succulents. In *Physiological ecology of North American plant communities,* ed. B. F. Chabot and H. A. Mooney, pp. 181–197. New York: Chapman and Hall.

Odening, W. R., B. R. Strain, and W. C. Oechel. 1974. The effect of decreasing water potential on net CO_2 exchange of intact desert shrubs. *Ecology* 55:1086–1095.

Oppenheimer, H. R. 1960. Adaptation to drought: xerophytism. In *Plant-water relationships in arid and semi-arid conditions, reviews of research,* pp. 105–138. Paris: UNESCO.

Orians, G. H. and O. T. Solbrig. 1977. *Convergent evolution in warm deserts.* Stroudsburg, PA: Dowden, Hutchinson, and Ross.

Pearson, L. C. 1966. Primary productivity in a northern desert area. *Oikos* 15:211–228.

Rickard, W. H. and J. R. Murdock. 1963. Soil moisture and temperature survey of a desert vegetation mosaic. *Ecology* 44:821–824.

Rzedowski, J. 1978. *Vegetacion de Mexico.* Mexico City: Editorial Limusa.

Shantz, H. L. and R. L. Piemeisel. 1924. Indicator significance of the natural vegetation of the southwestern desert region. *J. of Agricultural Research* 28:721–802.

Shreve, F. 1942. The desert vegetation of North America. *Botanical Review* 8:195–246.

Shreve, F. and A. L. Hinkley. 1937. Thirty years of change in desert vegetation. *Ecology* 18:463–478.

Shreve, F. and I. L. Wiggins. 1964. *Vegetation and flora of the Sonoran desert.* Stanford, CA: Stanford University Press.

Stebbins, R. C. 1974. Off-road vehicles and the fragile desert. *The American Biology Teacher* 36:203–208, 294–304.

Steenbergh, W. F. and C. H. Lowe. 1969. Critical factors during the first years of life of the saguaro (*Cereus giganteus*) at Saguaro National Monument, Arizona. *Ecology* 50:825–834.

Stocker, O. 1960. Physiological and morphological changes in plants due to water deficiency. In *Plant-water relationships in arid and semi-arid conditions, reviews of research*, pp. 63–104. Paris: UNESCO.

Syvertsen, J. P., G. L. Nickell, R. W. Spellenberg, and G. L. Cunningham. 1976. Carbon reduction pathways and standing crop in three Chihuahuan desert plant communities. *Southwestern Naturalist* 21:311–320.

Turner, R. W. 1963. Growth in four species of Sonoran desert trees. *Ecology* 44:760–765.

Valentine, K. A. and J. J. Norris. 1964. A comparative study of soils of selected creosote bush sites in southern New Mexico. *J. of Range Management* 17:23–32.

Vasek, F. C. and M. G. Barbour. 1977. Mojave desert scrub vegetation. In *Terrestrial vegetation of California*, ed. M. G. Barbour and J. Major, pp. 835–867. New York: Wiley-Interscience.

Vasek, F. C., H. B. Johnson, and G. D. Brum. 1975. Effects of power transmission lines on vegetation of the Mojave desert. *Madroño* 23:114–130.

Vasek, F. C., H. B. Johnson, and D. H. Eslinger. 1975. Effects of pipeline construction on creosote bush scrub vegetation of the Mojave desert. *Madroño* 23:1–13.

Vogl, R. J. and L. T. McHargue. 1966. Vegetation of California fan palm oases on the San Andreas fault. *Ecology* 47:532–540.

Wallace, A. and E. M. Romney. 1972. *Radioecology and ecophysiology of desert plants at the Nevada Test Site*. U.S. Atomic Energy Commission TID-25954. Washington, D.C.: U.S. Atomic Energy Commission.

Went, F. W. 1942. The dependence of certain annual plants on shrubs in southern California deserts. *Bulletin of the Torrey Botanical Club* 69:100–114.

———. 1949. Ecology of desert plants. II. The effect of rain and temperature on germination and growth. *Ecology* 30:1–13.

Went, F. W. and M. Westergaard. 1949. Ecology of desert plants. III. Development of plants in the Death Valley National Monument, California. *Ecology* 30:26–38.

West, N. E. 1987. Intermountain deserts, shrubsteppes, and woodlands. In *North American terrestrial vegetation*, ed. M. G. Barbour and W. D. Billings, Chap. 7. New York: Cambridge University Press.

West, N. E. and M. M. Caldwell. 1983. Snow as a factor in salt desert shrub vegetation patterns in Curlew Valley, Utah. *American Midland Naturalist* 109:376–379.

West, N. E. and K. I. Ibrahim. 1968. Soil-vegetation relationships in the shadscale zone of southeastern Utah. *Ecology* 49:445–456.

North America, Overall

Barbour, M. G. and W. D. Billings (eds.). 1987. *North American terrestrial vegetation*. New York: Cambridge University Press.

Chabot, B. F. and H. A. Mooney (eds.). 1985. *Physiological ecology of North American plant communities*. New York: Chapman and Hall.

Daubenmire, R. F. 1978. *Plant geography, with special reference to North America*. New York: Academic Press.

Eyre, S. R. 1963. *Vegetation and soils*. Chicago: Aldine.

Küchler, A. W. 1964. *Potential natural vegetation of the conterminous United States*. American Geographical Society, Special Pub. 36.

Vankat, J. L. 1979. *The natural vegetation of North America*. New York: Wiley.

LITERATURE CITED

Aarssen, L. W. and R. Turkington. 1983. What is community evolution? *Evolutionary Theory* 6:211–217.

Abrahamson, W. G. 1984. Species responses to fire on the Florida Lake Wales Ridge. *American J. Botany* 71:35–43.

Abrahamson, W. G. and M. Gadgil. 1973. Growth form and reproductive effort in goldenrods (*Solidago,* Compositae). *American Naturalist* 107:651–661.

Ackerman, E. A. 1941. The Köppen classification of climates in North America. *Geographical Review* 31:105–111.

Adams, M. S. and B. R. Strain. 1968. Photosynthesis in stems and leaves of *Cercidium floridum*: spring and summer diurnal field response and relation to temperature. *Oecologia Plantarum* 3:285–297.

———. 1969. Seasonal photosynthetic rates in stems of *Cercidium floridum* Benth. *Photosynthetica* 3:55–62.

Adams, S., B. R. Strain, and M. S. Adams. 1970. Water-repellent soil, fire, and annual plant cover in a desert shrub community of southeastern California. *Ecology* 51:696–700.

Adams, S., B. R. Strain, and J. P. Ting. 1967. Photosynthesis in chlorophyllous stem tissue and leaves of *Cercidium floridum*: accumulation and distribution of ^{14}C from $^{14}CO_2$. *Plant Physiology* 42:1797–1799.

Agee, J. K. 1973. *Prescribed fire effects on physical and hydrologic properties of mixed-conifer forest floor and soil.* Water Resources Center Contribution Report no. 143. Davis, CA: University of California.

Ahlgren, I. F. 1974. The effect of fire on soil organisms. In *Fire and ecosystems,* ed. T. T. Kozlowski and C. E. Ahlgren, ch. 3. New York: Academic Press.

Alabock, P. B. 1982. Dynamics of understory biomass in Sitka spruce-western hemlock forests of southeast Alaska. *Ecology* 63:1932–1948.

Al-Ani, H. A., B. R. Strain, and H. A. Mooney. 1972. The physiological ecology of diverse populations of the desert shrub *Simmondsia chinesis. J. Ecology* 60:41–57.

Albert, R. 1975. Salt regulation in halophytes. *Oecologia* 21:57–71.

Alexander, M. E. 1982. Calculating and interpreting forest fire intensities. *Canadian J. Botany* 60:349–375.

Allaway, W. G. and F. L. Milthorpe. 1976. Structure and functioning of stomata. In *Water deficits and plant growth*, vol. IV, ed. T. T. Kozlowski, pp. 57–102. New York: Academic Press.

Allen, E. A. and R. T. T. Forman. 1976. Plant species removals and old-field community structure and stability. *Ecology* 57:1233–1243.

Anderson, D. J. 1967. Studies on structure in plant communities. III. Data on pattern in colonizing species. *J. of Ecology* 55:397–404.

Anderson, J. E. and S. J. McNaughton. 1973. Effects of low soil temperature on transpiration, photosynthesis, leaf relative water content, and growth among elevationally diverse plant populations. *Ecology* 54:1220–1233.

Anderson, J. M. 1973. The breakdown and decomposition of sweet chestnut (*Castanea sativa* Mill.) and beech (*Fagus sylvatica* L.) leaf litter in two deciduous woodland soils. I. Breakdown, leaching, and decomposition. *Oecologia* 12:251–274.

Anderson, K. L. 1965. Fire ecology—some Kansas prairie forbs. In *Proc. Tall Timbers Fire Ecol. Conf.*, no. 4, pp. 153–159. Tallahassee, FL: Tall Timbers Research Station.

Anderson, M. C. 1973. Solar radiation and carbon dioxide in plant communities—conclusions. In *Photosynthesis and productivity in different environments*, ed. J. P. Cooper, pp. 345–354. London: Cambridge University Press.

Andrews, F., D. C. Coleman, J. E. Ellis, and J. S. Singh. 1974. Energy flow relationships in a shortgrass prairie ecosystem. In *Proc. First Int. Congr. Ecol.*, pp. 22–28. Wageningen, The Netherlands: Center for Agricultural Publishing and Documentation.

Appleby, R. F. and W. J. Davies. 1983. A possible evaporation site in the guard cell wall and the influence of leaf structure on the humidity response by stomata of woody plants. *Oecologia* 56:30–40.

Armstrong, W. P. 1977. Fire followers in San Diego County. *Fremontia* 4:3–9.

Art, H. W., F. II. Bormann, G. K. Voigt, and G. M. Woodwell. 1974. Barrier island forest ecosystem: role of meteorologic nutrient inputs. *Science* 184:60–62.

Ashton, D. H. 1975. Studies of litter in *Eucalyptus regnans* forest. *Australian J. of Botany* 23:413–433.

Atkinson, I. A. E. 1970. Successional trends in the coastal and lowland forest of Mauna Loa and Kilauea volcanoes, Hawaii. *Pacific Science* 24:387–400.

Attiwill, P. M. 1979. Nutrient cycling in a *Eucalyptus obliqua* (L'Herit.) forest. III: growth, biomass and net primary production. *Australian J. of Botany* 27:439–458.

Auclair, A. N. and F. G. Goff. 1971. Diversity relations of upland forests in the western Great Lakes area. *American Naturalist* 105:499–528.

Avery, T. E. 1964. To stratify or not to stratify. *J. Forestry* 62:106–108.

Axelrod, D. I. 1966. A method for determining the altitudes of Tertiary floras. *The Paleobotanist* 14:144–171.

———. 1977. Outline history of California vegetation. In *Terrestrial vegetation of California*, ed. M. G. Barbour and J. Major, pp. 139–193. New York: Wiley-Interscience.

Axelrod, D. I. and H. P. Bailey. 1969. Paleotemperature analysis of Tertiary floras. *Palaeogeography, Palaeoclimatology, Palaeoecology* 6:163–195.

Ayala, F. J. 1969. Experimental invalidation of the principle of competitive exclusion. *Nature* 224:176–179.

Ayyad, M. A. G. and R. L. Dix. 1964. An analysis of a vegetation-microenvironmental complex on prairie slopes in Saskatchewan. *Ecological Monographs* 34:421–442.

Azevedo, J. and D. L. Morgan. 1974. Fog precipitation in coastal California forests. *Ecology* 55:1135–1141.

Bakeless, J. 1961. *The eyes of discovery.* New York: Dover.

Baker, R. L. and C. E. Thomas. 1983. A point frame for circular plots in southern forest-ranges. *J. Range Management* 36:121–123.

Baldwin, I. T. and J. C. Schultz. 1983. Rapid changes in tree leaf chemistry induced by damage: evidence for communication between plants. *Science* 221:277–279.

Baldwin, M., C. E. Kellogg, and J. Thorp. 1938. Classification of soils on the basis of their characteristics. In *Soils and man: 1938 yearbook of agriculture,* pp. 979–1001. Washington, D.C.: U.S. Dept. of Agriculture.

Bamberg, S. A., G. E. Kleinkopf, A. Wallace, and A. T. Vollmer. 1975. Comparative photosynthetic production of Mojave Desert shrubs. *Ecology* 56:732–736.

Bamberg, S. A., A. T. Vollmer, G. E. Kleinkopf, and T. L. Ackerman. 1976. A comparison of seasonal primary production of Mojave Desert shrubs during wet and dry years. *American Midland Naturalist* 95:398–405.

Bannister, P. 1976. *Introduction to physiological plant ecology.* New York: Halsted Press.

Barbour, M. G. 1970a. Seedling ecology of *Cakile maritima* along the California coast. *Bulletin of the Torrey Botanical Club* 97:280–289.

———. 1970b. Is any angiosperm an obligate halophyte? *American Midland Naturalist* 84:106–119.

———. 1973a. Chemistry and community composition. In *Air pollution damage to vegetation,* Advances in Chemistry no. 122, ed. M. G. Barbour, pp. 85–100. Washington, D.C.: American Chemical Society.

———. 1973b. Desert dogma re-examined: root/shoot productivity and plant spacing. *American Midland Naturalist* 89:41–57.

Barbour, M. G., R. B. Craig, F. R. Drysdale, and M. T. Ghiselin. 1973. *Coastal ecology: Bodega Head.* Berkeley, CA: University of California Press.

Barbour, M. G., G. L. Cunningham, W. C. Oechel, and S. A. Bamberg. 1977. Growth and development, form and function. In *Creosote bush,* ed. T. J. Mabry, J. H. Hunziker, and D. R. Difeo, Jr., ch. 4. Stroudsburg, PA: Dowden, Hutchinson, and Ross.

Barbour, M. G., D. V. Diaz, and R. W. Breidenbach. 1974. Contributions to the biology of *Larrea* species. *Ecology* 55:1199–1215.

Barbour, M. G., J. A. MacMahon, S. A. Bamberg, and J. A. Ludwig. 1977. The structure and function of *Larrea* communities. In *Creosote bush,* ed. T. J. Mabry, J. H. Hunziker, and D. R. Difeo, Jr., ch. 7. Stroudsburg, PA: Dowden, Hutchinson, and Ross.

Barclay-Estrup, P. and C. H. Gingham. 1969. The description and interpretation of cyclic processes in a heath community. I. Vegetational change in relation to the *Calluna* cycle. *J. of Ecology* 57:737–758.

Bard, G. 1952. Secondary succession on the Piedmont of New Jersey. *Ecological Monographs* 22:195–215.

Barrett, E. C., and L. F. Curtis. 1982. *Introduction to environmental remote sensing* (2nd Ed.). New York: Chapman and Hall.

Barrs, H. D. 1968. Determination of water deficits in plant tissues. In *Water deficits and plant growth,* vol. I, ed. T. T. Kozlowski, pp. 235–268. New York: Academic Press.

Bartholomew, B. 1970. Bare zone between California shrub and grassland communities: the role of animals. *Science* 170:1210–1212.

Bartholomew, G. A. 1972. Body temperature and energy metabolism. In *Animal physiology: principles and adaptations,* ed. M. S. Gordon, pp. 298–368. New York: Macmillan.

Baskin, J. M. and C. C. Baskin. 1973. Plant population differences in dormancy and germination characteristics of seeds: heredity or environment? *American Midland Naturalist* 90:493–498.

Bates, H. W. 1962. *The naturalist on the River Amazon.* Berkeley, CA: University of California Press.

Bates, L. M. and A. E. Hall. 1981. Stomatal closure with soil water depletion not associated with changes in bulk leaf water status. *Oecologia* 50:62–65.

Bauer, H. L. 1943. The statistical analysis of chaparral and other plant communities by means of transect samples. *Ecology* 24:45–60.

Bauer, M. E. 1985. Spectral inputs to crop identification and condition assessment. *Proceedings IEEE* 73:1071–1085.

Beard, J. S. 1946. The mora forests in Trinidad, British West Indies. *J. of Ecology* 33:173–192.

Beardsell, M. F., P. G. Jarvis, and B. Davidson. 1972. A null-balance diffusion porometer suitable for use with leaves of many shapes. *J. of Applied Ecology* 9:677–690.

Beatley, J. C. 1966. Winter annual vegetation following a nuclear detonation in the northern Mojave Desert (Nevada Test Site). *Radiation Botany* 6:69–82.

———. 1967. Survival of winter annuals in the northern Mojave Desert. *Ecology* 48:745–750.

———. 1969. Biomass of desert winter annual plant populations in southern Nevada. *Oikos* 20:261–273.

———. 1974. Phenological events and their environmental triggers in Mojave Desert ecosystems. *Ecology* 55:856–863.

Beattie, A. J., D. E. Breedlove, and P. R. Ehrlich. 1973. The ecology of the pollinators and predators of *Frasera speciosa. Ecology* 54:81–91.

Beauchamp, J. J. and J. S. Olson. 1973. Corrections for bias in regression estimates after logarithmic transformation. *Ecology* 54:1403–1407.

Becking, R. W. 1957. The Zurich-Montpellier school of phytosociology. *Botanical Review* 23:411–488.

Bell, K. L. and L. C. Bliss. 1979. Autecology of *Kobresia bellardii:* why winter snow accumulation limits local distribution. *Ecological Monographs* 49:377–402.

Bender, M. M. 1971. Variation in the $^{13}C/^{12}C$ ratios of plants in relation to pathway of photosynthetic carbon fixation. *Phytochemistry* 10:1239–1244.

Bender, M. M., I. Rougani, H. M. Vines, and C. C. Black, Jr. 1973. $^{13}C/^{12}C$ ratio changes in crassulacean acid metabolism plants. *Plant Physiology* 52:427–430.

Bennert, W. H. and H. A. Mooney. 1979. The water relations of some desert plants in Death Valley, California. *Flora* 168:405–427.

Bernard, J. M. and J. G. McDonald. 1974. Primary production and life history of *Carex lacustris. Canadian J. of Botany* 52:117–123.

Bernard, J. M. and B. A. Solsky. 1977. Nutrient cycling in a *Carex lacustris* wetland. *Canadian J. of Botany* 55:630–638.

Berry, J. and O. Björkman. 1980. Photosynthetic response and adaptation to temperature in higher plants. *Annual Review of Plant Physiology* 31:491–543.

Berry, J. A. and G. D. Farquhar. 1978. The CO_2 concentrating function of C_4 photosynthesis: a biochemical model. In *Proceedings of the 4th International Congress on Photosynthesis,* ed. D. Hall, J. Coombs, and T. Goodwin, pp. 119–131. London: Biochemical Society.

Berthet, P. 1960. La mesure de la temperature par determination de la vitesse d'inversion du saccharose. *Vegetatio* 9:197–207.

Bidwell, R. G. S. 1974. *Plant physiology.* New York: Macmillan.

Billings, W. D. 1938. The structure and development of old-field short-leaf pine stands and certain associated physical properties of the soil. *Ecological Monographs* 8:437–499.

———. 1949. The shadscale vegetation zone of Nevada and eastern California in relation to climate and soils. *American Midland Naturalist* 42:87–109.

———. 1952. The environmental complex in relation to plant growth and distribution. *Quarterly Review of Biology* 27:251–265.

———. 1970. *Plants, man, and the ecosystem.* 2nd ed. Belmont, CA: Wadsworth.

———. 1985. The historical development of physiological plant ecology. In *Physiological ecology of North American plant communities,* ed. B. F. Chabot and H. A. Mooney, pp. 1–15. New York: Chapman and Hall.

Billings, W. D. and L. C. Bliss. 1959. An alpine snowbank environment and its effects on vegetation, plant development, and productivity. *Ecology* 40:388–397.

Billings, W. D., E. E. C. Clebsch, and H. A. Mooney. 1966. Photosynthesis and respiration rates of Rocky Mountain alpine plants under field conditions. *American Midland Naturalist* 75:34–44.

Billings, W. D., P. J. Godfrey, B. F. Chabot, and D. P. Bourque. 1971. Metabolic acclimation to temperature in arctic and alpine ecotypes of *Oxyria digyna*. *Arctic and Alpine Research* 3:277–289.

Biswell, H. H. 1958. Prescribed burning in Georgia and California compared. *J. of Range Management* 11:293–298.

———. 1963. Research in wildland fire ecology in California. In *Proc. Tall Timbers Fire Ecol. Conf.,* no. 2, pp. 63–97. Tallahassee, FL: Tall Timbers Research Station.

———. 1967. Forest fire in perspective. In *Proc. Tall Timbers Fire Ecol. Conf.,* no. 7, pp. 43–63. Tallahassee, FL: Tall Timbers Research Station.

———. 1974. Effects of fire on chaparral. In *Fire and ecosystems,* ed. T. T. Kozlowski and C. E. Ahlgren, ch. 10. New York: Academic Press.

———. 1977. *Giant sequoia fire ecology.* University Extension Course X417.4. Davis, CA: University of California.

Björkman, O. 1968*a*. Carboxydismutase activity in shape-adapted species of higher plants. *Physiologia Plantarum* 21:1–10.

———. 1968*b*. Further studies on differentiation of photosynthetic properties of sun and shade ecotypes of *Solidago virgaurea*. *Physiologia Plantarum* 21:84–99.

Björkman, O., H. A. Mooney, and J. Ehleringer. 1975. Photosynthetic characteristics of plants from habitats with contrasting thermal regimes: comparisons of photosynthetic responses of intact plants. *Carnegie Institution of Washington Yearbook* 74:743–748.

Black, C. C. 1973. Photosynthetic carbon fixation in relation to net CO_2 uptake. *Annual Review of Plant Physiology* 24:253–286.

Blackman, G. E. 1935. A study by statistical methods of the distribution of species in grassland associations. *Annals of Botany* 49:749–778.

Bleak, A. T. 1970. Disappearance of plant material under a winter snow cover. *Ecology* 51:915-917.

Bloom, Arnold J., F. Stuart Chapin, III, and Harold A. Mooney. 1985. Resource limitation in plants—an economic analogy. *Annual Review of Ecology and Systematics* 16:363–392.

Blum, U. and E. L. Rice. 1969. Inhibition of symbiotic nitrogen-fixation by gallic and tannic acid, and possible roles in old-field succession. *Bulletin of the Torrey Botanical Club* 96:531–544.

Bocock, K. L., O. J. W. Gilbert, C. K. Capstick, D. C. Twinn, J. S. Waid, and M. J. Woodman. 1960. Changes in the leaf litter when placed on the surface of soils with contrasting humus types. I. Losses in dry weight of oak and ash leaf litters. *J. of Soil Science* 11:1–9.

Bohm, W. 1979. *Methods of studying root systems*. Berlin: Springer-Verlag.

Bond, R. D. 1964. The influence of the microflora on the physical properties of soils. II. Field studies on water-repellent sands. *Australian J. of Soil Research* 2:123–131.

Bonner, J. 1940. On the growth factor requirements of isolated roots. *American J. of Botany* 27:696–701.

Bonnier, G. 1895. Recherches experimentales sur l'adaptation des plantes au climat alpin. *Annales des Sciences Naturelles Botanique*. 7th series 20:217–358.

Booth, W. E. 1941. Revegetation of abandoned fields in Kansas and Oklahoma. *American J. of Botany* 28:415–422.

Bormann, F. H. 1953. The statistical efficiency of sample plot size and shape in forest ecology. *Ecology* 34:474–487.

Bormann, F. H. and G. E. Likens. 1967. Nutrient cycling. *Science* 155:424–429.

Bormann, F. H., G. E. Likens, and J. M. Melillo. 1977. Nitrogen budget for an aggrading northern hardwood forest ecosystem. *Science* 196:981–983.

Bormann, F. H., G. E. Likens, T. G. Siccama, R. S. Pierce, and J. S. Eaton. 1974. The export of nutrients and recovery of stable conditions following deforestation at Hubbard Brook. *Ecological Monographs* 44:255–277.

Bormann, F. H., T. G. Siccama, G. E. Likens, and R. H. Whittaker. 1970. The Hubbard Brook ecosystem study: composition and dynamics of the tree stratum. *Ecological Monographs* 40:373–388.

Botkin, D. B., J. F. Janak, and J. R. Wallis. 1972. Some ecological consequences of a computer model of forest growth. *J. of Ecology* 60:849–872.

Botting, D. 1973. *Humboldt and the cosmos*. New York: Harper and Row.

Bourdeau, P. F. 1953. A test of random vs. systematic ecological sampling. *Ecology* 34:499–512.

Bowers, W. S., T. Ohta, J. S. Cleere, and P. A. Marsella. 1976. Discovery of insect anti-juvenile hormones in plants. *Science* 193:542–547.

Bowman, W. D. and S. W. Roberts. 1985. Seasonal and diurnal water relations adjustments in three evergreen chaparral shrubs. *Ecology* 66:738–742.

Box, T. W. 1967. Brush, fire and west Texas rangeland. In *Proc. Tall Timbers Fire Ecol. Conf.*, no. 6, pp. 7–19. Tallahassee, FL: Tall Timbers Research Station.

Boyce, S. G. 1951. Source of atmospheric salts. *Science* 113:620–621.

———. 1954. The salt spray community. *Ecological Monographs* 24:29–67.

Boyer, J. S. 1976. Water deficits and photosynthesis. In *Water deficits and plant growth*, vol. IV, ed. T. T. Kozlowski, pp. 153–190. New York: Academic Press.

Bradshaw, A. D. 1965. Evolutionary significance of phenotypic plasticity in plants. *Advances in Genetics* 13:115–155.

———. 1969. An ecologist's viewpoint. In *Ecological aspects of the mineral nutrition of plants*, ed. I. H. Rorison, pp. 415–427. Oxford: Blackwell Scientific Publications.

Brady, N. C. 1974. *The nature and properties of soils*. 8th Ed. New York: Macmillan.

Braun, E. L. 1916. *The physiographic ecology of the Cincinnati region*. Ohio Biological Survey Bulletin 7.

———. 1950. *Deciduous forests of eastern North America*. Philadelphia, PA: Blakiston.

Braun-Blanquet, J. 1932. *Plant sociology: the study of plant communities*. New York: McGraw-Hill.

Bray, J. R. and J. T. Curtis. 1957. An ordination of the upland forest communities of southern Wisconsin. *Ecological Monographs* 27:325–349.

Bray, J. R. and E. Gorham. 1964. Litter production in forests of the world. *Advances in Ecological Research* 2:101–157.

Brian, P. W. 1957. Effects of antibiotics on plants. *Annual Review of Plant Physiology* 8:413–426.

Briggs, D. and S. M. Walters. 1969. *Plant variation and evolution*. New York: McGraw-Hill.

Brinson, M. B. 1977. Decomposition and nutrient exchange of litter in an alluvial swamp forest. *Ecology* 58:601–609.

Brower, L. P. and J. V. Z. Brower. 1964. Birds, butterflies, and plant poisons: a study in ecological chemistry. *Zoologica* 49:137.

Brown, A. A. and K. P. Davis. 1973. *Forest fire: control and use*. 2nd ed. New York: McGraw-Hill.

Brown, R. T. 1967. Influence of naturally occurring compounds on germination and growth of jack pine. *Ecology* 48:542–546.

Brown, R. W. 1970. *Measurement of water potential with thermocouple psychrometers: construction and applications*. U.S. Dept. of Agriculture Forest Service Research Paper INT-80. Washington, D.C.: U.S. Dept. of Agriculture.

Brown, R. W. and W. T. McDonough. 1977. Thermocouple psychrometer for *in situ* leaf water potential determinations. *Plant and Soil* 48:5–10.

Brown, R. W. and B. P. van Haveren, eds. 1972. *Psychrometry in water relations research*. Utah Agricultural Experiment Station. Logan, UT: Utah State University.

Brownell, P. F. and C. J. Crossland. 1972. The requirements of sodium as a micronutrient by species having the C_4 dicarboxylic acid photosynthetic pathway. *Plant Physiology* 49:794–797.

———. 1974. Growth responses to sodium by *Bryophyllum tubiflorum* under conditions inducing crassulacean acid metabolism. *Plant Physiology* 54:416–417.

Bryant, J. P., F. S. Chapin III, and D. R. Klein. 1983. Carbon/nutrient balance of boreal plants in relation to vertebrate herbivory. *Oikos* 40:357–368.

Bryson, R. A. 1974. A perspective on climatic change. *Science* 184:753–759.

Bunce, J. A. 1977. Nonstomatal inhibition of photosynthesis at low water potentials in intact leaves of species from a variety of habitats. *Plant Physiology* 59:348–350.

Burbanck, M. P. and R. B. Platt. 1964. Granite outcrop communities on the Piedmont Plateau in Georgia. *Ecology* 45:292–305.

Burbanck, M. P. and D. L. Phillips. 1983. Evidence of plant succession on granite outcrops of the Georgia Piedmont. *American Midland Naturalist* 109:94–104.

Burk, J. H. and W. A. Dick-Peddie. 1973. Comparative production of *Larrea divaricata* Cav. on three geomorphic surfaces in southern New Mexico. *Ecology* 54:1094–1102.

Burkholder, P. R. 1952. Cooperation and conflict among primitive organisms. *American Scientist* 40:601–631.

Burris, R. H. and C. C. Black. 1976. *CO₂ metabolism and plant productivity.* Baltimore, MD: University Park Press.

Burrows, F. J. and F. L. Milthorpe. 1976. Stomatal conductance in the control of gas exchange. In *Water deficits and plant growth*, vol. IV, ed. T. T. Kozlowski, pp. 103–152. New York: Academic Press.

Busby, J. R., L. C. Bliss, and C. D. Hamilton. 1978. Microclimate control of growth rates and habitats of the boreal forest mosses, *Tomenthypnum nitens and Hylocomium splendens. Ecological Monographs* 48:95–110.

Buttrick, S. C. 1977. The alpine flora of Teresa Island, Atlin Lake, British Columbia, with notes on its distribution. *Canadian J. of Botany* 55:1399–1409.

Byers, H. R. 1953. Coast redwoods and fog drip. *Ecology* 34:192–193.

Byram, G. M. 1959. Combustion of forest fuels. In *Forest fire: control and use.* eds. A. A. Brown and K. T. Davis, pp. 61–69. New York: McGraw-Hill.

Cable, D. R. 1975. Influence of precipitation on perennial grass production in the semi-desert southwest. *Ecology* 56:981–986.

Cain, S. A. and G. M. De O. Castro. 1959. *Manual of vegetation analysis.* New York: Harper.

Caldwell, M. M. and R. Robberecht. 1980. A steep latitudinal gradient of solar ultraviolet-B radiation in the arctic alpine life zone. *Ecology* 61:600–611.

Caldwell, M. M., R. S. White, R. T. Moore, and L. B. Camp. 1977. Carbon balance, productivity, and water use of cold-winter desert shrub communities dominated by C₃ and C₄ species. *Oecologia* 29:275–300.

Camacho-B, S. E., A. E. Hall, and M. R. Kaufmann. 1974. Efficiency and regulation of water transport in some woody and herbaceous species. *Plant Physiology* 54:169–172.

Campbell, C. J. and W. A. Dick-Peddie. 1964. Comparison of phreatophyte communities on the Rio Grande in New Mexico. *Ecology* 45:492–502.

Campbell, G. S. 1977. *An introduction to environmental biophysics.* New York: Springer-Verlag.

Campbell, G. S. and M. D. Campbell. 1974. Evaluation of a thermocouple hygrometer for measuring leaf water potential *in situ. Agronomy J.* 66:24–27.

Cannon, W. A. 1905a. On the transpiration of *Fouquieria splendens. Bulletin of the Torrey Botanical Club* 32:397–414.

———. 1905b. On the water-conducting systems of some desert plants. *Botanical Gazette* 39:397–408.

———. 1911. *The root habits of desert plants.*Carnegie Institution of Washington Publ. 131. Washington, D.C.: Carnegie Institution of Washington.

Capon, B. and W. Van Asdall. 1966. Heat pretreatment as a means of increasing germination of desert annual seeds. *Ecology* 48:305–306.

Carlson, C. E., R. D. Pfister, L. J. Theroux, and C. E. Fiedler. 1985. *Release of a thinned budworm-infested Douglas-fir/ponderosa pine stand*. Research Paper INT-349. United States Forest Service. Ogden, Utah.

Carlisle, A., A. H. F. Brown, and E. J. White. 1966. The organic matter and nutrient elements in the precipitation beneath a sessile oak (*Quercus petraea*) canopy. *J. of Ecology* 54:87–98.

Carlquist, S. 1975. *Ecological strategies of xylem evolution.* Berkeley, CA: University of California Press.

Carman, J. G. 1982. A non-destructive stain technique for investigating root growth dynamics. *Journal of Applied Ecology* 19:873–879.

Carneggie, D. M., B. J. Schrumpf, and D. A. Mouat. 1983. Rangeland applications. In *Manual of remote sensing,* Vol. II, ed. R. N. Colwell, pp. 2325–2384. Falls Church, VA: American Society of Photogrammetry.

Carpenter, D. E., M. G. Barbour, and C. J. Bahre. 1986. Old-field succession in Mojave Desert scrub. *Madroño* 33:111–122.

Caswell, H. 1976. Community structure: a neutral model analysis. *Ecological Monographs* 46:327–354.

Cates, R. G. 1975. The interface between slugs and wild ginger: some evolutionary aspects. *Ecology* 56:391–400.

Cates, R. G. and G. H. Orians. 1975. Successional status and the palatability of plants to generalized herbivores. *Ecology* 56:410–418.

Cates, R. G. and R. A. Redak. 1986. Between-year population variation in resistance of Douglas-fir to the western spruce budworm. In *Natural resistance of plants to pests,* ed. M. Green and P. Hedin, pp. 106–115. American Chemical Society Symposium No. 296, Washington, D.C.

Cavers, P. B. and J. L. Harper. 1967a. Germination polymorphism in *Rumex crispus* and *R. obtusifolius. J. of Ecology* 54:367–382.

———. 1967b. The comparative biology of closely related species living in the same area. IX. *Rumex:* the nature of adaptation to a sea-shore habitat. *J. of Ecology* 55:73–82.

Center, T. D. and C. D. Johnson. 1974. Coevolution of some seed beetles (Coleoptera: Bruchidae) and their hosts. *Ecology* 55:1096–1103.

Ceska, A. and H. Roemer. 1971. A computer program for identifying species-relevé groups in vegetation studies. *Vegetatio* 23:255–277.

Chadwick, H. W. and P. D. Dalke. 1965. Plant succession in dune stands in Fremont County, Idaho. *Ecology* 46:765–780.

Chaney, R. W. 1948. *The ancient forests of Oregon.* Condon Lectures. Eugene, OR: Oregon State System of Higher Education.

Chapin, F. S., III. 1980. The mineral nutrition of wild plants. *Annual Review of Ecology and Systematics* 11:233–260.

———. 1985. Individualistic growth response of tundra plant species to environmental manipulations in the field. *Ecology* 66:564–576.

Chapin, F. S., III, and G. R. Shaver. 1985. Arctic. In *Physiological ecology of North American plant communities,* ed. B. F. Chabot and H. M. Mooney, pp. 16–40. New York: Chapman and Hall.

Chapin, F. S., III, and K. Van Cleve. 1981. Plant nutrient absorption and retention under differing fire regimes. In *Fire regimes and ecosystem processes,* ed. H. A. Mooney, T. M. Bonnicksen, N. L. Christensen, J. E. Lotan and W. A. Reiner, pp. 301–321. General Technical Report WO-26, United States Forest Service, Washington, DC.

Chapman, S. B. 1976. Production ecology and nutrient budgets. In *Methods in plant ecology,* ed. S. B. Chapman, pp. 157–228. New York: Halsted Press.

Chatterton, N. J. 1970. *Physiological ecology of Atriplex polycarpa: growth, salt tolerance, ion accumulation and soil-plant-water relations.* Ph.D. dissertation. Riverside, CA: University of California.

Chazdon, R. L., and R. W. Pearcy. 1986*a*. Photosynthetic responses to light variation in rainforest species. I. Induction under constant and fluctuating light conditions *Oecologia* 69:571–523.

———. 1986*b*. Photosynthetic responses to light variation in rainforest species. I. Carbon gain and photosynthetic efficiency during lightflecks. *Oecologia* 69:524–531.

Chepurko, N. I. 1972. The biological productivity and the cycle of N and ash elements in the dwarf shrub tundra ecosystems of the Khibini Mountains (Kola Peninsula). In *Int. Biol. Programme, Tundra Biome, Proc. IV Int. Meet. on the Biol. Productivity of Tundra,* ed. F. L. Weilgolaski and T. Rosswall, pp. 236–247. Stockholm, Sweden.

Cheung, Y. N. S., M. T. Tyree, and J. Dainty. 1975. Water relations parameters on single leaves obtained in a pressure bomb and some ecological interpretations. *Canadian J. of Botany* 53:1342–1346.

Chew, R. M. 1974. Consumers as regulators of ecosystems: an alternative to energetics. *Ohio J. of Science* 74:359–370.

Christensen, N. and C. H. Muller. 1975. Effects of fire on factors controlling plant growth in *Adenostoma* chaparral. *Ecological Monographs* 45:29–55.

Churchill, G. B., H. H. John, D. P. Duncan, and A. C. Hodson. 1964. Long term effects of defoliation of aspen by the forest tent caterpillar. *Ecology* 45:630–633.

Clapham, A. R. 1932. The form of the observational unit in quantitative ecology. *J. of Ecology* 20:192–197.

Clark, D. A. and D. B. Clark. 1984. Spacing dynamics of a tropical rain forest tree: evaluation of the Janzen-Connell model. *American Naturalist* 124:769–788.

Clark, D. D. and J. H. Burk. 1980. Resource allocation patterns of two California Sonoran Desert ephemerals. *Oecologia* 46:86–91.

Clark, P. J. and F. C. Evans. 1954. Distance to nearest neighbor as a measure of spatial relationships in populations. *Ecology* 35:445-453.

Clausen, J. 1962. *Stages in the evolution of plant species.* New York: Hafner.

Clausen, J., D. D. Keck, and W. M. Hiesey. 1940. *Experimental studies on the nature of species. I. The effect of varied environments on western North American plants.* Carnegie Institution of Washington Publ. 520. Washington, D.C.: Carnegie Institution of Washington.

Clements, E. S. 1960. *Adventures in ecology.* New York: Pageant Press.

Clements, F. E. 1916. *Plant succession: an analysis of the development of vegetation.* Carnegie Institution of Washington Publ. 242. Washington, D.C.: Carnegie Institution of Washington.

———. 1920. *Plant indicators: the relation of plant communities to process and practice.* Carnegie Institution of Washington Publ. 290. Washington, D.C.: Carnegie Institution of Washington.

———. 1936. Nature and structure of the climax. *J. of Ecology* 24:252–284.

Clements, F. E. and G. W. Goldsmith. 1924. *The phytometer method in ecology.* Carnegie Institution of Washington Publ. 356. Washington, D.C.: Carnegie Institution of Washington.

Clements, F. E. and H. M. Hall. 1921. Experimental taxonomy. *Carnegie Institute, Washington Year Book* 20:395–396.

Cloudsley-Thompson, J. L. and M. J. Chadwick. 1964. *Life in deserts.* Philadelphia, PA: Dufour.

Coats, R. N., R. L. Leonard, and C. R. Goldman. 1976. Nitrogen uptake and release in a forested watershed, Lake Tahoe Basin. *Ecology* 57:995–1004.

Cody, M. L. 1974. Optimization in ecology. *Science* 183:1156–1164.

Cody, M. L. and H. A. Mooney. 1978. Convergence versus nonconvergence in mediterranean-climate ecosystems. *Annual Review Ecology Systematics* 9:265–321.

Cohen, D. 1966. Optimizing reproduction in a randomly varying environment. *J. of Theoretical Biology* 12:119–129.

Cole, C. V., R. S. Innis, and J. W. B. Stewart. 1977. Simulation of phosphorus cycling in semiarid grasslands. *Ecology* 58:1–15.

Cole, D. W., S. P. Gessel, and S. F. Dice. 1967. Distribution and cycling of nitrogen, phosphorus, potassium and calcium in a secondary growth Douglas-fir ecosystem. In *Symp. primary productivity and mineral cycling in natural ecosystems*, pp. 196–232. Orono, ME: University of Maine Press.

Cole, K. 1985. Past rates of change, species richness, and a model of vegetational inertia in the Grand Canyon, Arizona. *American Naturalist* 125:289–303.

Coleman, D. C., C. P. P. Reid, and C. V. Cole. 1983. Biological strategies of nutrient cycling in soil ecosystems. *Advances in Biological Research* 13:1–55.

Colinvaux, P. A. 1973. *Introduction to ecology.* New York: Wiley.

Collier, B. D., G. W. Cox, A. W. Johnson, and P. C. Miller. 1973. *Dynamic ecology.* Englewood Cliffs, NJ: Prentice-Hall.

Collins, S. L. and D. E. Adams. 1983. Succession in grasslands: thirty-two years of change in a central Oklahoma tallgrass prairie. *Vegetatio* 51:181–190.

Collins, W., S. H. Chang, and J. T. Kuo. 1981. *Detection of hidden mineral deposits by airborne spectral analysis of forest canopies.* Final Rept., NASA contract NSG 5222.

Collins, W., S. H. Chang, G. Raines, F. Canney, and R. Ashley. 1983. Airborne biogeochemical mapping of hidden mineral deposits. *Economic Geology* 78:737–749.

Colwell, R. N. (ed.). 1983. *Manual of remote sensing*, 2nd ed. Falls Church, VA: American Society of Photogrammetry.

Conard, H. S. 1935. The plant associations on central Long Island: a study in descriptive sociology. *American Midland Naturalist* 16:433–516.

Connell, J. H. 1971. On the role of natural enemies in preventing competitive exclusion in some marine animals and in rain forest trees. In *Dynamics of population*, ed. P. J. den Boer and G. R. Gradwell, pp. 298–310. Wageningen, Netherlands: Center for Agricultural Publishing and Documentation.

Connell, J. H. and R. O. Slatyer. 1977. Mechanism of succession in natural communities and their role in community stability and organization. *American Naturalist* 111:1119–1144.

Cook, R. E. 1980. Germination and size dependent mortality in *Viola blanda*. *Oecologia* 47:115–117.

Cooper, C. F. 1957. The variable plot method for estimating shrub density. *J. Range Management* 19:111–115.

———. 1961. The ecology of fire. *Scientific American* 204:150–160.

Cooper, W. S. 1913. The climax forest of Isle Royale, Lake Superior, and its development. *Botanical Gazette* 55:1–44, 115–140, 189–235.

———. 1919. Ecology of the strand vegetation of the Pacific coast of North America. *Carnegie Institution of Washington Yearbook* 18:96–99.

———. 1922. *The broad-sclerophyll vegetation of California*. Carnegie Institution of Washington Publ. 319. Washington, D.C.: Carnegie Institution of Washington.

———. 1923. The recent ecological history of Glacier Bay, Alaska. *Ecology* 4:93–128, 223–246, 355–365.

———. 1931. A third expedition to Glacier Bay, Alaska. *Ecology* 12:61–95.

———. 1936. The strand and dune flora of the Pacific coast of North America: a geographic study. In *Essays in geobotany*, ed. T. H. Goodspeed, pp. 141–187. Berkeley, CA: University of California Press.

———. 1957. Sir Arthur Tansley and the science of ecology. *Ecology* 38:658–659.

Cornforth, I. S. 1970. Leaf fall in a tropical rain forest. *J. Applied Ecology* 7:603–608.

Cottam, G. and J. T. Curtis. 1956. Use of distance measures in phytosociological sampling. *Ecology* 37:451–460.

Cottam, W. P., J. M. Tucker, and R. Drobnick. 1959. Some clues to Great Basin postpluvial climates provided by oak distributions. *Ecology* 40:361–377.

Court, A. 1974. Water balance estimates for the United States. *Weatherwise* 27:252–256.

Courtney, S. P. 1985. Apparency in coevolving relationships. *Oikos* 44:91–98.

Cowles, H. C. 1901. The physiographic ecology of Chicago and vicinity: a study of the origin, development, and classification of plant societies. *Botanical Gazette* 31:73–108, 145–182.

———. 1909. Present problems in plant ecology: the trend of ecological philosophy. *American Naturalist* 43:356–368.

———. 1911. The causes of vegetative cycles. *Botanical Gazette* 51:161–183.

Cowling, R. M. and B. M. Campbell. 1980. Convergence in vegetation structure in the mediterranean communities of California, Chile, and South Africa. *Vegetatio* 43:191–197.

Cox, G. W. 1985. *Laboratory manual of general ecology*. Dubuque, IA: William C. Brown.

Crawley, M. J. 1983. *Herbivory*. Berkeley, CA: University of California Press.

Critchfield, W. B. 1971. *Profiles of California vegetation*. U.S. Dept. of Agriculture Forest Service Research Paper PSW-76, Berkeley, CA. Washington, D.C.: U.S. Dept. of Agriculture.

Crocker, R. L. and J. Major. 1955. Soil development in relation to vegetation and surface age at Glacier Bay, Alaska. *J. of Ecology* 43:427–448.

Cromack, K., Jr. 1973. Litter production and litter decomposition in a mixed hardwood watershed and in a white pine watershed at Coweeta Hydrologic Station, North Carolina. Dissertation, University of Georgia, Athens, GA.

Crosby, J. S. 1961. *Litter and duff fuel in shortleaf pine stands in southeast Missouri.* Central States Forest Experiment Station Technical Paper 178. Columbus, OH: U.S. Forest Service.

Cunningham, G. L. and B. R. Strain. 1969. Irradiance and productivity in a desert shrub. *Photosynthetica* 3:69–71.

Curran, P. J. 1985. *Principles of remote sensing.* New York: Longman.

Currey, D. B. 1965. An ancient bristlecone pine stand in eastern Nevada. *Ecology* 46:564–566.

Curtis, J. T. and G. Cottam. 1962. *Plant ecology workbook.* Minneapolis, MN: Burgess.

Curtis, J. T. and R. P. McIntosh. 1950. The interrelations of certain analytic and synthetic phytosociological characteristics. *Ecology* 31:434–455.

———. 1951. An upland forest continuum in the prairie-forest border region of Wisconsin. *Ecology* 32:476–498.

Cutler, J. M. and D. W. Rains. 1978. Effects of water stress and hardening on the internal water relations and osmotic constituents of cotton leaves. *Physiologia Plantarum* 42:261–268.

Cutler, J. M., D. W. Rains, and R. S. Loomis. 1977. The importance of cell size in the water relations of plants. *Physiologia Plantarum* 40:255–260.

Dahlman, R. C. and C. L. Kucera. 1965. Root productivity and turnover in native prairie. *Ecology* 46:84–89.

Dansereau, P. 1951. Description and recording of vegetation upon a structural basis. *Ecology* 32:172–229.

———. 1961. Essai de representation cartographique des elements structuraux de la vegetation. In *Methods de la cartographie de la vegetation,* 97th Colloquium, ed. H. Gaussen, pp. 233–255. Paris: Centre National de la Recherche Scientifique.

Darwin, C. R. 1859. *On the origin of species by means of natural selection, or the preservation of favoured races in the struggle for life.* London: John Murray.

Dasmann, R. F. 1972. *Environmental conservation,* 3rd ed. New York: Wiley.

Daubenmire, R. F. 1938. Merriam's life zones of North America. *Quarterly Review of Biology* 13:327–332.

———. 1952. Forest vegetation of northern Idaho and adjacent Washington and its bearing on concepts of vegetation classification. *Ecological Monographs* 22:301–330.

———. 1959. A canopy-coverage method of vegetational analysis. *Northwest Science* 33:43–66.

———. 1968a. *Plant communities.* New York: Harper and Row.

———. 1968b. Ecology of fire in grasslands. *Advances in Ecological Research* 5:209–266.

———. 1974. *Plants and environment.* 3rd ed. New York: Wiley.

Daubenmire, R., and D. Prusso. 1963. Studies of the decomposition rates of tree litter. *Ecology* 44:589–592.

Davis, P. H. and V. H. Heywood. 1963. *Principles of angiosperm taxonomy.* London: Oliver and Boyd.

Davis, R. B. 1966. Spruce-fir forests of the coast of Maine. *Ecological Monographs* 36:79–94.

Davis, T. A. W. and P. W. Richards. 1933. The vegetation of Moraballi Creek, British Guiana; an ecological study of a limited area of tropical rain forest. *J. of Ecology* 21:350–385.

DeBano, L. F., P. H. Dunn, and C. E. Conrad. 1977. Fire's effect on physical and chemical properties of chaparral soils. In *Proc. Symp. Env. Cons. Fire and Fuel Mangmt. in Medit. Ecosyst.*, pp. 65–74. Washington, D.C.: U.S. Dept. of Agriculture Forest Service.

DeBano, L. F. and J. Letey, eds. 1969. *Proc. Symp. on Water-repellent Soils.* Riverside, CA: University of California.

de Bolos, O. 1982. Josias Braun-Blanquet. [In French.] *Vegetatio* 48:193–196.

De Candolle, A. L. 1855. *Geógraphie botanique raisoné.* Paris: V. Masson.

———. 1874. Constitution dans le régne végétale des groupes physiologiques applicables à la geógraphie botanique ancienne et moderne. *Arch. Sci. Phys. Nat.*

Deevey, E. S., Jr. 1947. Life tables for natural populations. *The Quarterly Review of Biology* 22:283–314.

———. 1970. Mineral cycles. *Scientific American* 223(3):148–160.

DeJong, T. M. 1975. A comparison of three diversity indices based on their components of richness and evenness. *Oikos* 26:222–227.

de la Cruz, A. A. and B. C. Gabriel. 1974. Caloric, elemental, and nutritive changes in decomposing *Juncus roemerianus* leaves. *Ecology* 55:882–886.

del Moral, R. 1983. Competition as a control mechanism in subalpine meadows. *American J. Botany* 70:232–245.

del Moral, R. and C. H. Muller. 1969. Fog drip: a mechanism of toxin transport from *Eucalyptus globulus. Bulletin of the Torrey Botanical Club* 96:467–475.

Detling, J. K. and M. I. Dyer. 1981. Evidence for potential plant growth regulators in grasshoppers. *Ecology* 62:485–488.

Devlin, R. M. and F. H. Witham. 1983. *Plant physiology.* Boston: PWS Publishers.

de Wit, C. D. 1960. *On competition.* Versl. Landbouwkl. Onderzoek. no. 66.8. Wageningen, The Netherlands: Centre for Agricultural Publications and Documentation.

———. 1961. Space relationships within populations of one or more species. In *Mechanisms in biological competition*, ed. F. L. Milthorpe, pp. 314–429. New York: Cambridge University Press.

Dickinson, R. E. 1982. Modeling climate changes due to carbon dioxide increases. In *Carbon dioxide review: 1982*, ed. W. C. Clark, pp. 103–142. New York: Oxford University Press.

Dietvortst, P., E. van der Maarel, and H. van der Putten. 1982. A new approach to the minimal area of a plant community. *Vegetatio* 50:77–91.

Dilworth, J. R. and J. F. Bell. 1978. *Variable probability sampling.* Oregon State Univ. Book Stores, Corvallis.

Dittrich, P. and W. Huber. 1974. Carbon dioxide metabolism in members of hte Chlamydospermae. In *Proceedings of the Third International Congress on Photosynthesis*, pp. 1573–1578. Amsterdam: Elsevier.

Dix, R. L. 1957. Sugar maple in forest succession at Washington, D.C. *Ecology* 38:663–665.

———. 1961. An application of the point-centered quarter method to the sampling of grassland vegetation. *J. Range Management* 14:63–69.

Doman, E. R. 1967. Prescribed burning and brush type conversion in California National Forests. In *Proc. Tall Timbers Fire Ecol. Conf.*, no. 7, pp. 225–243. Tallahassee, FL: Tall Timbers Research Station.

Donald, C. M. 1951. Competition among pasture plants. I. Intra-specific competition among annual pasture plants. *Australian J. of Agricultural Research* 2:355–376.

Dorf, I. 1960. Climate changes of past and present. *American Scientist* 48:341–364.

Douglas, L. A. and J. C. F. Tedrow. 1959. Organic matter decomposition rates in arctic soils. *Soil Science* 88:305–312.

Downes, R. W. and J. D. Hesketh. 1968. Enhanced photosynthesis at low O_2 concentrations: differential response of temperate and tropical grasses. *Planta* 78:79–84.

Downton, W. J. S. 1975. Checklist of C_4 species. *Photosynthetica* 9:96–105.

Downton, W. J. S. and E. B. Tregunna. 1968. Carbon dioxide compensation—its relation to photosynthetic carboxylation reaction, systematics of the Graminae, and leaf anatomy. *Canadian J. Botany* 46:207–215.

Drude, O. 1890. *Handbuch der Pflanzengeographie.* Stuttgart, Germany: Englemann.

———. 1896. *Deutschlands Pflanzengeographie.* Stuttgart: Engelhorn.

Drury, W. H., Jr. 1956. Bog flats and physiographic processes in the upper Kuskokwin River region, Alaska. *Gray Herbarium Contributions* 178:1–30.

Drury, W. H., Jr. and I. C. T. Nisbet. 1973. Succession. *J. of the Arnold Arboretum* 54:31–368.

Dunn, P. H. and L. F. DeBano. 1977. Fire's effect on biological and chemical properties of chaparral soils. In *Proc. Symp. Env. Cons. Fire and Fuel Mangmt. in Medit. Ecosyst.,* pp. 75–84. Washington, D.C.: U.S. Dept. of Agriculture Forest Service.

DuRietz, G. E. 1931. Life forms of terrestrial flowering plants. *Acta Phytogeographica Suecica,* vol. III. Uppsala.

Duvdevani, S. 1964. Dew in Israel and its effect on plants. *Soil Science* 98:14–21.

Dyer, M. I. 1975. The effects of red-winged blackbirds (*Agelaius phoeniceus* L.) on biomass production of corn grain (*Zea mayes* L.). *J. of Applied Ecology* 12:719–726.

Dyer, M. I. and U. G. Bokhari. 1976. Plant-animal interactions: studies on the effects of grasshopper grazing on blue grama grass. *Ecology* 57:762–772.

Edwards, C. A. and G. W. Heath. 1963. The role of soil animals in break-down of leaf material. In *Soil organisms,* ed. J. Docksen and J. Van Der Drift, pp. 76–84. Amsterdam: North Holland Publishing Co.

Edwards, P. J. and S. D. Wratten. 1985. Induced plant defenses against insect grazing: fact or artifact? *Oikos* 44:70–74.

Egerton, F. N. 1976. Ecological studies and observations before 1900. In *Evolution of issues, ideas and events in America: 1776-1976,* ed. B. J. Taylor, pp. 311–351. Norman, OK: University of Oklahoma Press.

Egler, F. E. 1954. Vegetation science concepts. I. Initial floristic composition, a factor in old-field vegetation development. *Vegetatio* 4:412–417.

Ehleringer, J. R. 1978. Implications of quantum yield differences on the distributions of C_3 and C_4 grasses. *Oecologia* 31:255–267.

———. 1983. Ecophysiology of *Amaranthus palmeri,* a Sonoran Desert summer ephemeral. *Oecologia* 57:107–112.

———. 1985. Annuals and perennials of warm deserts. In *Physiological ecology of North American plant communities,* ed. B. F. Chabot and H. A. Mooney, pp. 162–180. New York: Chapman and Hall.

Ehleringer, J. and O. Bjorkman. 1978. A comparison of photosynthetic characteristics of *Encelia* species possessing glabrous and pubescent leaves. *Plant Physiology* 62:185–190.

Ehleringer, J., O. Björkman, and H. A. Mooney. 1976. Leaf pubescence: effects on absorptance and photosynthesis in a desert shrub. *Science* 192:376–377.

Ehleringer, J. and C. S. Cook. 1980. Measurements of photosynthesis in the field: utility of the CO_2 depletion technique. *Plant, Cell, and Environment* 3:479–482.

Ehleringer, J. and I. Forseth. 1980. Solar tracking by plants. *Science* 210:1094–1098.

Ehrlich, P. R. and P. H. Raven. 1964. Butterflies and plants: a study in coevolution. *Evolution* 18:568–608.

Eickmeier, W. G. 1978. Photosynthetic pathway distributions along an aridity gradient in Big Bend National Park, and implications for enhanced resource partitioning. *Photosynthetica* 12:290–297.

Ellenberg, H. 1958. Bodenreaktion (einschlieblich Kaltfrage). In *Handbuch der Pflanzenphysiologie*, vol. 4, ed. W. Ruhland, pp. 638–708. Berlin: Springer-Verlag.

England, C. B. and E. H. Lesesne. 1962. Evapotranspiration research in western North Carolina. *Agricultural Engineering* 43:526–528.

Engler, A. 1903. Untersuchungen uber das Wurzelwachstum der Holzarten. *Mitteilungen der schweizer Zentralanstalt fur das forstliche Versuchswesen* 7:247–317.

Epstein, E. 1972. *Mineral nutrition of plants: principles and perspectives.* New York: Wiley.

Esau, K. 1965. *Plant anatomy.* New York: Wiley.

Eshel, Y. and Y. Waisel. 1965. The salt relations of *Prosopis farcta* (Banks et Sol.) Eig. *Israel J. of Botany* 14:50–51 (abstract).

Etherington, J. R. 1975. *Environment and plant ecology.* New York: Wiley.

Evenari, M., L. Shanan, and N. H. Tadmor. 1971. *The Negev, the challenge of a desert.* Cambridge, MA: Harvard University Press.

Evans, G. C. 1976. A sack of uncut diamonds. *J. of Ecology* 64:1–39.

Evans, G. C. and D. E. Coombe. 1959. Hemispherical and woodland canopy photography and the light climate. *J. Ecology* 47:103–113.

Evans, L. S. 1984. Botanic aspects of acidic precipitation. *Botanical Review* 50:449–490.

Eyre, S. R. 1963. *Vegetation and soils.* Chicago: Aldine.

Faegri, K. and L. van der Pijl. 1971. *The principles of pollination ecology.* New York: Pergamon Press.

Farquhar, G. D. 1983. On the nature of carbon isotope discrimination in C_4 species. *Australian J. of Plant Physiology* 10:205–226.

Farquhar, G. D., M. C. Ball, S. von Caemmerer, and Z. Roksandic. 1982. Effect of salinity and humidity on the $\delta\,^{13}C$ value of halophytes—evidence for diffusional isotope fractionalization determined by the ratio of intercellular/atmospheric partial pressure of CO_2 under different environmental conditions. *Oecologia* 52:121–124.

Ferguson, C. W. 1968. Bristlecone pine: science and esthetics. *Science* 159:839–846.

Fernandez, O. A. and M. M. Caldwell. 1975. Phenology and dynamics of root growth of three cool semi-desert shrubs under field conditions. *J. of Ecology* 63:703–714.

Finck, A. 1969. *Pflanzenernahrung in Stichworten.* Kiel, Germany: F. Hirt.

Flint, R. F. 1957. *Glacial and pleistocene geology.* New York: Wiley.

Flock, J. A. W. 1978. Lichen-Bryophyte distributions along a snow-cover-soil-moisture gradient, Niwot Ridge, Colorado. *Arctic and Alpine Research* 10:31–47.

Flowers, T. J., P. F. Troke, and A. R. Yeo. 1977. The mechanism of salt tolerance in halophytes. *Annual Review of Plant Physiology* 28:89–121.

Fonda, R. W. 1974. Forest succession in relation to river terrace development in Olympic National Park, Washington. *Ecology* 55:927–942.

Fonteyn, P. J. and B. E. Mahall. 1978. Competition among desert perennials. *Nature* 275:544–545.

———. 1981. An experimental analysis of structure in a desert plant community. *J. Ecology* 69:883–896.

Forcier, L. K. 1975. Reproductive strategies and the co-occurrence of climax tree species. *Science* 189:808–810.

Ford, J. M. and J. E. Monroe. 1971. *Living systems.* San Francisco, CA: Canfield Press.

Forman, R. T. T. 1975. Canopy lichens with blue-green algae: a nitrogen source in a Columbian rain forest. *Ecology* 56:1176–1184.

Forseth, I. and J. Ehleringer. 1982. Ecophysiology of two solar tracking desert winter annuals. *Oecologia* 54:41–49.

Fortescue, J. A. C. and G. G. Marten. 1970. Analysis of temperate forest ecosystems. *Ecological studies,* vol. 1, ed. D. E. Reichle, pp. 173–198. New York: Springer.

Foster, N. W. and I. K. Morrison. 1976. Distribution and cycling of nutrients in a natural *Pinus banksiana* ecosystem. *Ecology* 57:110–120.

Fowells, H. A., ed. 1965. *Silvics of forest trees of the United States.* Agriculture Handbook no. 271. Washington, D.C.: U.S. Dept. of Agriculture Forest Service.

Franklin, J. F. and C. T. Dyrness. 1969. *Natural vegetation of Oregon and Washington.* U.S. Dept. of Agriculture Forest Service General Technical Report PNW-8. Portland, OR: U.S. Dept. of Agriculture Forest Service.

Freas, K. E. and P. R. Kemp. 1983. Some relationships between environmental reliability and seed dormancy in desert annuals. *J. of Ecology* 71:221–217.

Fryer, J. H. and F. T. Ledig. 1972. Microevolution of the photosynthetic temperature optimum in relation to an elevational complex gradient. *Canadian J. of Botany* 50:1231–1235.

Gadgil, M. and W. H. Bossert. 1970. Life historical consequences of natural selection. *American Naturalist* 104:1–24.

Gadgil, M. and O. T. Solbrig. 1972. The concept of *r*- and *K*-selection: evidence from wildflowers and some theoretical considerations. *The American Naturalist* 106:14–31.

Gambell, A. W. and D. W. Fisher. 1966. *Chemical composition of rainfall, eastern North Carolina and southeastern Virginia.* U.S. Geological Survey Water Supply Paper 1535-K.

Garrison, G. A. 1949. Uses and modifications for the "moosehorn" crown closure estimator. *J. of Forestry* 47:733–735.

Garth, R. E. 1964. The ecology of Spanish moss (*Tillandsia usneoides*). *Ecology* 45:470–481.

Gates, D. M. 1962. *Energy exchange in the biosphere.* New York: Harper and Row.

———. 1965a. Heat transfer in plants. *Scientific American* 213:76–83.

———. 1965b. Radiant energy, its receipt and disposal. *Meteorological Monographs* 6:1–26.

———. 1970. Physical and physiological properties of plants. In *Remote sensing with special reference to agriculture and forestry,* pp. 224–252. Washington, DC: National Academy of Sciences.

———. 1972. *Man and his environment: climate.* New York: Harper and Row.

———. 1980. *Biophysical ecology.* New York: Springer-Verlag.

Gates, D. M., H. J. Keegan, J. C. Schleter, and V. R. Weidner. 1965. Spectral properties of plants. *Applied Optics* 4:11–20.

Gauch, H. G., Jr. 1982. *Multivariate analysis in community ecology.* New York: Cambridge University Press.

Gauch, H. G., Jr. and R. H. Whittaker. 1981. Hierarchical classification of community data. *J. Ecology* 69:537–557.

Gause, G. F. 1934. *The struggle for existence.* Baltimore, MD: Williams and Wilkins.

Gausman, H. W., D. E. Escobar, and E. B. Knipling. 1977. Relation of *Peperomia obtusifolia*'s anomalous leaf reflectance to its leaf anatomy. *Photogrammetric Engineering and Remote Sensing* 43:487–491.

Gausman, H. W., E. E. Escobar, J. H. Everitt, A. J. Richardson, and R. R. Rodiguez. 1978. Distinguishing succulent plants from crop and woody plants. *Photogrammetric Engineering and Remote Sensing* 44:487–491.

Gill, A. M. 1977. Plant traits adaptive to fires in Mediterranean land ecosystems. In *Proc. Symp. Env. Cons. Fire and Fuel Mangmt. in Medit. Ecosyst.*, pp. 17–26. Washington, D.C.: U.S. Dept. of Agriculture Forest Service.

Gilmour, J. S. L. and J. Heslop-Harrison. 1954. The -deme terminology and the units of micro-evolutionary change. *Genetica* 27:146–161.

Gleason, H. A. 1926. The individualistic concept of the plant association. *Bulletin of the Torrey Botanical Club* 53:1–20.

———. 1953. Dr. H. A. Gleason, distinguished ecologist. *Bulletin of the Ecological Society of America* 34(2):40–42.

Goel, N. S. and R. L. Thompson. 1984. Inversion of vegetation canopy reflectance models for estimating agronomic variables. *Remote Sensing of the Environment* 16:69–85.

———. 1985. Optimal solar/viewing geometry for an accurate estimation of leaf area index and leaf angle distribution from bidirectional canopy reflectance data. *International J. Remote Sensing* 6:1493–1520.

Goetz, A. F. H., B. N. Rock, and L. C. Rowan. 1983. Remote sensing for exploration: an overview. *Economic Geology* 78:573–590.

Goetz, A. F. H., G. Vane, J. E. Solomon, and B. N. Rock. 1985*a*. Imaging spectrometry for earth remote sensing. *Science* 228:1147–1153.

Goetz, A. F. H., J. B. Wellman, and W. L. Barnes. 1985*b*. Optimal remote sensing of the earth. *Proceedings Institute of Electrical and Electronics Engineers* 73:950–969.

Goldberg, D. E. 1982. The distribution of evergreen and deciduous trees relative to soil type: an example from the Sierra Madre, Mexico and a general model. *Ecology* 63:942–951.

Goldberg, D. E. and P. A. Werner. 1983. Equivalence of competitors in plant communities: a null hypothesis and a field experimental approach. *American J. Botany* 70:1098–1104.

Goldsmith, F. B. and C. M. Harrison. 1976. Description and analysis of vegetation. In *Methods in plant ecology,* ed. S. B. Chapman, pp. 85–155. New York: Halsted Press.

Goldsworthy, A. 1976. *Photorespiration.* Carolina Biology Readers, no. 80. Burlington, NC: Carolina Biological Suppliers.

Golley, F. B. 1961. Energy values of ecological materials. *Ecology* 42:581–584.

———. ed. 1977. *Ecological succession*. Stroudsburg, PA: Dowden, Hutchinson, and Ross.

Good, R. E. 1931. A theory of plant geography. *The New Phytologist* 30:149–203.

———. 1953. *The geography of the flowering plants*. 2nd ed. New York: Longmans, Green, and Co.

Good, R. E. and N. F. Good. 1975. Growth characteristics of two populations of *Pinus ridgida* Mill, from the pine barrens of New Jersey. *Ecology* 56:1215–1220.

Goodall, D. W. 1953. Objective methods for the classification of vegetation. I. The use of positive interspecific correlation. *Australian J. of Botany* 1:39–63.

———. 1957. Some considerations in the use of point quadrat methods for the analysis of vegetation. *Australian J. of Biological Science* 5:1–41.

———. 1973. Sample similarity and species correlation. In *Handbook of vegetation science. V. Ordination and classification of communities*, ed. R. H. Whittaker, pp. 107–156. The Hague: Junk.

Goodland, R. J. 1975. The tropical origin of ecology: Eugen Warming's jubilee. *Oikos* 26:240–245.

Goodman, L. A. 1969. The analysis of population growth when the birth and death rates depend upon several factors. *Biometrics* 25:659–681.

Gorham, E. 1979. Shoot height, weight, and standing crop in relation to density of mono-specific plant stands. *Nature* 279:148–150.

Goss, R. W. 1960. Mycorrhizae of ponderosa pine in Nebraska grassland soils. *Nebraska Agricultural Experiment Station Research Bulletin* 192:1–47.

Gosz, J. R., G. E. Likens, and F. H. Bormann. 1973. Nutrient release from decomposing leaf and branch litter in the Hubbard Brook Forest, New Hampshire. *Ecological Monographs* 43:173–191.

Gouyon, P. H., P. Fort, and G. Caraux. 1983. Selection of seedlings of *Thymus vulgaris* by grazing slugs. *J. Ecology* 71:299–306.

Graham, B. F., Jr. and F. H. Bormann. 1966. Natural root grafts. *Botanical Review* 32:255–292.

Gray, A. 1889. *Scientific papers,* selected by C. S. Sargent. New York: Houghton Mifflin.

Gray, J. T. 1983. Nutrient use by evergreen and deciduous shrubs in southern California I. *J. Ecology* 71:21–41.

Gray, J. T. and W. H. Schlesinger. 1983. Nutrient use by evergreen and deciduous shrubs in southern California II. *J. Ecology* 71:43–56.

Green, D. S. 1983. The efficacy of dispersal in relation to safe site density. *Oecologia* 56:356–358.

Greenland, D., N. Caine, and O. Pollak. 1984. The summer water budget and its importance in the alpine tundra of Colorado. *Physical Geography* 5:221–239.

Greenway, H. and R. Munns. 1980. Mechanisms of salt tolerance in non-halophytes. *Annual Review of Plant Physiology* 31:149–190.

Greenway, H. and D. A. Thomas. 1965. Plant response to saline substrates. V. Chloride regulation in the individual organs of *Hordeum vulgare* during treatment with sodium chloride. *Australian J. of Biological Science* 18:505–524.

Gregor, J. W. 1946. Ecotypic differentiation. *The New Phytologist* 45:254–270.

Greig-Smith, P. 1964. *Quantitative plant ecology.* London: Butterworths.

Greig-Smith, P. and K. A. Kershaw. 1958. The significance of pattern in vegetation. *Vegetatio* 8:189–192.

Grier, C. C. and S. W. Running. 1977. Leaf area of mature northwestern coniferous forests: relation to site water balance. *Ecology* 58:893–899.

Griffin, J. R. 1973. Xylem sap tension in three woodland oaks of central California. *Ecology* 54:152–159.

Griffin, J. R. and W. B. Critchfield. 1972. *The distribution of forest trees in California.* U.S. Dept. of Agriculture Forest Service Research Paper PSW-82/1972. Washington, D.C.: U.S. Dept. of Agriculture Forest Service.

Griffin, K. O. 1975. Vegetation studies and modern pollen spectra from the Red Lake peatland, northern Minnesota. *Ecology* 56:531–546.

Grime, J. P. 1977. Evidence for the existence of three primary strategies in plants and its relevance to ecological and evolutionary theory. *American Naturalist* 111:1169–1194.

Grime, J. P. 1979. *Plant strategies and vegetation processes.* New York: Wiley.

———. 1982. The concept of strategies: use and abuse. *J. of Ecology* 70:863–865.

Grime, J. P. and K. Thompson. 1976. An apparatus for measurement of the effect of amplitude of temperature fluctuation upon the germination of seeds. *Annals of Botany* 40:795–799.

Grisebach, A. H. R. 1872. *Die Vegetation der Erde nach ihrer klimatischen Anordnung.* Leipzig, Germany: Engelmann.

Grosenbaugh, L. R. 1952. Plotless timber estimates—new, fast, easy. *J. of Forestry* 50:32–37.

Gross, K. L. 1980. Colonization by *Verbascum thapsus* (mullein) of an old-field in Michigan: Experiments on the effects of vegetation. *J. of Ecology* 68:919–927.

Grubb, P. J., H. E. Green, and R. C. J. Merrifield. 1969. The ecology of chalk heath: its relevance to the calcicole-calcifuge and soil acidification problems. *J. of Ecology* 57:175–212.

Gupta, S. R. and J. S. Singh. 1977. Decomposition of litter in a tropical grassland. *Pedobiologia* 17:330–333.

Gutterman, Y. 1972. Delayed seed dispersal and rapid germination as survival mechanisms of the desert plant *Blepharis persica* (Burm.) Kuntze. *Oecologia* 10:145–149.

Gutterman, Y., T. H. Thomas, and W. Heydecker. 1975. Effect on the progeny of applying different day length and hormone treatments to parent plants of *Lactuca scariola.* *Physiologia Plantarum* 34:30–38.

Hacskaylo, E., ed. 1971. *Proc. of the First North American Conf. on Mycorrhizae, 1969.* U.S. Dept. of Agriculture Forest Service Misc. Publ. 1189. Washington, D.C.: U.S. Dept. of Agriculture Forest Service.

Hadley, E. B. 1970. Net productivity and burning responses of native eastern North Dakota prairie communities. *American Midland Naturalist* 84:121–135.

Hadley, J. L. and W. K. Smith. 1983. Influence of wind exposure on needle desiccation and mortality for timberline conifers in Wyoming, USA. *Arctic and Alpine Research* 15:127–135.

Haeckel, E. H. P. A. 1866. *Generelle Morphologie der Organismen.* Berlin: Reimer.

Hagan, R. M., H. R. Haise, and T. W. Edminster, eds. 1967. *Irrigation of agricultural lands.* Madison, WI: American Society of Agronomy.

Hall, A. E. and M. R. Kaufmann. 1975. The regulation of water transport in the soil-plant-atmosphere continuum. In *Perspectives in biophysical ecology,* ed. D. M. Gates and R. B. Schmerl, pp. 187–202. Berlin: Springer-Verlag.

Hall, A. E., E. D. Schulze, and O. L. Lange. 1976. Current perspectives of steady-state stomatal responses to environment. In *Water and plant life: problems and modern approaches,* ed. O. L. Lange, L. Kappen, and E. D. Schulze, pp. 169–188. Berlin: Springer-Verlag.

Halligan, J. P. 1973. Bare areas associated with shrub stands in grassland: the case of *Artemesia californica. BioScience* 23:429–432.

Halm, G. J. 1972. *The phosphorus cycle in a grassland ecosystem.* Ph.D. dissertation. Saskatoon, Canada: University of Saskatchewan.

Hanes, T. L. 1977. California chaparral. In *Terrestrial vegetation of California,* ed. M. G. Barbour and J. Major, pp. 417–469. New York: Wiley-Interscience.

Hannapel, R. J., W. H. Fuller, S. Bosma, and J. S. Bullock. Phosphorus movement in a calcareous soil: I. Predominance of organic forms of phosphorus in phosphorus movement. *Soil Science* 97:350–357.

Hanscom, Z., III and P. Ting. 1978. Responses of succulents to plant water stress. *Plant Physiology* 61:327–330.

Hanson, H. C. 1917. Leaf-structure as related to environment. *American J. of Botany* 4:553–560.

———. 1962. *Dictionary of ecology.* New York: Philosophical Library.

Harley, J. L. 1969. *The biology of mycorrhizae.* 2nd ed. London: Leonard Hill.

Harper, J. L. 1961. Approaches to the study of plant competition. *Symposium of the Society for Experimental Biology* 15:1–39.

———. 1964. The individual in the population. *J. of Ecology* 52 (Supplement):149–158.

———. 1967. A Darwinian approach to plant ecology. *J. of Ecology* 55:247–270.

———. 1977. *Population biology of plants.* New York: Academic Press.

Harper, J. L., and A. D. Bell. 1979. The population dynamics of growth form in organisms with modular construction. In *Population dynamics,* ed. R. M. Anderson, B. C. Turner, and L. R. Taylor, pp. 29–52. Oxford, England: Blackwell.

Harper, J. L. and J. Ogden. 1970. The reproductive strategy of higher plants. I. The concept of strategy with special reference to *Senecio vulgaris. J. of Ecology* 58:681–698.

Harper, J. L. and J. White. 1974. The demography of plants. *Annual Review of Ecology and Systematics* 5:419–463.

Harris, G. A. 1967. Some competitive relationships between *Agropyron spictum* and *Bromus tectorum. Ecological Monographs* 37:89–111.

Harris, P. 1974. A possible explanation of plant yield increases following insect damage. *Agro-Ecosystems* 1:219–225.

Harrison, A. T., E. Small, and H. A. Mooney. 1971. Drought relationships and distribution of two Mediterranean-climate California plant communities. *Ecology* 52:869–875.

Harshberger, J. W. 1911. *Photogeographic survey of North America.* New York: Stechert.

———. 1911. An hydrometric investigation of the influence of sea water on the distribution of salt water and estuarine plants. *Proc. of the American Philosophical Society* 50:457–497.

———. 1916. *The vegetation of the New Jersey pine-barrens.* Philadelphia, PA: Sower.

Hartesveldt, R. J. 1964. The fire-ecology of the giant sequoias. *Natural History* 73:12–19.

Hartesveldt, R. J. and H. T. Harvey. 1967. The fire ecology of sequoia regeneration. In *Proc. Tall Timbers Fire Ecol. Conf.,* no. 7, pp. 65–78. Tallahassee, FL: Tall Timbers Research Station.

Hartesveldt, R. J., H. T. Harvey, H. S. Shellhammer, and R. E. Stecker. 1975. *The giant sequoia of the Sierra Nevada.* U.S. Dept. of the Interior National Park Service Publ. no. 120. Washington, D.C.: U.S. Dept. of the Interior, National Park Service.

Hartsock, T. L. and P. S. Nobel. 1976. Watering converts a CAM plant to daytime CO_2 uptake. *Nature* 262:574–576.

Harvey, H. T., H. S. Shellhammer, and R. E. Stecker. 1980. *Giant sequoia ecology: fire and reproduction.* U.S. Dept. of the Interior National Park Service Sci. Mono. 12. Washington, D.C.: U.S. Dept. of the Interior, National Park Service.

Hastings, J. R. and R. M. Turner. 1965. *The changing mile.* Tucson, AZ: University of Arizona Press.

Hatch, A. B. 1937. The physical basis of mycotrophy in *Pinus. Black Rock Forest Bulletin 6.*

Hatch, M. D. and C. R. Slack. 1966. Photosynthesis by sugar-cane leaves: a new carboxylation reaction and the pathway of sugar formation. *Biochemical J.* 101:103–111.

Hatch, M. D. and C. B. Osmond. 1976. Compartmentation and transport in C_4 photosynthesis. In *Transport in plants. III. Intracellular interactions and transport processes,* ed. C. R. Stocking and U. Heber, pp. 144–184. Berlin-Heidelberg: Springer-Verlag.

Heine, R. W. 1971. Hydraulic conductivity in trees. *J. of Experimental Botany* 22:503–511.

Heller, R. C., and J. J. Ulliman. 1983. Forest resource assessments. In *Manual remote sensing,* ed. R. N. Colwell, pp. 2229–2324. Falls Church, VA: American Society of Photogrammetry.

Hellkvist, J., G. P. Richards, and P. G. Jarvis. 1974. Vertical gradients of water potential and tissue water relations in Sitka spruce trees measured with the pressure chamber. *J. of Applied Ecology* 11:637–667.

Hellmers, H. 1964. An evaluation of the photosynthetic efficiency of forests. *Quarterly Review of Biology* 39:249–257.

———. 1966. Growth response of redwood seedlings to thermoperiodism. *Forest Science* 12:276–283.

Henderson, G. S., W. F. Harris, D. E. Todd, Jr., and T. Grizzard. 1977. Quantity and chemistry of throughfall as influenced by forest type and season. *J. of Ecology* 65:365–374.

Heslop-Harrison, J. 1964. Forty years of genecology. *Advances in Ecological Research* 2:159–247.

Heyward, F. 1939. The relation of fire to stand composition of longleaf pine forests. *Ecology* 20:287–304.

Heywood, V. H. 1967. *Plant taxonomy.* New York: St. Martin's Press.

Hickman, J. C. 1975. Environmental unpredictability and plastic energy allocation strategies in the annual *Polygonum cascadense. J. of Ecology* 63:689–701.

———. 1977. Energy allocation and niche differentiation in four coexisting species of *Polygonum* in western North America. *J. of Ecology* 65:317–326.

Hickman, J. C. and L. F. Pitelka. 1975. Dry weight indicates energy allocation in ecological strategy analyses of plants. *Oecologia* 21:117–121.

Higgins, K. F. 1984. Lightning fires in North Dakota grasslands and in pine savanna lands of South Dakota. *J. Range Management* 37:100–103.

Highkin, H. R. 1958. Transmission of phenotypic variability within a pure line. *Nature* 182:1460.

Hill, M. O. 1973. Diversity and evenness: a unifying notation and its consequences. *Ecology* 54:427–432.

————. 1979. DECORANA—a FORTRAN program for detrended correspondence analysis and reciprocal averaging. Ithaca, NY: Cornell University.

Hill, M. O. and H. G. Gauch, Jr. 1980. Detrended correspondence analysis: an improved ordination technique. *Vegetatio* 42:47–58.

Hils, M. H. and J. L. Vankat. 1982. Species removals from a first-year old field plant community. *Ecology* 63:705–711.

Holdridge, L. R. 1967. *Life zone ecology*, rev. ed. San José, Costa Rica: Tropical Science Center.

Hopkins, A. D. 1938. *Bioclimatics: a science of life and climatic relations*. U.S. Dept. of Agriculture Misc. Publ. 280. Washington, D.C.: U.S. Dept. of Agriculture.

Hopkins, B. 1957. Pattern in the plant community. *J. of Ecology* 45:451–463.

Horn, H. S. 1971. *The adaptive geometry of trees*. Princeton, NJ: Princeton University Press.

————. 1974. The ecology of secondary succession. *Annual Review of Ecology and Systematics* 5:25–37.

————. 1975. Forest succession. *Scientific American* 232(5):90–98.

Horton, J. S. 1977. The development and perpetuation of the permanent Tamarisk type in the phreatophyte zone of the southwest. In *Importance, preservation, and management of riparian habitat: symp. proc.*, pp. 124–127. U.S. Dept. of Agriculture Forest Service General Technical Report RM-43. Washington, D.C.: U.S. Dept. of Agriculture Forest Service.

Horton J. S. and C. J. Campbell. 1974. Management of phreatophyte and riparian vegetation maximum multiple-use values. USDA Forest Service Research Paper Rm-117. 23 pp.

Houkioja, E. and S. Neuvonen. 1985. Induced long-term resistance of birch foliage against defoliators: defensive or incidental? *Ecology* 66:1303–1308.

Howe, H. F. and J. Smallwood. 1982. Ecology of seed dispersal. *Annual Review of Ecology Systematics* 13:201–228.

Hrapko, J. O. and G. A. La Roi. 1978. The alpine tundra vegetation of Signal Mountain, Jasper National Park. *Canadian J. of Botany* 56:309–332.

Hsiao, T. C. 1973. Plant responses to water stress. *Annual Review of Plant Physiology* 24:519–570.

————. 1976. Stomatal ion transport. In *Encyclopaedia of plant physiology*, vol. 2, part B, ed. A. Pirson and M. H. Zimmerman, pp. 195–217. Berlin: Springer-Verlag.

Hubbell, S. P. 1980. Seed predation and the coexistence of tree species in tropical forests. *Oikos* 35:214–229.

Huey, R. B. and E. R. Pianka. 1977. Patterns of niche overlap among broadly sympatric versus narrowly sympatric Kalahari lizards (Scinidae: Maybuya). *Ecology* 58:119–128.

Hull, R. J. and O. A. Leonard. 1964. Physiological aspects of parasitism in mistletoes (*Arceuthobium* and *Phoradendron*), parts I and II. *Plant Physiology* 39:996–1017.

Humboldt, A. von and A. Bonpland. 1807. *Essai sur la geographie des plantes*. Paris.

————. 1807–1834. *Voyage aux régions equinoxiales du nouveau continent*. 30 vols.

Hume, L. and P. B. Cavers. 1981. A methodological problem in genecology. Seeds versus clones as source material for uniform gardens. *Canadian J. Botany* 59:763–768.

Humphrey, R. R. 1974. Fire in the deserts and desert grassland of North America. In *Fire and ecosystems*, ed. T. T. Kozlowski and C. E. Ahlgren, ch. 11. New York: Academic Press.

Hunt, W. H. 1977. A simulation model for decomposition in grasslands. *Ecology* 58: 469–484.

Hurlbert, S. H. 1971. The nonconcept of species diversity: a critique and alternative parameters. *Ecology* 52:577–586.

Hutchings, M. J. and J. P. Barkham. 1976. An investigation of shoot interactions in *Mercurialis perennis* L., a rhizomatous perennial herb. *J. of Ecology* 64:723–743.

Hutchison, B. A. and D. R. Matt. 1977. The distribution of solar radiation within a deciduous forest. *Ecological Monographs* 47:185–207.

Idso, S. B. 1984. An empirical evaluation of Earth's surface air temperature response to radiative forcing, including feedback, as applied to the CO_2-climate problem. *Archives for Meteorology, Geophysics, and Bioclimatology* 34:1–19.

Ingram, M. 1957. Microorganisms resisting high concentrations of sugars and salts. In *Seventh Symposium of the Society for General Microbiology*, pp. 90–133. Cambridge, England: Cambridge University Press.

Ivimey-Cook, R. B. and M. C. F. Proctor. 1966. The application of association-analysis to phytosociology. *J. of Ecology* 54:179–192.

Iwaki, H., M. Monsi, and B. Midorikawa. 1966. Dry matter production of some herb communities in Japan. *The Eleventh Pacific Science Congress*, pp. 1–15. Tokyo, Japan: Science Council of Japan.

Jackson, J. R. and R. W. Willemsen. 1976. Allelopathy in the first stages of secondary succession on the piedmont of New Jersey. *American J. of Botany* 63:1015–1023.

Jackson, M. T. 1966. Effects of microclimate on spring flowering phenology. *Ecology* 47:407–415.

Jackson, R. D. 1985. Evaluating evapotranspiration at local and regional scales. *Proceedings IEEE* 73:1086–1096.

Jackson, R. D., and C. E. Ezra. 1985. Spectral response of cotton to suddenly induced water stress. *International Journal of Remote Sensing* 6:177–185.

Jackson, R. D., P. N. Slater, and P. J. Pinter, Jr. 1983. Discrimination of growth and water stress in wheat by various vegetation indices through clear and turbid atmospheres. *Remote Sensing of the Environment* 13:187–208.

Janzen, D. H. 1970. Herbivores and the number of tree species in tropical forests. *American Naturalist* 104:501–528.

———. 1973. Community structure of secondary compounds in plants. *Pure and Applied Chemistry* 34:529–538.

———. 1974. Tropical black water rivers, animals, and mast fruiting by the Dipterocarpaceae. *Biotropica* 6:69–103.

———. 1975. *Ecology of plants in the tropics*. London: Edward Arnold.

———. 1976. Why bamboos wait so long to flower. *Annual Review of Ecology and Systematics* 7:347–391.

Jarvis, P. G. and T. A. Mansfield. 1981. *Stomatal physiology*. Cambridge, England: Cambridge University Press.

Jeffree, E. P. 1960. Some long term means from the phenological reports (1891–1948) of the Royal Meteorological Society. *Quarterly J. of the Royal Meteorological Society* 86:95–103.

Jennings, D. H. 1968. Halophytes, succulence, and sodium in plants--a unified theory. *New Phytologist* 67:899–911.

Jenny, H. 1941. *The factors of soil formation*. New York: McGraw-Hill.

Jenny, H., R. J. Arkley, and A. M. Schultz. 1969. The pygmy forest-podsol ecosystem and its dune associates of the Mendocino Coast. *Madroño* 20:60–74.

Jenny, H., S. P. Gessel, and F. T. Bingham. 1949. Comparative study of decomposition rates of organic matter in temperate and tropical regions. *Soil Science* 68:419–432.

Jensen, W. A. and F. B. Salisbury. 1972. *Botany: an ecological approach.* Belmont, CA: Wadsworth.

Johnson, D. A. and M. M. Caldwell. 1975. Gas exchange of four arctic and alpine tundra plant species in relation to atmospheric and soil moisture stress. *Oecologia* 21:93–108.

Johnson, D. W., and D. W. Cole. 1980. Anion mobility in soils: relevance to nutrient transport from terrestrial ecosystems. *Environment International* 3:79–90.

Johnson, H. B. 1976. Vegetation and plant communities of southern California deserts— a functional view. In *Symposium proc., plant communities of southern California,* Special Publ. no. 2 of the California Native Plant Society, ed. J. Latting, pp. 125–164. Berkeley, CA: California Native Plant Society.

Johnson, N. M. 1971. Mineral equilibria in ecosystem geochemistry. *Ecology* 52:529–531.

Johnson, P. L., ed. 1977. *An ecosystem paradigm for ecology.* Oak Ridge, TN: Oak Ridge Associated Universities.

Johnson, P. L. and W. D. Billings. 1962. The alpine vegetation of the Beartooth Plateau in relation to cryopedogenic processes and patterns. *Ecological Monographs* 32:105–135.

Jones, C. E. and S. L. Buchmann. 1974. Ultra-violet floral patterns as functional orientation cues in Hymenopteran pollination systems. *Animal Behavior* 22:481–485.

Jones, E. W. 1945. The structure and reproduction of the virgin forests of the north temperate zone. *The New Phytologist* 44:130–148.

Jones, M. M. and N. C. Turner. 1978. Osmotic adjustment in leaves of sorghum in response to water deficits. *Plant Physiology* 61:122–126.

Jordan, C. F. 1971. A world pattern in plant energetics. *American Scientist* 59:425–433.

———. 1982. The nutrient balance of an Amazonian rain forest. *Ecology* 63:647–654.

Jordan, D. F. and J. R. Kline. 1972. Mineral cycling: some basic concepts and their application in a tropical rain forest. *Annual Review of Ecology and Systematics* 3:33–50.

Justice, C. O., J. R. G. Townshend, B. N. Holben, and C. J. Tucker. 1985. Analysis of the phenology of global vegetation using meteorological satellite data. *International J. Remote Sensing* 6:1271–1318.

Kalle, K. 1958. In *Handbuch der Pflanzenphysiologie,* Bd. IV, ed. W. Ruhland, pp. 170–178. Berlin-Gottingen-Heidelberg: Springer-Verlag.

Kaminsky, R. 1981. The microbial origin of the allelopathic potential of *Adenostoma fasciculatum. Ecological Monographs* 51:365–382.

Kanemasu, E. T., G. W. Thurtell, and C. B. Tanner. 1969. Design, calibration and field use of a stomatal diffusion porometer. *Plant Physiology* 44:881–885.

Kaplanis, J. N., M. J. Thompson, W. E. Robbins, and B. M. Bryce. 1967. Insect hormones: alpha ecdysone and 20-hydroxyecdysone in bracken fern. *Science* 157:1436–1438.

Keeley, J. E. 1977. Fire-dependent strategies in *Arctostaphylos* and *Ceanothus.* In *Proc. Symp. Env. Cons. Fire and Fuel Mangmt. in Medit. Ecosyst.,* pp. 391–396. Washington, D.C.: U.S. Dept. of Agriculture Forest Service.

Keeley, J. E., B. A. Morton, A. Pedrosa and P. Trotter. 1985. Role of allelopathy, heat and charred wood in the germination of chaparral herbs and suffrutescents. *J. of Ecology* 73:445–458.

Keeley, S. 1977. The relationship of precipitation to post-fire succession in southern California chaparral. In *Proc. Symp. Env. Cons. Fire and Fuel Mangmt. in Medit. Ecosyst.*, pp. 387–390. Washington, D.C.: U.S. Dept. of Agriculture Forest Service.

Keeling, C. D., W. G. Mook, and P. P. Tans. 1979. Recent trends in the $^{13}C/^{12}C$ ratio of atmospheric carbon dioxide. *Nature* (London) 277:121–123.

Keever, C. 1950. Causes of succession on old fields of the piedmont, North Carolina. *Ecological Monographs* 20:229–250.

———. 1983. Retrospective view of old-field succession after 35 years. *American Midland Naturalist* 110:397–404.

Kelly, G. J., E. Latzko, and M. Gibbs. 1976. Regulatory aspects of photosynthetic carbon metabolism. *Annual Review of Plant Physiology* 27:181–205.

Kelly, J. M. 1975. Dynamics of root biomass in two eastern Tennessee old-field communities. *American Midland Naturalist* 94:54–61.

Kenoyer, L. A. 1927. A study of Raunkiaer's Law of Frequency. *Ecology* 8:341–349.

Kerner, A. 1863. *The plant life of the Danube Basin.* Translated by H. S. Conrad. Republished. Ames, IA: Iowa State College Press. 1951.

———. 1895. *The natural history of plants, their forms, growth, reproduction and distribution.* Translated by F. W. Oliver. London: Blackie.

Kershaw, K. A. 1973. *Quantitative and dynamic plant ecology.* 2nd ed. New York: American Elsevier.

Kilgore, B. M. 1972. The role of fire in giant sequoia-mixed conifer forest. In *Proc. American Association for the Advancement of Science Symp.*, 1971, pp. 1–38.

Kilgore, B. M. and H. H. Biswell. 1971. Seedling germination following fire in a giant sequoia forest. *California Agriculture Magazine* 25:8–10.

Kira, T. 1975. Primary production of forests. In *Photosynthesis and productivity in different environments*, ed. J. P. Cooper, pp. 5–40. Cambridge, England: Cambridge University Press.

Kira, T., H. Ogawa, and K. Shinozaki. 1953. Intraspecific competition among higher plants. I. Competition, density-yield interrelationships in regularly dispersed populations. *J. Institute Polytechnic*, Osaka City University D. 4:1–16.

Kira, T., H. Ogawa, K. Yoda, and K. Ogina. 1967. Comparative ecological studies on three main types of forest vegetation in Thailand. IV. Dry matter production, with special reference to the Khao Chong rain forest. *Nature and Life in Southeast Asia* 5:149–174.

Kirkpatrick, Mark. 1984. Demographic models based on size, not age, for organisms with indeterminate growth. *Ecology* 65:1874–1884.

Kittredge, J. 1944. Estimation of the amount of foliage on trees and stands. *J. of Forestry* 42:905–912.

Kluge, M. 1974. Metabolism of carbohydrates and organic acids. In *Progress in botany*, ed. H. Ellenberg et al., vol. 36, pp. 90–98. Berlin: Springer-Verlag.

Knapp, R. 1957. Über den Einfluss der Temperatur während der Keimung auf die spatere Entweicklung einiger annueller Pflanzenarten. *Zeit. Naturforsch* 126:564–568.

Knipling, E. B. 1967. Measurement of leaf water potential by the dye method. *Ecology* 48:1038–1041.

Knipling, E. B. and P. J. Kramer. 1967. Comparison of the dye method with the thermo-couple psychrometer for measuring leaf water potentials. *Plant Physiology* 42:1315–1320.

Koelling, M. R. and C. L. Kucera. 1965. Production and turnover relationships in native tall-grass prairie. *Iowa State J. of Science* 39:387–392.

Komarek, E. V., Sr. 1963. Fire, research, and education. In *Proc. Tall Timbers Fire Ecol. Conf.*, no. 2, pp. 181–187. Tallahassee, FL: Tall Timbers Research Station.

———. 1964. The natural history of lightning. In *Proc. Tall Timbers Fire Ecol. Conf.*, no. 3, pp. 139–184. Tallahassee, FL: Tall Timbers Research Station.

———. 1965. Fire ecology—grasslands and man. In *Proc. Tall Timbers Fire Ecol. Conf.*, no. 4, pp. 169–220. Tallahassee, FL: Tall Timbers Research Station.

———. 1967. Fire ecology—and the ecology of man. In *Proc. Tall Timbers Fire Ecol. Conf.*, no. 6, pp. 143–170. Tallahassee, FL: Tall Timbers Research Station.

———. 1968. Lightning and lightning fires as ecological forces. In *Proc. Tall Timbers Fire Ecol. Conf.*, no. 8, pp. 169–197. Tallahassee, FL: Tall Timbers Research Station.

———. 1969. Fire and animal behavior. In *Proc. Tall Timbers Fire Ecol. Conf.*, no. 9, pp. 161–207. Tallahassee, FL: Tall Timbers Research Station.

———. 1974. Effects of fire on temperate forests and related ecosystems: southeastern United States. In *Fire and ecosystems*, ed. T. T. Kozlowski and C. E. Ahlgren, ch. 8. New York: Academic Press.

Komarkova, V. 1979. *Alpine vegetation of the Indian Peaks area, Front Range, Colorado Rocky Mountains* (2 vols.). Liechtenstein: Cramer, Vaduz.

Komarkova, V. and P. J. Webber. 1978. An alpine vegetation map of Niwot Ridge, Colorado. *Arctic and Alpine Rsearch* 10:1–30.

Kortshak, H. P., C. E. Hartt, and G. O. Burr. 1965. Carbon dioxide fixation in sugar cane leaves. *Plant Physiology* 40:209–213.

Kovacic, D. A., T. V. St. John, and M. I. Dyer. 1984. Lack of vesicular-arbuscular mycor-rhizal inoculum in a ponderosa pine forest. *Ecology* 65:1755–1759.

Kozlovsky, D. G. 1968. A critical evaluation of the trophic level concept. *Ecology* 49:48–60.

Kramer, P. J. 1969. *Plant and soil water relationships: a modern synthesis.* New York: McGraw-Hill.

Krammes, J. S. and L. F. DeBano. 1965. Soil wettability: a neglected factor in watershed management. *Water Resources Research* 1:283–286.

Krammes, J. S. and J. Osborn. 1969. Water-repellent soils and wetting agents as factors influencing erosion. In *Proc. Symp. on Water-repellent Soils*, ed. L. F. DeBano and J. Letey, pp. 117–187. Riverside, CA: University of California.

Krebs, C. J. 1972. *Ecology: the experimental analysis of distribution and abundance.* New York: Harper and Row.

———. 1985. *Ecology: the experimental analysis.* New York: Harper and Row.

Krenzer, E. G., Jr., D. N. Moss, and R. K. Crookston. 1975. Carbon dioxide compensation points of flowering plants. *Plant Physiology* 56:194–206.

Krieger, R. I., P. P. Feeny, and C. F. Wilkinson. 1971. Detoxication enzymes in the guts of caterpillars: an evolutionary answer to plant defenses? *Science* 172:579–581.

Kruckeberg, A. R. 1954. The ecology of serpentine soils. III. Plant species in relation to serpentine soils. *Ecology* 35:267–274.

Kucera, C. L., R. C. Dahlman, and M. R. Koelling. 1967. Total net productivity and turnover on an energy basis for tallgrass prairie. *Ecology* 48:536–541.

Küchler, A. W. 1964. *The potential natural vegetation of the conterminous United States.* American Geographical Society Special Publ. no. 36. New York: American Geographical Society.

———. 1967. *Vegetation mapping.* New York: Ronald Press.

———. 1974. A new vegetation map of Kansas. *Ecology* 55:586–604.

———. 1977. Potential natural vegetation of California. In *Terrestrial vegetation of California,* ed. M. G. Barbour and J. Major, pp. 909–938 and map. New York: Wiley-Interscience.

Küchler, A. W. and J. McCormick. 1965. *International bibliography of vegetation maps. Volume I: North America.* Lawrence, KS: University of Kansas Press.

Kuhlman, H. 1977. *Regeneration of a knobcone pine forest after a fire.* San Jose State University Biology 160 (General Ecology). Unpublished.

Kuijt, J. 1969. *The biology of parasitic flowering plants.* Berkeley, CA: University of California Press.

Kummerow, J., D. Krause, and W. Jow. 1978. Seasonal changes of fine root density in the southern Californian chaparral. *Oecologia* 37:201–212.

Kyriakopoulos, E. and H. Richter. 1976. A comparison of methods for the determination of water status in *Quercus ilex* L. *Zhurnal Pflanzenphysiologie* 82:14–27.

Laetsch, W. M. 1974. The C_4 syndrome: a structural analysis. *Annual Review of Plant Physiology* 25:1974.

LaMarche, V. C., Jr. 1974. Paleoclimatic inferences from long tree-ring records. *Science* 183:1043–1048.

Landsberg, H. E., H. Lippmann, K. H. Paffen, and C. Troll. 1966. *World maps of climatology.* Berlin: Springer-Verlag.

Lang, G. E. 1974. Litter dynamics in a mixed oak forest on the New Jersey Piedmont. *Bulletin of the Torrey Botanical Club* 101:277–286.

Lange, O. L. and L. Kappen. 1972. Photosynthesis of lichens from Antarctica. In *Antarctic terrestrial biology,* ed. G. A. Llano, pp. 83–95. Antarctic Research Series, vol. 20. Washington, D.C.: American Geophysical Union.

Lange, O. L., R. Lösch, E. D. Schulze, and L. Kappen. 1971. Responses of stomata to changes in humidity. *Planta* 100:76–86.

Langlet, O. 1959. A cline or not a cline—a question of scots pine. *Sylvae Genetica* 8:13–22.

Larcher, W. 1975. *Physiological plant ecology.* Translated by M. A. Bierdman-Thorson. New York: Springer-Verlag.

La Roi, G. H. 1967. Ecological studies in the boreal spruce-fir forests of the North American taiga. I. Analysis of the vascular flora. *Ecological Monographs* 37:229–253.

Larsen, J. A. 1922. Effect of removal of the virgin white pine stand upon the physical factors of site. *Ecology* 3:302–305.

Laudermilk, J. D. and P. A. Munz. 1938. *Plants in the dung of Nothrotherium from Gypsum Cave, Nevada.* In Carnegie Institution of Washington Publ. 453, pp. 29–37. Washington, D.C.: Carnegie Institution of Washington.

Law, R. 1979. The cost of reproduction in annual meadow grass. *American Naturalist* 113:3–16.

Lawrey, J. D. 1983. Lichen herbivore preference: a test of two hypotheses. *American J. Botany* 70:1188–1194.

Lawson, E. R. 1967. Throughfall and stemflow in a pine-hardwood stand in the Ouachita Mountains of Arkansas. *Water Resources Research* 3:731–735.

Lee, D. W. and J. B. Lowry. 1980. Young leaf anthocyanin and solar ultraviolet. *Biotropica* 12:75–76.

Lee, J. J. and D. L. Inman. 1975. The ecological role of consumers—an aggregated systems view. *Ecology* 56:1455–1458.

Lee, R. 1969. Chemical temperature determination. *J. of Applied Meteorology* 8:423–430.

Lefkovitch, L. P. 1965. The study of population growth in organisms grouped by stages. *Biometrics* 21:1–18.

Lemon, E. R. 1960. Photosynthesis under field conditions. II. An aerodynamic method for determining the turbulent carbon dioxide exchange between the atmosphere and a corn field. *Agronomy J.* 52:697–703.

Lemon, P. C. 1968. Fire and wildlife grazing on an African plateau. In *Proc. Tall Timbers Fire Ecol. Conf.*, no. 8, pp. 71–88. Tallahassee, FL: Tall Timbers Research Station.

Leonard, R. E. 1961. Net precipitation in a northern hardwood forest. *J. of Geophysical Research* 66:2417–2421.

Leopold, A. 1966. *A sand county almanac.* San Francisco/New York: Sierra Club/Ballantine Books.

Leopold, A. C. and P. E. Kriedemann. 1975. *Plant growth and development.* 2nd ed. New York: McGraw-Hill.

Leps, J., J. Osbornova-Kosinova, and M. Rejmanek. 1982. Community stability, complexity, and species life-history strategies. *Vegetatio* 50:53–63.

Leslie, P. H. 1945. On the use of matrices in certain population mathematics. *Biometrika* 33:183–212.

Levin, S. A. 1976. Alkaloid-bearing plants: an ecological perspective. *The American Naturalist* 110:261–284.

Lewis, E. R. 1972. Delay-line models of population growth. *Ecology* 53:797–807.

Liebig, J. 1840. *Chemistry in its agriculture and physiology.* London: Taylor and Walton.

Lieth, H. 1968. The determination of plant dry-matter production with special emphasis on the underground parts. In *Proc. Copenhagen Symp: Functioning of terrestrial ecosystems at the primary production level*, ed. F. E. Eckard. *Natural Resources Research* 5:179–186. Paris: UNESCO.

———. 1973. Primary production: terrestrial ecosystems. *Human Ecology* 1:303–332.

Lieth, H. and R. H. Whittaker, eds. 1975. *The primary production of the biosphere.* New York: Springer-Verlag.

Likens, G. E. and F. H. Bormann. 1972. Nutrient cycling in ecosystems. In *Ecosystem structure and function*, ed. J. A. Weins, pp. 25–67. Corvallis, OR: Oregon State University Press.

Likens, G. E., F. H. Bormann, N. M. Johnson, W. D. Fisher, and R. S. Pierce. 1970. Effects of forest cutting and herbicide treatment on nutrient budgets in the Hubbard Brook watershed-ecosystem. *Ecological Monographs* 40:23–47.

Likens, G. E., F. H. Bormann, N. M. Johnson, and R. S. Pierce. 1967. The calcium, magnesium, potassium, and sodium budgets for a small forested ecosystem. *Ecology* 48:772–785.

Likens, G. E., F. H. Bormann, R. S. Pierce, J. S. Eaton, and N. M. Johnson. 1977. *Biogeochemistry of forested ecosystems.* New York: Springer–Verlag.

Lindberg, S. E., G. M. Lovett, D. D. Richter, and D. W. Johnson. 1986. Atmospheric deposition and canopy interactions of major ions in a forest. *Science* 231:141–145.

Lindsey, A. A. 1955. Testing and line-strip method against full tallies in diverse forest types. *Ecology* 36:485–495.

———. 1956. Sampling methods and community attributes in forest ecology. *Forest Science* 2:287–296.

Lindsey, A. A., J. D. Barton, and S. R. Miles. 1958. Field efficiencies of forest sampling methods. *Ecology* 39:428–444.

Lindsey, A. A. and J. O. Sawyer, Jr. 1971. Vegetation-climate relationships in the eastern United States. *Proc. Indiana Academy of Science* 80:210–214.

List, R. L. 1951. *Smithsonian meteorological tables.* (6th ed.) Smithsonian Institute Publication No. 4014. Washington, D.C.

Livingstone, D. A. 1968. Some interstadial and postglacial pollen diagrams from eastern Canada. *Ecological Monographs* 38:87–125.

Lloyd, M. G. 1961. The contribution of dew to the summer water budget of northern Idaho. *Bulletin of the American Meteorological Society* 42:572–580.

Loach, K. 1967. Shade tolerance in tree seedlings. I. Leaf photosynthesis and respiration in plants raised under artificial shade. *New Phytologist* 66:607–621.

Logan, K. T. 1970. Adaptations of the photosynthetic apparatus of sun- and shade-grown yellow birch (*Betula alleghaniensis* Britt.). *Canadian J. of Botany* 48:1681–1688.

Longman, K. A. and J. Jenik. 1974. *Tropical forest and its environment.* London: Longman.

Lorimer, C. G. 1977. The presettlement forest and natural disturbance cycle of northeastern Maine. *Ecology* 58:139–148.

Lotan, J. E. 1974. Cone serotiny-fire relationships in lodgepole pine. In *Proc. Tall Timbers Fire Ecol. Conf.*, no. 14, pp. 267–278. Tallahassee, FL: Tall Timbers Research Station.

Lotka, A. J. 1925. *Elements of physical biology.* Baltimore, MD: Williams and Wilkins.

Loucks, O. L. 1970. Evolution of diversity, efficiency, and community stability. *American Zoologist* 10:17–25.

Louda, Svata M. 1983. Seed predation and seedling mortality in the recruitment of a shrub, *Haplopappus venetus* (Asteraceae), along a climatic gradient. *Ecology* 64:511–521.

Lowry, W. P. 1969. *Weather and life.* New York: Academic Press.

Lutz, H. J. 1930. The vegetation of Heart's Content, a virgin forest in northwestern Pennsylvania. *Ecology* 11:1–29.

Luxmoore, R. J. 1983. Water budget of an eastern deciduous forest stand. *Soil Science Society of America Journal* 47:785–791.

MacArthur, R. H. 1958. Population ecology of some warblers of northeastern coniferous forests. *Ecology* 39:599–619.

———. 1962. Some generalized theorems of natural selection. *Proc. National Academy of Sciences* 48:1893–1897.

———. 1972. *Geographical ecology: patterns in the distribution of species.* New York: Harper and Row.

MacArthur, R. H. and E. O. Wilson. 1967. *The theory of island biogeography*. Princeton, NJ: Princeton University Press.

Machta, L. 1972. Mauna Loa and global trends in air quality. *American Meteorological Society Bulletin* 53:402–420.

———. 1983. The atmosphere. In *Changing climate: report of the carbon dioxide assessment committee*, pp. 242–251. Washington, DC: National Academy Press.

Macior, L. W. 1973. The pollination ecology of *Pedicularis* on Mount Ranier. *American J. of Botany* 60:863–871.

Mack, R. N. and J. L. Harper. 1977. Interference in dune annuals: spatial pattern and neighborhood effects. *J. of Ecology* 65:345–363.

MacMahon, J. A. 1980. Ecosystems over time: succession and other types of change. In *Forests: fresh perspectives from ecosystem analysis*, ed. R. H. Waring, pp. 27–58. Proceedings 40th Annual Biology Colloquium. Corvallis, OR: Oregon State University Press.

———. 1981. Successional processes: comparisons among biomes with special reference to probable roles of and influences on animals. In *Forest succession: concept and application*, ed. H. H. Shugart, D. B. Botkin, and S. D. West, pp. 207–304. New York: Springer-Verlag.

Madgwick, H. A. I. 1968. Seasonal changes in biomass and annual production of an old-field *Pinus virginiana* stand. *Ecology* 49:149–152.

Madgwick, H. A. I. and T. Satoo. 1975. On estimating the aboveground weights of tree stands. *Ecology* 56:1446–1450.

Maguire, D. A. and R. T. T. Forman. 1983. Herb cover effects on tree seedling patterns in a mature hemlock-hardwood forest. *Ecology* 64:1367–1380.

Mahall, B. E. and F. H. Bormann. 1978. A quantitative description of the vegetative phenology of herbs in a northern hardwood forest. *Botanical Gazette* 139:467–481.

Major, J. 1969. Historical development of the ecosystem concept. In *The ecosystem concept in natural resource management*, ed. G. M. Van Dyne, pp. 9–22. New York: Academic Press.

———. 1974. Kinds and rates of changes in vegetation and chronofunctions. In *Handbook of vegetation science, part VIII, vegetation dynamics*, ed. R. Knapp, pp. 9–18. The Hague: Junk.

———. 1977. California climate in relation to vegetation. In *Terrestrial vegetation of California*, ed. M. G. Barbour and J. Major, pp. 11–74. New York: Wiley-Interscience.

Malcolm, W. M. 1966. Biological interactions. *Botanical Review* 32:243–254.

Margalef, R. 1968. *Perspectives in ecological theory*. Chicago: University of Chicago Press.

Marks, G. C. and T. T. Kozlowski, eds. 1973. *Ectomycorrhizae*. New York: Academic Press.

Marsh, G. P. 1864. *Man and nature*. Reprint 1965. Ed. D. Lowenthal. Cambridge, MA: Harvard University Press.

Marshall, D. R. and S. K. Jain. 1969. Interference in pure and mixed populations of *Avena fatua* and *A. barbata*. *J. of Ecology* 57:251–270.

Matthaei, G. L. C. 1905. Experimental researches on vegetable assimilation and respiration. III. On the effect of temperature on carbon-dioxide assimilation. *Philosophical Transactions of the Royal Society of London, Series B* 197:47–105.

Mattson, W. J., Jr. 1980. Herbivory in relation to plant nitrogen content. *Annual Review Ecology Systematics* 11:119–161.

Mattson, W. J. and N. D. Addy. 1975. Phytophagous insects as regulators of forest primary production. *Science* 190:515–522.

Maycock, P. F. 1967. Jozef Paczoski: founder of the science of phytosociology. *Ecology* 48:1031–1034.

Mayr, E. 1970. *Populations, species, and evolution.* Cambridge, MA: Belknap Press of Harvard University Press.

McBride, J. and E. C. Stone. 1976. Plant succession on the sand dunes of the Monterey Peninsula, California. *American Midland Naturalist* 96:118–132.

McClaugherty, C. A. and J. D. Aber. 1982. The role of fine roots in the organic matter and nitrogen budgets of two forested ecosystems. *Ecology* 63:1481–1490.

McClaugherty, C. A., J. Pastor, J. D. Aber, and J. M. Melillo. 1985. Forest litter decomposition in relation to soil nitrogen dynamics and litter quality. *Ecology* 66:266–275.

McCormick, J. 1968. Succession. *Via* 1:22–35, 131–132. Philadelphia, PA: Graduate School of Fine Arts, University of PA.

McCormick, J. and M. F. Buell. 1968. The plains: pygmy forests of the New Jersey pine barrens, a review and annotated bibliography. *Bulletin New Jersey Academy of Science* 13:20–34.

McCormick, J. and P. A. Harcombe. 1968. Phytograph: useful tool or decorative doodle? *Ecology* 49:13–20.

McCormick, J. F. and R. B. Platt. 1980. Recovery of an Appalachian forest following the chestnut blight, or Catherine Keever—you were right! *American Midland Naturalist* 104:264–273.

McCune, B. and J. A. Antos. 1981. Correlations between forest layers in the Swan Valley, Montana. *Ecology* 62:1196–1204.

McCutchan, M. H. 1977. Climatic features as a fire determinant. In *Proc. Symp. Env. Cons. Fire and Fuel Mangmt. in Medit. Ecosyst.*, pp. 1–11. Washington, D.C.: U.S. Dept. of Agriculture Forest Service.

McDonald, C. D. and G. H. Hughes. 1964. *Studies of consumptive use of water by phreatophytes and hydrophytes near Yuma, Arizona.* Geological Survey Professional Paper 486-F. Washington, D.C.: U.S. Geological Survey.

McDougall, W. B. 1922. Symbiosis in a deciduous forest, part I. *Botanical Gazette* 73:200–212.

———. 1925. Symbiosis in a deciduous forest, part II. *Botanical Gazette* 79:95–102.

———. 1927. *Plant Ecology.* Philadelphia: Lea and Febiger.

McGraw, J. B. 1985a. Experimental ecology of *Dryas octopetala* ecotypes: relative response to competitors. *New Phytologist* 100:233–241.

———. 1985b. Experimental ecology of *Dryas octopetala* ecotypes. III. Environmental factors and plant growth. *Arctic and Alpine Research* 17:229–239.

McGraw, J. B. and J. Antonovics. 1983. Experimental ecology of *Dryas octopetala* ecotypes. I. Ecotypic differentiation and life-cycle stages of selection. *J. Ecology* 71:879–897.

McIntosh, R. P. 1962. Raunkiaer's "Law of Frequency." *Ecology* 43:533–555.

———. 1974. Plant ecology, 1947–1972. *Annals of the Missouri Botanical Garden* 61:132–165.

———. 1976. Ecology since 1900. In *Evolution of issues, ideas and events in America: 1776–1976,* ed. B. J. Taylor, pp. 353–372. Norman, OK: University of Oklahoma Press.

———. 1980. The relationship between succession and the recovery process in ecosystems. In *The recovery process in damaged ecosystems,* ed. J. Cairns, Jr., pp. 11–62. Ann Arbor, MI: Ann Arbor Science.

————. 1983a. Excerpts from the work of L. G. Ramensky. *Bulletin Ecological Society America* 64(1):7–12.

————. 1983b. Pioneer support for ecology. *BioScience* 33:107–112.

McIntyre, L. 1985. Humboldt's way. *National Geographic* 168(3):318–351.

McMillan, C. 1956. The edaphic restriction of *Cupressus* and *Pinus* in the Coast Ranges of central California. *Ecological Monographs* 26:177–212.

McNaughton, S. J. 1966. Thermal inactivation properties of enzymes from *Typha latifolia* L. ecotypes. *Plant Physiology* 41:1736–1738.

————. 1967. Photosynthetic system II: racial differentiation in *Typha latifolia*. *Science* 156:1363.

————. 1972. Enzymatic thermal adaptations: the evolution of homeostasis in plants. *The American Naturalist* 106:165–172.

————. 1973. Comparative photosynthesis of Quebec and California ecotypes of *Typha latifolia*. *Ecology* 54:1260–1270.

————. 1983. Compensatory plant growth as a response to herbivory. *Oikos* 40:329–336.

McPherson, J. K. and C. H. Muller. 1969. Allelopathic effects of *Adenostoma fasciculatum*, "chamise," in the California chaparral. *Ecological Monographs* 39:177–198.

Meentemeyer, V. 1978. Macroclimate and lignin control of litter decomposition rates. *Ecology* 59:465–472.

Meidner, H. and T. A. Mansfield. 1968. *Physiology of stomata*. New York: McGraw-Hill.

Meidner, H. and D. W. Sheriff. 1976. *Water and plants*. New York: Wiley.

Melillo, J. M., J. D. Abner, and J. F. Muratore. 1982. Nitrogen and lignin control of hardwood leaf litter decomposition dynamics. *Ecology* 63:621–626.

Melin, E. 1930. Biological decomposition of some types of litter from North American forests. *Ecology* 11:72–101.

Merriam, C. H. 1890. Results of a biological survey of the San Francisco mountain region and desert of the Little Colorado, Arizona. *North American Fauna* 3:1–136.

————. 1894. Laws of temperature control of the geographic distribution of animals and plants. *National Geographic* 6:229–238.

————. 1898. Life zones and crop zones of the United States. *U.S. Dept. of Agriculture Biological Survey Division Bulletin* 10:9–79.

Meyer, F. H. 1973. Ectomycorrhizae. In *Ectomycorrhizae*, eds. G. C. Marks and T. T. Kozlowski. New York: Academic Press.

Meyers, V. (Ed.). 1983. Remote sensing applications in agriculture. In *Manual of remote sensing* (2nd Ed.), ed. R. N. Colwell, pp. 2111–2228. American Society of Photogrammetrics. Falls Church, VA: Sheridan Press.

Milewski, A. V. 1983. A comparison of ecosystems in mediterranean Australia and southern Africa. *Annual Review Ecology and Systematics* 14:57–76.

Miller, A. and J. C. Thompson. 1975. *Elements of meteorology*. 2nd ed. Columbus, OH: Merrill.

Miller, C. E. 1938. *Plant physiology*. New York: McGraw-Hill.

Miller, P. C. and L. Tieszen. 1972. A preliminary model of processes affecting primary production in the arctic tundra. *Arctic and Alpine Research* 4:1–18.

Miller, P. C., D. K. Poole and P. M. Miller. 1983. The influence of annual precipitation, topography, and vegetative cover on soil moisture and summer drought in southern California. *Oecologia* 56:385–391.

Miller, P. R. 1969. Air pollution and the forests of California. *California Air Environment* 1:1–3.

Milthorpe, F. L. 1955. The significance of the measurements made by the cobalt chloride paper method. *J. of Experimental Botany* 6:17–19.

Minchin, P. R. 1986. A comparative evaluation of ordination techniques. *Vegetatio* (In press).

Moldenke, A. R. 1975. Niche specialization and species diversity along a California transect. *Oecologia* 21:219–242.

Monson, R. K. and S. D. Smith. 1982. Seasonal water potential components of Sonoran Desert plants. *Ecology* 63:113–123.

Monteith, J. L. 1973. *Principles of environmental physics.* London: Arnold.

Mooney, H. A. 1972. The carbon balance of plants. *Annual Review of Ecology and Systematics* 3:315–346.

———. 1977a. Southern coastal scrub. In *Terrestrial vegetation of California,* ed. M. G. Barbour and J. Major, pp. 471–489. New York: Wiley-Interscience.

———. ed. 1977b. *Convergent evolution in Chile and California: Mediterranean climate ecosystems.* Stroudsburg, PA: Dowden, Hutchinson, and Ross.

Mooney, H. A. and W. D. Billings. 1961. Comparative physiological ecology of arctic and alpine populations of *Oxyria digyna. Ecological Monographs* 31:1–29.

Mooney, H. A. and E. L. Dunn. 1970a. Convergent evolution of Mediterranean-climate evergreen sclerophyll shrubs. *Evolution* 24:292–303.

———. 1970b. Photosynthetic systems of Mediterranean-climate shrubs and trees of California and Chile. *The American Naturalist* 104:447–453.

Mooney, H. A. and J. R. Ehleringer. 1978. The carbon gain benefits of solar tracking in a desert annual. *Plant Cell and Environment* 1:307–311.

Mooney, H. A., J. Ehleringer, and J. Berry. 1976. High photosynthetic capacity of a winter annual in Death Valley. *Science* 194:322–323.

Mooney, H. A., J. Ehleringer, and O. Björkman. 1977. The energy balance of leaves of the evergreen desert shrub *Atriplex hymenelytra. Oecologia* 29:301–310.

Mooney, H. A. and A. T. Harrison. 1970. The influence of conditioning temperature on subsequent temperature-related photosynthetic capacity in higher plants. In *Prediction and measurement of photosynthetic productivity,* ed. C. T. de Wit, pp. 411–417. Wageningen, The Netherlands: Center for Agricultural Publishing and Documentation.

Mooney, H. A. and M. West. 1964. Photosynthetic acclimation of plants of diverse origin. *American J. of Botany* 51:825–827.

Moore, R. M., ed. 1970. *Australian grasslands.* Canberra, Australia: Australian National University Press.

Morrow, P. A. and V. C. LaMarche. 1978. Tree-ring evidence for chronic suppression of productivity in subalpine Eucalyptus. *Science* 201:1244–1246.

Mueller-Dombois, D. and H. Ellenberg. 1974. *Aims and methods of vegetation ecology.* New York: Wiley.

Muir, J. 1878. The new sequoia forest of California. *Harpers* 57:813–827.

———. 1894. *The mountains of California.* Garden City, NY: Doubleday.

Muir, P. S. and J. E. Lotan. 1985. Serotiny and life history of *Pinus contorta* var. *latifolia. Canadian J. Botany* 63:938–945.

Muller, C. H. 1953. The association of desert annuals with shrubs. *American J. of Botany* 40:53–60.

———. 1966. The role of chemical inhibition (allelopathy) in vegetational composition. *Bulletin of the Torrey Botanical Club* 93:332–351.

Muller, C. H., R. B. Hanawalt, and J. K. McPherson. 1968. Allelopathic control of herb growth in the fire cycle of California chaparral. *Bulletin of the Torrey Botanical Club* 95:225–231.

Muller, W. H. and C. H. Muller. 1956. Association patterns involving desert plants that contain toxic products. *American J. of Botany* 43:354–361.

Mulroy, T. W. and P. W. Rundel. 1977. Annual plants: adaptations to desert environments. *BioScience* 27:109–114.

Mutch, R. W. 1970. Wildland fires and ecosystems—a hypothesis. *Ecology* 51:1046–1051.

Nadelhoffer, K. J., J. D. Aber, and J. M. Melillo. 1985. Fine roots, net primary production, and soil nitrogen availability: a new hypothesis. *Ecology* 66:1377–1390.

Nash, R. 1973. *Wilderness and the American mind.* Revised ed. New Haven, CT: Yale University Press.

Naveh, Z. and R. H. Whittaker. 1979. Structural and floristic diversity of shrublands and woodlands in northern Israel and other mediterranean areas. *Vegetatio* 41:171–190.

Neales, T. F. 1973. The effect of night temperatures on CO_2 assimilation, transpiration, and water use efficiency in *Agave americana* L. *Australian J. of Biological Science* 26:705–714.

Newbould, P. J. 1967. *Methods for estimating the primary production of forests.* IBP Handbook no. 2. Oxford: Blackwell Scientific Publ.

———. 1968. Methods of estimating root production. In *Proc. Copenhagen Symp.: Functioning of terrestrial ecosystems at the primary production level*, ed. F. E. Eckhard. *Natural Resources Research* 5:187–190. Paris: UNESCO.

Newell, S. J. and E. J. Tramer. 1978. Reproductive strategies in herbaceous plant communities during succession. *Ecology* 59:228–234.

Ng, E. and P. C. Miller. 1980. Soil moisture relations in the southern California chaparral. *Ecology* 61:98–107.

Niering, W. A., R. H. Whittaker, and C. H. Lowe. 1963. The saguaro: a population in relation to its environment. *Science* 142:15–23.

Nixon, S. W., C. A. Oviatt, J. Garber, and V. Lee. 1976. Diel metabolism and nutrient dynamics in a salt marsh embayment. *Ecology* 57:740–750.

Nobel, P. S. 1974a. *Introduction to biophysical plant physiology.* San Francisco: W. H. Freeman.

———. 1974b. Boundary layers of air adjacent to cylinders: estimation of effective thickness and measurements on plant material. *Plant Physiology* 54:177–181.

———. 1976. Water relations and photosynthesis of a desert CAM plant, *Agave deserti.* *Plant Physiology* 58:576–582.

———. 1977. Water relations and photosynthesis of barrel cactus, *Ferocactus acanthodes*, in the Colorado Desert. *Oecologia* 27:117–133.

———. 1980a. Morphology, surface temperatures, and northern limits of columnar cacti in the Sonoran desert. *Ecology* 61:1–7.

———. 1980b. Influences of minimum stem temperatures on ranges of cacti in southwestern United States and central Chile. *Oecologia* 47:101–115.

———. 1982. Low temperature tolerance and cold hardening of cacti. *Ecology* 63:1650–1656.

———. 1982*a*. Orientation of terminal cladodes of platyopuntias. *Botanical Gazette* 143:219–224.

———. 1982*b*. Interaction between morphology, PAR interception, and nocturnal acid accumulation in cacti. In *Crassulacean acid metabolism,* ed. I. P. Ting and M. Gibbs, pp. 260–277. Rockville, MD: American Society of Plant Physiology.

Noy-Meir, I. 1975. Stability of grazing systems: an application of predator-prey graphs. *J. of Ecology* 63:459–481.

Nye, P. H. 1961. Organic matter and nutrient cycles under moist tropical forests. *Plant and Soil* 13:333–346.

Oberlander, G. T. 1956. Summer fog precipitation on the San Francisco peninsula. *Ecology* 37:851–852.

Odening, W. R., B. R. Strain, and W. C. Oechel. 1974. The effect of decreasing water potential on net CO_2 exchange of intact desert shrubs. *Ecology* 55:1086–1095.

Odum, E. P. 1963. Limits of remote ecosystems containing man. *American Biology Teacher* 25:429–443.

———. 1969. The strategy of ecosystem development. *Science* 164:262–270.

———. 1971. *Fundamentals of ecology.* 3rd ed. Philadelphia, PA: W. B. Saunders.

Odum, S. 1965. Germination of ancient seeds: floristic observations and experiments with archeologically dated soil samples. *Dansk Botanik Arkiven* 24(2):1–70.

Oechel, W. C., B. R. Strain, and W. R. Odening. 1972. Tissue water potential, photosynthesis, ^{14}C-labeled photosynthetic utilization and growth in the desert shrub *Larrea divaricata. Ecological Monographs* 42:127–141.

O'Leary, M. H. and C. B. Osmond. 1980. Diffusional contributions to carbon isotope fractionation during dark CO_2 fixation in CAM plants. *Plant Physiology* 66:931–934.

Olmsted, C. E. III and E. L. Rice. 1970. Relative effects of known plant inhibitors on species from the first two stages of old-field succession. *Southwestern Naturalist* 15:165–173.

Olson, J. S. 1958. Rates of succession and soil changes on southern Lake Michigan sand dunes. *Botanical Gazette* 119:125–170.

———. 1963. Energy storage and balance of producers and decomposers in ecological systems. *Ecology* 44:322–331.

Oosting, H. J. 1942. An ecological analysis of the plant communities of Piedmont, North Carolina. *American Midland Naturalist* 28:1–126.

———. 1956. *The study of plant communities.* 2nd ed. San Francisco, CA: Freeman.

Oosting, H. J. and W. D. Billings. 1951. A comparison of virgin spruce-fir forest in the northern and southern Appalachian system. *Ecology* 32:84–103.

Oppenheimer, H. R. 1960. Adaptation to drought: xerophytism. In *Plant-water relationships in arid and semi-arid conditions,* pp. 105–138. Paris: UNESCO.

Orians, G. H. and O. T. Solbrig. 1977. A cost-income model of leaves and roots with special reference to arid and semiarid areas. *American Naturalist* 111:677–690.

Orloci, L. 1966. Geometric models in ecology. I. The theory and application of some ordination methods. *J. of Ecology* 54:193–215.

Orloci, L. and N. C. Kenkel. 1985. *Introduction to data analysis with examples from population and community ecology.* Burtonsville, MD: International Co-operative Publishing House.

Orshan, G. 1972. Morphological and physiological plasticity in relation to drought. In *Wildland shrubs—their biology and utilization,* ed. C. M. McKell, J. P. Blaisdell, and J. R. Goodin, pp. 245–254. U.S. Dept. of Agriculture Forest Service General Technical Report INT-1. Ogden, UT: U.S. Dept. of Agriculture Forest Service.

Osmond, C. B. 1976. CO_2 assimilation and dissimilation in the light and dark in CAM plants. In *CO_2 metabolism and plant productivity,* ed. R. H. Burris and C. C. Black, pp. 217–233. Baltimore, MD: University Park Press.

Osmond, C. B., O. Bjorkman and D. J. Anderson. 1980. Physiological processes in plant ecology: toward and synthesis with *Atriplex. Ecological Studies,* Series #36. Berlin: Springer/Verlag.

Osonubi, O. and W. J. Davies. 1978. Solute accumulation in leaves and roots of woody plants subjected to water stress. *Oecologia* 32:323–332.

Ovington, J. D., D. Heitkamp, and D. B. Lawrence. 1963. Plant biomass and productivity of prairie, savanna, oakwood and maize field ecosystems in central Minnesota. *Ecology* 44:52–63.

Ownbey, R. S. and B. E. Mahall. 1983. Salinity and root conductivity: differential responses of a coastal succulent halophyte, *Salicornia virginica* and a weedy glycophyte, *Raphanus sativus. Physiologia Plantarum* 57:189–195.

Paczoski, J. 1891. Stadii razwitija flory. (In Polish.) *Westnik Estestboznania* No. 8.

———. 1896. Zycie gromadne roslin. (In Polish.) *Wszechswiat,* No. 16, Warsaw.

———. 1921. *Osnowy fitosocjologji.* Cherson, Izd. Stud. Comitet Tech.

Palta, J. 1983. Photosynthesis, transpiration, and leaf duffusive conductance of the cassava leaf in response to water stress. *J. of Botany* 61:373–375.

Parker, G. G. 1983. Throughfall and stemflow in the forest nutrient cycle. *Advances in Ecological Research* 13:58–134.

Parker, G. R. and D. J. Leopold. 1983. Replacement of *Ulmus americana* L. in a mature east-central Indiana woods. *Bulletin Torrey Botanical Club* 110:482–488.

Parkinson, K. J. and B. J. Legg. 1972. A continuous flow porometer. *J. of Applied Ecology* 9:669-675.

Parmeter, J. R. Jr. 1977. Effects of fire on pathogens. In *Proc. Symp. Env. Cons. Fire and Fuel Mangmt. in Medit. Ecosyst.,* pp. 58–64. Washington, D.C.: U.S. Dept. of Agriculture Forest Service.

Parrish, J. A. D. and F. A. Bazzaz. 1976. Underground niche separation in successional plants. *Ecology* 57:1281–1288.

Parsons, D. J. 1976. Vegetation structure in the Mediterranean scrub communities of California and Chile. *J. of Ecology* 64:435–447.

Parsons, D. J. and A. R. Moldenke. 1975. Convergence in vegetation structure along analogous climatic gradients in California and Chile. *Ecology* 56:950–957.

Pastor, J. and J. G. Bockheim. 1984. Distribution and cycling of nutrients in an aspen-mixed-hardwood-spodosol ecosystem in northern Wisconsin. *Ecology* 65:339–353.

Paysen, T. E., J. A. Derby, and C. E. Conrad. 1982. *A vegetation classification system for use in California.* USDA Forest Service, General Technical Report PSW-63, Berkeley, CA.

Pearcy, R. W. 1976. Temperature responses of growth and photosynthetic CO_2 exchange rates in coastal and desert races of *Atriplex lentiformis. Oecologia* 26:245–255.

———. 1983. The light environment and growth of C_3 and C_4 tree species in the under-story of a Hawaiian forest. *Oecologia* 58:19–25.

Pearcy, R. W., and H. W. Calkin. 1983. Carbon dioxide exchange of C_3 and C_4 tree species in the understory of a Hawaiian forest. *Oecologia* 58:26–32

Pearcy, R. W. and A. T. Harrison. 1974. Comparative photosynthetic and respiratory gas exchange characteristics of *Atriplex lentiformis* (Torr.) Wats. in coastal and desert habitats. *Ecology* 55:1104–1111.

Pearson, G. A. 1942. Herbaceous vegetation a factor in natural regeneration of ponderosa pine in the southwest. *Ecological Monographs* 12:315–338.

Peet, R. K. 1974. The measurement of species diversity. *Annual Review of Ecology and Systematics* 5:285–307.

———. 1975. Relative diversity indices. *Ecology* 56:496–498.

Penfound, W. T. 1964. Effects of denudation on the productivity of grassland. *Ecology* 45:838–845.

Penman, H. L. 1950. Evaporation over the British Isles. *Quarterly J. of the Royal Meteoro-logical Society* 76:372–383.

Perino, J. V. and P. G. Risser. 1972. Some aspects of structure and function in Oklahoma old-field succession. *Bulletin of the Torrey Botanical Club* 99:233–239.

Peters, G. A. 1978. Blue-green algae and algal associations. *BioScience* 28:580–585.

Petrusewicz, K. and W. L. Grodzinski. 1975. The role of herbivore consumers in various ecosystems. In *Productivity of world ecosystems*, pp. 64–70. Washington, D.C.: National Academy of Sciences.

Phillips, D. L. and J. A. MacMahon. 1981. Competition and spacing patterns in desert shrubs. *J. Ecology* 69:97–115.

Phillips, E. A. 1959. *Methods of vegetation study.* New York: Holt, Rinehart and Winston.

Phillips, H. 1965. Fire—as master and servant: its influence in the bioclimatic regions of Trans-Saharan Africa. In *Proc. Tall Timbers Fire Ecol. Conf.*, no. 4, pp. 7–109. Tal-lahassee, FL: Tall Timbers Research Station.

———. 1974. Effects of fire in forest and savanna ecosystems of Sub-Saharan Africa. In *Fire and ecosystems*, ed. T. T. Kozlowski and C. E. Ahlgren, ch. 13. New York: Academic Press.

Philpot, C. W. 1977. Vegetative features as determinants of fire frequency and intensity. In *Proc. Symp. Env. Cons. Fire and Fuel Mangmt. in Medit. Ecosyst.*, pp. 12–16. Washington, D. C.: U.S. Dept. of Agriculture Forest Service.

Pianka, E. R. 1970. On r- and K-selection. *American Naturalist* 104:592–597.

———. 1983. *Evolutionary ecology.* New York: Harper and Row.

Pickett, S. T. A. and F. A. Bazzaz. 1976. Divergence of two co-occurring successional annuals on a soil moisture gradient. *Ecology* 57:169–176.

Pielou, E. C. 1961. Segregation and symmetry in two-species populations as studied by nearest neighbor relations. *J. of Ecology* 49:255–269.

———. 1969. *An introduction to mathematical ecology.* New York: Wiley-Interscience.

———. 1977. *Mathematical ecology.* New York: Wiley.

———. 1981. The usefulness of ecological models. *Quarterly Review of Biology* 56:17–31.

———. 1984. *The interpretation of ecological data. A primer on classification and ordination.* New York: Wiley.

Pike, L. H., D. M. Tracy, M. A. Sherwood, and D. Nielsen. 1972. Estimates of biomass and fixed nitrogen of epiphytes from old-growth Douglas-fir. In *Proc.—research on coniferous forests ecosystems—a symp.*, ed. J. F. Franklin, L. J. Dempster, and R. H. Waring, pp. 177–187. Portland, OR: Pacific Northwest Forest and Range Experiment Station. Washington, D.C.: U.S. Dept. of Agriculture Forest Service.

Pimm, S. L. 1984. The complexity and stability of ecosystems. *Nature* 307:321–326.

Pinder, J. E., III. 1975. Effects of species removal on an old-field plant community. *Ecology* 56:747–751.

Pinero, D., J. Sarukhan, and P. Alberdi. 1982. The costs of reproduction in a tropical palm, *Astrocaryum mexicanum*. *J. of Ecology* 70:473–481.

Pomeroy, L. R. 1970. The strategy of mineral cycling. *Annual Review of Ecology and Systematics* 1:171–190.

Poore, M. E. D. 1955*a*. The use of phytosociological methods in ecological investigations. I. The Braun-Blanquet system. *J. of Ecology* 43:226–244.

———. 1955*b*. The use of phytosociological methods in ecological investigations. II. Practical issues involved in an attempt to apply the Braun-Blanquet system. *J. of Ecology* 43:245–269.

Porsild, A. E., C. R. Harington, and G. A. Mulligan. 1967. *Lupinus arcticus* Wats. grown from seeds of Pleistocene age. *Science* 158:113–114.

Post, L. J. 1970. Dry matter production of mountain maple and balsam fir in northwestern New Brunswick. *Ecology* 51:548–550.

Potzger, J. E., M. E. Potzger, and J. McCormick. 1956. The forest primeval of Indiana as recorded in the original land surveys and an evaluation of previous interpretations of Indiana vegetation. *Butler University Botanical Studies* 13:95–111.

Powles, S. B. and C. B. Osmond. 1978. Inhibition of the capacity and efficiency of photosynthesis in bean leaflets illuminated in a CO_2-free atmosphere at low oxygen. *Australian J. of Plant Physiology* 5:619–629.

Preston, F. W. 1948. The commonness and rarity of species. *Ecology* 29:254–283.

Prose, C. V. and S. K. Metzger. 1985. *Recovery of soils and vegetation in World War II military base camps, Mojave Desert*. Open File Report, US Geological Survey, Menlo Park, CA.

Quick, C. R. 1959. Ceanothus seeds and seedlings on burns. *Madroño* 15:79–81.

Quinn, J. A. and J. C. Colosi. 1977. Separating genotype from environment in germination ecology studies. *American Midland Naturalist* 97:484–489.

Rabotnov, T. A. 1953. L. G. Ramensky. [In Russian.] *Botanisheskii Zhurnal* 38:5.

———. 1969. On coenopopulations of perennial herbaceous plants in natural coenoses. *Vegetatio* 19:87–95.

———. 1978. Structure and method of studying coenotic populations of perennial herbaceous plants. *Soviet J. of Ecology* 9:99–105.

———. 1978. Concepts of ecological individuality of plant species and of the continuum of the plant cover in the works of L. G. Ramensky. *Ekologiya* 5:25–32. [Translation in *Soviet J. Ecol.* 9:417–422.]

Radosevich, S. R. and J. S. Holt. 1984. *Weed ecology*. New York: Wiley.

Rains, D. W. 1976. Mineral metabolism. In J. Bonner and J. E. Varner (eds.), *Plant biochemistry* (3rd ed.). New York: Academic Press.

Raison, J. K., J. A. Berry, P. A. Armond, and C. S. Pike. 1980. Membrane properties in relation to the adaptation of plants to high and low temperature stress. In *Adaptations of plants to water and high temperature stress*, ed. N. C. Turner and P. J. Kramer, pp. 261–277. New York: Wiley/Interscience.

Ranwell, D. S. 1972. *Ecology of salt marshes and sand dunes.* London: Chapman and Hall.

Raschke, K. 1976. How stomata resolve the dilemma of opposing priorities. *Philosophical Transactions of the Royal Society of London, Series B* 273:551–560.

Raunkiaer, C. 1918. Récherches statistiques sur les formations végétales. *Kgl. Danske Videnskabernes Selskab Biologiske Meddelelser* 1:1–47.

———. 1934. *The life forms of plants and statistical plant geography.* Oxford: Clarendon Press.

Raven, P. H. and D. I. Axelrod. 1978. *Origin and relationships of the California flora.* Berkeley: University of California Press.

Raven, P. H. and H. Curtis. 1970. *Biology of plants.* New York: Worth.

Rawlins, S. L. 1976. Measurement of water content and the state of water in soils. In *Water deficits and plant growth*, vol. IV, ed. T. T. Kozlowski, pp. 1–55. New York: Academic Press.

Real, L. (ed.). 1983. *Pollination ecology.* New York: Academic Press.

Reardon, P. O., C. L. Leinweber, and L. B. Merrill. 1972. The effect of bovine saliva on grasses. *J. of Animal Science* 34:897–898.

———. 1974. Response of sideoats grama to animal saliva and thiamine. *J. of Range Management* 27:400–401.

Redmann, R. E. 1975. Production ecology of grassland plant communities in western North Dakota. *Ecological Monographs* 45:83–106.

Reed, H. S. 1942. *A short history of the plant sciences.* New York: Ronald Press.

Reichle, D. E., R. A. Goldstein, R. I. Van Hook, Jr., and G. J. Dodson. 1973. Analysis of insect consumption in a forest canopy. *Ecology* 54:1076–1084.

Reimold, R. J. and W. H. Queen. 1974. *Ecology of halophytes.* New York: Academic Press.

Reiners, W. A. 1972. Structure and energetics of three Minnesota forests. *Ecological Monographs* 42:71–94.

Reiners, W. A. and N. M. Reiners. 1970. Energy and nutrient dynamics of forest floors in three Minnesota forests. *J. of Ecology* 58:497–519.

Rejmanek, M. 1977. The concept of structure in phytosociology with references to classification of plant communities. *Vegetatio* 35:55–61.

———. 1984. Perturbation-dependent coexistence and species diversity in ecosystems. In *Stochastic phenomena and chaotic behavior in complex systems*, ed. P. Schuster, pp. 220–230. New York: Springer-Verlag.

Reuss, J. O. and G. S. Innis. 1977. A grassland nitrogen flow simulation model. *Ecology* 58:379–388.

Rhoades, D. F. 1979. Evolution of plant chemical defense against herbivores. In *Herbivores: their interaction with secondary plant metabolites*, ed. G. A. Rosenthal and D. H. Janzen, pp. 3–54. New York: Academic Press.

———. 1983a. Herbivore population dynamics and plant chemistry. In *Variable plants and herbivores in natural and managed systems*, ed. R. F. Denno and M. S. McClure, pp. 155–220. New York: Academic Press.

———. 1983*b*. Responses of alder and willow to attack by tent caterpillars and webworms: evidence for pheromonal sensitivity of willows. *American Chemical Society Symposium Series* 208:55–68.

———. 1985. Offensive-defensive interactions between herbivores and plants: their relevance in herbivore population dynamics and ecological theory. *American Naturalist* 125:205–238.

Rhoades, D. H. and R. G. Cates. 1976. Toward a general theory of plant antiherbivore chemistry. *Recent Advances in Phytochemistry* 10:168–213.

Rice, E. L. 1964. Inhibition of nitrogen-fixing and nitrifying bacteria by seed plants, part I. *Ecology* 45:824–837.

———. 1965. Inhibition of symbiotic nitrogen-fixing and nitrifying bacteria by seed plants, part II. *Physiologia Plantarum* 18:255–268.

———. 1974. *Allelopathy.* New York: Academic Press.

Rice, E. L. and R. W. Kelting. 1955. The species-area curve. *Ecology* 36:7–11.

Rice, E. L. and R. L. Parenti. 1978. Causes of decreases in productivity in undisturbed tall grass prairie. *American J. of Botany* 65:1091–1097.

Richards, L. A. and G. Ogata. 1958. Thermocouple for vapor pressure measurements in biological and soil systems at high humidity. *Science* 128:1089–1090.

Richards, P. W. 1936. Ecological observations on the rain forest of Mount Dulit, Sarawak. *J. of Ecology* 24:1–37 and 340–360.

———. 1973. The tropical rain forest. *Scientific American* 229(6):58–67.

Richardson, J. L. 1977. *Dimensions of ecology.* Baltimore, MD: Williams and Wilkins.

Ricklefs, R. E. 1973. *Ecology.* Newton, MA: Chiron Press.

———. 1979. *Ecology.* 2nd ed. New York: Chiron Press.

Riley, D. and A. Young. 1974. *World vegetation.* London: Cambridge University Press.

Risser, P. G. and P. H. Zedler. 1968. An evaluation of the grassland quarter method. *Ecology* 49:1006–1009.

Ritchie, G. A. and T. M. Hinckley. 1975. The pressure chamber as an instrument for ecological research. *Advances in Ecological Research* 9:166–254.

Robberecht, R. and M. M. Caldwell. 1980. Leaf ultraviolet optical properties along a latitudinal gradient in the arctic alpine life zone. *Ecology* 61:612–619.

Roberts, S. W. and K. R. Knoerr. 1977. Components of water potential estimated from xylem pressure measurements in five tree species. *Oecologia* 28:191–202.

Roberts, S. W., B. R. Strain, and K. R. Knoerr. 1980. Seasonal patterns of leaf water relations in four co-occurring forest tree species: parameters from pressure volume curves. *Oecologia* 46:330–337.

Robertson, G. P. and P. M. Vitousek. 1981. Nitrification potentials in primary and secondary succession. *Ecology* 62:376–386.

Robichaud, R. H. and M. F. Buell. 1973. *Vegetation of New Jersey.* New Brunswick, NJ: Rutgers University Press.

Robinson, T. 1967. *The organic constituents of higher plants.* 2nd ed. Minneapolis, MN: Burgess.

Rochow, J. J. 1974. Litter fall relations in a Missouri forest. *Oikos* 25:80–85.

Rockwood, L. L. 1974. Seasonal changes in the susceptibility of *Crescentia alata* leaves to the flea beetle, *Oedionyehus* sp. *Ecology* 55:142–148.

———. 1976. Plant selection and foraging patterns in two species of leaf cutting ants (*Atta*). *Ecology* 57:48–61.

Rodin, L. E. and N. I. Basilevic. 1967. *Production and mineral cycling in terrestrial vegetation.* Edinburgh, Scotland: Oliver and Boyd.

———. 1968. World distribution of plant biomass. In *Proc. Copenhagen Symp.: functioning of terrestrial ecosystems at the primary production level,* ed. F. E. Eckhard. *Natural Resources Research* 5:45–52. Paris: UNESCO.

Rogers, H. H., J. F. Thomas, and G. E. Bingham. 1983. Responses of agronomic and forest species to elevated atmospheric carbon dioxide. *Science* 220:428–429.

Rogers, R. S. 1980. Hemlock stands from Wisconsin to Nova Scotia: transitions in understory composition along a floristic gradient. *Ecology* 61:178–193.

———. 1981. Mature mesophytic hardwood forest: community transitions, by layer, from east-central Minnesota to southeastern Michigan. *Ecology* 62:1634–1647.

———. 1982. Early spring herb communities in mesophytic forests of the Great Lakes region. *Ecology* 63:1050–1063.

———. 1985. Local coexistence of deciduous forest ground layer species growing in different seasons. *Ecology* 66:701–707.

Rook, D. A. 1969. The influence of growing temperature on photosynthesis and respiration of *Pinus radiata* seedlings. *New Zealand J. of Botany* 7:43–55.

Rorison I. H., ed. 1969. *Ecological aspects of the mineral nutrition of plants.* Oxford: Blackwell.

Rosenberg, N. J. 1974. *Micro-climate: the biological environment.* New York: Wiley.

Rosenzweig, M. L. 1968. Net primary productivity of terrestrial communities: prediction from climatological data. *American Naturalist* 102:67–74.

Rowe, J. S. 1964. Environmental preconditions with special reference to forestry. *Ecology* 45:399–403.

Rübel, E. 1930. *Pflanzengesellschaften der Erde.* Berlin: H. Huber.

Rundel, P. W. 1971. Community structure and stability in the giant Sequoia groves of the Sierra Nevada, California. *American Midland Naturalist* 85:478–492.

———. 1974. Water relations and morphological variation in *Ramalina menziesii* Tayl. *The Bryologist* 77:23–32.

Rundel, P. W., D. J. Parsons, and D. T. Gordon. 1977. Montane and subalpine vegetation of the Sierra Nevada and Cascade Ranges. In *Terrestrial vegetation of California,* ed. M. G. Barbour and J. Major, ch. 17. New York: Wiley-Interscience.

Rundel, P. W. and R. E. Stecker. 1977. Morphological adaptations of tracheid structure to water stress gradients in the crown of *Sequoiadendron giganteum. Oecologia* 27:135–139.

Russell, E. W. B. 1983. Indian-set fires in the forests of the northeastern United States. *Ecology* 64:78–88.

Rutter, A. J. 1975. The hydrological cycle in vegetation. In *Vegetation and the atmosphere,* vol. 1, ed. J. L. Monteith, pp. 111–154. New York: Academic Press.

Saeki, T. 1973. Distribution of radiant energy and CO_2 in terrestrial communities. In *Photosynthesis and productivity in different environments,* ed. J. P. Cooper, pp. 297–322. London: Cambridge University Press.

Salisbury, E. 1952. *Downs and dunes.* London: G. Bell and Sons.

Salisbury, F. B. and C. Ross. 1969. *Plant physiology.* Belmont, CA: Wadsworth.

Salthe, S. N. 1985. *Evolving hierarchical systems.* New York: Columbia University Press.

Sampson, A. W. 1944. Plant succession on burned chaparral lands in northern California. *California Agriculture Experiment Station Bulletin* 685:1–144.

Sarukhan, J. and M. Gadgil. 1974. Studies on plant demography: *Ranunculus repens* L., *R. bulbosus* L., and *R. acris* L: III. A mathematical model incorporating multiple modes of reproduction. *J. of Ecology* 62:921–936.

Sarukhan, J. and J. L. Harper. 1973. Studies on plant demography: *Ranunculus repens* L., *R. bulbosa* L., and *R. acris* L. *J. of Ecology* 61:675–716.

Saunier, R. E. and R. F. Wagle. 1965. Root grafting in *Quercus turbinella* Greene. *Ecology* 46:749–750.

Savage, J. M. 1963. *Evolution*. New York: Holt, Rinehart and Winston.

Schaffer, W. M. and M. D. Gadgil. 1975. Selection for optimal life histories in plants. In *Ecology and evolution of communities*, ed. M. L. and J. M. Diamond, pp. 142–157. Cambridge, MA: Harvard University Press.

Schimper, A. F. W. 1898. *Pflanzengeographie auf physiologischer Grundlage*. Jena, Germany: Fischer.

———. 1903. Plant geography upon a physiological basis. Translated by W. R. Fisher. Oxford: Clarendon Press.

Schlesinger, W. H. 1985. Decomposition of chaparral shrub foliage. *Ecology* 66:1353–1359.

Schlesinger, W. H. and M. M. Hasey. 1981. Decomposition of chaparral shrub foliage: losses of organic and inorganic constituents from deciduous and evergreen leaves. *Ecology* 62:762–774.

Schlesinger, W. H. and P. L. Marks. 1977. Mineral cycling and the niche of Spanish moss, *Tillandsia usneoides* L. *American J. of Botany* 64:1254–1262.

Schlesinger, W. H. and W. A. Reiners. 1974. Deposition of water and cations on artificial foliar collectors in fir *krummholz* of New England mountains. *Ecology* 55:378–386.

Schlesinger, W. H., J. T. Gray, and F. S. Gilliam. 1982. Atmospheric deposition processes and their importance as sources of nutrients in a chaparral ecosystem of southern California. *Water Resources Research* 18:623–629.

Scholander, P. F. 1968. How mangroves desalinate seawater. *Physiologia Plantarum* 21:251–261.

Scholander, P. F., H. T. Hammel, E. A. Hemmingsen, and E. D. Bradstreet. 1964. Hydrostatic pressure and osmotic potential in leaves of mangroves and some other plants. *National Academy of Sciences* 52:119–125.

Schouw, J. F. 1822. *Grundtraek til en almindelig plantegeographie*. Copenhagen, Denmark: Glydendal.

Schultz, A. M. 1969. A study of an ecosystem: the arctic tundra. In *The ecosystem concept in natural resource management*, ed. G. M. Van Dyne, pp. 77–93. New York: Academic Press.

Schultz, A. M., R. P. Gibbens, and L. DeBano. 1961. Artificial populations for teaching and testing range techniques. *J. of Range Management* 14:237–242.

Schulze, E. D., O. L. Lange, U. Buschbom, L. Kappen, and M. Evenari. 1972. Stomatal responses to changes in humidity in plants growing in the desert. *Planta* 108:259–270.

———. 1973. Stomatal responses to changes in temperature at increasing water stress. *Planta* 110:29–42.

Schulze, E. D., H. Ziegler, and W. Stichler. 1976. Environmental control of crassulacean acid metabolism in *Welwitschia mirabilis* Hook. Fil. in its range of natural distribution in the Namib desert. *Oecologia* 24:323–334.

Scott, D. and W. D. Billings. 1964. Effects of environmental factors on standing crop and productivity of an alpine tundra. *Ecological Monographs* 34:243–270.

Scott, F. M. 1932. Some features of the anatomy of *Fouquieria splendens*. *American J. of Botany* 19:673–678.

Scott, G. D. 1969. *Plant symbiosis*. New York: St. Martin's Press.

Seigler, D. and P. W. Price. 1976. Secondary compounds in plants: primary functions. *The American Naturalist* 110:101–105.

Sellers, P. J. 1985. Canopy reflectance, photosynthesis and transpiration. *International J. Remote Sensing* 6:1335–1372.

Selter, C. M., W. D. Pitts, and M. G. Barbour. 1986. Site microenvironment and seedling survival of Shasta red fir. *American Midland Naturalist* 115:288–300.

Semenza, R. J., J. A. Young, and R. A. Evans. 1978. Influence of light and temperature on the germination and seed bed ecology of common Mullein (*Verbascum thapsus*). *Weed Science* 26:577–581.

Semikhatova, O. A. 1960. The after-effect of temperature on photosynthesis. *Botanisheskii Zhurnal* 45:1488–1501.

Sestak, Z., J. Catsky, and P. G. Jarvis. 1971. *Plant photosynthetic production: manual of methods*. The Hague: Junk.

Shanks, R. E. and J. S. Olson. 1961. First year breakdown of leaf litter in southern Appalachian forests. *Science* 134:194–195.

Shannon, C. E. and W. Weaver. 1949. *The mathematical theory of communication*. Urbana, IL: University of Illinois Press.

Shantz, H. L. 1945. Frederick Edward Clements (1874–1945). *Ecology* 26:317–319.

Sharitz, R. R. and J. F. McCormick. 1973. Population dynamics of two competing annual plant species. *Ecology* 54:723–739.

Shelford, V. E. 1913. *Animal communities in temperate America*. Chicago: University of Chicago Press.

Shelton, J. S. 1966. *Geology illustrated*. San Francisco, CA: W. H. Freeman.

Shimshi, D. 1969. A rapid field method for measuring photosynthesis with labeled carbon dioxide. *J. Experimental Botany* 20:381–401.

Shimwell, D. W. 1971. *The description and classification of vegetation*. Seattle, WA: University of Washington Press.

Shirley, H. L. 1945. Reproduction of upland conifers in the Lake States as affected by root competition and light. *American Midland Naturalist* 33:537–612.

Short, N. M. 1982. *The Landsat tutorial workbook*. NASA Ref. Publ. #1078.

Shmida, A. 1984. Whittaker's plant diversity sampling method. *Israel J. Botany* 33:41–46.

Shmida, A. and M. G. Barbour. 1982. A comparison of two types of mediterranean scrub in Israel and California. In *Proceedings of Symposium on dynamics and management of mediterranean-type ecosystems*, ed. C. E. Conrad and W. C. Oechel, pp. 100–106. US Forest Service, Pacific SW Forest and Range Exp. Sta. Gen. Tech. Rep. PSW-58.

Shreve, E. B. 1923. Seasonal changes in the water relations of desert plants. *Botanical Gazette* 39:397–408.

Shreve, F. 1914. The role of winter temperatures in determining the distribution of plants. *American J. of Botany* 1:194–202.

———. 1917. The establishment of desert perennials. *J. of Ecology* 5:210–216.

————. 1942. The desert vegetation of North America. *Botanical Review* 8:195–246.

Shugart, H. H., Jr., D. E. Reichle, N. T. Edwards, and J. R. Kercher. 1976. A model of calcium-cycling in an east Tennessee *Liriodendron* forest. *Ecology* 57:99–109.

Shure, D. J. and A. J. Lewis. 1973. Dew formation and stem flow on common ragweed (*Ambrosia artemisiafolia*). *Ecology* 54:1152–1155.

Siccama, T. G., F. H. Bormann, and G. E. Likens. 1970. The Hubbard Brook ecosystem study: productivity, nutrients, and phytosociology of the herbaceous layer. *Ecological Monographs* 40:389–402.

Silvertown, J. W. 1982. *Introduction to plant population ecology.* London: Longman.

Simpson, E. H. 1949. Measurement of diversity. *Nature* 163:688.

Singer, S. F., ed. 1970. *Global effects of environmental pollution.* New York: Springer-Verlag.

Singh, J. S. and S. R. Gupta. 1977. Plant decomposition and soil respiration in terrestrial ecosystems. *Botanical Review* 43:449–528.

Singh, J. S., W. K. Lauenroth, and R. K. Steinhorst. 1975. Review and assessment of various techniques for estimating net aerial primary productivity in grasslands from harvest data. *Botanical Review* 41:181–232.

Singh, J. S., W. K. Lauenroth, H. W. Hunt, and D. M. Swift. 1984. Bias and random errors in estimators of net root production: a simulation approach. *Ecology* 65:1760–1764.

Slatyer, R. O. 1967. *Plant-water relationships.* New York: Academic Press.

Slobodkin, L. B. 1974. Comments from a biologist to a mathematician. In *Proc. SIAM-SIMS Conference*, Alta, UT, ed. S. A. Leven, pp. 318–329.

Slobodkin, L. B. and A. Rapoport. 1974. An optimal strategy of evolution. *The Quarterly Review of Biology* 49:181–200.

Slobodkin, L. B., F. E. Smith, and N. G. Hairston. 1967. Regulation in terrestrial ecosystems, and the implied balance of nature. *American Naturalist* 101:109–124.

Smith, A. D. 1940. A discussion of the application of a climatological diagram, the hythergraph, to the distribution of natural vegetation types. *Ecology* 21:184–191.

Smith, B. N. and S. Epstein. 1971. Two categories of $^{13}C/^{12}C$ ratios for higher plants. *Plant Physiology* 47:380–384.

Smith, C. C. 1970. The coevolution of pine squirrels (*Tamiasciurus*) and conifers. *Ecological Monographs* 40:349–371.

Smith, H. B. 1927. Annual vs. biennial growth habit and its inheritance in *Melilotus alba*. *American J. of Botany* 14:129–146.

Smith, P. F. 1966. Leaf analysis of citrus. In *Fruit nutrition*, ed. N. F. Childers. New Brunswick, NJ: Horticultural Publications.

Smith, S. S. E. 1980. Mycorrhizas of autotrophic plants. *Biological Review* 55:475–510.

Smith, W. H. 1976. Character and significance of forest tree root exudates. *Ecology* 57:324–331.

Smith, W. K. and P. S. Nobel. 1977a. Temperature and water relations for sun and shade leaves of a desert broadleaf, *Hyptis emoryi*. *J. of Experimental Botany* 28:169–183.

————. 1977b. Influences of seasonal changes in leaf morphology on water use efficiency for three desert broadleaf shrubs. *Ecology* 58:1033–1043.

Sobolev, L. N. and V. D. Utekhin. 1978. Russian (Ramensky) approaches to community systematization. In *Ordination of plant communities*, ed. R. H. Whittaker, pp. 71–97. The Hague, Netherlands: Junk.

Soil Survey Staff. 1960. *Soil classification, a comprehensive system (7th Approximation).* Washington, D.C.: U.S. Dept. of Agriculture Soil Conservation Service.

———. 1975. *Soil taxonomy.* Agriculture Handbook no. 436. Washington, D.C.: U.S. Dept. of Agriculture Forest Service.

Sondheimer, E. and J. B. Simeone, eds. 1970. *Chemical ecology.* New York: Academic Press.

Soo Hoo, C. F. and G. Fraenkel. 1964. The resistance of ferns to the feeding of *Prodenia eridania* larvae. *Annals of the Entomological Society of America* 57:788–790.

Southwood, T. R. E. 1985. Interactions of plants and animals: patterns and processes. *Oikos* 44:5–11.

Spanner, P. C. 1951. The Peltier effect and its use in the measurement of suction pressure. *J. of Experimental Botany* 2:145–168.

Specht, R. L. 1969. A comparison of the sclerophyllous vegetation . . . in France, California, and southern Australia, parts I and II. *Australian J. Botany* 17:277–308.

Spurr, S. H. and B. V. Barnes. 1973. *Forest ecology.* New York: Ronald Press Co.

Stalter, R. and W. T. Batson. 1969. Transplantation of salt marsh vegetation, Georgetown, South Carolina. *Ecology* 50:1087–1089.

Stanton, M. L. 1984. Developmental and genetic sources of seed weight variation in *Raphanus raphanistrum* L. (Brassicaceae). *American J. of Botany* 71:1090–1098.

Stearn, W. T., ed. 1968. *Humboldt, Bonplant, Kunth and tropical American botany, a miscellany on the Nova Genera et Species Plantarum.* Cramer.

Stearns, S. C. 1977. The evolution of life history traits: a critique of the theory and a review of the data. *Annual Review of Ecology and Systematics* 8:145–171.

Steenbergh, W. F. and C. H. Lowe. 1969. Critical factors during the first years of life of the saguaro (*Cereus giganteus*) at Saguaro National Monument. *Ecology* 50:825–834.

Stewart, O. C. 1963. Barriers to understanding the influence of fire by aborigines on vegetation. In *Proc. Tall Timbers Ecol. Conf.,* no. 2, pp. 117–126. Tallahassee, FL: Tall Timbers Research Station.

Stone, E. C. 1963. The ecological importance of dew. *Quarterly Review of Biology* 38:328–341.

Stowe, L. G. and J. A. Teeri. 1978. The geographic distribution of C_4 species of the Dicotyledonae in relation to climate. *American Naturalist* 112:609–623.

Strahler, A. N. and A. H. Strahler. 1973. *Environmental geosciences: interaction between natural systems and man.* Santa Barbara, CA: Hamilton Publishing Co.

———. 1974. *Introduction to environmental science.* Santa Barbara, CA: Hamilton Publishing Co.

Strain, B. R. and V. C. Chase. 1966. Effect of past and prevailing temperatures on the carbon dioxide exchange capacities of some woody desert perennials. *Ecology* 47:1043–1045.

Sweeney, J. R. 1956. Responses of vegetation to fire. *University of California Publications in Botany* 28:143–206.

———. 1967. Ecology of some "fire type" vegetations in northern California. *Proc. Tall Timbers Fire Ecol. Conf.,* no. 7, pp. 111–125. Tallahassee, FL: Tall Timbers Research Station.

———. 1977. *Fire ecology.* Paper given to California Botanical Society, 17 March 1977 at University of California, Berkeley.

Sydes, C. and J. P. Grime. 1981. Effects of tree leaf litter on herbaceous vegetation in deciduous woodland. *J. Ecology* 69:237–262.

Sykes, J. M., A. D. Horrill, and M. D. Mountford. 1983. Use of visual cover assessments as quantitative estimators of some British woodland taxa. *J. Ecology* 71:437–450.

Syvertsen, J. P., G. L. Cunningham, and T. V. Feather. 1975. Anomalous diurnal patterns of stem xylem water potentials in *Larrea tridentata*. *Ecology* 56:1423–1428.

Syvertsen, J. P., G. L. Nickell, R. W. Spellenberg, and G. L. Cunningham. 1976. Carbon reduction pathways and standing crop in three Chihuahuan desert plant communities. *Southwestern Naturalist* 21:311–320.

Tadros, T. M. 1957. Evidence of the presence of an edaphobiotic factor in the problem of serpentine tolerance. *Ecology* 38:14–23.

Taha, F. K., H. G. Fisser, and R. E. Ries. 1983. A modified 100-point frame for vegetation inventory. *J. Range Management* 36:124–125.

Tall Timbers Fire Ecol. Confs. 1962–1979. *Proc. Tall Timbers Fire Ecol. Conf.* Tallahassee, FL: Tall Timbers Research Station.

Tansley, A. G. 1914. Presidential address to the first annual general meeting of the British Ecological Society. *J. of Ecology* 2:194–202.

———. 1935. The use and abuse of vegetation concepts and terms. *Ecology* 16:284–307.

———. 1939. The plant community and the ecosystem. *J. of Ecology* 27:513–530.

———. 1940. Henry Chandler Cowles, 1869–1939. *J. of Ecology* 28:450–452.

———. 1947a. Frederic Edward Clements, 1874–1945. *J. of Ecology* 34:194–196.

———. 1947b. The early history of modern plant ecology in Britain. *J. of Ecology* 35:130–137.

Tappeiner, J. C. and A. A. Alm. 1975. Undergrowth vegetation effects on the nutrient content of litterfall and soils in red pine and birch stands in northern Minnesota. *Ecology* 56:1193–1200.

Taylor, O. R., Jr. and D. W. Inouye. 1985. Synchrony and periodicity of flowering in *Frasera speciosa* (Gentianaceae). *Ecology* 66:521–527.

Taylor, R. J. and R. W. Pearcy. 1976. Seasonal patterns of CO_2 exchange characteristics of understory plants from a deciduous forest. *Canadian J. of Botany* 54:1094–1103.

Taylor, S. E. 1975. Optimal leaf form. In *Perspectives in biophysical ecology,* ed. D. M. Gates and R. B. Schmerl, pp. 73–86. New York Springer-Verlag.

Teeri, J. A. and L. G. Stowe. 1976. Climatic patterns and the distribution of C_4 grasses in North America. *Oecologia* 23:1–12.

Templeton, A. R. and D. A. Levin. 1979. Evolutionary consequences of seed pools. *American Naturalist* 114:232–249.

Tevis, L., Jr. 1958a. Germination and growth of ephemerals induced by sprinkling a sandy desert. *Ecology* 39:681–687.

———. 1958b. A population of desert ephemerals germinated by less than one inch of rain. *Ecology* 39:688–695.

Thomas, R. W., L. H. Beck, C. E. Brown, and S. L. Wall. 1984. Development of a satellite-aided crop acreage estimation and mapping system for the state of California. In *Eighteenth International Symposium of Remote Sensing of Environment,* pp. 1283–1294. Environmental Research Institute of Michigan, Ann Arbor.

Thompson, D. Q. and R. H. Smith. 1970. The forest primeval in the Northeast—A great myth? *Proc. Tall Timbers Ecol. Conf.,* no. 10, pp. 255–265. Tallahassee, FL: Tall Timbers Research Station.

Thompson, K. and J. P. Grime. 1979. Seasonal variation in the seed banks of herbaceous species in ten contrasting habitats. *J. of Ecology* 67:893–921.

Thompson, L. M. and F. R. Troeh. 1973. *Soils and soil fertility.* New York: McGraw-Hill.

Thomson, J. W. 1974. Lichenology in North America 1947–1972. *Annals of the Missouri Botanical Garden* 61:45–55.

Thorne, J. F. and S. P. Hamburg. 1985. Nitrification potentials of an old-field chronosequence in Campton, New Hampshire. *Ecology* 66:1333–1338.

Thornthwaite, C. W. 1931. The climates of North America, according to a new classification. *Geographical Review* 21:633–655.

Tieszen, L. L. 1978. Photosynthesis in the principal Barrow, Alaska species: A summary of field and laboratory responses. In *Vegetation and production ecology of an Alaskan arctic tundra,* ed. L. L. Tieszen, pp. 241–268. New York: Springer-Verlag.

Tikhomirov, B. A. 1963. *Contributions to the biology of Arctic plants.* Leningrad, USSR: Acad. Nauk.

Tilman, D. 1982. Resource competition and community structure. *Monographs in population biology,* No. 17. Princeton, NJ: Princeton University Press.

———. 1985. The resource-ratio hypothesis of plant succession. *American Naturalist* 125:827–852.

Ting, I. P., and M. Gibbs (eds.). 1982. *Crassulacean acid metabolism.* Rockville, MD: American Society of Plant Physiology.

Tobey, R. C. 1981. *Saving the prairies: the life cycle of the founding school of American plant ecology, 1895–1955.* Berkeley: Univ. California Press.

Tobiessen, P., and M. B. Werner. 1980. Hardwood seedling survival under plantations of scotch pine and red pine in central New York. *Ecology* 61:25–29.

Torrey, J. G. 1978. Nitrogen fixation by actinomycete-nodulated angiosperms. *BioScience* 28:586–592.

Transeau, E. N. 1926. The accumulation of energy by plants. *Ohio J. of Science* 26:1–10.

Triska, F. J. and J. R. Sedell. 1976. Decomposition of four species of leaf litter in response to nitrate manipulation. *Ecology* 57:783–792.

Troughton, J. and L. A. Donaldson. 1972. *Probing plant structure.* New York: McGraw-Hill.

Tucker, C. J., J. R. G. Townshend, and T. E. Goff. 1985. African land-cover classification using satellite data. *Science* 227:369–375.

Tukey, H. B., Jr. 1966. Leaching of metabolites from aboveground plant parts and its implications. *Bulletin of the Torrey Botanical Club* 93:385–401.

Tukey, H. B., Jr. and R. A. Mecklenberg. 1964. Leaching of metabolites from foliage and subsequent reabsorption and redistribution of the leachate in plants. *American J. of Botany* 51:737–742.

Turesson, G. 1922a. The species and variety as ecological units. *Hereditas* 3:100–113.

———. 1922b. The genotypical response of the plant species to the habitat. *Hereditas* 3:211–350.

———. 1925. The plant species in relation to habitat and climate. *Hereditas* 6:147–236.

———. 1930. The selective effect of climate upon the plant species. *Hereditas* 14:99–152.

Turitzin, S. N. 1982. Nutrient limitations to plant growth in a California serpentine grassland. *American Midland Naturalist* 107:95–99.

Turner, R. M., S. M. Alcorn, G. Olin, and J. A. Booth. 1966. The influence of shade, soil, and water on saguaro seedling establishment. *Botanical Gazette* 127:95–102.

Turrill, W. B. 1946. The ecotype concept, a consideration with appreciation and criticism especially of recent trends. *The New Phytologist* 45:34–43.

Tyler, G. 1971. Distribution and turnover of organic matter and minerals in a shore meadow ecosystem. *Oikos* 22:265–291.

Tyree, M. T. 1976. Negative turgor pressure in plant cells: fact or fallacy? *Canadian J. of Botany* 54:2738–2746.

Tyree, M. T. and H. T. Hammel. 1972. The measurement of the turgor pressure and the water relations of plants by the pressure-bomb technique. *J. of Experimental Botany* 23:267–282.

UNESCO. 1973. *International classification and mapping of vegetation.* Paris: UNESCO.

Ungar, I. A. 1962. Influence of salinity on seed germination in succulent halophytes. *Ecology* 43:763–764.

———. 1977. The relationship between soil water potential and plant water potential in two inland halophytes under field conditions. *Botanical Gazette* 138:498–501.

Ustin, S. L:, R. A. Woodward, M. G. Barbour, and J. L. Hatfield. 1985. Relationships between sunfleck dynamics and red fir seedling distributions. *Ecology* 65:1420–1428.

Vaadia, Y. and Y. Waisel. 1963. Water absorption by the aerial organs of plants. *Physiologia Plantarum* 16:44–51.

van Bavel, C. H. M. 1966. Potential evaporation: the combination concept and its experimental verification. *Water Resources Research* 2:445–467.

Vanderbilt, V. C. 1985. Measuring plant canopy structure. *Remote Sensing Environment* 18:281–294.

Vanderbilt, V. C., L. Grant, and C. S. T. Daughtry. 1985. Polarization of light scattered by vegetation. *Proceedings Institute Electrical and Electronics Engineers* 73:1012–1024.

van der Maarel, L. Orloci, and S. Pignatti (eds.). 1980. *Modern summary of European techniques used in sampling and data analysis.* The Hague: Junk.

Vandermeer, J. 1980. Saguaros and nurse trees: a new hypothesis to account for population fluctuations. *Southwestern Naturalist* 25:357–360.

van der Valk, A. G. 1974. Mineral cycling in coastal foredune plant communities in Cape Hatteras National Seashore. *Ecology* 55:1349–1358.

van Dobben, W. H. and R. R. Lowe-McConnell, eds. 1972. *Unifying concepts in ecology.* Wageningen, The Netherlands: Centre for Agricultural Publishing and Documentation.

Van Dyne, G. M., W. G. Vogel, and H. G. Fisser. 1963. Influence of small plot size and shape on range herbage production estimates. *Ecology* 44:746–759.

Van Valen, L. 1975. Life, death, and energy of a tree. *Biotropica* 7:260–269.

Varley, G. C. 1967. The effects of grazing by animals on plant productivity. In *Secondary productivity of terrestrial ecosystems (principles and methods)*, vol. II, ed. K. Petrusewicz, pp. 773–777. Warsaw: Polish Academy of Sciences.

Vasek, F. C. 1979. Early successional stages in Mojave Desert scrub vegetation. *Israel J. Botany* 28:133–148.

Vasek, F. C. and L. J. Lund. 1980. Soil characteristics associated with a primary plant succession on a Mojave Desert dry lake. *Ecology* 61:1013–1018.

Vasek, F. C., H. B. Johnson, and G. D. Brum. 1975*a*. Effects of power transmission lines on vegetation of the Mojave Desert. *Madroño* 23:114–130.

Vasek, F. C., H. B. Johnson, and D. H. Eslinger. 1975*b*. Effects of pipeline construction on creosote bush scrub vegetation of the Mojave Desert. *Madroño* 23:1–13.

Venable, D. L. and L. Lawlor. 1980. Delayed germination and dispersal in desert annuals: escape in space and time. *Oecologia* 46:272–282.

Viereck, L. A. 1966. Plant succession and soil development on gravel outwash of the Muldrow Glacier, Alaska. *Ecological Monographs* 36:181–199.

Viro, P. J. 1974. Effects of forest fire on soil. In *Fire and ecosystems*, ed. T. T. Kozlowski and C. E. Ahlgren, ch. 2. New York: Academic Press.

Vitousek, P. M. 1982. Nutrient cycling and nutrient use efficiency. *American Naturalist* 119:553–572.

Vitousek, P. M. and P. A. Matson. 1984. Mechanisms of nitrogen retention in forest ecosystems: a field experiment. *Science* 225:51–52.

———. 1985*a*. Causes of delayed nitrate production in two Indiana forests. *Forest Science* 31:122–131.

———. 1985*b*. Disturbance, nitrogen availability, and nitrogen losses in an intensively managed loblolly pine plantation. *Ecology* 66:1360–1376.

Vitousek, P. M., J. R. Gosz, C. C. Grier, J. M. Melillo, and W. A. Reiners. 1982. A comparative analysis of potential nitrification and nitrate mobility in forest ecosystems. *Ecological Monographs* 52:155–177.

Vitousek, P. M., J. R. Gosz, C. C. Grier, J. M. Melillo, W. A. Reiners, and R. L. Todd. 1979. Nitrate losses from disturbed ecosystems. *Science* 204:469–474.

Vogelmann, H. W., T. G. Siccama, D. Leedy, and D. C. Ovitt. 1968. Precipitation from fog moisture in the Green Mountains of Vermont. *Ecology* 49:1205–1207.

Vogl, R. J. 1972. Effects of fire on southeastern grasslands. *Proc. Tall Timbers Fire Ecol. Conf.*, no. 12, pp. 175–198. Tallahassee, FL: Tall Timbers Research Station.

———. 1973. Ecology of knobcone pine in the Santa Ana Mountains, California. *Ecological Monographs* 43:125–143.

———. 1974. Effects of fire on grasslands. In *Fire and ecosystems*, ed. T. T. Kozlowski and C. E. Ahlgren, ch. 5. New York: Academic Press.

Vogl, R. J., W. P. Armstrong, K. L. White, and K. L. Cole. 1977. The closed-cone pines and cypress. In *Terrestrial vegetation of California*, ed. M. G. Barbour and J. Major, pp. 295–358. New York: Wiley-Interscience.

Vogl, R. J. and L. T. McHargue. 1966. Vegetation of California fan palm oases on the San Andreas Fault. *Ecology* 47:532–540.

Vogt, K. A., C. C. Grier, C. E. Meier, and R. L. Edmonds. 1982. Mycorrhizal role in net primary production and nutrient cycling in *Abies amabilis* ecosystems in western Washington. *Ecology* 63:370–380.

Voigt, G. K. 1965. Nitrogen recovery from decomposing tree leaf tissue and forest humus. *Soil Science Society of America, Proc.* 29:756–759.

Volterra, V. 1926. Fluctuations in the abundance of a species considered mathematically. *Nature* 118:558–560.

von Willert, D. J., B. M. Eller, E. Brinckmann, and R. Baash. 1982. CO_2 gas exchange and transpiration of *Welwitschia mirabilis* Hook. Fil. in the central Namib desert. *Oecologia* 55:21–29.

Waisel, Y. 1972. *Biology of halophytes.* New York: Academic Press.

Wakimoto, R. H. 1977*a*. Chaparral growth and fuel assessment in southern California. In *Proc. Symp. Env. Cons. Fire and Fuel Mangmt. in Medit. Ecosyst.*, pp. 412–418. Washington, D.C.: U.S. Dept. of Agriculture Forest Service.

———. 1977*b*. Presentation at giant sequoia fire ecology extension course, October 1977, University of California, Davis.

Walker, D. 1970. Direction and rate in some British postglacial hydroseres. In *The vegetational history of the British Isles,* ed. D. Walker and R. West, pp. 117–139. Cambridge, England: Cambridge University Press.

Walker, R. B. 1954. The ecology of serpentine soils. II. Factors affecting plant growth on serpentine soils. *Ecology* 35:259–266.

Wallace, J. W. and R. L. Mansell, eds. 1976. *Biochemical interaction between plants and insects.* New York: Plenum Press.

Walter, H. 1979. *Vegetation of the earth and the ecological systems of the geobiosphere.* 2nd ed. New York: Springer-Verlag.

Walter, H. 1979. *Vegetation of the earth in relation to climate and the eco-physiological conditions.* New York: Springer-Verlag.

Wang, J. Y. 1979. *Instruments for physical environmental measurement,* vol. 1. San Jose, CA: Milieu Information Service.

Wanner, H. 1970. Soil respiration, litter fall, and productivity of tropical rain forest. *J. of Ecology* 58:543–547.

Waring, R. H. 1983. Estimating forest growth and efficiency in relation to canopy leaf area. *Advances in Ecological Research* 13:327–354.

Waring, R. H. and B. D. Cleary. 1967. Plant moisture stress: evaluation by pressure bomb. *Science* 155:1248–1254.

Waring, R. H. and G. B. Pitman. 1985. Modifying lodgepole pine stands to change susceptibility to mountain pine beetle attack. *Ecology* 66:889–897.

Warming, J. E. B. 1892. *Lagoa Santa.* Copenhagen, Denmark: Lun.

———. 1895. *Plantesamfund, grundtraek af den okologiske plantegeografi.* Copenhagen, Denmark: Philipsen.

———. 1909. *Oecology of plants.* London: Oxford University Press.

Watt, A. S. 1947. Pattern and process in the plant community. *J. of Ecology* 35:1–22.

———. 1955. Bracken versus heather, a study in plant sociology. *J. of Ecology* 43:490–506.

Weaver, J. E. 1920. *Root development in the grassland formation.* Carnegie Institution of Washington Publ. 286. Washington, D.C.: Carnegie Institution of Washington.

Weaver, J. E. and F. E. Clements. 1929. *Plant ecology.* New York: McGraw-Hill.

Weaver, J. E. and T. J. Fitzpatrick. 1934. The prairie. *Ecological Monographs* 4:109–295.

Webb, W. L., W. K. Lauenroth, S. R. Szarek, and R. S. Kinerson. 1983. Primary production and abiotic controls in forests, grasslands, and desert ecosystems in the United States. *Ecology* 64:134–151.

Webb, W., S. Szarek, W. Lauenroth, R. Kinerson, and M. Smith. 1978. Primary productivity and water use in native forest, grassland, and desert ecosystems. *Ecology* 59:1239–1247.

Webley, D. M., D. J. Eastwood, and C. H. Gimingham. 1952. Development of a soil microflora in relation to plant succession on sand dunes, including the "rhizosphere" flora associated with colonising species. *J. of Ecology* 40:168–178.

Wells, P. V. 1961. Succession in desert vegetation on streets of a Nevada ghost town. *Science* 134:670–671.

Wells, P. V. and R. Berger. 1967. Late Pleistocene history of coniferous woodland in the Mojave Desert. *Science* 155:1640–1647.

Went, F. W. 1942. The dependence of certain annual plants on shrubs in southern California deserts. *Bulletin of the Torrey Botanical Club* 69:100–114.

———. 1948. Ecology of desert plants. I. Observations on germination in Joshua Tree National Monument. *Ecology* 29:242–253.

Went, F. W. and N. Stark. 1968. Mycorrhiza. *BioScience* 18:1035–1039.

Went, F. W. and M. Westergaard. 1949. Ecology of desert plants. III. Development of plants in the Death Valley National Monument, California. *Ecology* 30:26–38.

Wentworth, T. R. 1981. Vegetation on limestone and granite in the Mule Mountains, Arizona. *Ecology* 62:469–482.

———. 1983. Distributions of C^4 plants along environmental and composition gradients in southeastern Arizona. *Vegetatio* 52:21–34.

Werner, P. A. 1975. Predictions of fate from rosette size in teasel (*Dipsacus fullonum* L.). *Oecologia* 20:197–201.

———. 1976. Ecology of plant populations in successional environments. *Systematic Botany* 1:246–268.

Werner, P. A. and H. Caswell. 1977. Population growth rates and age versus stage-distribution models for teasel (*Dipsacus sylvestris* Huds.). *Ecology* 58:1103–1111.

West, N. E. and P. T. Tueller. 1971. Special approaches to studies of competition and succession in shrub communities. In *Wildland shrubs*, U.S. Dept. of Agriculture Forest Service General Technical Report INT-1, ed. C. M. McKell, J. P. Blaisdell, and J. R. Goodin, pp. 165–171. Logan, UT: U.S. Dept. of Agriculture Forest Service.

Westman, W. E. 1975. Edaphic climax pattern of the pygmy forest region of California. *Ecological Monographs* 45:109–135.

Westman, W. E., and R. K. Peet. 1982. Robert H. Whittaker (1920–1980): the man and his work. *Vegetatio* 48:97–122.

Westman, W. E. and R. H. Whittaker. 1975. The pygmy forest region of northern California: studies on biomass and primary productivity. *J. of Ecology* 63:493–520.

Westoby, M. 1981. The place of the self-thinning rule in population dynamics. *American Naturalist* 118:581–587.

———. 1984. The self-thinning rule. *Advances in Ecological Research* 14:167–225.

White, D. 1941. Prairie soil as a medium for tree growth. *Ecology* 22:399–407.

White, J. 1979. The plant as a metapopulation. *Annual Review of Ecology and Systematics* 10:109–145.

———. 1980. Demographic factors in populations of plants. In *Demography and evolution in plant populations*, ed. O. T. Solbrig, pp. 21–48. Oxford, England: Blackwell.

———. 1981. The allometric interpretation of the self-thinning rule. *J. of Theoretical Biology* 89:475–500.

Whitford, W. G., V. Meentemeyer, T. R. Seastedt, K. Cromack, Jr., D. A. Crossley, Jr., P. Santox, R. L. Todd, and J. B. Waide. 1981. Exceptions to the AET model: deserts and clear-cut forests. *Ecology* 62:275–277.

Whitney, G. G. 1986. The presettlement pine forests of northern lower Michigan as related to substrate and disturbance history. *Ecology* (In press).

Whittaker, R. H. 1953. A consideration of climax theory: the climax as a population and pattern. *Ecological Monographs* 23:41–78.

————. 1954. The ecology of serpentine soils. I. Introduction. *Ecology* 35:258–259.

————. 1956. Vegetation of the Great Smoky Mountains. *Ecological Monographs* 26:1–80.

————. 1960. Vegetation of the Siskiyou Mountains of California. *Ecological Monographs* 30:279–338.

————. 1962. Classification of natural communities. *Botanical Review* 28:1–239.

————. 1966. Forest dimensions and production in the Great Smoky Mountains. *Ecology* 47:103–121.

————. 1967. Gradient analysis of vegetation. *Biological Review* 42:207–264.

————. 1973a. Climax concepts and recognition. *Handbook of Vegetation Science* 8:137–154.

————. ed. 1937b. *Ordination and classification of communities.* Gotingen, The Netherlands: Junk.

————. 1975. *Communities and ecosystems.* 2nd ed. New York: Macmillan.

————. 1978. Direct gradient analysis. In *Ordination of plant communities,* ed. R. H. Whittaker, pp. 7–50. Groningen, The Netherlands: Junk.

Whittaker, R. H. and P. P. Feeny. 1971. Allelochemics: chemical interactions between species. *Science* 171:1757–1770.

Whittaker, R. H., S. A. Levin, and R. B. Root. 1973. Niche, habitat, and ecotype. *The American Naturalist* 107:321–338.

Whittaker, R. H. and G. E. Likens. 1975. The biosphere and man. In *The primary production of the biosphere,* ed. H. Lieth and R. H. Whittaker, pp. 305–328. New York: Springer-Verlag.

Whittaker, R. H. and P. L. Marks. 1975. Methods of assessing terrestrial productivity. In *Primary productivity of the biosphere,* ed. H. Lieth and R. H. Whittaker, pp. 55–118. New York: Springer.

Whittaker, R. H. and W. A. Niering. 1975. Vegetation of the Santa Catalina Mountains, Arizona: V. Biomass, production and diversity along the elevation gradient. *Ecology* 56:771–790.

Whittaker, R. H. and G. M. Woodwell. 1968. Dimension and production relations of trees and shrubs in the Brookhaven Forest, New York. *J. of Ecology* 56:1–25.

————. 1969. Structure, production, and diversity of the oak-pine forest at Brookhaven, New York. *J. of Ecology* 57:155–174.

————. 1972. Evolution of natural communities. In *Ecosystem structure and function,* Oregon State University Annual Biology Colloquia, volume 31, ed. J. A. Wiens, pp. 137–156. Corvallis, OR: Oregon State University.

Wiegert, R. G. and F. C. Evans. 1964. Primary production and disappearance of dead vegetation on an old field in southeastern Michigan. *Ecology* 45:49–63.

Wiegert, R. G. and J. T. McGinnis. 1975. Annual production and disappearance of detritus on three South Carolina old fields. *Ecology* 56:129–140.

Wiegert, R. G. and D. F. Owen. 1971. Trophic structure, available resources and population density in terrestrial vs. aquatic ecosystems. *J. of Theoretical Biology* 30:69–81.

Wieland, N. K. and F. A. Bazzaz. 1975. Physiological ecology of three codominant successional annuals. *Ecology* 56:681–688.

Wilbur, H. M. 1976. Life history evolution in seven milkweeds of the genus *Asclepias. J. of Ecology* 64:223–240.

Williams, C. B. 1964. *Patterns in the balance of nature.* New York: Academic Press.

Williams, C. M. 1970. Hormonal interactions between plants and insects. In *Chemical ecology,* ed. E. Sondheimer and J. B. Simeone, pp. 103–132. New York: Academic Press.

Williams, W. T. and J. M. Lambert. 1959. Multivariate methods in plant ecology. I. Association analysis in plant communities. *J. of Ecology* 47:83–101.

———. 1961. Multivariate methods in plant ecology. III. Inverse association-analysis. *J. of Ecology* 49:717–729.

Willis, A. J. 1963. Braunton burrows: the effects on the vegetation of the addition of mineral nutrients to the dune soils. *J. of Ecology* 51:353–374.

Willmot, A. and P. D. Moore. 1973. Adaptation to light intensity in *Silene alba* and *S. dioica. Oikos* 24:458–464.

Wilson, D. S. 1980. *The natural selection of populations and communities.* Menlo Park, CA: Benjamin/Cummings.

Wilson, R. E. and E. L. Rice. 1968. Allelopathy as expressed by *Helianthus annuus* and its role in old-field succession. *Bulletin of the Torrey Botanical Club* 95:432–448.

Winn, C. 1977. *Decomposition in southern California chaparral.* Master's Thesis. Fullerton, CA: California State University.

Winter, K. 1975. *Die Rolle des Crassulacean-säurestoffwechsels als Biochemishe grundlage zur anpassung von Halophyten an Standorte hoher Salinität.* Dissertation. Darmstadt, Germany.

Winter, K. and U. Lüttge. 1976. Balance between C_3 and CAM pathway of photosynthesis. *Ecological Studies,* 19:323–334.

Woods, F. W. and K. Brock. 1964. Interspecific transfer of Ca_{45} and P_{32} by root systems. *Ecology* 45:886–889.

Woods, F. W. and R. E. Shanks. 1959. Natural replacement of chestnut by other species in the Great Smoky Mountains National Park. *Ecology* 40:349–361.

Woods, K. D. 1984. Patterns of tree replacement: canopy effects on understory pattern in hemlock-northern hardwood forests. *Vegetatio* 56:87–107.

Woodwell, G. M., J. E. Hobbie, R. A. Houghton, J. M. Melillo, B. Moore, B. J. Peterson, and G. R. Shaver. 1983. Global deforestation: contribution to atmospheric carbon dioxide. *Science* 222:1081–1086.

Woodwell, G. M. and H. H. Smith, eds. 1969. *Diversity and stability in ecological systems.* Brookhaven Symposia in Biology, no. 22. Upton, NY: Brookhaven National Laboratories.

Woodwell, G. M., R. H. Whittaker, and R. S. Houghton. 1975. Nutrient concentrations in plants in the Brookhaven oak-pine forest. *Ecology* 56(2):318–332.

Workman, C. 1980. A new chemical method for measurement of mean temperatures with special applicability to cold environments. *Oikos* 35:365–372.

Workman, J. P. and N. E. West. 1967. Germination of *Eurotia lanata* in relation to temperature and salinity. *Ecology* 48:659–661.

Wright, E. and R. F. Tarrant. 1957. *Microbial soil properties after logging and slash burning.* U.S. Forest Service Pacific Northwest Forest Range Experimental Station Research Notes 157. Washington, D.C.: U.S. Dept. of Agriculture Forest Service.

Wright, H. A. and A. W. Bailey. 1982. *Fire ecology.* New York: John Wiley.

Wright, R. D. 1974. Rising atmospheric CO_2 and photosynthesis of San Bernardino Mountain plants. *American Midland Naturalist* 91:360–370.

Yavitt, J. B. and E. L. Smith, Jr. 1983. Spatial patterns of mesquite and associated herbaceous species in an Arizona desert grassland. *American Midland Naturalist* 109:89–93.

Yeaton, R. I. 1978. A cyclic relationship between *Larrea tridentata* and *Opuntia leptocaulis* in the northern Chihuahuan Desert. *J. of Ecology* 66:651–656.

Yemm, E. W. and R. G. S. Bidwell. 1969. Carbon dioxide exchanges in leaves. I. Discrimination between $^{14}CO_2$ and $^{12}CO_2$ in photosynthesis. *Plant Physiology* 44:1328–1334.

Yoda, K., T. Kira, H. Ogawa, and K. Hozumi. 1963. Intraspecific competition among higher plants. XI. Self-thinning in overcrowded pure stands under cultivated and natural conditions. *J. of Biology*, Osaka City University 14:107–129.

Young, J. A., R. A. Evans, and P. T. Tueller. 1975. Great Basin plant communities—pristine and grazed. In *Holocene climates in the Great Basin*, Occasional Paper, Nevada Archeological Survey, ed. R. Elston, pp. 186–215. Reno, NV: Nevada Archeological Survey.

Zanstra, P. E. and F. Hagenzieker. 1977. Comments on the psychrometric determination of leaf water potentials *in situ*. *Plant and Soil* 48:347–367.

Zavitkovski, J. and M. Newton. 1968. Ecological importance of snowbrush, *Ceanothus velutinus*, in the Oregon Cascades agricultural ecology. *Ecology* 49:1134–1145.

Zinke, P. J. 1977. The redwood forest and associated north coast forests. In *Terrestrial vegetation of California*, ed. M. G. Barbour and J. Major, pp. 679–698. New York: Wiley-Interscience.

Zucker, W. V. 1983. Tannins: does structure determine function? An ecological perspective. *American Naturalist* 121:335–365.

INDEX

Page numbers given in bold refer to definitions

624   *Index*